W. Greiner · J. A. Maruhn

NUCLEAR MODELS

Springer
Berlin
Heidelberg
New York
Barcelona
Budapest
Hong Kong
London
Milan
Paris
Santa Clara
Singapore
Tokyo

Greiner
Quantum Mechanics
An Introduction 3rd Edition

Greiner
Quantum Mechanics
Special Chapters

Greiner · Müller
Quantum Mechanics
Symmetries 2nd Edition

Greiner
Relativistic Quantum Mechanics
Wave Equations

Greiner · Reinhardt
Field Quantization

Greiner · Reinhardt
Quantum Electrodynamics
2nd Edition

Greiner · Schäfer
Quantum Chromodynamics

Greiner · Maruhn
Nuclear Models

Greiner · Müller
Gauge Theory of Weak Interactions
2nd Edition

Greiner
Mechanics I
(in preparation)

Greiner
Mechanics II
(in preparation)

Greiner
Electrodynamics
(in preparation)

Greiner · Neise · Stöcker
Thermodynamics
and Statistical Mechanics

Walter Greiner · Joachim A. Maruhn

NUCLEAR MODELS

With a Foreword by
D. A. Bromley

With 50 Figures,
and 39 Worked Examples and Problems

Springer

Professor Dr. Walter Greiner
Professor Dr. Joachim A. Maruhn

Institut für Theoretische Physik der
Johann Wolfgang Goethe-Universität Frankfurt
Postfach 111932
D-60054 Frankfurt am Main
Germany

Street address:

Robert-Mayer-Strasse 8–10
D-60325 Frankfurt am Main
Germany

email:
greiner@th.physik.uni-frankfurt.de (W. Greiner)
maruhn@th.physik.uni-frankfurt.de (J. A. Maruhn)

Title of the original German edition: *Theoretische Physik,* Ein Lehr- und Übungsbuch,
Band 11: Kernmodelle © Verlag Harri Deutsch, Thun 1995

Die Deutsche Bibliothek – CIP-Einheitsaufnahme

Greiner, Walter:
Nuclear models : with 39 worked examples and problems / Walter Greiner ; Joachim A. Maruhn. With a foreword by D. A. Bromley. – Berlin ; Heidelberg ; New York ; Barcelona ; Budapest ; Hong Kong ; London ; Milan ; Paris ; Santa Clara ; Singapore ; Tokyo : Springer, 1996
Einheitssacht.: Kernmodelle <engl.>
ISBN 3-540-59180-X
NE: Maruhn, Joachim:

ISBN 3-540-59180-X Springer-Verlag Berlin Heidelberg New York

Printed and bound by Hamilton Printing Company, Rensselaer, NY.
Printed in the United States of America.

9 8 7 6 5 4 3 2

Typesetting: Camera ready copy from the authors using a Springer T_EX macro package
Cover design; Design Concept, Emil Smejkal, Heidelberg
Copy Editor: V. Wicks
Production Editor: P. Treiber
SPIN 10469337 56/3144 - 5 4 3 2 1 - Printed on acid-free paper

Foreword to Earlier Series Editions

More than a generation of German-speaking students around the world have worked their way to an understanding and appreciation of the power and beauty of modern theoretical physics – with mathematics, the most fundamental of sciences – using Walter Greiner's textbooks as their guide.

The idea of developing a coherent, complete presentation of an entire field of science in a series of closely related textbooks is not a new one. Many older physicists remember with real pleasure their sense of adventure and discovery as they worked their ways through the classic series by Sommerfeld, by Planck and by Landau and Lifshitz. From the students' viewpoint, there are a great many obvious advantages to be gained through use of consistent notation, logical ordering of topics and coherence of presentation; beyond this, the complete coverage of the science provides a unique opportunity for the author to convey his personal enthusiasm and love for his subject.

The present five-volume set, *Theoretical Physics*, is in fact only that part of the complete set of textbooks developed by Greiner and his students that presents the quantum theory. I have long urged him to make the remaining volumes on classical mechanics and dynamics, on electromagnetism, on nuclear and particle physics, and on special topics available to an English-speaking audience as well, and we can hope for these companion volumes covering all of theoretical physics some time in the future.

What makes Greiner's volumes of particular value to the student and professor alike is their completeness. Greiner avoids the all too common "it follows that . . . " which conceals several pages of mathematical manipulation and confounds the student. He does not hesitate to include experimental data to illuminate or illustrate a theoretical point and these data, like the theoretical content, have been kept up to date and topical through frequent revision and expansion of the lecture notes upon which these volumes are based.

Moreover, Greiner greatly increases the value of his presentation by including something like one hundred completely worked examples in each volume. Nothing is of greater importance to the student than seeing, in detail, how the theoretical concepts and tools under study are applied to actual problems of interest to a working physicist. And, finally, Greiner adds brief biographical sketches to each chapter covering the people responsible for the development of the theoretical ideas and/or the experimental data presented. It was Auguste Comte (1798–1857) in his *Positive Philosophy* who noted, "To understand a science it is necessary to know its history". This is all too often forgotten in modern physics teaching and the bridges that Greiner builds to the pioneering figures of our science upon whose work we build are welcome ones.

Greiner's lectures, which underlie these volumes, are internationally noted for their clarity, their completeness and for the effort that he has devoted to making physics an integral whole; his enthusiasm for his science is contagious and shines through almost every page.

These volumes represent only a part of a unique and Herculean effort to make all of theoretical physics accessible to the interested student. Beyond that, they are of enormous value to the professional physicist and to all others working with quantum phenomena. Again and again the reader will find that, after dipping into a particular volume to review a specific topic, he will end up browsing, caught up by often fascinating new insights and developments with which he had not previously been familiar.

Having used a number of Greiner's volumes in their original German in my teaching and research at Yale, I welcome these new and revised English translations and would recommend them enthusiastically to anyone searching for a coherent overview of physics.

Yale University
New Haven, CT, USA
1989

D. Allan Bromley
Henry Ford II Professor of Physics

Preface

Theoretical physics has become a many-faceted science. For the young student it is difficult enough to cope with the overwhelming amount of new scientific material that has to be learned, let alone to obtain an overview of the entire field, which ranges from mechanics through electrodynamics, quantum mechanics, field theory, nuclear and heavy-ion science, statistical mechanics, thermodynamics, and solid-state theory to elementary-particle physics. And this knowledge should be acquired in just 8–10 semesters during which, in addition, a Diploma or Master's thesis has to be worked on or examinations prepared for. All this can be achieved only if the university teachers help to introduce the student to the new disciplines as early on as possible, in order to create interest and excitement that in turn set free essential new energy. Naturally, all inessential material must simply be eliminated.

At the Johann Wolfgang Goethe University in Frankfurt we therefore confront the student with theoretical physics immediately in the first semester. Theoretical Mechanics I and II, Electrodynamics, and Quantum Mechanics I – an Introduction are the basic courses during the first two years. These lectures are supplemented with many mathematical explanations and much support material. After the fourth semester of studies, graduate work begins and Quantum Mechanics II – Symmetries, Statistical Mechanics and Thermodynamics, Relativistic Quantum Mechanics, Quantum Electrodynamics, the Gauge Theory of Weak Interactions, and Quantum Chromodynamics are obligatory. Apart from these, a number of supplementary courses on special topics are offered, such as Hydrodynamics, Classical Field Theory, Special and General Relativity, Many-Body Theories, Nuclear Models, Models of Elementary Particles, and Solid-State Theory. Some of them, for example the two-semester courses on Theoretical Nuclear Physics and Theoretical Solid-State Physics, are obligatory.

This volume is devoted to the Theory of Nuclear Models, which forms a two-semester cycle together with a course on Nuclear Reactions. For this field it appeared to be especially important to present a relatively short textbook actually suitable for accompanying a lecture, since while there are excellent and comprehensive treatises on nuclear models, the wealth of material presented in those tends to overwhelm the students initially. In this connection we mention preferentially the three-volume work by Eisenberg and Greiner,[1] on which the present treatment

[1] J. M. Eisenberg and W. Greiner, *Nuclear Theory*, 3 Volumes, Third Edition (North Holland, Amsterdam 1973–1987).

is based in many respects, and the textbook by Ring and Schuck,[2] which puts more emphasis on many-body approaches.

A textbook for direct use with a lecture has to concentrate on the most essential points, emphasize the explanation of ideas and methods, and forego the presentation of a wealth of individual results, which cannot be shown in a lecture anyway if it is not to degenerate into a slide show. Another characteristic that makes the theory of nuclear models different from the classical fields of theoretical physics is the scarcity of examples that can be calculated from start to finish without the use of computers.

For all of these reasons the focus is on the discussion of the most important types of models and the requisite mathematical methods. Since experience shows that most students have not really mastered the crucial methods of angular-momentum coupling and second quantization, we have not relegated these topics to an appendix but treated them at the start of the book. Of course these chapters can be ignored if desired. Even in these chapters the material was carefully restricted to what is actually used in the rest of the book. Following this there is a short discussion of group-theoretical methods, which are essential, for example, for the IBA model.

The fifth chapter treats the theory of the radiation field up to the definition of multipole transition probabilities. Again with a view to brevity the magnetic transitions are only dicussed in general terms. The sixth chapter presents the classical collective models, which because of their didactic value and their fundamental importance for introducing concepts form the centerpiece of the book. A short overview of the phenomenological properties of nuclear matter is followed by a treatment of the geometric collective model (surface vibrations, the rotation–vibration model, etc.) in the various limiting cases, the IBA model, and the collective theory of giant resonances.

Only a little less space is devoted to microscopic models in Chapter 7. The most important concepts, from Hartree–Fock theory via phenomenological single-particle models to the relativistic mean-field model, are introduced successively. The next chapter treats the coupling of single-particle and collective motion both with respect to the particle-plus-core model and to the microscopic description of collective vibrations.

The final chapter presents large-amplitude collective motion, concentrating on ways to describe nuclear fission and similar processes. This includes two-center models, the general problem of collective mass parameters, time-dependent Hartree–Fock, the generator-coordinate method, and an elementary overview of high-spin states.

In addition to the classical syllabus of nuclear models that still form the basic equipment of the nuclear theorist, short discussions of topics of present-day interest are interspersed in many places – such as superheavy elements, high-spin states, and the relativistic mean-field model. These should give young physicists an impression of the continuing vitality of this science. The reader will also note in various places that the book is based on repeated practical experience with such a course and offers many explanations and illustrations motivated by typical student questions.

[2] P. Ring and P. Schuck, *The Nuclear Many-Body Problem* (Springer-Verlag, New York 1980).

We express our sincere thanks to Dr. Dirk Troltenier, Michael Bender, Christian Spieles, and Klemens Rutz for their efficient help in the formulation, formatting, and editing of the text as well as to Ms. Astrid Steidl for support in producing some of the graphics.

Finally we wish to thank Springer-Verag; in particular Dr. H. J. Kölsch for his encouragement and patience, and Petra Treiber for the production and Dr. Victoria Wicks for expertly copy-editing the English edition.

Frankfurt am Main,
November 1995

Walter Greiner
Joachim Maruhn

Contents

Contents of Examples and Exercises

1. Introduction

1.1 Nuclear Structure Physics

The nuclear models discussed in this book belong to the realm of *nuclear structure theory*. In present usage, nuclear structure physics is devoted to the study of the properties of nuclei at low excitation energies, where individual energy levels can be resolved. This means that typically quantum effects are predominant and the states of the nucleus have a very complicated structure that depends on the intricate interrelations of all the many nucleons involved.

In contrast, at higher energies and especially for heavy-ion reactions, quantum mechanics becomes less important and the preeminent place is instead given to methods of statistical mechanics. Theories then typically employ bulk properties of nuclear matter such as the equation of state or the dissipation coefficients, or are even based on purely classical many-body physics like the cascade models.

Of course it is impossible to give an exact energy boundary between these types of theories. The theories presented here, however, are typically employed for excitation energies up to 2–3 MeV. Usually only the lowest few energy levels can be described well by a theoretical model, and the number of levels increases so rapidly above that energy range that it becomes impossible to make any sensible comparison with experiment (for nuclei with an odd number of neutrons or protons or both this is even more dramatic – most nuclear models prefer even–even nuclei with their relatively simple spectra). Also one should remember that in experimental spectra only a relatively small number of states can be identified as to spin and parity, and that to really test a model transitions, i.e., essentially overlaps between the wave functions, are needed, which again are often not known even for the most interesting states.

It is thus not surprising that the models presented in this book usually explain a relatively small number of low-lying states and to a modest accuracy, and even this is a considerable achievement. To esteem that, remember that we are dealing with a system of particles whose number is neither small enough to allow direct solution nor large enough to make statistical methods highly accurate, and which interact through an interaction that has still not been pinned down to any definite form. It is this extraordinary difficulty and the freedom with which methods and ideas from many other branches are applied here that make nuclear structure physics so fascinating and so much alive.

1.2 The Basic Equation

To find the proper theoretical starting point some more ballpark estimates of the relevent physical quantitities have to be introduced. Let us first recall a few numbers from elementary experimental nuclear physics.

The elements known at the time of this writing have nuclei consisting of (at present) $Z = 1, \ldots, 111$ protons and $N = 0, \ldots, 161$ neutrons, giving a total number of A nucleons. The radii of nuclei follow the empirical law

$$R(A) = r_0 A^{1/3} \tag{1.1}$$

with $r_0 \approx 1.2\,\mathrm{fm}$. Nuclear radii thus range up to about 7.5 fm. The formula also implies that the nuclear volume is proportional to the number of particles in the nucleus, indicating the near incompressibility of nuclear matter (the true density profile observed by electron scattering is a bit more complicated). The least-bound nucleon has a binding energy of the order of 8 MeV and a kinetic energy close to 40 MeV.

This information is already sufficient to form some rough ideas about what is essential in the theories. Since a nucleon has a mass of $mc^2 \approx 938\,\mathrm{MeV}$, the kinetic energy is quite negligible by comparison, so that a *nonrelativistic* approach appears quite sufficient, and this assumption is made in the vast majority of nuclear structure models. More recently, however, relativistic approaches have become important – this theme is taken up in Sect. 7.4 in connection with the relativistic mean-field model, and we also explain there why relativistic effects can be important in spite of the simple estimate given above.

The velocity of a nucleon with a kinetic energy of $T = 40\,\mathrm{MeV}$ is given by

$$v = \sqrt{\frac{2T}{m}} = c\sqrt{\frac{2T}{mc^2}} \approx 0.3\,c \quad , \tag{1.2}$$

and the associated de Broglie wavelength by

$$\lambda = \frac{2\pi\hbar}{mv} = \frac{2\pi(\hbar c)}{(mc^2)(v/c)} \approx 4.5\ \mathrm{fm} \quad . \tag{1.3}$$

Here the useful constant $\hbar c \approx 197.32$ MeV fm was used. The result shows that quantum effects are certainly not negligible, as λ is by no means small compared to the nuclear radii. This is even more pronounced for the more tightly bound nucleons, which have a smaller kinetic energy.

Taking these considerations into account, the starting point for a theory of nuclear eigenstates should be a stationary Schrödinger equation very generally given by

$$\hat{H}\,\psi = E\,\psi \quad . \tag{1.4}$$

The rest of this book is about what to write for $\hat{H}$ and which degrees of freedom to use in the wave functions.

1.3 Microscopic versus Collective Models

The most natural selection for the degrees of freedom is, of course, to use the nucleonic ones, i.e., the A sets of positions $\boldsymbol{r}_i$, spins $\boldsymbol{s}_i$, and isospins $\boldsymbol{\tau}_i$. The wave function then takes the general form

$$\psi(\boldsymbol{r}_1, \boldsymbol{s}_1, \boldsymbol{\tau}_1, \boldsymbol{r}_2, \boldsymbol{s}_2, \boldsymbol{\tau}_2, \ldots, \boldsymbol{r}_A, \boldsymbol{s}_A, \boldsymbol{\tau}_A) \quad , \tag{1.5}$$

while for the Hamiltonian we would try the natural expression

$$\hat{H} = -\sum_{i=1}^{A} \frac{\hbar^2}{2m} \nabla^2 + \frac{1}{2} \sum_{i,j} v(i,j) \quad . \tag{1.6}$$

This is the principal starting point for *microscopic models*, in which the degrees of freedom are those of the constituent particles of the nucleus. Here $v(i,j)$ is the nucleon–nucleon interaction, which may depend on all the degrees of freedom of a pair of nucleons. Clearly it is impractical even with modern computers to solve the many-particle Schrödinger equation directly for A larger than three or four, so that the search for suitable approximations is the overriding concern in this type of model.

It is one of the important features of nuclear theory that *there is no a priori theory for $v(i,j)$*. Instead, various parametrizations are employed which are good for different purposes – one may be adapted for describing nucleon–nucleon scattering while another may be suitable for Hartree–Fock calculations of heavy nuclei. It is not even clear how important three-body forces not included in the above Hamiltonian might be. Until the nucleon–nucleon interaction can be derived from a more fundamental theory such as quantum chromodynamics, we will have to live with this situation. Contrast this with the situation in atomic physics, where the fundamental interaction theory – QED – is very well known and it is "only" a matter of approximation methods to find solutions to the problem.

A typical microscopic model thus depends on a nucleon–nucleon interaction which necessarily contains parameters fitted to reproduce some experimental data. This justifies the name "model" even for the microscopic approaches: the lacking knowledge about the fundamental interaction is replaced by the proposal of a reasonable functional form with a limited number of parameters, which cannot be determined from an underlying theory.

The proposal of suitable functional forms for the nucleon–nucleon interaction depends to a large extent on symmetry arguments. These considerations and an overview of some interactions used in practice will be given in Sect. 7.1.

A complementary and important role is also played by *collective models*. These are based on degrees of freedom that do not refer to individual nucleons, but instead indicate some bulk property of the nucleus as a whole. Simple but trivial examples for *collective coordinates* are the center-of-mass vector for the nucleus and the quadrupole moment,

$$\boldsymbol{R} = \frac{1}{A} \sum_{i=1}^{A} \boldsymbol{r}_i \quad , \quad Q_{20} = \sum_{i=1}^{A} r_i^2 Y_{2\mu}(\Omega_i) \quad . \tag{1.7}$$

Note that in these cases it is possible to express the collective coordinates in terms of the microscopic ones, so that at least theoretically one should be able to go back and forth between both descriptions. In practice, however, one often introduces coordinates defined without any reference to microscopic physics, for example, by expanding the surface of the nucleus in spherical harmonics and using the coefficients as coordinates; let us just give an impression of the issues involved and refer the reader to Chap. 6 for details.

In a special case of the geometric model the nuclear surface is given in spherical coordinates by

$$R(\Omega) = R_0\big(1 + \alpha(t) Y_{20}(\Omega)\big) \quad . \tag{1.8}$$

This will be seen to describe shapes close to ellipsoids a deformation α that is not too large. A time-dependent $\alpha(t)$ then describes oscillations in the shape of the nucleus, and if the nucleus is a sphere in its equilibrium state, it appears natural to assume harmonic vibrations around $\alpha = 0$, leading to a classical Lagrangian

$$L = \tfrac{1}{2} B \alpha^2 - \tfrac{1}{2} C \alpha^2 \tag{1.9}$$

with a *mass parameter* B and a *stiffness parameter* C, which are as yet undetermined. To develop this into, and exploit it as, a full-fledged model the following steps may be taken.

1. Quantize the Lagrangian. For the harmonic oscillator this is easy, but if the collective coordinate is defined in a more complicated way, quantization may be a problem. An example is given by the rotation–vibration model discussed in Sect. 6.4.
2. Determine the eigenstates. This is mostly a matter of mathematical prowess.
3. Set up expressions for other observables. It is not sufficient to determine the energy spectrum; calculating transition probabilities also requires operators such as the quadrupole operator, which has to be given in terms of the collective coordinates. Again, a physical model has to be employed to find such expressions.
4. Compare with experiment. Since the models always involve some undetermined parameters, one should concentrate on the *characteristic structure*. For the harmonic oscillator, for example, we would expect a spectrum with equally spaced levels (in the realistic case, one does indeed find an equidistant spectrum, but a rich angular-momentum structure is superimposed). If an experimental spectrum comes close to this, the spacing between the levels determines the oscillator frequency $\omega = \sqrt{C/B}$, and it turns out that a single transition probability is sufficient to determine B and C independently. This defines this simple model completely, and any additional quantity in experiment agreeing well with the model supports the model as a reasonable interpretation of the data. At this stage one might claim to have "gained an understanding" of the structure of a certain nucleus.
5. Finally, if possible, check whether the parameters B and C can be understood from a slightly deeper description of the physics; for example, the mass parameter B might be calculated from a model about the movement of the matter inside the nucleus during a surface vibration.

The final goal in this context is to unify microscopic and collective models by explaining the latter in terms of the former: collective models would be used to classify spectra and explain their structure whereas microscopic models should explain why collective coordinates of a certain type should lead to a viable model.

1.4 The Role of Symmetries

As was discussed in this chapter, many of the foundations of nuclear structure theory are still unclear. In this situation is is very important to use symmetry arguments to restrict the freedom inherent in setting up new models for the nucleon–nucleon interaction, the collective Hamiltonian, etc. A deep understanding of symmetry arguments in quantum physics is therefore essential for anyone interested in nuclear structure. The most important parts of this subject are repeated in the next chapter, where we give only as much detail as is actually needed for this book. Active work in nuclear structure theory requires much more than can be presented in the available space and suitable works on angular-momentum theory should be consulted.

2. Symmetries

2.1 General Remarks

In this chapter the principal symmetries of interest to nuclear physics and their mathematical treatment will be introduced. The importance of a good understanding of these mathematical developments cannot be overestimated; many very useful concepts, such as isospin, become quite obscure if one does not see the formal analogies, in this case to angular momentum.

Symmetry operations generally correspond to a mathematical group, i.e., they fulfil the following four properties.

1. If $\hat{S}$ and $\hat{S}'$ are two symmetry operations one can always define a *product* $\hat{T} = \hat{S} \cdot \hat{S}'$ such that $\hat{T}$ is also a symmetry operation of the same type (belongs to the same group). Generally one defines the product as the application of first $\hat{S}'$ and then $\hat{S}$ to the physical system. For example, for rotations the group property means that the result of carrying out two rotations one after the other should be representable by a single rotation.
2. Multiplication is associative, i.e., for all $\hat{S}$, $\hat{S}'$, and $\hat{S}''$ we have
$$\hat{S} \cdot (\hat{S}' \cdot \hat{S}'') = (\hat{S} \cdot \hat{S}') \cdot \hat{S}'' \quad . \tag{2.1}$$
3. There is an identity operation $\mathbb{1}$ which has the property that
$$\mathbb{1} \cdot \hat{S} = \hat{S} \quad , \quad \hat{S} \cdot \mathbb{1} = \hat{S} \quad . \tag{2.2}$$
Generally the identity is realized by the operation of "doing nothing", for example, the rotation about an angle of zero.
4. For each $\hat{S}$ there is an *inverse* $\hat{S}^{-1}$ such that
$$\hat{S} \cdot \hat{S}^{-1} = \mathbb{1} \quad , \quad \hat{S}^{-1} \cdot \hat{S} = \mathbb{1} \quad . \tag{2.3}$$
For example, the inverse of a rotation about some axis through an angle ϕ is a rotation about the same axis through an angle $-\phi$.

The groups of interest to us can be divided into two quite different types: groups whose elements depend on continuous parameters such as the translation and rotation groups, which are parametrized in terms of rotation angles and the translation vector, respectively, and groups consisting of discrete operations, for example, space or time inversion.

2.2 Translation

2.2.1 The Operator for Translation

Translational invariance provides a symmetry which is not too useful in nuclear physics, but which serves as a very simple case for illustrating many of the methods that will be used for the much more complicated case of rotation. Translational invariance can normally be taken into account quite simply, but there are cases in nuclear physics where it plays a more subtle role; for example, a phenomenological single-particle model with a prescribed potential violates translational invariance, because the potential has to be located at a fixed position in space, and special considerations have to be made to correct for this.

The translation of a point particle located at $\boldsymbol{r}$ with momentum $\boldsymbol{p}$ and spin $\boldsymbol{s}$ is defined by the operation

$$\boldsymbol{r} \rightarrow \boldsymbol{r}' = \boldsymbol{r} + \boldsymbol{a} \quad , \quad \boldsymbol{p} \rightarrow \boldsymbol{p}' = \boldsymbol{p} \quad , \quad \boldsymbol{s} \rightarrow \boldsymbol{s}' = \boldsymbol{s} \quad . \tag{2.4}$$

(Here, as in the following, the transformation of the spin $\boldsymbol{s}$ is included for completeness, although spin will be defined formally only in the later sections on angular momentum.) The vector $\boldsymbol{a}$ is the constant vector by which the particle is moved. This is the *active* view of the transformation; the *passive* view can be achieved by moving the coordinate system instead, which corresponds to moving the particle by $-\boldsymbol{a}$. These two formulations of transformations are equivalent but may lead to different signs in the formulas. In this book the active view will be used throughout.

If the particle is described by a wave function $\psi(\boldsymbol{r}, \boldsymbol{p}, \boldsymbol{s})$, the translated wave function may be defined simply by letting the value be carried along with the particle, i.e., its value at the new position $\boldsymbol{r}'$ is the same as that of the original wave function at $\boldsymbol{r}$:

$$\psi'(\boldsymbol{r}') = \psi'(\boldsymbol{r} + \boldsymbol{a}) = \psi(\boldsymbol{r}) \quad . \tag{2.5}$$

Its value at the point $\boldsymbol{r}$ is then given by moving both points by $-\boldsymbol{a}$:

$$\psi'(\boldsymbol{r}) = \psi(\boldsymbol{r} - \boldsymbol{a}) \quad . \tag{2.6}$$

In quantum mechanics the action of transforming $\psi(\boldsymbol{r})$ into $\psi'(\boldsymbol{r})$ has to be expressed as an operator $\hat{U}(\boldsymbol{a})$. This can be done by expanding (2.6) in a Taylor series:

$$\begin{aligned} \psi'(\boldsymbol{r}) &= \psi(\boldsymbol{r}) - \boldsymbol{a} \cdot \nabla\psi(\boldsymbol{r}) + \frac{1}{2!}(-\boldsymbol{a} \cdot \nabla)^2\psi(\boldsymbol{r}) - \ldots \\ &= \sum_{n=0}^{\infty} \frac{(-\boldsymbol{a} \cdot \nabla)^n}{n!}\psi(\boldsymbol{r}) \quad . \end{aligned} \tag{2.7}$$

Formally the sum can be written as an exponential function and the operator ∇ may be expressed through the momentum operator $\hat{\boldsymbol{p}} = -\mathrm{i}\hbar\nabla$, so that

$$\psi'(\boldsymbol{r}) = \exp(-\boldsymbol{a} \cdot \nabla)\psi(\boldsymbol{r}) = \exp\left(-\tfrac{\mathrm{i}}{\hbar}\boldsymbol{a} \cdot \hat{\boldsymbol{p}}\right)\psi(\boldsymbol{r}) \quad . \tag{2.8}$$

Thus the momentum operator is directly connected with translations; in fact, an expansion for small displacements $\boldsymbol{a}$ yields

$$\psi'(\boldsymbol{r}) \approx \left(1 - \tfrac{\mathrm{i}}{\hbar}\boldsymbol{a}\cdot\hat{\boldsymbol{p}}\right)\psi(\boldsymbol{r}) \quad , \tag{2.9}$$

so that one may call $\hat{\boldsymbol{p}}$ the *operator* or, alternatively, the *generator* for *infinitesimal* translations.

We have thus obtained the operator for translations

$$\hat{U}(\boldsymbol{a}) = \exp\left(-\tfrac{\mathrm{i}}{\hbar}\boldsymbol{a}\cdot\hat{\boldsymbol{p}}\right) \quad , \tag{2.10}$$

transforming wave functions according to

$$\psi'(\boldsymbol{r}) = \psi(\boldsymbol{r}-\boldsymbol{a}) = \hat{U}(\boldsymbol{a})\,\psi(\boldsymbol{r}) \quad . \tag{2.11}$$

To find out how position-dependent operators $\hat{A}(\boldsymbol{r})$ have to be transformed, just keep in mind that $\hat{A}(\boldsymbol{r})\psi(\boldsymbol{r})$ must transform like a wave function, so that

$$\begin{aligned}\hat{A}'(\boldsymbol{r})\psi'(\boldsymbol{r}) &= \hat{A}(\boldsymbol{r}-\boldsymbol{a})\,\psi(\boldsymbol{r}-\boldsymbol{a})\\ &= \hat{U}(\boldsymbol{a})\left(\hat{A}(\boldsymbol{r})\,\psi(\boldsymbol{r})\right)\\ &= \hat{U}(\boldsymbol{a})\,\hat{A}(\boldsymbol{r})\,\hat{U}^{-1}(\boldsymbol{a})\,\hat{U}(\boldsymbol{a})\,\psi(\boldsymbol{r}) \quad .\end{aligned} \tag{2.12}$$

So operators are transformed according to

$$\hat{A}' = \hat{U}\hat{A}\hat{U}^{-1} \quad , \tag{2.13}$$

and obviously this is a general result holding analogously for any transformation group.

If we use the power series expansion it is seen immediately that for the exponential of an operator

$$\exp\left(\hat{T}\right)^{\dagger} = \exp\left(\hat{T}^{\dagger}\right) \quad , \quad \exp\left(\hat{T}\right)^{-1} = \exp\left(-\hat{T}\right) \quad . \tag{2.14}$$

Since $\hat{\boldsymbol{p}}$ is a Hermitian operator, its product with an imaginary number changes sign under Hermitian conjugation and we have

$$\hat{U}^{\dagger}(\boldsymbol{a}) = \hat{U}^{-1}(\boldsymbol{a}) = \hat{U}(-\boldsymbol{a}) \quad . \tag{2.15}$$

The operator $\hat{U}(\boldsymbol{a})$ thus is *unitary*, so that it conserves the norm of wave functions as well as the matrix elements between them. That the inverse translation is the same as the translation by $-\boldsymbol{a}$ just states formally what is intuitively obvious.

2.2.2 Translational Invariance

These arguments are still applicable to arbitrary one-particle systems; the formulas derived simply express the action of translation on wave functions without assuming any invariance property. A physical system is *translationally invariant* if the Hamiltonian does not change under translations (note that all other arguments of $\hat{H}$ such as spin and momentum are omitted for brevity). For arbitrary $\boldsymbol{a}$ it must be true that

$$\hat{H}'(\boldsymbol{r}) = \hat{H}(\boldsymbol{r}-\boldsymbol{a}) = \hat{H}(\boldsymbol{r}) \quad . \tag{2.16}$$

This implies

$$\hat{H}(\boldsymbol{r}) = \hat{U}(\boldsymbol{a})\,\hat{H}(\boldsymbol{r})\,\hat{U}^{-1}(\boldsymbol{a}) \tag{2.17}$$

or, by multiplying with $\hat{U}(\boldsymbol{a})$ from the right,

$$\hat{H}(\boldsymbol{r})\,\hat{U}(\boldsymbol{a}) = \hat{U}(\boldsymbol{a})\,\hat{H}(\boldsymbol{r}) \quad , \tag{2.18}$$

i.e.,

$$\left[\hat{H}(\boldsymbol{r}), \hat{U}(\boldsymbol{a})\right] = 0 \quad , \tag{2.19}$$

so that the Hamiltonian commutes with the operator of translation for arbitrary $\boldsymbol{a}$. At this point it becomes advantageous to use (2.10): clearly $\hat{U}(\boldsymbol{a})$ will commute with $\hat{H}$ independent of the specific displacement $\boldsymbol{a}$, provided the momentum operator does. This leads to the simpler condition

$$\left[\hat{H}, \hat{p}\right] = 0 \quad . \tag{2.20}$$

Summarizing the above considerations we may conclude that *the properties of a physical system with respect to translation can all be expressed in terms of the momentum operator.*

2.2.3 Many-Particle Systems

Translation of a many-particle system leads naturally to the concept of total momentum, again giving a simple introduction to what will be more complex for angular momentum. Translating a system of N particles by the displacement $\boldsymbol{a}$ is expressed as

$$(\boldsymbol{r}_1, \boldsymbol{r}_2, \ldots, \boldsymbol{r}_N) \rightarrow (\boldsymbol{r}_1 + \boldsymbol{a}, \boldsymbol{r}_2 + \boldsymbol{a}, \ldots, \boldsymbol{r}_N + \boldsymbol{a}) \tag{2.21}$$

(the momenta and spins are not affected, as before). For the many-body wave function the transformation is given by

$$\psi'(\boldsymbol{r}_1, \boldsymbol{r}_2, \ldots, \boldsymbol{r}_N) = \psi(\boldsymbol{r}_1 - \boldsymbol{a}, \boldsymbol{r}_2 - \boldsymbol{a}, \ldots, \boldsymbol{r}_N - \boldsymbol{a}) \tag{2.22}$$

and one may apply the translational operators separately for each coordinate. Since they refer to different degrees of freedom, these operators commute and one may choose any ordering.

$$\psi'(\boldsymbol{r}_1, \boldsymbol{r}_2, \ldots, \boldsymbol{r}_N) = \hat{U}_1(\boldsymbol{a})\hat{U}_2(\boldsymbol{a})\cdots\hat{U}_N(\boldsymbol{a})\,\psi(\boldsymbol{r}_1, \boldsymbol{r}_2, \ldots, \boldsymbol{r}_N) \quad . \tag{2.23}$$

Here $\hat{U}_i(\boldsymbol{a})$ acts on coordinate number i:

$$\hat{U}_i(\boldsymbol{a}) = \exp(-\boldsymbol{a} \cdot \nabla_i) = \exp\left(-\tfrac{\mathrm{i}}{\hbar}\boldsymbol{a} \cdot \hat{\boldsymbol{p}}_i\right) \quad . \tag{2.24}$$

Again because of the commutation of the $\hat{\boldsymbol{p}}_i$ the arguments of the exponentials may be combined to yield

$$\psi'(\boldsymbol{r}_1, \boldsymbol{r}_2, \ldots, \boldsymbol{r}_N) = \exp\left(-\tfrac{\mathrm{i}}{\hbar}\boldsymbol{a} \cdot \hat{\boldsymbol{P}}\right)\,\psi(\boldsymbol{r}_1, \boldsymbol{r}_2, \ldots, \boldsymbol{r}_N) \quad , \tag{2.25}$$

where $\hat{\boldsymbol{P}}$ is the *total momentum* operator

$$\hat{P} = \sum_{i=1}^{N} \hat{p}_i \quad . \tag{2.26}$$

So the total momentum operator appears as the operator for simultaneous infinitesimal translation of all particles.

This also makes it easy to understand what translational invariance of a many-particle system implies in practice: the Hamiltonian should be invariant under *simultaneous* translation of *all* the particles. For example, for two-body interactions a Hamiltonian of standard form like

$$\hat{H} = \sum_{i=1}^{N} \frac{\hat{p}_i^2}{2m_i} + \frac{1}{2} \sum_{\substack{i,j=1 \\ i \neq j}}^{N} V(\boldsymbol{r}_i, \boldsymbol{r}_j) \tag{2.27}$$

will be invariant if the potential depends only on the relative positions $\boldsymbol{r}_i - \boldsymbol{r}_j$.

The canonically conjugate coordinate to the total momentum is the center-of-mass vector

$$\boldsymbol{R} = \frac{\sum_{i=1}^{N} m_i \, \boldsymbol{r}_i}{\sum_{i=1}^{N} m_i} \quad . \tag{2.28}$$

This can be proved easily by checking that the cartesian components $\hat{R}_k$ and $\hat{P}_{k'}$ fulfil

$$\left[\hat{R}_k, \hat{P}_{k'}\right] = \mathrm{i}\hbar \, \delta_{kk'} \quad . \tag{2.29}$$

2.3 Rotation

2.3.1 The Angular Momentum Operators

For the case of rotation we will proceed along the same lines as for translation, as far as is possible for this more complicated case. Let us make a few general remarks first, though. In this book only the fundamental definitions and methods will be discussed. Active research in nuclear theory requires a much deeper knowledge of angular-momentum theory, for which a number of textbooks can be recommended [Ro57, Ed60, Br68c]. These should also be consulted for possible differences in definitions, which can affect signs, notation, and even additional factors, for example, in the Wigner–Eckart theorem. The textbooks cited all use the set of definitions that seems to be almost universally accepted in nuclear theory nowadays, but the reader should carefully check which conventions are used, especially when consulting older papers. A comprehensive modern collection of angular-momentum formulas is given in the book by Varshalovich et al. [Va88]. To simplify the introductory developments, regard first a two-dimensional rotation (Fig. 2.1) in polar coordinates. The point $\boldsymbol{r} = (r, \phi)$ is carried into $\boldsymbol{r}' = (r, \phi + \theta)$ by a rotation through the angle θ. The rotation will be denoted by $\mathcal{R}(\theta)$. Similarly to the case of translations, we can define the rotated wave function ψ' by

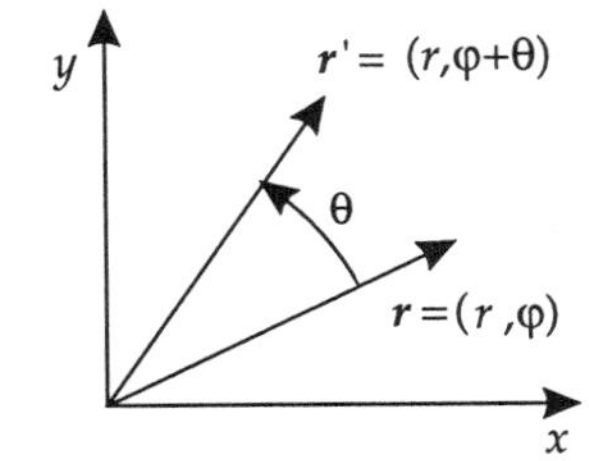

Fig. 2.1 Rotation in the plane through an angle θ, carrying the vector $\boldsymbol{r}$ into the vector $\boldsymbol{r}'$

$$\psi'(\boldsymbol{r}') = \psi(\boldsymbol{r}) \quad , \tag{2.30}$$

and its value at $\boldsymbol{r}$ is determined by the value of ψ at that point which is carried into $\boldsymbol{r}$ by the rotation:

$$\mathcal{R}(\theta)\, \psi(r, \phi) = \psi'(r, \phi) = \psi(r, \phi - \theta) \quad . \tag{2.31}$$

The shift in ϕ can also be expressed by a Taylor series expansion

$$\begin{aligned}\psi'(r, \phi) &= \sum_{n=0}^{\infty} \frac{(-\theta)^n}{n!} \frac{\partial^n}{\partial \phi^n} \psi(r, \phi) \\ &= \exp\left(-\theta \frac{\partial}{\partial \phi}\right) \psi(r, \phi) \\ &= \exp\left(-\tfrac{\mathrm{i}}{\hbar} \theta \hat{J}_z\right) \psi(r, \phi) \quad .\end{aligned} \tag{2.32}$$

The operator $\hat{J}_z$ for infinitesimal rotations can also be written in cartesian coordinates,

$$\hat{J}_z = -\mathrm{i}\hbar \frac{\partial}{\partial \phi} = -\mathrm{i}\hbar \left(x \frac{\partial}{\partial y} - y \frac{\partial}{\partial x}\right) \quad , \tag{2.33}$$

and thus is identical with the angular-momentum operator.

EXERCISE

2.1 Cartesian Form of the Angular-Momentum Operator $\hat{J}_z$

Problem. Derive the cartesian form of the angular-momentum operator $\hat{J}_z$.

Solution. One way is to simply transform the expression in cylindrical coordinates to cartesian ones. It is more instructive, however, to go back to the definition of the rotated wave function. A rotation through an angle $-\theta$ is given by $x \to x\cos\theta + y\sin\theta$ and $y \to y\cos\theta - x\sin\theta$, so that

$$\mathcal{R}(\theta)\, \psi(x, y) = \psi(x\cos\theta + y\sin\theta, y\cos\theta - x\sin\theta) \quad , \tag{1}$$

which for small θ reduces to

$$\begin{aligned}\mathcal{R}(\theta)\, \psi(x, y) &\approx \psi(x + y\theta, y - x\theta) \\ &= \left(1 + y\frac{\partial}{\partial x} - x\frac{\partial}{\partial y}\right) \psi(x, y) \quad .\end{aligned} \tag{2}$$

Comparison with the corresponding small-angle result

$$\mathcal{R}(\theta) = \exp\left(-\tfrac{\mathrm{i}}{\hbar}\theta \hat{J}_z\right) \approx \left(1 - \tfrac{\mathrm{i}}{\hbar}\theta \hat{J}_z\right) \tag{3}$$

immediately yields the above result for $\hat{J}_z$.

The rotation of the vector $\boldsymbol{r}$ itself in cartesian coordinates may be written in matrix form as

$$\begin{pmatrix} x' \\ y' \end{pmatrix} = \begin{pmatrix} \cos\theta & -\sin\theta \\ \sin\theta & \cos\theta \end{pmatrix} \begin{pmatrix} x \\ y \end{pmatrix} \quad . \tag{2.34}$$

If the rotation matrix is expanded to first order for small θ, we also get a *matrix representation for* $\hat{J}_z$:

$$\begin{pmatrix} x' \\ y' \end{pmatrix} \approx \begin{pmatrix} 1 & -\theta \\ \theta & 1 \end{pmatrix} \begin{pmatrix} x \\ y \end{pmatrix} = \left[\mathbb{1} - \theta \begin{pmatrix} 0 & 1 \\ -1 & 0 \end{pmatrix} \right] \begin{pmatrix} x \\ y \end{pmatrix} \quad , \tag{2.35}$$

so that

$$J_z = -\mathrm{i}\hbar \begin{pmatrix} 0 & 1 \\ -1 & 0 \end{pmatrix} \quad . \tag{2.36}$$

We now show that the finite rotation can be recovered from this matrix. This evaluation of the exponential function of a matrix is quite instructive in itself, because the trick used always recurs in such cases: it can be applied if some power of the matrix in the exponent is proportional to the identity matrix. In the present case we have

$$\begin{aligned} \exp\left(-\tfrac{\mathrm{i}}{\hbar}\theta J_z\right) &= \sum_{n=0}^{\infty} \frac{(-\theta)^n}{n!} \begin{pmatrix} 0 & 1 \\ -1 & 0 \end{pmatrix}^n \\ &= \sum_{n=0}^{\infty} \frac{(-\theta)^{2n}}{(2n)!} \begin{pmatrix} 0 & 1 \\ -1 & 0 \end{pmatrix}^{2n} + \sum_{n=0}^{\infty} \frac{(-\theta)^{2n+1}}{(2n+1)!} \begin{pmatrix} 0 & 1 \\ -1 & 0 \end{pmatrix}^{2n+1} \\ &= \begin{pmatrix} 1 & 0 \\ 0 & 1 \end{pmatrix} \sum_{n=0}^{\infty} \frac{(-1)^n \theta^{2n}}{(2n)!} - \begin{pmatrix} 0 & 1 \\ -1 & 0 \end{pmatrix} \sum_{n=0}^{\infty} \frac{(-1)^n \theta^{2n+1}}{(2n+1)!} \\ &= \cos\theta \begin{pmatrix} 1 & 0 \\ 0 & 1 \end{pmatrix} - \sin\theta \begin{pmatrix} 0 & 1 \\ -1 & 0 \end{pmatrix} \\ &= \begin{pmatrix} \cos\theta & -\sin\theta \\ \sin\theta & \cos\theta \end{pmatrix} \quad , \end{aligned} \tag{2.37}$$

where we have made use of

$$\begin{pmatrix} 0 & 1 \\ -1 & 0 \end{pmatrix}^2 = -\begin{pmatrix} 1 & 0 \\ 0 & 1 \end{pmatrix} \quad , \tag{2.38}$$

so that in general for even powers

$$\begin{pmatrix} 0 & 1 \\ -1 & 0 \end{pmatrix}^{2n} = (-1)^n \begin{pmatrix} 1 & 0 \\ 0 & 1 \end{pmatrix} \tag{2.39}$$

and for odd powers

$$\begin{pmatrix} 0 & 1 \\ -1 & 0 \end{pmatrix}^{2n+1} = (-1)^n \begin{pmatrix} 0 & 1 \\ -1 & 0 \end{pmatrix} \quad . \tag{2.40}$$

This allowed the matrices to be taken out of the sums and the sums to be replaced by trigonometric functions.

The representation of finite transformations through the exponential function of the operator for infinitesimal transformations, which can in turn be expressed either as a differential operator or as a matrix, is called the *exponential representation.*

An alternative way of deriving the angular-momentum operator rotations is in terms of derivatives of finite rotations. For both the matrix and the differential operator versions we can write

$$\hat{J}_z = \mathrm{i}\hbar \left. \frac{\partial \mathcal{R}(\theta)}{\partial \theta} \right|_{\theta=0} \quad , \tag{2.41}$$

which simply results from its definition as the first-order coefficient in the Taylor series.

In the case of three dimensions there are three degrees of freedom for rotations. Examining rotations around the three cartesian axes leads to a simple generalization of the results for two dimensions. For rotations about the z axis

$$\begin{pmatrix} x' \\ y' \\ z' \end{pmatrix} = \begin{pmatrix} \cos\theta_z & -\sin\theta_z & 0 \\ \sin\theta_z & \cos\theta_z & 0 \\ 0 & 0 & 1 \end{pmatrix} \begin{pmatrix} x \\ y \\ z \end{pmatrix} \quad , \tag{2.42}$$

so that according to (2.41) the matrix for the angular momentum is

$$J_z = -\mathrm{i}\hbar \begin{pmatrix} 0 & 1 & 0 \\ -1 & 0 & 0 \\ 0 & 0 & 0 \end{pmatrix} \quad . \tag{2.43}$$

Similarly for rotations about the y axis

$$\begin{pmatrix} x' \\ y' \\ z' \end{pmatrix} = \begin{pmatrix} \cos\theta_y & 0 & \sin\theta_y \\ 0 & 1 & 0 \\ -\sin\theta_y & 0 & \cos\theta_y \end{pmatrix} \begin{pmatrix} x \\ y \\ z \end{pmatrix} \quad . \tag{2.44}$$

The signs in the matrix keep account of the fact that a positive rotation about the y axis turns the x axis into the *negative* z direction. The associated angular-momentum matrix is

$$J_y = -\mathrm{i}\hbar \begin{pmatrix} 0 & 0 & -1 \\ 0 & 0 & 0 \\ 1 & 0 & 0 \end{pmatrix} \quad . \tag{2.45}$$

Finally for a rotation about the x axis

$$\begin{pmatrix} x' \\ y' \\ z' \end{pmatrix} = \begin{pmatrix} 1 & 0 & 0 \\ 0 & \cos\theta_x & -\sin\theta_x \\ 0 & \sin\theta_x & \cos\theta_x \end{pmatrix} \begin{pmatrix} x \\ y \\ z \end{pmatrix} \tag{2.46}$$

and

$$J_x = -\mathrm{i}\hbar \begin{pmatrix} 0 & 0 & 0 \\ 0 & 0 & 1 \\ 0 & -1 & 0 \end{pmatrix} \quad . \tag{2.47}$$

The angular-momentum matrices fulfil the familiar commutation relations for angular momentum

$$[J_x, J_y] = \mathrm{i}\hbar J_z \quad , \quad [J_y, J_z] = \mathrm{i}\hbar J_x \quad , \quad [J_z, J_x] = \mathrm{i}\hbar J_y \quad . \tag{2.48}$$

The same commutation relations hold for the representation of the angular momenta as differential operators

$$\begin{aligned} \hat{J}_x &= -\mathrm{i}\hbar\left(y\frac{\partial}{\partial z} - z\frac{\partial}{\partial y}\right) \quad , \\ \hat{J}_y &= -\mathrm{i}\hbar\left(z\frac{\partial}{\partial x} - x\frac{\partial}{\partial z}\right) \quad , \\ \hat{J}_z &= -\mathrm{i}\hbar\left(x\frac{\partial}{\partial y} - y\frac{\partial}{\partial x}\right) \quad . \end{aligned} \tag{2.49}$$

The angular-momentum operators form a *Lie algebra*, whose properties are determined by the commutation relations. We will see that these commutation relations by themselves determine the properties of finite rotations to a large extent, and usually it is much simpler mathematically to study the Lie algebra instead of the rotation group itself. Formally a Lie algebra is a set of operators which is closed under linear combination and under the commutation relations: a commutator of two elements of the Lie algebra must be expressible as a linear combination of elements of the algebra. The commutation relations of the angular-momentum algebra clearly have this property.

Finite rotations about any of these axes may be written in the exponential representation as

$$\mathcal{R}(\theta_k) = \exp\left(-\tfrac{\mathrm{i}}{\hbar}\theta_k \hat{J}_k\right) \quad , \quad k \in \{x, y, z\} \quad . \tag{2.50}$$

There is a problem, however, in expressing arbitrary rotations, because rotations about the different axes do not commute. If one has a composite rotation denoted formally by $\boldsymbol{\theta} = (\theta_x, \theta_y, \theta_z)$, one should specify precisely in which order the rotations about the three coordinate axes should be performed. Furthermore, it is not clear whether a given finite rotation can be parametrized *uniquely* by a $\boldsymbol{\theta}$ and how such a parametrization could be determined in practice. We will later see that for this purpose it is better to use the *Euler angles* and not the rotations about the three axes, and for the formal developments in this chapter we merely assume that finite rotations $\mathcal{R}(\boldsymbol{\theta})$ can somehow be defined uniquely.

A final remark concerning the group terminology: rotations in three dimensions are represented by real 3×3 matrices and conserve the scalar product between vectors. The condition that

$$\boldsymbol{a} \cdot \boldsymbol{b} = \boldsymbol{a}' \cdot \boldsymbol{b}' = \left(\mathcal{R}(\boldsymbol{\theta})\, \boldsymbol{a}\right) \cdot \left(\mathcal{R}(\boldsymbol{\theta})\, \boldsymbol{b}\right) \tag{2.51}$$

can be rewritten in matrix notation as

$$a^{\mathrm{T}} b = \left(\mathcal{R}(\boldsymbol{\theta})\, a\right)^{\mathrm{T}} \mathcal{R}(\boldsymbol{\theta})\, b = a^{\mathrm{T}} \mathcal{R}^{\mathrm{T}}(\boldsymbol{\theta})\, \mathcal{R}(\boldsymbol{\theta})\, b \quad , \tag{2.52}$$

so that the matrix for $\mathcal{R}(\boldsymbol{\theta})$ must fulfil the orthogonality condition

$$\mathcal{R}^{\mathrm{T}}(\boldsymbol{\theta})\, \mathcal{R}(\boldsymbol{\theta}) = \mathbb{1} \quad . \tag{2.53}$$

Since

$$\det \mathcal{R}^{\mathrm{T}}(\boldsymbol{\theta}) = \det \mathcal{R}(\boldsymbol{\theta}) \tag{2.54}$$

orthogonal matrices can have a determinant of either +1 or −1. Those with a negative determinant should not be included, because they change the handedness of the coordinate system and are thus not true rotations (they are rotations combined with a space inversion).

Thus we may conclude that rotations in three dimensions are represented by matrices belonging to the *special orthogonal group* SO(3), i.e., the group consisting of all 3×3 matrices $\mathcal{R}$ fulfilling

$$\mathcal{R}^{\mathrm{T}} \mathcal{R} = 1\!\!1 \quad , \quad \det \mathcal{R} = 1\!\!1 \quad . \tag{2.55}$$

2.3.2 Representations of the Rotation Group

We have seen that the angular-momentum operators may be represented by matrices. In the same way, any abstract rotation $\mathcal{R}(\boldsymbol{\theta})$ acting on a wave function can be represented by matrices via a basis expansion. Expanding

$$\psi'(\boldsymbol{r}) = \mathcal{R}(\boldsymbol{\theta})\, \psi(\boldsymbol{r}) \tag{2.56}$$

in a complete orthonormal basis $\phi_i(\boldsymbol{r})$ and taking the overlap with ϕ_i on both sides yields

$$\begin{aligned} \langle \phi_i | \psi' \rangle &= \langle \phi_i | \mathcal{R}(\boldsymbol{\theta}) | \psi \rangle \\ &= \sum_j \langle \phi_i | \mathcal{R}(\boldsymbol{\theta}) | \phi_j \rangle \langle \phi_j | \psi \rangle \quad . \end{aligned} \tag{2.57}$$

So the abstract rotation $\mathcal{R}(\boldsymbol{\theta})$ is "represented" by the matrix with elements

$$R_{ij}(\boldsymbol{\theta}) = \langle \phi_i | \mathcal{R}(\boldsymbol{\theta}) | \phi_j \rangle \quad , \tag{2.58}$$

and clearly the representation must reproduce the group structure in the sense that products and inverses in the group are represented by products and inverses of matrices:

$$\begin{aligned} \mathcal{R}(\boldsymbol{\theta}'') = \mathcal{R}(\boldsymbol{\theta})\, \mathcal{R}(\boldsymbol{\theta}') \quad &\rightarrow \quad R_{ij}(\boldsymbol{\theta}'') = \sum_k R_{ik}(\boldsymbol{\theta})\, R_{kj}(\boldsymbol{\theta}') \quad , \\ \mathcal{R}(\boldsymbol{\theta}') = \mathcal{R}^{-1}(\boldsymbol{\theta}) \quad &\rightarrow \quad R_{ij}(\boldsymbol{\theta}') = R^{-1}{}_{ij}(\boldsymbol{\theta}) \quad . \end{aligned} \tag{2.59}$$

The set of matrices is called a *representation* of the rotation group. It need not be *faithful*, i.e., different rotations may be represented by the same matrix – a simple example is a scalar wave function, for which all rotations correspond to the identity transformation. The number of basic functions ϕ_i, which is also the dimension of the matrices, is called the dimension of the representation.

In many cases representations can be reduced to simpler ones. For example, the total space of wave functions may be decomposable into two invariant subspaces, with the wave functions of each subspace mixing among themselves only under rotations. With a suitable choice of basis the matrices of the representation then all take the form

$$R_{ij}(\boldsymbol{\theta}) = \left(\begin{array}{c|c} R_{ij}^{(1)}(\boldsymbol{\theta}) & 0 \\ \hline 0 & R_{ij}^{(2)}(\boldsymbol{\theta}) \end{array}\right) \quad , \tag{2.60}$$

where both $R_{ij}^{(1)}(\boldsymbol{\theta})$ and $R_{ij}^{(2)}(\boldsymbol{\theta})$ are representations of smaller dimension. Of special interest are the *irreducible* representations, which cannot be decomposed in this sense. For the rotation group it can be shown that all representations can be built up from finite-dimensional irreducible representations.

In practice it is usually much easier to construct the representations of the Lie algebra, in the sense of constructing matrices with the correct commutation properties, and then to obtain those of the group by the exponential formula. For the rotation group this means that we need to determine only the matrices representing the three operators $\hat{J}_x$, $\hat{J}_y$, and $\hat{J}_z$. Because all of these operators commute with $\hat{J}^2$, the eigenvalue of $\hat{J}^2$ cannot be changed by any rotation and it must be the same within one irreducible representation. As the components of the angular momentum vector do not commute amongst themselves, only one of them can be chosen to be diagonal in addition to $\hat{J}^2$.

Let us try then to use a basis diagonal in both $\hat{J}^2$ and $\hat{J}_z$,

$$\hat{J}^2|jm\rangle = \hbar^2 \Lambda_j |jm\rangle \quad , \quad \hat{J}_z|jm\rangle = \hbar m|jm\rangle \quad , \tag{2.61}$$

with j constant and m varying in a range still to be determined (elementary quantum mechanics suggests that although the eigenvalue of $\hat{J}_z$ will turn out to be $\hbar m$, that of $\hat{J}^2$ will not be as simple, so that we tentatively write Λ_j). Because $\hat{J}_x$ and $\hat{J}_y$ do not commute with $\hat{J}_z$, they cannot simultaneously be chosen to be diagonal. Instead of studying the action of $\hat{J}_x$ and $\hat{J}_y$ on these wave functions, it is simpler to replace them by *shift operators*

$$\hat{J}_+ = \hat{J}_x + \mathrm{i}\hat{J}_y \quad , \quad \hat{J}_- = \hat{J}_x - \mathrm{i}\hat{J}_y \quad . \tag{2.62}$$

The meaning of the term “shift operator” becomes clear through the commutation relations

$$\left[\hat{J}_z, \hat{J}_\pm\right] = \pm\hbar\hat{J}_\pm \quad , \tag{2.63}$$

which allows us to compute the eigenvalue corresponding to $\hat{J}_\pm|jm\rangle$:

$$\begin{aligned} \hat{J}_z\left(\hat{J}_\pm|jm\rangle\right) &= \hat{J}_\pm\hat{J}_z|jm\rangle \left[\hat{J}_\pm, \hat{J}_z\right]|jm\rangle \\ &= \hbar(m \pm 1)|jm\rangle \quad . \end{aligned} \tag{2.64}$$

This shows that $\hat{J}_\pm$ shifts the eigenvalue of $\hat{J}_z$ by $\pm\hbar$, or the value of m by ± 1.

The underlying idea, which will be used several times in this book, is that if two operators A and B have a commutation relation of the form

$$\left[A, B\right] = \beta B \tag{2.65}$$

with β some number, then the same calculation as above shows that B shifts the eigenvalue of A by β.

Given a particular basis state $|jm\rangle$, by using the operators $\hat{J}_\pm$ we can successively construct the states $|jm \pm 1\rangle$, $|jm \pm 2\rangle$, etc. Since we are looking for representations of finite dimension, this process must end somewhere (in fact it can be shown that "all irreducible unitary representations of a connected simple compact group" are finite-dimensional – details may be found in textbooks on group theory . Let us only remark that this theorem applies to the rotation group). Use μ for the largest value attainable by m. Then we must have

$$\hat{J}_+|j\mu\rangle = 0 \quad , \tag{2.66}$$

because anything else would imply the existence of a vector with eigenvalue $\mu+1$. Now compute the action of $\hat{J}^2$ on $|j\mu\rangle$ using

$$\hat{J}^2 = \tfrac{1}{2}\left(\hat{J}_+\hat{J}_- + \hat{J}_-\hat{J}_+\right) + \hat{J}_z^2 \quad . \tag{2.67}$$

This may be rewritten using the commutation relation

$$\left[\hat{J}_+, \hat{J}_-\right] = 2\hbar\hat{J}_z \tag{2.68}$$

to yield

$$\hat{J}^2 = \hat{J}_-\hat{J}_+ + \hat{J}_z^2 + \hbar\hat{J}_z \quad , \tag{2.69}$$

and the first term on the right produces zero on $|j\mu\rangle$, while $\hat{J}_z$ has the eigenvalue $\hbar\mu$, so that we get the eigenvalue of $\hat{J}^2$:

$$\hat{J}^2|j\mu\rangle = \hbar^2\mu(\mu+1)|j\mu\rangle \tag{2.70}$$

as

$$\Lambda_j = \hbar^2\mu(\mu+1) \quad . \tag{2.71}$$

Up to now the variable j had no direct physical meaning; it only enumerated the eigenvalues of $\hat{J}^2$. The preceding equation suggests we use the maximal eigenvalue of $\hat{J}_z$ instead. Thus we identify j with μ and keep the letter j. Then the states of the representation fulfil

$$\hat{J}^2|jm\rangle = \hbar^2 j(j+1)|jm\rangle \quad , \quad \hat{J}_z|jm\rangle = \hbar m|jm\rangle \quad , \tag{2.72}$$

and the one with the largest eigenvalue of $\hat{J}_z$ is $|jj\rangle$.

Now the same arguments can be applied to the lowest possible value of m. Assume that for some positive number n we have

$$\hat{J}_-|j, j-n\rangle = 0 \tag{2.73}$$

to stop the generation of states on the lower end. The angular momentum squared of (2.67) can also be expressed as

$$\hat{J}^2 = \hat{J}_+\hat{J}_- + \hat{J}_z^2 - \hat{J}_z \quad , \tag{2.74}$$

and again the first term will yield zero, so that

$$\hat{J}^2|j, j-n\rangle = \hbar^2\left[(j-n)^2 - j + n\right]|j, j-n\rangle \quad . \tag{2.75}$$

Since the eigenvalue of $\hat{J}^2$ must still be the same, it must be true that

$$j(j+1) = (j-n)^2 - (j-n) \quad , \tag{2.76}$$

which is a quadratic equation for n with the only positive solution $n = 2j$. This makes the lowest projection equal to $-j$.

The representation we have constructed thus has the basis states

$$|jm\rangle \quad , \quad m = -j, -j+1, \ldots, j \quad , \tag{2.77}$$

which are $2j+1$ in number and are eigenstates of the angular momentum squared and of the z projection according to

$$\hat{J}^2 |jm\rangle = \hbar^2 j(j+1) |jm\rangle \quad , \quad \hat{J}_z |jm\rangle = \hbar m |jm\rangle \quad . \tag{2.78}$$

The construction of the basis with the shift operators did not explicitly normalize the states, but this can be done easily if we note that $\hat{J}_+$ is just the Hermitian conjugate of $\hat{J}_-$, so that the norm of the state $\hat{J}_\pm |jm\rangle$ is given by

$$\begin{aligned} \langle jm|\hat{J}_\mp \hat{J}_\pm |jm\rangle &= \langle jm|\hat{J}^2 - \hat{J}_z^2 \mp \hat{J}_z|jm\rangle \\ &= \hbar^2 \left[j(j+1) - m^2 \mp m \right] \\ &= \hbar^2 (j \pm m + 1)(j \mp m) \quad . \end{aligned} \tag{2.79}$$

The reciprocal square root of this expression will be the normalizing factor for the state $|j, m \pm 1\rangle$, and the matrix elements of the shift operators also result as

$$\langle j, m \pm 1 | \hat{J}_\pm | jm \rangle = \hbar \sqrt{(j \pm m + 1)(j \mp m)} \quad . \tag{2.80}$$

These matrix elements define the matrix representations of the shift operators and hence of the operators $\hat{J}_x$ and $\hat{J}_y$ via

$$\hat{J}_x = \tfrac{1}{2} \left(\hat{J}_+ + \hat{J}_- \right) \quad , \quad \hat{J}_y = -\tfrac{\mathrm{i}}{2} \left(\hat{J}_+ - \hat{J}_- \right) \tag{2.81}$$

as

$$\begin{aligned} \langle j, m'|\hat{J}_x|jm\rangle &= \tfrac{\hbar}{2} \Big(\sqrt{(j+m+1)(j-m)}\, \delta_{m',m+1} \\ &\qquad + \sqrt{(j-m+1)(j+m)}\, \delta_{m',m-1} \Big) \quad , \\ \langle j, m'|\hat{J}_y|jm\rangle &= -\tfrac{\mathrm{i}\hbar}{2} \Big(\sqrt{(j+m+1)(j-m)}\, \delta_{m',m+1} \\ &\qquad - \sqrt{(j-m+1)(j+m)}\, \delta_{m',m-1} \Big) \quad , \\ \langle jm'|\hat{J}_z|jm\rangle &= \hbar m\, \delta_{m'm} \quad , \end{aligned} \tag{2.82}$$

where for completeness the matrix elements of $\hat{J}_z$ are also given. Thus we have represented the full Lie algebra by $(2j+1)\times(2j+1)$ matrices. There is a freedom of choice in the phases of the matrix elements, since multiplying them by an arbitrary phase would not change the normalization argument. The phase chosen here is the *Condon and Shortley phase* and is the usual choice. Another choice of phase will be used in the BCS model for pairing (Chap. 7.5).

2.3.3 The Rotation Matrices

Finite rotations are quite a bit more complicated. It is important to find a unique parametrization of these rotations in terms of rotations performed about certain axes in a prescribed order. The order is important because rotations about different axes do not in general commute. The most familiar set of angles are the *Euler angles* $\boldsymbol{\theta} = (\theta_1, \theta_2, \theta_3)$, which are defined as follows (all rotations performed counterclockwise).

1. Rotate the system through an angle θ_1 about the z axis, yielding new axes x', y', and z'.
2. The second rotation is through an angle θ_2 about the *new* axis y'. This yields (x'', y'', z'').
3. Finally, rotate through θ_3 about the *new* axis z'' produced in the first two steps.

What is the advantage of using the Euler angles compared with the rotations about the three Cartesian axes that we have considered with up to now? It is much easier to determine the Euler angles needed to orient a body uniquely in space by using the first two angles to orient its body-fixed z axis correctly and then to turn it about that axis into the correct position. Also, for the Euler angles two of the rotations involve a diagonal $\hat{J}_z$. On the other hand, for the infinitesimal rotations the Euler angles are useless because for $\theta_2 \approx 0$, θ_1 and θ_3 rotate about the same axis and do not lead to independent angular-momentum operators.

To write these rotations in terms of angular-momentum operators, we have to define $\hat{J}_{y'}$ as the operator for infinitesimal rotations about the rotated y' axis needed for θ_2, and $\hat{J}_{z''}$ as that for the final z'' axis needed for θ_3. The rotation operator is then given by

$$\mathcal{R}(\boldsymbol{\theta}) = \exp\left(-\tfrac{\mathrm{i}}{\hbar}\theta_3\hat{J}_{z''}\right)\exp\left(-\tfrac{\mathrm{i}}{\hbar}\theta_2\hat{J}_{y'}\right)\exp\left(-\tfrac{\mathrm{i}}{\hbar}\theta_1\hat{J}_z\right) \quad . \tag{2.83}$$

This is very difficult to use as it stands, but can fortunately be transformed into a form with rotations about fixed axes. Instead of rotating the system about y', we can clearly first go back to the original axes, rotate about the *original* y axis, and then rotate back to the primed coordinate system. Expressing this by operators replaces this part of the rotation operator by

$$\exp\left(-\tfrac{\mathrm{i}}{\hbar}\theta_2\hat{J}_{y'}\right) = \exp\left(-\tfrac{\mathrm{i}}{\hbar}\theta_1\hat{J}_z\right)\exp\left(-\tfrac{\mathrm{i}}{\hbar}\theta_2\hat{J}_y\right)\exp\left(\tfrac{\mathrm{i}}{\hbar}\theta_1\hat{J}_z\right) \quad , \tag{2.84}$$

and a similar argument for θ_3 yields

$$\begin{aligned}\exp\left(-\tfrac{\mathrm{i}}{\hbar}\theta_3\hat{J}_{z''}\right) &= \exp\left(-\tfrac{\mathrm{i}}{\hbar}\theta_2\hat{J}_{y'}\right)\exp\left(-\tfrac{\mathrm{i}}{\hbar}\theta_1\hat{J}_z\right)\\ &\quad\times\exp\left(-\tfrac{\mathrm{i}}{\hbar}\theta_3\hat{J}_z\right)\exp\left(\tfrac{\mathrm{i}}{\hbar}\theta_1\hat{J}_z\right)\exp\left(\tfrac{\mathrm{i}}{\hbar}\theta_2\hat{J}_{y'}\right)\end{aligned} \tag{2.85}$$

with the two other rotations reversed and then reapplied. Inserting this result into (2.83), cancelling the operators as far as possible, and then using the formula (2.84) for the remaining $\hat{J}_{y'}$ term finally yields

$$\mathcal{R}(\boldsymbol{\theta}) = \exp\left(-\tfrac{\mathrm{i}}{\hbar}\theta_1\hat{J}_z\right)\exp\left(-\tfrac{\mathrm{i}}{\hbar}\theta_2\hat{J}_y\right)\exp\left(-\tfrac{\mathrm{i}}{\hbar}\theta_3\hat{J}_z\right) \quad , \tag{2.86}$$

i.e., the interesting fact that one produces the same rotation by rotating about the *fixed original axes* with the order of the Euler angles reversed.

The matrices for these rotations in the irreducible representation with angular momentum j are defined by

$$\mathcal{D}^{(j)}_{m'm}(\boldsymbol{\theta}) = \langle jm'|\mathcal{R}(\boldsymbol{\theta})|jm\rangle \quad , \tag{2.87}$$

which implies that a state $|jm\rangle$ transforms into

$$|jm\rangle' = \sum_{m'} |jm'\rangle \mathcal{D}^{(j)}_{m'm}(\boldsymbol{\theta}) \quad . \tag{2.88}$$

We will not need many explicit properties of the matrices $\mathcal{D}^{(j)}_{m'm}(\boldsymbol{\theta})$; special forms will be given when needed. It is useful, however, to reduce the matrix to simpler form by noting that in (2.86) the first and last operator is diagonal in the basis $|jm\rangle$, so that one may simplify the matrix to

$$\mathcal{D}^{(j)}_{m'm}(\boldsymbol{\theta}) = \exp\left[-\tfrac{\mathrm{i}}{\hbar}(\theta_1 m' + \theta_3 m)\right] d^{(j)}_{m'm}(\theta_2) \quad . \tag{2.89}$$

In this way the dependence on two of the angles has become trivial, and the only complicated function is the *reduced rotation matrix* $d^{(j)}_{m'm}(\theta_2)$.

2.3.4 SU(2) and Spin

Let us return to the question of which values of j are actually possible. The construction of the representations of the angular-momentum algebra led to the condition that $2j$ should be integer, a consequence of (2.76). So the somewhat surprising result is that j may be integer or half-integer, leading to the natural appearance of spin in the latter case. The representations with half-integer angular momentum cannot correspond to normal rotations of classical objects. To see this, simply examine a rotation for an angle of 2π. For the z axis, for example, this is given by

$$\mathcal{R}(\theta_z = 2\pi) = \exp\left(-2\pi\tfrac{\mathrm{i}}{\hbar}\hat{J}_z\right) \quad , \tag{2.90}$$

and for a state with angular-momentum projection $\frac{1}{2}$, for example, it will be multiplied by a factor of $\exp(-\pi\mathrm{i}) = -1$. For wave functions there is nothing wrong with this, as all measurable quantities lead to matrix elements containing the square of this factor. But if a classical object such as a vector is rotated by 2π, it should always be transformed into itself. Certainly the group SO(3) that we started with has the property that all rotations about an angle of 2π revert to the identity matrix, so how can it be that its representations do not have this property in general? The answer is that the half-integer representations are not representations of SO(3).

Remember that we constructed the representations from the angular-momentum operators. These operators together with their commutation relations form the Lie algebra associated with the Lie group of the rotations, for which they determine the infinitesimal rotations. Now it turns out that the Lie algebra does not completely determine the associated Lie group, and in this special case the groups SO(3) and SU(2) have the same Lie algebra. Because an understanding of SU(2) and its relation to spin is very important for more general applications such as isospin, it is worth explaining SU(2) in some more detail.

The group SU(2) is the *special unitary* group of all 2×2 unitary matrices with determinant 1, i.e., matrices U fulfilling

$$U^\dagger U = 1\!\!1 \quad , \quad \det U = 1 \quad . \tag{2.91}$$

Its relation to rotations has already been used in classical mechanics in connection with the *Cayley–Klein* parameters, and here we sketch the derivation. Examine a general 2×2 complex matrix

$$U = \begin{pmatrix} a & b \\ c & d \end{pmatrix} \quad . \tag{2.92}$$

Requiring the determinant to be 1 yields the condition

$$ad - bc = 1 \quad , \tag{2.93}$$

whereas the unitarity condition reads

$$\begin{pmatrix} a^* & c^* \\ b^* & d^* \end{pmatrix} \begin{pmatrix} a & b \\ c & d \end{pmatrix} = \begin{pmatrix} 1 & 0 \\ 0 & 1 \end{pmatrix} \quad , \tag{2.94}$$

or, explicitly,

$$a^*a + c^*c = 1 \quad , \tag{2.95a}$$
$$b^*b + d^*d = 1 \quad , \tag{2.95b}$$
$$a^*b + c^*d = 0 \quad , \tag{2.95c}$$
$$b^*a + d^*c = 0 \quad . \tag{2.95d}$$

The last equation is simply the complex conjugate of the preceding one and can thus be omitted. These conditions allow one to reduce the number of degrees of freedom in the matrices. From (2.95c) we have

$$d = -\frac{a^*b}{c^*} \quad , \tag{2.96}$$

and inserting this into (2.93) yields

$$1 = -\frac{a^*ab}{c^*} - cb = -(a^*a + c^*c)\frac{b}{c^*} = -\frac{b}{c^*} \quad , \tag{2.97}$$

where the content of the parentheses is equal to 1 because of (2.95a). So we must have $b = -c^*$ and, inserting this again into (2.96), also $d = a^*$. Then (2.95b) is automatically fulfilled and the matrix can be written in the more specific form

$$U = \begin{pmatrix} a & b \\ -b^* & a^* \end{pmatrix} \tag{2.98}$$

with the subsidiary condition

$$a^*a + b^*b = 1 \quad . \tag{2.99}$$

Of the four complex numbers only two are left and because of the subsidiary condition three real degrees of freedom remain, identical in number to the three degrees of freedom of rotations in three dimensions.

To see the relation to these rotations, associate with a vector $\boldsymbol{r} = (x, y, z)$ a matrix

$$P(\boldsymbol{r}) = \begin{pmatrix} z & x - \mathrm{i}y \\ x + \mathrm{i}y & -z \end{pmatrix} \quad . \tag{2.100}$$

This matrix is hermitian and has trace zero. Conversely, any 2×2 hermitian matrix with trace zero defines a vector in three dimensions, which can be obtained simply by reading the components from the real and imaginary parts of the matrix elements. Now investigate transformations of the form

$$P' = UPU^{\dagger} \quad , \tag{2.101}$$

where U is any matrix in SU(2). Does this describe a rotation of the associated vector? The matrix P' is also hermitian because

$$P'^{\dagger} = \left(UPU^{\dagger}\right)^{\dagger} = UP^{\dagger}U^{\dagger} = UPU^{\dagger} = P' \tag{2.102}$$

and it also has trace zero. To see the latter, we write the trace of a product of matrices in component form and check that the matrices in the product can be permuted cyclically without changing the trace:

$$\begin{aligned} \mathrm{Tr}\{ABC \ldots Z\} &= \sum_{ijkl\cdots n} A_{ij} B_{jk} C_{kl} \cdots Z_{ni} \\ &= \sum_{jkl\cdots ni} B_{jk} C_{kl} \cdots Z_{ni} A_{ij} \\ &= \mathrm{Tr}\{BC \cdots ZA\} \quad . \end{aligned} \tag{2.103}$$

Applying this to our transformation formula yields

$$\mathrm{Tr}\{P'\} = \mathrm{Tr}\left\{UPU^{\dagger}\right\} = \mathrm{Tr}\left\{PU^{\dagger}U\right\} = \mathrm{Tr}\{P\} = 0 \tag{2.104}$$

since $U^{\dagger}U = 1$. So P' also defines a vector $\boldsymbol{r}'$, and to prove that $\boldsymbol{r}'$ is obtained from $\boldsymbol{r}$ by a rotation, we only have to check its length, which is given by the determinant:

$$\det P = -z^2 - (x - \mathrm{i}y)(x + \mathrm{i}y) = -\left(x^2 + y^2 + z^2\right) = -r^2 \quad . \tag{2.105}$$

The determinant of a product of matrices is equal to the product of the determinants, so that

$$\det P' = \det \left(UPU^{\dagger}\right) = \det U \, \det P \, \det U^{\dagger} = \det P \tag{2.106}$$

and $r'^2 = r^2$. Now a linear transformation which preserves the lengths of all vectors must be a rotation, possibly combined with reflections.

To complete the construction, the rotations about the three coordinate axes have to be constructed in this representation (see Exercise 2.2). The result for rotations about the z axis is

$$U(\theta_z) = \begin{pmatrix} \mathrm{e}^{-\mathrm{i}\theta_z/2} & 0 \\ 0 & \mathrm{e}^{\mathrm{i}\theta_z/2} \end{pmatrix} \quad , \tag{2.107}$$

and for those about the x axis

$$U(\theta_x) = \begin{pmatrix} \cos(\theta_x/2) & -\mathrm{i}\sin(\theta_x/2) \\ -\mathrm{i}\sin(\theta_x/2) & \cos(\theta_x/2) \end{pmatrix} \quad , \tag{2.108}$$

and finally for the y axis

$$U(\theta_y) = \begin{pmatrix} \cos(\theta_y/2) & \sin(\theta_y/2) \\ -\sin(\theta_y/2) & \cos(\theta_y/2) \end{pmatrix} \quad . \tag{2.109}$$

The form of these expressions allows two important conclusions. First, the angles are halved in the arguments of the trigonometric functions, and all of these matrices reduce to *minus* the unit matrix for an angle of 2π. This does not cause problems in this case, because in the rotation law (2.101) U appears twice and the signs cancel. Second, expanding the rotation matrices for small angles yields the angular momentum operators

$$\hat{J}_x = \frac{1}{2}\begin{pmatrix} 0 & 1 \\ 1 & 0 \end{pmatrix} \quad , \quad \hat{J}_y = \frac{1}{2}\begin{pmatrix} 0 & -\mathrm{i} \\ \mathrm{i} & 0 \end{pmatrix} \quad , \quad \hat{J}_z = \frac{1}{2}\begin{pmatrix} 1 & 0 \\ 0 & -1 \end{pmatrix} \quad , \tag{2.110}$$

which can also be expressed in terms of the Pauli matrices as

$$\hat{\boldsymbol{J}} = \tfrac{1}{2}\boldsymbol{\sigma} \tag{2.111}$$

with

$$\sigma_x = \begin{pmatrix} 0 & 1 \\ 1 & 0 \end{pmatrix} \quad , \quad \sigma_y = \begin{pmatrix} 0 & -\mathrm{i} \\ \mathrm{i} & 0 \end{pmatrix} \quad , \quad \sigma_z = \begin{pmatrix} 1 & 0 \\ 0 & -1 \end{pmatrix} \quad , \tag{2.112}$$

and fulfil the usual angular momentum commutation relations. This shows that SU(2) and SO(3) have the same Lie algebras.

Although the Cayley–Klein formulation as described above will not be used further in this book, this discussion should have made the intimate connection between rotations and the group SU(2) clear. Since SU(2) describes arbitrary unitary transformations with determinant 1 in a two-dimensional space of wave functions, angular-momentum methods can be used *by mathematical analogy* whenever one is dealing with symmetries in such a space. Isospin will be the principal application of this idea.

EXERCISE

2.2 Cayley–Klein Representation of the Rotation Matrix

Problem. Derive the transformation matrix in the Cayley–Klein representation for rotations about the z axis.

Solution. We have to compare the usual transformation

$$x' = x\cos\theta - y\sin\theta \quad , \quad y' = x\sin\theta + y\cos\theta \quad , \quad z' = z \tag{1}$$

with the one given by (2.101). With the matrix set in the form of (2.98), this leads to the matrix equation

$$\begin{pmatrix} z' & x' - \mathrm{i}y' \\ x' + \mathrm{i}y' & -z' \end{pmatrix} = \begin{pmatrix} a & b \\ -b^* & a^* \end{pmatrix} \begin{pmatrix} z & x - \mathrm{i}y \\ x + \mathrm{i}y & -z \end{pmatrix} \begin{pmatrix} a^* & -b \\ b^* & a \end{pmatrix} \quad . \tag{2}$$

Exercise 2.2

Carrying through the matrix multiplications and inserting the expressions for the primed coordinates yields four complex equations, in which the coefficients of x, y, and z can be compared separately. In the diagonal, the coefficients of x and y must be zero and that of z must be 1 or -1, leading to

$$ab^* = a^*b = 0 \quad , \quad a^*a - b^*b = 1 \quad . \tag{3}$$

In the off-diagonal terms the coefficient of z must vanish, requiring $ab = 0$. Now split the lower off-diagonal equation into real and imaginary parts; defining $a = a_r + \mathrm{i}a_i$ and $b = b_r + \mathrm{i}b_i$ one obtains

$$\begin{aligned} \cos\theta &= -a_i^2 + a_r^2 + b_i^2 - b_r^2 - \sin\theta = 2a_i a_r + 2b_i b_r \quad , \\ \cos\theta &= -a_i^2 + a_r^2 + b_i^2 - b_r^2 \sin\theta = -2a_i a_r + 2b_i b_r \quad , \end{aligned} \tag{4}$$

from which $b_i = b_r = 0$, i.e., $b = 0$, and

$$2a_i a_r = -\sin\theta \quad , \quad a_r^2 - a_i^2 = \cos\theta \tag{5}$$

follow, which can be solved yielding

$$a_r = -\cos\left(\tfrac{\theta}{2}\right) \quad , \quad a_i = -\sin\left(\tfrac{\theta}{2}\right) \quad , \quad a = \exp\left(-\mathrm{i}\tfrac{\theta}{2}\right) \quad . \tag{6}$$

This completes the construction of the matrix.

2.3.5 Coupling of Angular Momenta

In a system of two particles, there is for each of the particles an angular-momentum operator that infinitesimally rotates that particle about the origin of the coordinate system. Let us designate these operators by $\hat{\boldsymbol{J}}_1$ and $\hat{\boldsymbol{J}}_2$. If the particles interact, the energy of the system will not be invariant if only one of the particles is rotated, but if both are rotated simultaneously, their relative position, relative spin orientation, etc., will not be changed and the physics should remain the same. Thus it makes sense to study the operator for a common infinitesimal rotation of both particles, which is the *total angular-momentum operator*

$$\hat{\boldsymbol{J}} = \hat{\boldsymbol{J}}_1 + \hat{\boldsymbol{J}}_2 \quad , \tag{2.113}$$

(cf. the construction of the total momentum in Sect. 2.2.3) and for the two-particle system the eigenfunctions of $\hat{J}^2$ and $\hat{J}_z$ must be sought. Normally these are obtained by *angular-momentum coupling* from the products of eigenfunctions of the separate angular momenta.

Let two sets of eigenfunctions $|j_1 m_1\rangle$ and $|j_2 m_2\rangle$ be given such that

$$\begin{aligned} \hat{J}_1^2 |j_1 m_1\rangle &= \hbar^2 j_1(j_1+1)|j_1 m_1\rangle \quad , \quad \hat{J}_{1z}|j_1 m_1\rangle = \hbar m_1 |j_1 m_1\rangle \quad , \\ \hat{J}_2^2 |j_2 m_2\rangle &= \hbar^2 j_2(j_2+1)|j_2 m_2\rangle \quad , \quad \hat{J}_{2z}|j_2 m_2\rangle = \hbar m_2 |j_2 m_2\rangle \quad . \end{aligned} \tag{2.114}$$

A basis for the system of two particles may then be built out of the products of these states, forming the so-called uncoupled basis

$$|j_1 m_1 j_2 m_2\rangle = |j_1 m_1\rangle |j_2 m_2\rangle \quad . \tag{2.115}$$

One sees immediately that such a state is an eigenstate of the z component of the total angular momentum with eigenvalue $m_1 + m_2$, since

$$\hat{J}_z |j_1 m_1 j_2 m_2\rangle = \left(\hat{J}_{1z} + \hat{J}_{2z}\right) |j_1 m_1 j_2 m_2\rangle = \hbar(m_1 + m_2) |j_1 m_1 j_2 m_2\rangle \quad . \tag{2.116}$$

However, it cannot be an eigenstate of $\hat{J}^2$, and to see this we have to examine the commutation relations of the different angular momenta.

First note that all components of $\hat{\boldsymbol{J}}_1$ will commute with all components of $\hat{\boldsymbol{J}}_2$, since they refer to different particles. Thus we get immediately

$$\left[\hat{J}_z, \hat{J}_1^2\right] = \left[\hat{J}_z, \hat{J}_2^2\right] = \left[\hat{J}^2, \hat{J}_1^2\right] = \left[\hat{J}^2, \hat{J}_2^2\right] = 0 \quad , \tag{2.117}$$

as well as

$$\left[\hat{J}_z, \hat{J}_{1z}\right] = \left[\hat{J}_z, \hat{J}_{2z}\right] = 0 \quad , \tag{2.118}$$

while the square of the total angular momentum does not commute with the z projections of the individual angular momenta; for example,

$$\left[\hat{J}^2, \hat{J}_{1z}\right] = \left[\hat{J}_1^2, \hat{J}_{1z}\right] + \left[\hat{J}_2^2, \hat{J}_{1z}\right] + \left[2\hat{\boldsymbol{J}}_1 \cdot \hat{\boldsymbol{J}}_2, \hat{J}_{1z}\right] \quad . \tag{2.119}$$

Although the first two commutators on the right-hand side vanish, the third one does not and the result becomes

$$\left[\hat{J}^2, \hat{J}_{1z}\right] = 2\mathrm{i}\hbar\left(\hat{J}_{2y}\hat{J}_{1x} - \hat{J}_{2x}\hat{J}_{1y}\right) \quad . \tag{2.120}$$

A fully commuting set of operators is thus given by $\hat{J}^2$, $\hat{J}_z$, $\hat{J}_1^2$, and $\hat{J}_2^2$, and angular-momentum coupling essentially consists in replacing the quantum numbers m_1 and m_2 by j and m. The new basis vectors may be denoted by

$$|jmj_1 j_2\rangle \quad , \tag{2.121}$$

and the unitary matrix leading to this basis is simply defined by

$$|jmj_1 j_2\rangle = \sum_{m_1 m_2} |j_1 m_1 j_2 m_2\rangle \langle j_1 m_1 j_2 m_2 | jmj_1 j_2\rangle \quad . \tag{2.122}$$

Obviously for practical notation the repetition of j_1 and j_2 on both sides of the matrix element is superfluous, so that the transformation coefficient is more succinctly defined as

$$(j_1 j_2 j | m_1 m_2 m) = \langle j_1 m_1 j_2 m_2 | jmj_1 j_2\rangle \quad . \tag{2.123}$$

This is the *Clebsch–Gordan coefficient.* The transformation is now written as

$$|jmj_1 j_2\rangle = \sum_{m_1 m_2} |j_1 m_1 j_2 m_2\rangle (j_1 j_2 j | m_1 m_2 m) \quad . \tag{2.124}$$

The following properties of the Clebsch–Gordan coefficients are used in this book (for a derivation see the literature on angular momentum).

1. *Selection rules:* the coefficient is zero unless the quantum numbers fulfil the two conditions

$$m_1 + m_2 = m \tag{2.125}$$

and

$$|j_1 - j_2| \leq j \leq j_1 + j_2 \quad . \tag{2.126}$$

The first one merely repeats what we said above about the eigenvalue of $\hat{J}_z$, while the latter is known as the "triangular condition": the size of the total angular momentum is restricted to those values that are allowed by the rules of vector addition, where the vectors $\hat{\boldsymbol{J}}_1$, $\hat{\boldsymbol{J}}_2$, and $\hat{\boldsymbol{J}}$ form a triangle.

2. *The coefficients are real.* This is not a general condition, because the phases of the basis states are arbitrary in principle, but is assured for the standard choice of basis states due to *Condon and Shortley*. Note that a real transformation matrix is orthogonal so that the inverse transform corresponds to the transposed matrix:

$$\langle j_1 m_1 j_2 m_2 | j m j_1 j_2 \rangle = \langle j m j_1 j_2 | j_1 m_1 j_2 m_2 \rangle = (j_1 j_2 j \,|m_1 m_2 m) \quad , \tag{2.127}$$

and the transformation back to the uncoupled basis uses the same coefficients but the summation is over different indices:

$$|j_1 m_1 j_2 m_2\rangle = \sum_{jm} |j m j_1 j_2\rangle (j_1 j_2 j \,|m_1 m_2 m) \quad . \tag{2.128}$$

3. In the language of group representations, angular-momentum coupling corresponds to the reduction of the product of two representations:

$$\mathcal{D}^{(j_1)} \times \mathcal{D}^{(j_2)} = \mathcal{D}^{(j_1+j_2)} + \mathcal{D}^{(j_1+j_2-1)} + \dots + \mathcal{D}^{(|j_1-j_2|)} \quad . \tag{2.129}$$

The dimensions of the bases involved on both sides coincide, i.e.,

$$(2j_1 + 1)(2j_2 + 1) = \sum_{j=|j_1-j_2|}^{j_1+j_2} (2j + 1) \quad , \tag{2.130}$$

which may be checked easily by direct calculation.

4. *Special formulas* for the Clebsch–Gordan coefficients are to be found in angular momentum textbooks and special tables; the few that are needed in this book will always be quoted explicitly.

2.3.6 Intrinsic Angular Momentum

Up to now the rotational invariance of wave functions has been investigated in terms of their dependence on the space coordinates, leading to *orbital angular momentum*. There are also fields that have an *intrinsic angular momentum*, for example, vector fields. Figure 2.2 shows the situation for this case. A rotation not only moves the field to a different point $\boldsymbol{r}'$ in space, but also rotates the vector itself. So we must have

$$\boldsymbol{A}'(\boldsymbol{r}') = \mathcal{R}\,\boldsymbol{A}(\boldsymbol{r}) \quad , \tag{2.131}$$

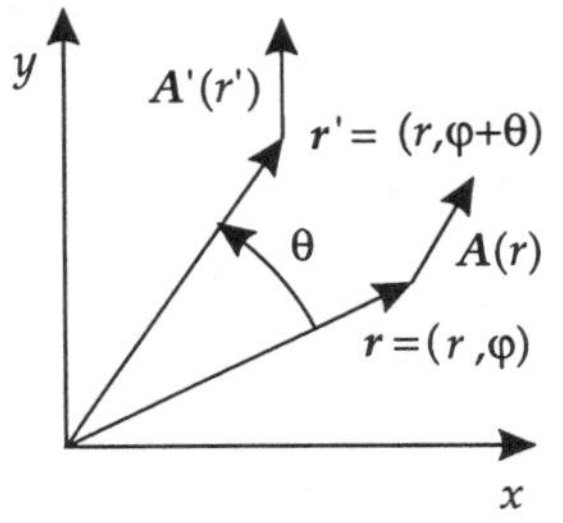

Fig. 2.2 Rotation of a vector field. In addition to the transformation of the base point the field vector itself is also rotated

where $\mathcal{R}$ is the appropriate rotation matrix for a vector in three dimensions, for example, for a rotation about the z axis

$$\mathcal{R}(\theta_z) = \begin{pmatrix} \cos\theta_z & -\sin\theta_z & 0 \\ \sin\theta_z & \cos\theta_z & 0 \\ 0 & 0 & 1 \end{pmatrix} \quad . \tag{2.132}$$

The rotated field at $\boldsymbol{r}$ is then given by

$$\boldsymbol{A}'(\boldsymbol{r}) = \mathcal{R}\,\boldsymbol{A}\big(\mathcal{R}^{-1}\boldsymbol{r}\big) \quad . \tag{2.133}$$

The infinitesimal rotation is easy to derive from this expression: for the dependence on the coordinates we get the usual angular-momentum operator and for the rotation of the vector itself the matrix $\mathcal{R}$ has to be expanded in the angles. For a rotation about the z axis this is

$$\mathcal{R}(\theta_z) \approx \begin{pmatrix} 1 & -\theta_z & 0 \\ \theta_z & 1 & 0 \\ 0 & 0 & 0 \end{pmatrix} = 1 - \tfrac{\mathrm{i}}{\hbar}\theta_z \hat{S}_z \quad , \tag{2.134}$$

where $\hat{S}_z$ is the matrix

$$\hat{S}_z = \hbar \begin{pmatrix} 0 & -\mathrm{i} & 0 \\ \mathrm{i} & 0 & 0 \\ 0 & 0 & 0 \end{pmatrix} \quad . \tag{2.135}$$

This is one component of the operator of *intrinsic* angular momentum, which is usually called spin, even if it does not refer to half-integral angular momentum. In this case, for a vector field, the angular momentum is unity, corresponding to the three-dimensional representation with $j = 1$. An example of this type of field is a photon.

The other angular-momentum components, obtained in the same way, are

$$\hat{S}_y = \hbar \begin{pmatrix} 0 & 0 & \mathrm{i} \\ 0 & 0 & 0 \\ -\mathrm{i} & 0 & 0 \end{pmatrix} \quad , \quad \hat{S}_x = \hbar \begin{pmatrix} 0 & 0 & 0 \\ 0 & 0 & -\mathrm{i} \\ 0 & \mathrm{i} & 0 \end{pmatrix} \quad . \tag{2.136}$$

These matrices fulfil, as indeed they must, the commutation relations for angular momentum.

For the field $\boldsymbol{A}(\boldsymbol{r})$ the infinitesimal rotation is given by

$$\boldsymbol{A}'(\boldsymbol{r}) = \left[1 - \tfrac{\mathrm{i}}{\hbar}\theta_z \left(\hat{L}_z + \hat{S}_z\right)\right] \boldsymbol{A}(\boldsymbol{r}) \quad . \tag{2.137}$$

Here the orbital angular momentum is denoted by $\hat{L}$, to reserve the letter J for the total angular momentum $\hat{\boldsymbol{J}}$ given by

$$\hat{\boldsymbol{J}} = \hat{\boldsymbol{L}} + \hat{\boldsymbol{S}} \quad . \tag{2.138}$$

Obviously $\hat{\boldsymbol{J}}$ generates rotations for the vector field, and eigenstates of $\hat{J}^2$ and $\hat{J}_z$ have to be constructed by angular-momentum coupling.

For this coupling, however, the coordinate system used for the vector is not adequate. As is clear from the form of the matrix $\hat{S}_z$, it is not diagonal in cartesian coordinates and one has to go over to *spherical coordinates* to obtain spin eigenstates. Formally we need to construct vectors fulfilling $\hat{S}_z \boldsymbol{e}_\mu = \hbar\mu \boldsymbol{e}_\mu$ or, if we assume components a, b, c for the unknown eigenvector,

$$\begin{pmatrix} 0 & -\mathrm{i} & 0 \\ \mathrm{i} & 0 & 0 \\ 0 & 0 & 0 \end{pmatrix} \begin{pmatrix} a \\ b \\ c \end{pmatrix} = \hbar\mu \begin{pmatrix} a \\ b \\ c \end{pmatrix} \quad . \tag{2.139}$$

This equation can have solutions only if the determinant $\det\left(\hat{S}_z - \hbar\mu\right)$ vanishes, which leads to the condition

$$\mu\left(\mu^2 - 1\right) = 0 \tag{2.140}$$

with solutions $\mu = -1, 0, 1$ as should be expected for the three projections of unit angular momentum. Inserting these eigenvalues into the equation, finding the required relations between a, b, and c, and normalizing the resulting vectors leads to

$$\begin{aligned} \boldsymbol{e}_{+1} &= -\tfrac{1}{\sqrt{2}}(\boldsymbol{e}_x + \mathrm{i}\boldsymbol{e}_y) \quad , \\ \boldsymbol{e}_0 &= \boldsymbol{e}_z \quad , \\ \boldsymbol{e}_{-1} &= \tfrac{1}{\sqrt{2}}(\boldsymbol{e}_x - \mathrm{i}\boldsymbol{e}_y) \quad . \end{aligned} \tag{2.141}$$

These complex vectors could in principle be multiplied by an arbitrary phase without affecting their eigenvector properties, and the special choice made here is partly conventional and also made such that the general property

$$\boldsymbol{e}_\mu^* = (-1)^\mu \boldsymbol{e}_{-\mu} \tag{2.142}$$

is fulfilled. This is analogous to spherical harmonics and corresponds to the *Condon–Shortley choice of phases* of the-angular momentum eigenstates.

Let us list a few properties of spherical coordinates. The basis vectors of (2.141) are orthogonal in the generalized version for complex vectors:

$$\boldsymbol{e}_\mu^* \boldsymbol{e}_{\mu'} = \delta_{\mu\mu'} \quad , \tag{2.143}$$

and if spherical components a_μ, $\mu = -1, 0, 1$, of a vector $\boldsymbol{a}$ are defined via

$$\sum_\mu a_\mu^* \boldsymbol{e}_\mu = \boldsymbol{a} \quad , \tag{2.144}$$

they are related to the cartesian components by

$$a_{\pm 1} = \mp\tfrac{1}{\sqrt{2}}(a_x \pm \mathrm{i}a_y) \quad , \quad a_0 = a_z \quad . \tag{2.145}$$

To understand the physical meaning of spherical components, regard the case of a plane-wave electromagnetic field with a polarization vector given by one of the spherical basis vectors:

$$\begin{aligned} \boldsymbol{A}(\boldsymbol{r}, t) &= \boldsymbol{e}_1 \exp(\mathrm{i}\boldsymbol{k}\cdot\boldsymbol{r} - \mathrm{i}\omega t) \\ &= -\tfrac{1}{\sqrt{2}}[\boldsymbol{e}_x \cos(\boldsymbol{k}\cdot\boldsymbol{r} - \omega t) - \boldsymbol{e}_y \sin(\boldsymbol{k}\cdot\boldsymbol{r} - \omega t)] \\ &\quad - \tfrac{\mathrm{i}}{\sqrt{2}}[\boldsymbol{e}_x \sin(\boldsymbol{k}\cdot\boldsymbol{r} - \omega t) + \boldsymbol{e}_y \cos(\boldsymbol{k}\cdot\boldsymbol{r} - \omega t)] \quad . \end{aligned} \tag{2.146}$$

The real part of this, corresponding to the physical field, and evaluated for simplicity at $\boldsymbol{r} = 0$, is

$$\mathrm{Re}\,\{\boldsymbol{A}(0,t)\} = -\frac{1}{\sqrt{2}}[\boldsymbol{e}_x \cos(\omega t) + \boldsymbol{e}_y \sin(\omega t)] \quad , \tag{2.147}$$

which describes a field rotating in the mathematically positive sense about the z axis and thus nicely conforming to the idea of positive intrinsic angular momentum about that axis.

A second important case of internal angular momentum is that of *spin fields*. In this case the wave function has two components at each point in space that transform under rotations according to the representation of the rotation group with angular momentum $j = \frac{1}{2}$. There are no classical analogies and in this case we do not have to worry about the choice of the basis: since there is no predefined physical set of basis states, we can immediately use the eigenstates of angular momentum $\hat{s}_z$. The two basic states may be called $|\frac{1}{2}\rangle$ and $|-\frac{1}{2}\rangle$ with

$$\hat{s}_z \,| \pm \tfrac{1}{2}\rangle = \pm\tfrac{1}{2}\hbar \,| \pm \tfrac{1}{2}\rangle \quad , \tag{2.148}$$

and the spin operator is denoted by the small letter s to distinguish it from the case of intrinsic angular momentum of unity, where capital $\hat{S}$ is commonly used. Its components are determined by

$$\hat{\boldsymbol{s}} = \tfrac{1}{2}\hbar\,\hat{\boldsymbol{\sigma}} \quad , \tag{2.149}$$

where $\hat{\boldsymbol{\sigma}}$ are the Pauli matrices as defined above in (2.112). For both operators $\hat{\boldsymbol{S}}$ and $\hat{\boldsymbol{s}}$ this agrees with the representation constructed in Sect. 2.3.2, as may be checked by calculating the matrix elements for $\hat{J}_z$ and $\hat{J}_\pm$ given there for these special cases and then obtaining $\hat{J}_x$ and $\hat{J}_y$ from them.

2.3.7 Tensor Operators

Up to now only the transformation of wavefunctions under rotations has been examined. Before considering operators in more detail, we first introduce the various types of transformation properties. A *scalar* is a quantity that does not change under rotations, whereas a *vector* or *tensor of rank* 1 is an object with three components that transforms like the position vector $\boldsymbol{r}$; as discussed in the last section this corresponds to unit angular momentum.

Higher-rank tensors can be constructed in two ways, reminiscent of the alternative of angular-momentum coupled or uncoupled bases. A *cartesian tensor of rank n* has components with n indices, each of them running from -1 to $+1$ (i.e., spherical components are assumed, but cartesian ones can be used as well), and with each index transforming according to the representation for unit angular momentum:

$$a'_{m_1 m_2 \ldots m_n} = \sum_{m'_1 m'_2 \ldots m'_n} a_{m'_1 m'_2 \ldots m'_n} \mathcal{D}^{(1)}_{m'_1 m_1} \mathcal{D}^{(1)}_{m'_2 m_2} \cdots \mathcal{D}^{(1)}_{m'_n m_n} \quad . \tag{2.150}$$

For the simplest case of a rank-2 tensor this reads

$$a'_{ij} = \sum_{i'j'} a_{i'j'} \mathcal{D}^{(1)}_{i'i} \mathcal{D}^{(1)}_{j'j} \tag{2.151}$$

and is essentially the same formula as that giving the rotations for an uncoupled basis built from two angular momenta of magnitude 1. Such a cartesian tensor has 3^n components.

Equivalently, and usually much more usefully in nuclear structure, one may employ spherical tensors (more precisely, *irreducible spherical tensors*), which are objects transforming under rotations like a wave function of good angular momentum. Thus an irreducible tensor with angular momentum $\hat{T}^k$ consists of components $\hat{T}^k_q$, $q = -k, \ldots, +k$, which under rotations transform according to

$$\hat{T}'^k_q = \sum_{q'=-k}^{k} \hat{T}^k_{q'} \mathcal{D}^{(k)}_{q'q}(\boldsymbol{\theta}) \quad . \tag{2.152}$$

The most useful application, of course, is the coupling of such tensors, for which all the considerations for wave functions are still applicable. Two tensors R^k and $\hat{S}^{k'}$ of angular momenta k and k', respectively, may be coupled to total angular momentum K using

$$\hat{T}^K_Q = \sum_{qq'} (kk'K|qq'Q)\, R^k_q \hat{S}^{k'}_{q'} \quad , \quad Q = -K, \ldots, K \quad . \tag{2.153}$$

This type of angular-momentum coupling is so important that we introduce a more concise notation,

$$\hat{T}^K_Q = \left[R^k \times \hat{S}^{k'}\right]^K_Q \quad \text{or} \quad \hat{T}^K = \left[R^k \times \hat{S}^{k'}\right]^K \quad . \tag{2.154}$$

The latter version is useful when one is not interested in the projections.

EXERCISE

2.3 Coupling of Two Vectors to Good Angular Momentum

Problem. What are the possible couplings of two vectors $\boldsymbol{a}$ and $\boldsymbol{b}$ to good total angular momentum?

Solution. According to the triangular rule, the coupling of the vectors is that of two tensors of unit angular momentum to a resulting angular momentum $K = 0, 1, 2$. The general formula is

$$T^K_Q = \sum_q (11K|q\; Q-q\; Q)\, a_q\, b_{Q-q} \quad , \tag{1}$$

involving the spherical components of the vectors. For $K = 0$ we can use $(110|q - q0) = -(-1)^q/\sqrt{3}$ to get

$$\begin{aligned} T^0_0 &= -\tfrac{1}{\sqrt{3}}(a_0 b_0 - a_1 b_{-1} - a_{-1} b_1) \\ &= -\tfrac{1}{\sqrt{3}}\left[a_z b_z + \tfrac{1}{2}(a_x + \mathrm{i}a_y)(b_x - \mathrm{i}b_y) + \tfrac{1}{2}(a_x - \mathrm{i}a_y)(b_x + \mathrm{i}b_y)\right] \\ &= -\tfrac{1}{\sqrt{3}}(a_x b_x + a_y b_y + a_z b_z) = -\tfrac{1}{\sqrt{3}}\boldsymbol{a} \cdot \boldsymbol{b} \quad , \end{aligned} \tag{2}$$

Exercise 2.3

i.e., the not very surprising result that the scalar built up from the two vectors is just the scalar product, aside from a constant factor.

For $K = 1$ let us do only one component explicitly: $Q = 1$. Here we have $(11 1|101) = -(11 1|011) = 1/\sqrt{2}$ and

$$\begin{aligned} T_1^1 &= \tfrac{1}{\sqrt{2}}(-a_0 b_1 + a_1 b_0) \\ &= \tfrac{1}{2}[a_z(b_x + \mathrm{i}b_y) - b_z(a_x + \mathrm{i}a_y)] \\ &= -\tfrac{\mathrm{i}}{2}[(a_y b_z - a_z b_y) + \mathrm{i}(a_z b_x - a_x b_z)] \\ &= \tfrac{\mathrm{i}}{\sqrt{2}}(\boldsymbol{a} \times \boldsymbol{b})_{+1} \quad , \end{aligned} \tag{3}$$

the spherical component of the vector product. The other components work out similarly and in general we can write

$$[\boldsymbol{a} \times \boldsymbol{b}]^1_Q = \frac{\mathrm{i}}{\sqrt{2}}(\boldsymbol{a} \times \boldsymbol{b})_Q \quad . \tag{4}$$

Again this result is intuitively clear. Finally, for $K = 2$, we write down the first three components explicitly:

$$\begin{aligned} T_2^2 &= a_1 b_1 = \tfrac{1}{2}(a_x b_x - a_y b_y + 2\mathrm{i}a_x b_y) \quad , \\ T_1^2 &= \tfrac{1}{\sqrt{2}}(a_1 b_0 + a_0 b_1) = -\tfrac{1}{2}[a_x b_z + a_z b_x + \mathrm{i}(a_y b_z + a_z b_y)] \quad , \\ T_0^2 &= \tfrac{1}{\sqrt{6}}(a_1 b_{-1} + a_{-1} b_1) + \sqrt{\tfrac{2}{3}}a_0 b_0 = -\tfrac{1}{\sqrt{6}}(a_x b_x + a_y b_y) + \sqrt{\tfrac{2}{3}}a_z b_z \quad . \end{aligned} \tag{5}$$

The negative Q-components can be obtained from the rule $T^K_{-Q} = (-1)^Q T^{*K}_{\ Q}$.

One may also view these results in terms of a decomposition of the cartesian tensor product

$$\begin{pmatrix} a_x b_x & a_x b_y & a_x b_z \\ a_y b_x & a_y b_y & a_y b_z \\ a_z b_x & a_z b_y & a_z b_z \end{pmatrix} \tag{6}$$

into a scalar $\boldsymbol{a} \cdot \boldsymbol{b}$, a vector $\boldsymbol{a} \times \boldsymbol{b}$, and a symmetric tensor T^2_Q with components like $a_x b_y + a_y b_x$.

An *irreducible spherical-tensor operator* is now simply an irreducible spherical tensor whose components are operators. This introduces a new condition, because an operator $\hat{T}^k_q$ can be transformed either by using (2.152) or via the general formula

$$\hat{T}'^k_q = \mathcal{D}(\boldsymbol{\theta})\, \hat{T}^k_q\, \mathcal{D}^{-1}(\boldsymbol{\theta}) \quad , \tag{2.155}$$

where $\mathcal{D}(\boldsymbol{\theta})$ is the appropriate operator for rotating the wave functions (which, for wave functions of good angular momentum, may again be a rotation matrix, though possibly for angular momentum other than k). What do these two different transformation laws imply? Although (2.155) holds for any operator and essentially says that one may apply the rotated operator by first rotating the wave function back, then applying the original operator, and finally rotating the result forward to the desired position, (2.152) is valid *only* for irreducible spherical-tensor operators. It requires that a set of operators corresponding to the components of the spherical-tensor operator transform under rotations like a set of spherical components in the way prescribed by (2.152).

EXAMPLE

2.4 The Position Operator as an Irreducible Spherical Tensor

The three operators x, y, and z, which multiply the wave function by the value of the corresponding coordinate, can be combined into an irreducible spherical-tensor operator r_q of unit angular momentum and with components following the usual definition of spherical components:

$$r_0 = z \quad , \quad r_{+1} = -\tfrac{1}{\sqrt{2}}(x + \mathrm{i}y) \quad , \quad r_{-1} = \tfrac{1}{\sqrt{2}}(x - \mathrm{i}y) \quad . \tag{1}$$

From the discussion of vector fields it should be clear that r_q indeed transforms according to (2.152) with $k = 1$.

Comparing the two ways of transforming an irreducible spherical-tensor operator, one would expect that, if infinitesimal rotations are considered, some condition on the commutation of $\hat{T}^k$ with the angular momentum operator should follow. If we assume an infinitesimal rotation about the α axis, (2.155) may be written up to second order as

$$\begin{aligned}\hat{T}'^k_q &\approx \left(1 - \tfrac{\mathrm{i}}{\hbar}\theta_\alpha \hat{J}_\alpha\right)\hat{T}^k_q\left(1 + \tfrac{\mathrm{i}}{\hbar}\theta_\alpha \hat{J}_\alpha\right)\\ &\approx \hat{T}^k_q - \tfrac{\mathrm{i}}{\hbar}\theta_\alpha\left[\hat{J}_\alpha, \hat{T}^k_q\right] \quad .\end{aligned} \tag{2.156}$$

On the other hand, the rotation matrix element in (2.152) may also be simplified for infinitesimal rotations:

$$\mathcal{D}^{(k)}_{q'q}(\theta_\alpha) \approx \langle kq'|1 - \tfrac{\mathrm{i}}{\hbar}\theta_\alpha \hat{J}_\alpha|kq\rangle = \delta_{qq'} - \tfrac{\mathrm{i}}{\hbar}\theta_\alpha\langle kq'|\hat{J}_\alpha|kq\rangle \quad , \tag{2.157}$$

so that comparing the two ways of rotation yields

$$\left[\hat{J}_\alpha, \hat{T}^k_q\right] = \sum_{q'} \hat{T}^k_{q'}\langle kq'|\hat{J}_\alpha|kq\rangle \quad . \tag{2.158}$$

Using the matrix elements of the angular- momentum operators given at the end of Sect. 2.3.2 we obtain

$$\left[\hat{J}_z, \hat{T}^k_q\right] = q\hat{T}^k_q \quad , \quad \left[\hat{J}_\pm, \hat{T}^k_q\right] = \hbar\sqrt{(k \pm q + 1)(k \mp q)}\,\hat{T}^k_{q\pm 1} \quad . \tag{2.159}$$

Note that this result essentially means that the angular-momentum operators act on the components of an irreducible spherical-tensor operator in the same way as they act on a wave function; for an operator the product merely has to be replaced by a commutator, because the $\hat{J}$ also needs to be "passed through" to act on the wave function.

EXAMPLE

2.5 Commutation Relations of the Position Operator

The position vector operator $\hat{\boldsymbol{r}}$ should fulfil

$$\begin{aligned} &[\hat{J}_z, \hat{r}_0] = 0 \quad , && [\hat{J}_z, \hat{r}_{\pm 1}] = \pm\hbar\hat{r}_{\pm 1} \quad , && \\ &[\hat{J}_\pm, \hat{r}_0] = \sqrt{2}\hbar\hat{r}_{\pm 1} \quad , && [\hat{J}_+, \hat{r}_{+1}] = 0 \quad , && [\hat{J}_-, \hat{r}_{-1}] = 0 \quad , \\ &[\hat{J}_+, \hat{r}_{-1}] = \sqrt{2}\hbar\hat{r}_0 \quad , && [\hat{J}_-, \hat{r}_{+1}] = \sqrt{2}\hbar\hat{r}_0 \quad , && \end{aligned} \tag{1}$$

which can be checked easily using the definition of $\hat{\boldsymbol{J}}$ in terms of differential operators. The angular-momentum operator itself is also a spherical tensor with unit angular momentum, but *it is important to realize that the components are not identical to the shift operators*. The spherical components are denoted by subscripts $+1$ and -1 and are given by

$$\begin{aligned} \hat{J}_0 &= \hat{J}_z \\ \hat{J}_{+1} &= -\tfrac{1}{\sqrt{2}}\left(\hat{J}_x + \mathrm{i}\hat{J}_y\right) = -\tfrac{1}{\sqrt{2}}\hat{J}_+ \\ \hat{J}_{-1} &= \tfrac{1}{\sqrt{2}}\left(\hat{J}_x - \mathrm{i}\hat{J}_y\right) = \tfrac{1}{\sqrt{2}}\hat{J}_- \quad . \end{aligned} \tag{2}$$

In general, any vector operator can be written as a spherical-tensor operator with unit angular momentum by using spherical components.

The applications of tensor operators are mostly concerned with the possibility of coupling them

- with wave functions: a new wave function with good angular momentum can be constructed through

$$|\beta KM\rangle = \sum_{qm} (jkK|mqM)\,\hat{T}^k_q\,|\alpha jm\rangle \quad , \tag{2.160}$$

  where α and β refer to any other quantum numbers affected by the operator. The resulting angular momenta are determined by the selection rules implicit in the Clebsch–Gordan coefficients.
- with other operators: this is similar to (2.154) above. Of course for operators one has to keep track of the ordering because of their possible noncommutation. In this way the scalar and vector products of operators can be defined as above as the coupling to zero or unit angular momentum.

The distinction between operators and fields is crucial for some considerations and the "same" mathematical expression may have to be treated differently depending on how it is used. We illustrate this point for the position vector.

Acting on wave functions, $\boldsymbol{r}$ is really composed of the three operators x, y, and z, which may be combined into the irreducible spherical-tensor operator r_μ,

$\mu = -1, 0, 1$, as described above. Its application to a wave function produces three new functions $r_\mu \psi(\boldsymbol{r})$, each of which may be rotated using the rotation operator:

$$\begin{aligned}\big(r_\mu \psi(\boldsymbol{r})\big)' &= \big(r''_\mu \psi(\boldsymbol{r}'')\big)_{(\boldsymbol{r}''=\mathcal{D}^{-1}(\boldsymbol{\theta}))} \\ &= \mathcal{D}(\boldsymbol{\theta})\big(r_\mu \psi(\boldsymbol{r})\big) \quad .\end{aligned} \tag{2.161}$$

Because $\boldsymbol{r}$ is a spherical-tensor operator, we can alternatively write

$$\begin{aligned}\big(r_\mu \psi(\boldsymbol{r})\big)' &= \left(\sum_{\mu'} r_{\mu'} \mathcal{D}^{(1)}_{\mu'\mu}\right) \mathcal{D}(\boldsymbol{\theta}) \psi(\boldsymbol{r}) \\ &= \left(\sum_{\mu'} r_{\mu'} \mathcal{D}^{(1)}_{\mu'\mu}\right) \psi\big(\mathcal{D}^{-1}(\boldsymbol{\theta}) \boldsymbol{r}\big) \quad .\end{aligned} \tag{2.162}$$

This last way of writing the rotation makes it clear how this works: the spherical operator $\boldsymbol{r}$ is rotated at a fixed point and thus multiplies the wave function with the values that pertain to the location $\mathcal{D}^{-1}(\boldsymbol{\theta})\,\boldsymbol{r}$.

The situation is different if $\boldsymbol{r}$ is regarded as a *vector field* with an intrinsic angular momentum of 1. Now we have a vector function $\boldsymbol{r}(x, y, z)$ which under rotations will have both its dependence on the coordinates and its direction at a fixed point affected. In fact, it can be written as

$$\boldsymbol{r} = -\sqrt{3} \sum_{\mu=-1}^{+1} (110|\mu - \mu 0)\, r_\mu\, \boldsymbol{e}_{-\mu} \tag{2.163}$$

and so appears explicitly as a scalar. This agrees with intuition since the field $\boldsymbol{r}$ is obviously spherically symmetric. Formally, the position-independent basic vectors $\boldsymbol{e}_\mu$ carry the intrinsic angular momentum and the coefficients r_μ carry the orbital angular momentum, which in this case just couple to zero total angular momentum.

2.3.8 The Wigner–Eckart Theorem

Since the states of different projection belonging to a multiplet of fixed angular momentum are in some sense trivially related to each other by application of the operators $\hat{J}_+$ or $\hat{J}_-$, one would expect that matrix elements should also in some sense depend "trivially" on the projections involved, i.e., simply through manipulations having to do with the rotation group and its representations. In fact, this is the contents of the *Wigner–Eckart theorem* which will now be derived.

Take a matrix element of an irreducible spherical-tensor operator $\hat{T}^k_q$ between angular-momentum states of the form $|\alpha j m\rangle$. As the operator may also change the nonrotational quantum numbers, they are explicitly represented by α and β. The matrix element we are interested in is

$$\langle \beta j' m' | \hat{T}^k_q | \alpha j m \rangle \quad . \tag{2.164}$$

The idea will now be to couple the right-hand state with the tensor operator to good angular momentum and then to exploit orthogonality with the left-hand state.

The coupling yields intermediate states

$$|\gamma j''m''\rangle = \sum_{mq} \left(jkj''|mqm''\right) \hat{T}_q^k \,|\alpha jm\rangle \quad , \tag{2.165}$$

with j'' and m'' taking all allowed values. This coupling can be inverted according to Sect. 2.3.5, yielding

$$\hat{T}_q^k|\alpha jm\rangle = \sum_{j''m''} \left(jkj''|mqm''\right) |\gamma j''m''\rangle \quad . \tag{2.166}$$

Taking the overlap with $|\beta j'm'\rangle$ produces

$$\begin{aligned}\langle\beta j'm'|\hat{T}_q^k|\alpha jm\rangle &= \sum_{j''m''} \left(jkj''|mqm''\right) \langle\beta j'm'|\gamma j''m''\rangle \\ &= \left(jkj'|mqm'\right) \langle\beta j'm'|\gamma j'm'\rangle \quad .\end{aligned} \tag{2.167}$$

The second matrix element on the right-hand side does not depend on m', which can be seen by examining

$$\begin{aligned}\langle\beta j\; m+1|\gamma jm+1\rangle &= \frac{\langle\beta jm|\hat{J}_-\hat{J}_+|\gamma jm\rangle}{\sqrt{\langle\beta jm|\hat{J}_-\hat{J}_+|\beta jm\rangle\langle\gamma jm|\hat{J}_-\hat{J}_+|\gamma jm\rangle}} \\ &= \langle\beta jm|\gamma jm\rangle \quad ,\end{aligned} \tag{2.168}$$

in which the operator combination $\hat{J}_-\hat{J}_+$ is diagonal and its eigenvalue is independent of any additional quantum numbers – in fact, this expression was used in Sect. 2.3.2 to derive the matrix element of $\hat{J}_+$. The same argument can be used for lowering the projection, of course.

The result is thus that the matrix element can be split up into a Clebsch–Gordan coefficient, which contains all the information about the projection dependence, and an as yet unknown matrix element $\langle\beta j'm'|\gamma j'm'\rangle$, which describes the physics of the operator $\hat{T}_q^k$. To make this fact clear, we define a *reduced matrix element* by writing

$$\langle\beta j'm'|\hat{T}_q^k|\alpha jm\rangle = (-1)^{2k}\langle\beta j'||\hat{T}^k||\alpha j\rangle\left(jkj'|mqm'\right) \quad . \tag{2.169}$$

The reduced matrix element $\langle\beta j'||\hat{T}^k||\alpha j\rangle$ is determined implicitly: to calculate its value simply calculate the left-hand side of (2.169) for any convenient combination of projections m, m', and q, for which it does not vanish, and then use the equation to obtain the reduced matrix element itself. The advantage is then that all other combinations of projections can be calculated easily, and that the reduced matrix element as defined does not explicitly contain any reference to the projections, thus providing a more general characterization of the operator's action. The factor $(-1)^{2k}$ is a conventional phase factor and different authors have employed different definitions for the reduced matrix element.

2.3.9 6j and 9j Symbols

The coupling of three or four angular momenta leads to the definition of 6*j* and 9*j* symbols. Their most important application is for the decomposition of reduced matrix elements; since they are used rarely in this book we give only a brief discussion and refer the reader to the books on angular momentum theory for their properties and evaluation.

Assume an uncoupled basis given by the individual angular momentum quantum numbers as $|j_1m_1, j_2m_2, j_3m_3\rangle$. The total angular momentum is given by $\hat{J} = \hat{\jmath}_1 + \hat{\jmath}_2 + \hat{\jmath}_3$ and both its square and the z component will yield good quantum numbers. Additional quantum numbers are provided by the squares of the individual angular momenta, and all these operators also commute with the squares of the sum of two angular momenta such as $\hat{J}_{12}^2 = (\hat{\jmath}_1 + \hat{\jmath}_2)^2$. But clearly we have $[\hat{J}_{12}, \hat{J}_{23}] \neq 0$, so that at most one of these intermediate coupling operators will be diagonal. If $\hat{J}_{12}^2$ is chosen to be diagonal, the construction proceeds by first coupling $\hat{\jmath}_1$ and $\hat{\jmath}_2$ with the standard method,

$$|j_1j_2J_{12}M_{12}\rangle = \sum_{m_1m_2}(j_1j_2J_{12}|m_1m_2M_{12})|j_1m_1\rangle|j_2m_2\rangle \quad , \tag{2.170}$$

and then performing the additional coupling

$$|jM, (j_1j_2)J_{12}j_3\rangle = \sum_{M_{12}m_3}(j_{12}j_3J\,|M_{12}m_3M)|j_1j_2J_{12}M_{12}\rangle|j_3m_3\rangle \quad . \tag{2.171}$$

An alternative is to first couple $\hat{\jmath}_2$ and $\hat{\jmath}_3$, leading to the following construction:

$$\begin{aligned}|j_2j_3J_{23}M_{23}\rangle &= \sum_{m_2m_3}(j_2j_3J_{23}|m_2m_3M_{23})|j_2m_2\rangle|j_3m_3\rangle \quad ,\\ |jM, j_1(j_2j_3)J_{23}\rangle &= \sum_{M_{23}m_1}(j_1J_{23}J\,|m_1M_{23}M)|j_1m_1\rangle|j_2j_3J_{23}M_{23}\rangle \quad .\end{aligned} \tag{2.172}$$

Thus we have two alternative sets of basis states for the system of three coupled angular momenta, which must be related by a transformation. Conventionally two definitions are used. One is the *Racah coefficient* W given by

$$\begin{aligned}&W(j_1j_2Jj_3; J_{12}J_{23})\\ &\quad = \sqrt{(2J_{12}+1)(2J_{23}+1)}\,\langle J, (j_1j_2)J_{12}j_3|J, j_1(j_2j_3)J_{23}\rangle \quad .\end{aligned} \tag{2.173}$$

Note that M was omitted in the wave functions since it turns out that W is independent of M. The normalization factor is useful for making W more symmetric. The transformation reads

$$\begin{aligned}|jM, (j_1j_2)J_{12}j_3\rangle = \sum_{J_{23}}&\sqrt{(2J_{12}+1)(2J_{23}+1)}\\ &\times W(j_1j_2Jj_3; J_{12}J_{23})|jM, j_1(j_2j_3)J_{23}\rangle \quad .\end{aligned} \tag{2.174}$$

An even higher degree of symmetry is possessed by the 6*j* *symbol* defined as

$$\begin{Bmatrix} j_1 & j_2 & J_{12} \\ j_3 & J & J_{23} \end{Bmatrix} = (-1)^{j_1+j_2+J+j_3}\,W(j_1j_2Jj_3; J_{12}J_{23}) \quad . \tag{2.175}$$

In the case of four angular momenta the ideas are developed quite similarly. One may for example couple j_1 and j_2 to J_{12} as well as j_3 and j_4 to J_{34}, and then combine these to the total J, or obtain J from a coupling of J_{13} and J_{24}. The transformation is given by the $9j$ *symbol* defined through

$$\begin{aligned}&\langle J,(j_1j_2)J_{12}(j_3j_4)J_{34}|J,(j_1j_3)J_{13}(j_2j_4)J_{24}\rangle\\ &\quad=\sqrt{(2J_{12}+1)(2J_{34}+1)(2J_{13}+1)(2J_{24}+1)}\begin{Bmatrix} j_1 & j_2 & J_{12}\\ j_3 & j_4 & J_{34}\\ J_{13} & J_{24} & J\end{Bmatrix}\quad. \end{aligned}\tag{2.176}$$

Note that each column or row describes an angular-momentum coupling. The $9j$ symbol is also known by the name of *Fano X coefficient* which is denoted by $X(j_1j_2J_{12}, j_3j_4J_{34}, J_{13}J_{24}J)$.

As mentioned above, these symbols are very useful for the decomposition of matrix elements of tensor operators, which will briefly be discussed now.

One often has to deal with wave functions and tensor operators referring to different subsystems. For example, the wave function may be built up from an orbital and a spin part, and the operator is also a coupling of a spin and an orbital contribution. Another example is that of two particles combining their angular momenta and an operator consisting of a product of two operators each acting on one of the particles only. We are then forced to consider matrix elements of the type $\langle j_1j_2J||\hat{T}^k||j_1'j_2'J'\rangle$, where

$$\hat{T}^k=\left[\hat{R}^{k_1}\times\hat{S}^{k_2}\right]^k\quad,\tag{2.177}$$

and $\hat{R}$ acts on particle 1 characterized by j_1 and j_1', whereas $\hat{S}$ acts only on particle 2 with j_2 and j_2'. The following general formula allows the expression of the total matrix element in terms of a product of matrix elements of the subsystems:

$$\begin{aligned}\langle j_1j_2J||\hat{T}^k||j_1'j_2'J'\rangle&=\sqrt{(2J'+1)(2k+1)(2j_1+1)(2j_2+1)}\\ &\quad\times\begin{Bmatrix} J & J' & k\\ j_1 & j_1' & k_1\\ j_2 & j_2' & k_2\end{Bmatrix}\langle j_1||\hat{R}^{k_1}||j_1'\rangle\langle j_2||\hat{S}^{k_2}||j_2'\rangle\quad. \end{aligned}\tag{2.178}$$

In the special case where one of the operators is the identity, so that, for example, $\hat{T}=\hat{S}$, the matrix element must be diagonal for particle 2, but the result still depends on the angular-momentum coupling present in the wave functions. The formula is

$$\begin{aligned}\langle j_1j_2J||\hat{R}^k||j_1'j_2'J'\rangle&=(-1)^{J+j_1'-k-j_2}\sqrt{(2J'+1)(2j_1+1)}\\ &\quad\times W\left(j_1j_1'JJ';kj_2\right)\langle j_1||\hat{R}^k||j_1'\rangle\delta_{j_2j_2'}\quad. \end{aligned}\tag{2.179}$$

2.4 Isospin

As has been shown in Sect. 2.3.4, the spin operators are intimately related to the generators of the group SU(2), which describes unitary transformations (with unit determinant) in a two-dimensional space of wave functions. Thus it should not come as a surprise that whenever one considers such a space – no matter what its physical interpretation – the group SU(2) and the spin representations will play a role.

In nuclear physics the interesting application is that of *isospin*. The proton and neutron appear to be two states of the same particle, which are indistinguishable under the strong interactions. If the Coulomb force is neglected in a first approximation, the physics of nuclei should therefore be invariant under transformations which transform proton and neutron states into each other. Mathematically, if we assign the basis states "proton", $|\mathrm{p}\rangle$, and "neutron", $|\mathrm{n}\rangle$, as the unit basis in a two-dimensional space,

$$|\mathrm{p}\rangle = \begin{pmatrix} 1 \\ 0 \end{pmatrix} \quad , \quad |\mathrm{n}\rangle = \begin{pmatrix} 0 \\ 1 \end{pmatrix} \quad . \tag{2.180}$$

A transformation in SU(2) will then mix proton and neutron states and we can take over all the results from angular momentum theory because of the identical group structure. The name "isospin" is chosen such as to show both the mathematical similarity and the different physical meaning (the term is derived from "isotopic" spin).

The generators of the group are the *isospin operators* $\hat{\boldsymbol{t}} = \frac{1}{2}\hat{\boldsymbol{\tau}}$, where $\hat{\boldsymbol{\tau}}$ is a vector of matrices that is identical to the Pauli vector $\hat{\boldsymbol{\sigma}}$ and just has a different notation to make clear that it refers to a different physical degree of freedom. The proton and neutron must be the eigenstates of $\hat{t}_z$ in the two-dimensional representation corresponding to isospin $\frac{1}{2}$; whether one chooses the positive eigenvalue for the proton or the neutron is a matter of choice. In this book – and in most nuclear physics publications – the proton gets the positive isospin projection, so that

$$\hat{t}_z|\mathrm{p}\rangle = \tfrac{1}{2}|\mathrm{p}\rangle \quad , \quad \hat{t}_z|\mathrm{n}\rangle = -\tfrac{1}{2}|\mathrm{n}\rangle \quad . \tag{2.181}$$

The operator $\hat{t}^2$ has the eigenvalue $\frac{1}{2}\left(\frac{1}{2}+1\right) = \frac{3}{4}$, and $\hat{t}_x$ and $\hat{t}_y$ transform protons into neutrons and vice versa; for example

$$\hat{t}_x|\mathrm{p}\rangle = \frac{1}{2}\begin{pmatrix} 0 & 1 \\ 1 & 0 \end{pmatrix}\begin{pmatrix} 1 \\ 0 \end{pmatrix} = \frac{1}{2}\begin{pmatrix} 0 \\ 1 \end{pmatrix} = \tfrac{1}{2}|\mathrm{n}\rangle \quad . \tag{2.182}$$

A Hamiltonian is isospin invariant if it commutes with the operators $\hat{\boldsymbol{t}}$. This is actually slightly more general than the starting consideration: the Hamiltonian will be invariant under *any* finite rotation in isospin, i.e., any mixture of proton and neutron state, whereas in nature only the eigenstates of $\hat{t}_z$, the proton and neutron, are realized. However, requiring this more general invariance is quite successful and there is no need to consider special transformations *exchanging* proton and neutron only. Because isospin is not an angular momentum, the dimensional factor $\hbar$ is not necessary and is dropped from all the formulas.

Of course the formalism of isospin would be pretty useless if only this single representation for isospin $\frac{1}{2}$ were considered. There is, however, a useful extension of the concept of angular-momentum coupling to isospin. Mathematically, this works exactly the same, using the same Clebsch–Gordan coefficients. If a system consists of A nucleons, its *total isospin* is defined by

$$\hat{\boldsymbol{T}} = \sum_{i=1}^{A} \hat{\boldsymbol{t}}_i \quad . \tag{2.183}$$

Isospin invariance of the Hamiltonian implies that it commutes with $\hat{\boldsymbol{T}}$ so that its eigenstates can also be classified in terms of representations analogous to $\mathcal{D}^j$ with j being replaced by t. Representing all other quantum numbers of the nucleus by λ, the eigenstates will fulfil

$$\hat{T}^2|tt_z\lambda\rangle = t(t+1)|tt_z\lambda\rangle \quad , \quad \hat{T}_z|tt_z\lambda\rangle = t_z|tt_z\lambda\rangle \quad . \tag{2.184}$$

As for angular momentum, the eigenvalue of the projection is simply the sum of the projections for the individual nucleons; since each of these contributes $+\frac{1}{2}$ for protons and $-\frac{1}{2}$ for neutrons, it is given by

$$t_z = \tfrac{1}{2}(Z - N) \quad . \tag{2.185}$$

The eigenstates for different t_z thus correspond to nuclei with the same number of nucleons, but different proton-to-neutron ratios. If the Hamiltonian is isospin invariant, these must all have the same energy, since the isospin shift operators commute with $\hat{H}$, and we get the prediction that there should exist *analog states* in a chain of isobaric nuclei that have the same internal structure. Such states are in fact found in light nuclei where Coulomb effects are negligible. These are the well-known isobaric analog states.

Of what kind of multiplet such a state may be a member depends on the Z and N of the nucleus. A symmetric nucleus, for example, has an isospin projection of zero and may thus be a member of any isospin multiplet, the simplest one being the singlet with zero isospin, for which case there are no analogue states in neighbouring isobars. In general a state in a nucleus (Z, N) requires an isospin

$$t \geq t_z = \tfrac{1}{2}(Z - N) \quad . \tag{2.186}$$

Another advantage of isospin is that all nucleons can be treated in the same way. If proton and neutron were kept distinct, one would have to antisymmetrize the wave functions separately for each type of particle, no antisymmetry being imposed upon exchanges of protons with neutrons because they are different particles. Isospin allows us to regard them as the same particle if the requirement is added that the wave function *including isospin dependence* is antisymmetric.

EXAMPLE

2.6 The Two-Nucleon System

For the isospin part of the wave functions we have the uncoupled basis consisting of the four states (the indices 1 and 2 refer to the two particles)

$$|\mathrm{p}_1\rangle|\mathrm{p}_2\rangle \quad , \quad |\mathrm{p}_1\rangle|\mathrm{n}_2\rangle \quad , \quad |\mathrm{n}_1\rangle|\mathrm{p}_2\rangle \quad , \quad |\mathrm{n}_1\rangle|\mathrm{n}_2\rangle \quad . \tag{1}$$

Applying angular momentum results in the coupled basis

$$\begin{aligned}
|t=0,t_z=0\rangle &= \frac{1}{\sqrt{2}}\left(|\mathrm{p}_1\rangle|\mathrm{n}_2\rangle - |\mathrm{n}_1\rangle|\mathrm{p}_2\rangle\right) \quad , \\
|t=1,t_z=0\rangle &= \frac{1}{\sqrt{2}}\left(|\mathrm{p}_1\rangle|\mathrm{n}_2\rangle + |\mathrm{n}_1\rangle|\mathrm{p}_2\rangle\right) \quad , \\
|t=1,t_z=1\rangle &= |\mathrm{p}_1\rangle|\mathrm{p}_2\rangle \quad , \\
|t=1,t_z=-1\rangle &= |\mathrm{n}_1\rangle|\mathrm{n}_2\rangle \quad .
\end{aligned} \tag{2}$$

The states with total isospin $t = 1$ are obviously symmetric under exchange of the two particles, whereas the state with isospin 0 is antisymmetric, and consequently the spin and orbital parts of the wave function must have the opposite symmetry in order to make the total wave function antisymmetric. Isospin invariance of the strong interaction implies that it should not depend on the isospin projection, so that the systems p–p and n–n should have similar scattering behavior. For the p–n system only the symmetric component with isospin 1 should act the same, whereas the antisymmetric component with isospin 0 may have totally different scattering properties. Thus isospin invariance does *not* predict identical behavior of the p–n system to p–p and n–n (modifications due to Coulomb effects are in any case neglected in this discussion).

Finally we should mention the useful formula expressing the charges of nucleons or nuclei in terms of isospin. Since the proton has a charge of $+e$ and the neutron has no charge, this can be achieved simply by adding $\frac{1}{2}$ to the eigenvalue of isospin and multiplying by e, so that the charge operator is

$$\hat{q} = e\left(\hat{t}_z + \tfrac{1}{2}\right) \tag{2.187}$$

for one nucleon and analogously

$$\hat{Q} = e\left(\hat{T}_z + \tfrac{1}{2}A\right) \tag{2.188}$$

for a nucleus.

2.5 Parity

2.5.1 Definition

The operation of spatial inversion is defined by

$$\boldsymbol{r} \longrightarrow -\boldsymbol{r} \quad , \quad \boldsymbol{p} \longrightarrow -\boldsymbol{p} \quad , \quad \boldsymbol{s} \longrightarrow \boldsymbol{s} \quad , \quad t \longrightarrow t \quad . \tag{2.189}$$

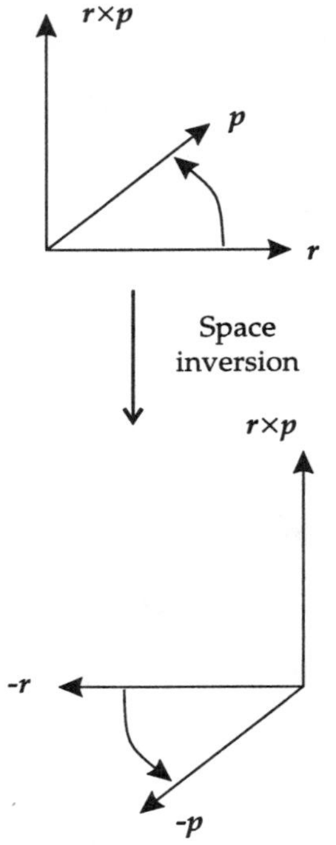

Fig. 2.3 The behaviour of axial vectors under space inversion: the vector product $\boldsymbol{r} \times \boldsymbol{p}$ is unchanged if both $\boldsymbol{r}$ and $\boldsymbol{p}$ are inverted

The reason why the spin vector is not inverted is that as an angular momentum it corresponds to a vector product. If the two vectors forming a vector product are inverted, the right-hand convention for the vector product makes the resulting vector point in the same direction as before (Fig. 2.3). Thus there are two types of vectors: the elementary vectors like $\boldsymbol{r}$ and $\boldsymbol{p}$, which are called *polar vectors*, and the *axial vectors* such as the angular momenta, which are defined in terms of rotation or vector products and do not change sign under space inversion.

Calling the operator of spatial inversion $\hat{P}$, we can also write (2.189) as

$$\hat{P}\boldsymbol{r} = -\boldsymbol{r} \quad , \quad \hat{P}\boldsymbol{p} = -\boldsymbol{p} \quad , \quad \hat{P}\boldsymbol{s} = \boldsymbol{s} \quad , \quad \hat{P}t = t \quad . \tag{2.190}$$

The operator $\hat{P}$ operates on wave functions in a similar spirit as the continuous transformations of the previous sections did: the transformed wave function at the new point takes the value of the old wave function at the old point,

$$\left(\hat{P}\psi\right)\left(\hat{P}\boldsymbol{r}\right) = \psi(\boldsymbol{r}) \quad , \tag{2.191}$$

and since $\hat{P}$ must be its own inverse (trivially $\hat{P}^2 = 1$), we also have

$$\hat{P}\psi(\boldsymbol{r}) = \psi(-\boldsymbol{r}) \quad . \tag{2.192}$$

Again, as for translation and rotation, if the Hamiltonian is invariant under spatial inversion, it must commute with $\hat{P}$, and the wave functions can be chosen as eigenstates of both $\hat{H}$ and $\hat{P}$. If we denote the eigenvalue of $\hat{P}$ by π, we have

$$\psi(-\boldsymbol{r}) = \hat{P}\psi(\boldsymbol{r}) = \pi\psi(\boldsymbol{r}) \quad , \tag{2.193}$$

and, applying $\hat{P}$ again,

$$\psi(\boldsymbol{r}) = \hat{P}^2\psi(\boldsymbol{r}) = \pi^2\psi(\boldsymbol{r}) \quad . \tag{2.194}$$

Thus the *parity quantum number* can only take the values ± 1.

2.5.2 Vector Fields

Similar to the case of angular momentum, the definition of parity has to be considered carefully for vector fields. If we deal with a single vector, such as the position or momentum vector, then it is clear that it is inverted if it is a polar vector and remains the same for an axial vector. For a vector field, we have to do the same *at each point* and also examine its dependence on the coordinates.

For example, for a polar vector field $\boldsymbol{A}(\boldsymbol{r})$ to have good parity requires all the components to satisfy

$$\begin{aligned} A_x(\boldsymbol{r}) &= \pi_r\, A_x(-\boldsymbol{r}) \quad , \\ A_y(\boldsymbol{r}) &= \pi_r\, A_y(-\boldsymbol{r}) \quad , \\ A_z(\boldsymbol{r}) &= \pi_r\, A_z(-\boldsymbol{r}) \quad , \end{aligned} \tag{2.195}$$

where π_r is the *orbital parity*, while the total parity also contains an additional factor of -1 describing the *intrinsic parity* of the polar vector field. Its total parity π is then given by $-\pi_r$, whereas that for an axial vector field is given by π_r, so that it may be defined by the conditions

$$-\boldsymbol{A} = \boldsymbol{A}(-\boldsymbol{r}) \tag{2.196}$$

for polar fields and

$$\boldsymbol{A} = \boldsymbol{A}(-\boldsymbol{r}) \tag{2.197}$$

for axial vector fields. As an example, consider the position vector field $\boldsymbol{r}$. Its components $x(\boldsymbol{r})$, $y(\boldsymbol{r})$, and $z(\boldsymbol{r})$ have negative orbital parity, so that, because of the additional negative parity due to the polar character of the vector $\boldsymbol{r}$, the total parity is $+1$. This agrees with the reflection-invariant character of the field when plotted. The example also shows clearly that the distinction between single vectors and vector fields is very important: the vector $\boldsymbol{r}$ by itself has only the intrinsic parity of -1. Remember the analogous discussion concerning the angular momentum of the field $\boldsymbol{r}$ in Sect. 2.3.7.

Note that some authors use π_r as the total parity, which of course leads to the same final results in physical calculations if used consistently. Usually the parity of scalar products is needed, and if the vector fields are of the same type their intrinsic parities will cancel, so that the parity of the product is in fact determined by only the orbital parity.

2.6 Time Reversal

Although both Coulomb and strong interactions are invariant under time reversal, this symmetry plays only a minor role in nuclear structure physics (it is much more important for reactions, where time-reversed processes are of natural interest). Time reversal is defined by

$$\boldsymbol{r} \longrightarrow \boldsymbol{r} \quad , \quad \boldsymbol{p} \longrightarrow -\boldsymbol{p} \quad , \quad \boldsymbol{s} \longrightarrow -\boldsymbol{s} \quad , \quad t \longrightarrow -t \quad . \tag{2.198}$$

One application is to restrict functional forms of Hamiltonians; for example, a term such as $\boldsymbol{r} \cdot \boldsymbol{p}$ cannot be allowed. In this book, however, the principal application will be in the theory of pairing (Sect. 7.3), so that the reader may skip this section until the need arises.

Time reversal has unusual properties compared with the other symmetries discussed. Using the defining properties, the commutator of position and momentum should be transformed according to

$$\mathrm{i}\hbar = \left[x, p_x\right] \quad \longrightarrow \quad \left[x, -p_x\right] = -\mathrm{i}\hbar \quad . \tag{2.199}$$

This implies that the constant $i\hbar$ changes sign under time reversal! The only way to achieve this is to assume that time reversal includes complex conjugation. This also appears reasonable if we consider the time-dependent Schrödinger equation

$$i\hbar\frac{\partial}{\partial t}\psi(\boldsymbol{r},t) = \hat{H}\,\psi(\boldsymbol{r},t) \quad . \tag{2.200}$$

As $\hat{H}$ should not change under time reversal, this equation retains its form if a change in sign in t is balanced by a sign change in $i\hbar$.

The consequence is that the operator for time reversal, $\hat{\mathcal{T}}$, cannot be unitary, because it is not even a linear operator. A linear operator commutes with an arbitrary c number, so that one should have

$$\hat{\mathcal{T}} i\hbar\hat{\mathcal{T}}^{-1} = i\hbar\hat{\mathcal{T}}\hat{\mathcal{T}}^{-1} = i\hbar \quad , \tag{2.201}$$

which would not allow the complex conjugation that is needed. Operators that have the property $\hat{\mathcal{T}}\alpha = \alpha^*\hat{\mathcal{T}}$ are called *antilinear operators*.

One has to be very careful when constructing the eigenstates of $\hat{\mathcal{T}}$, because many of the usual operator properties do not hold. Assume that $|A\rangle$ is an eigenstate of $\hat{\mathcal{T}}$ with eigenvalue A. Applying $\hat{\mathcal{T}}^2$ then yields

$$\hat{\mathcal{T}}^2|A\rangle = \hat{\mathcal{T}}A|A\rangle = A^*\hat{\mathcal{T}}|A\rangle = |A|^2|A\rangle \quad , \tag{2.202}$$

and because the definition of time reversal shows that it still satisfies $\hat{\mathcal{T}}^2 = 1$, we must have

$$A = e^{i\phi_A} \tag{2.203}$$

with some angle ϕ_A. The angle depends on the choice of phase for the eigenstate. If $|A\rangle$ is replaced by a new state

$$|A'\rangle = e^{\frac{1}{2}i\phi_A}|A\rangle \quad , \tag{2.204}$$

the eigenvalue also changes:

$$\hat{\mathcal{T}}|A'\rangle = \hat{\mathcal{T}}e^{\frac{1}{2}i\phi_A}|A\rangle = e^{-\frac{1}{2}i\phi_A}\hat{\mathcal{T}}|A\rangle = e^{-\frac{1}{2}i\phi_A}e^{i\phi_A}|A\rangle = |A'\rangle \quad . \tag{2.205}$$

It is thus possible to make the eigenvalue of $\hat{\mathcal{T}}$ equal to 1 by a change of phase, and the reader should keep in mind that the properties of time reversal are intimately linked with the phases of wave functions.

For nuclear physics the interplay between angular momentum and time reversal is interesting. Since $\hat{\mathcal{T}}$ inverts the sign of $\hat{\boldsymbol{J}}$, it commutes with $\hat{J}^2$ but not with $\hat{J}_z$. If it is combined with a rotation $\mathcal{R}$ that inverts the direction of the z axis, it should commute with both:

$$\left[\mathcal{R}\hat{\mathcal{T}},\hat{J}^2\right] = 0 \quad , \quad \left[\mathcal{R}\hat{\mathcal{T}},\hat{J}_z\right] = 0 \quad , \tag{2.206}$$

and $\mathcal{R}\hat{\mathcal{T}}$ can be diagonalized in the usual angular-momentum basis. The possible eigenvalues, by the same argument as for $\hat{\mathcal{T}}$, must have the form $\exp(i\phi_A)$ and can again be made equal to 1 by a change of phase of the wave function. Denoting any additional quantum numbers by α, we can thus assume a system of eigenfunctions $|\alpha JM\rangle$ such that

$$\mathcal{R}\hat{\mathcal{T}}|\alpha JM\rangle = |\alpha JM\rangle \tag{2.207}$$

in addition to the usual angular momentum eigenvalue relations. This requires a special choice of phase for the basic functions.

The rotation $\mathcal{R}$ is arbitrary except for the condition of inverting the z axis. It is customary to choose a rotation by an angle π about the y axis, $\mathcal{R} = R_y(\pi)$.

The action of $R_y(\pi)$ on the wave functions can be evaluated from the rotation matrices:

$$\langle\alpha JM'|R_y(\pi)|\alpha' JM\rangle = d^J_{M'M}(\pi)\,\delta_{\alpha\alpha'} = (-1)^{J-M}\delta_{M\,-M'}\delta_{\alpha\alpha'} \quad , \tag{2.208}$$

so that

$$R_y^{-1}(\pi)|\alpha JM\rangle = (-1)^{J-M}|\alpha J \; -M\rangle \quad , \tag{2.209}$$

and because $\mathcal{R}\hat{\mathcal{T}}$ has an eigenvalue of 1 this implies

$$\hat{\mathcal{T}}|\alpha JM\rangle = R_y^{-1}(\pi)\big(\mathcal{R}\hat{\mathcal{T}}\,|\alpha JM\rangle\big) = (-1)^{J-M}|\alpha J \; -M\rangle \quad . \tag{2.210}$$

Let us now examine the action of $(\mathcal{R}\hat{\mathcal{T}})^2$. On the one hand, since its eigenvalue is 1, we have

$$\big(\mathcal{R}\hat{\mathcal{T}}\big)^2|\alpha JM\rangle = |\alpha JM\rangle \quad . \tag{2.211}$$

On the other hand, $\mathcal{R}$ commutes with $\hat{\mathcal{T}}$, because the finite rotation $R_y(\pi)$ is given by $\exp(\mathrm{i}\pi\hat{J}_y)$ and $\hat{\mathcal{T}}$ inverts the signs of both $\hat{J}_y$ and the imaginary factor $\mathrm{i}\pi$. Thus we can also write

$$\big(\mathcal{R}\hat{\mathcal{T}}\big)^2|\alpha JM\rangle = \mathcal{R}^2\hat{\mathcal{T}}^2|\alpha JM\rangle = R_y(2\pi)\hat{\mathcal{T}}^2|\alpha JM\rangle \quad . \tag{2.212}$$

Now $R_y(2\pi)$ is $+1$ for integral spin particles and -1 for fractional spin. We must thus have

$$\hat{\mathcal{T}}^2 = \begin{cases} +1 & \text{for integral spin,} \\ -1 & \text{for fractional spin} \end{cases} \quad . \tag{2.213}$$

On the other hand, $\hat{\mathcal{T}}^2$ can only have eigenvalues of $+1$, because from the eigenvalues of $\hat{\mathcal{T}}$ it follows that if $\hat{\mathcal{T}}|A\rangle = \exp(\mathrm{i}\phi_A)\,|A\rangle$ then

$$\hat{\mathcal{T}}^2|A\rangle = \hat{\mathcal{T}}\mathrm{e}^{\mathrm{i}\phi_A}|A\rangle = \mathrm{e}^{-\mathrm{i}\phi_A}\hat{\mathcal{T}}|A\rangle = |A\rangle \quad . \tag{2.214}$$

So for integral spin there is no problem: both arguments show that for eigenvectors of $\hat{\mathcal{T}}$ the eigenvalues of $\hat{\mathcal{T}}^2$ are $+1$. For fractional spin on the other hand there is a conflict in sign. The resolution is that in this case there are no eigenvectors of $\hat{\mathcal{T}}$, or, in other words, $\hat{\mathcal{T}}|A\rangle$ is always linearly independent of $|A\rangle$. As the Hamiltonian is invariant under time reversal and thus commutes with $\hat{\mathcal{T}}$, $|A\rangle$ and $\hat{\mathcal{T}}|A\rangle$ are two linearly independent but degenerate eigenstates of the Hamiltonian.

This general result can be restated as follows. If the Hamiltonian for a system consisting of an odd number of fermions (and thus having a fractional spin) is invariant under time reversal, its eigenstates will always show twofold degeneracy with the two states being time-reversed with respect to each other. This is called *Kramers degeneracy*. A typical application is in microscopic nuclear models, where single-particle wave functions with opposite values of the angular-momentum projection are always degenerate.

EXERCISE

2.7 The Time-Reversal Operator for Spinors with Spin $\frac{1}{2}$

Problem. Give an explicit form for the time-reversal operator for spinors with spin $\frac{1}{2}$.

Solution. From (2.210) time reversal acts on a spinor $|\sigma\rangle$ with $\sigma = \pm\frac{1}{2}$ according to

$$\hat{\mathcal{T}}|\sigma\rangle = (-1)^{(\frac{1}{2}-\sigma)}|-\sigma\rangle \quad , \tag{1}$$

or

$$\hat{\mathcal{T}}|\tfrac{1}{2}\rangle = |-\tfrac{1}{2}\rangle \quad , \quad \hat{\mathcal{T}}|-\tfrac{1}{2}\rangle = -|\tfrac{1}{2}\rangle \quad . \tag{2}$$

This can easily be reformulated as a matrix operator

$$\hat{\mathcal{T}} = \begin{pmatrix} 0 & -1 \\ 1 & 0 \end{pmatrix} = -\mathrm{i}\sigma_y \quad . \tag{3}$$

In view of the definition of the phases using rotations about the y axis, it is not surprising that the Pauli matrix σ_y appears in this result.

3. Second Quantization

3.1 General Formalism

3.1.1 Motivation

The mathematical apparatus of second quantization is extremely useful for dealing with many-body systems. Although in nuclear physics the total number of particles is fixed in most problems (the principal particles that can be created and destroyed in low-energy nuclear physics are photons and – taking the concept of "particle" in a broader sense – phonons), the formalism takes the required symmetry properties into account more elegantly and allows the formulation of the extremely useful *particle-hole picture* of nuclear excitations.

Here we will first motivate the mathematical structures by examining what kinds of operators are needed to create and destroy particles, derive the most appropriate commutation relations, and then develop the formalism on the basis of these commutations as far as is necessary for later needs.

The simplest starting point is a system of noninteracting particles. "Noninteracting" means that the Hamiltonian of the total system is simply the sum of the Hamiltonians for each individual particle: there are no interaction terms depending on the coordinates of more than one particle. For such a Hamiltonian,

$$\hat{H} = \sum_{i=1}^{A} \hat{h}(\boldsymbol{r}_i, \hat{\boldsymbol{p}}_i) \quad , \tag{3.1}$$

a solution can be found easily in terms of a product of single particle states. If we have a set of one-particle wave functions $\psi_k(\boldsymbol{r})$ fulfilling $\hat{h}(\boldsymbol{r}, \hat{\boldsymbol{p}})\, \psi_k(\boldsymbol{r}) = \varepsilon_k\, \psi_k(\boldsymbol{r})$, a product state like

$$\Psi(\boldsymbol{r}_1, \ldots, \boldsymbol{r}_A) = \psi_{k_1}(\boldsymbol{r}_1) \cdots \psi_{k_A}(\boldsymbol{r}_A) \tag{3.2}$$

will be an eigenstate of $\hat{H}$ with

$$\hat{H}\Psi = E\Psi \quad , \quad E = \sum_{k=1}^{A} \varepsilon_k \quad . \tag{3.3}$$

This is not yet the correct solution, though, because the wave function must still be symmetrized for bosons and antisymmetrized for fermions in order to fulfil the requirement that the wave function take the same value (bosons) or change its sign (fermions) under the exchange of two particles. Thus we end up with an expression like the Slater determinant

$$\Psi(\boldsymbol{r}_1, \boldsymbol{r}_2, \ldots, \boldsymbol{r}_A) = \frac{1}{\sqrt{A!}} \sum_\pi (-1)^\pi \prod_{k=1}^{A} \psi_k(\boldsymbol{r}_{k_\pi}) \quad , \tag{3.4}$$

for the case of A fermions, where π is a permutation of the indices $i = 1, \ldots, A$ and $(-1)^\pi$ is its sign, i.e., $+1$ for even and -1 for odd permutations. The permutation changes the index i into i_π. For bosons this sign is left out. The sum is over all $A!$ permutations of the A indices and the normalization factor is obtained simply from this number of terms, each of which is normalized. Clearly dealing with such wave functions directly is cumbersome, to say the least.

However, it is also clear that there is a lot of superfluous information in these expressions. Quantum mechanics insists that the information about which particle occupies which wave function is meaningless because of the indistinguishability of the particles: therefore, a better formalism should not explicitly contain this information. The *only* meaningful information is *how many particles populate each state* $\psi_i(\boldsymbol{r})$. Calling this the *occupation number* n_i, we can define the many-particle state as an abstract (normalized) vector in the *occupation-number representation.*

$$|\Psi\rangle = |n_1, n_2, \ldots, n_A\rangle \quad . \tag{3.5}$$

The space of these abstract vectors characterized by varying particle numbers is called *Fock space.* Each of the occupation numbers n_i can take on values of 0 or 1 for fermions, and $0, \ldots, \infty$ for bosons.

To simplify matters, look at the simple case that only one single-particle wave function is available for occupation. When we leave out the index k the many-particle states will be characterized by only the occupation number n of this wave function, and we can simply denote it by $|n\rangle$. We need an operator that changes the number n; let us try the definition

$$|n-1\rangle = \hat{c}|n\rangle \quad . \tag{3.6}$$

Normalization of the states then requires

$$1 = \langle n-1|n-1\rangle = \langle n|\hat{c}^\dagger\hat{c}|n\rangle \quad , \tag{3.7}$$

and this is the only nonvanishing matrix element of the operator $\hat{c}^\dagger\hat{c}$, so that this operator must be diagonal in the states $|n\rangle$ with an eigenvalue of 1. Unfortunately this causes trouble for $n = 0$: applying $\hat{c}$ to $|0\rangle$ we then get a normalized state $|-1\rangle$ with a negative occupation number, which is clearly useless physically. One should rather require $\hat{c}|0\rangle = 0$, so that the construction of states with lower particle numbers is stopped at $n = 0$. But then $\hat{c}^\dagger\hat{c}$ must have an eigenvalue of 0 for the state $|0\rangle$, and our tentative definition of the operator $\hat{c}$ does not work.

A natural way out is to assume that the eigenvalue of $\hat{c}^\dagger\hat{c}$ is not 1 but n, so that this difficulty does not appear (and we also obtain a simple operator determining the number of particles in the wave function). The state $\hat{c}|n\rangle$ cannot then be normalized to unity and instead we have for its norm

$$\langle n|\hat{c}^\dagger\hat{c}|n\rangle = n \quad . \tag{3.8}$$

So we try the new definition

$$\hat{c}|n\rangle = \sqrt{n}|n-1\rangle \quad , \tag{3.9}$$

or, written as a matrix element,

$$\langle n-1|\hat{c}|n\rangle = \sqrt{n} \quad ; \tag{3.10}$$

all other matrix elements of this operator being assumed to vanish. Taking the hermitian conjugate of this matrix element yields

$$\langle n|\hat{c}^\dagger|n-1\rangle = \sqrt{n} \tag{3.11}$$

and it is clear that while $\hat{c}$ lowers the number of particles by 1, $\hat{c}^\dagger$ raises it by 1. To complete this development it only remains for us to examine the commutation relation between $\hat{c}$ and $\hat{c}^\dagger$. For any n we must have

$$\begin{aligned}\langle n|\left[\hat{c},\hat{c}^\dagger\right]|n\rangle &= \langle n|\hat{c}\hat{c}^\dagger|n\rangle - \langle n|\hat{c}^\dagger\hat{c}|n\rangle \\ &= (n+1) - n \\ &= 1 \quad ,\end{aligned} \tag{3.12}$$

so that the commutation relations must read

$$\left[\hat{c},\hat{c}^\dagger\right] = 1 \quad , \quad \left[\hat{c},\hat{c}\right] = 0 \quad , \quad \left[\hat{c}^\dagger,\hat{c}^\dagger\right] = 0 \quad . \tag{3.13}$$

The latter two are, of course, trivial.

Leaving the simple case of one single-particle state, let us return to a set of single-particle wave functions $\psi_k(\boldsymbol{r})$. The operators must then also receive this index, so that $\hat{c}_k$ lowers the number of particles in wave function $\psi_k(\boldsymbol{r})$ by 1. For bosons the order in which the particles are inserted into the state is unimportant, so that operators with different indices should commute and we get

$$\left[\hat{c}_i,\hat{c}_j^\dagger\right] = \delta_{ij} \quad , \quad \left[\hat{c}_i,\hat{c}_j\right] = 0 \quad , \quad \left[\hat{c}_i^\dagger,\hat{c}_j^\dagger\right] = 0 \quad . \tag{3.14}$$

These developments make it appear that the commutation relations could be used in the same way for bosons and fermions, at least for the case of one single-particle wave function. This is not true, because as set up the operator $\hat{c}^\dagger$ would allow the construction of states with an arbitrarily large number of particles in the same single-particle wave function. Instead, the Pauli principle requires that n should take only the values 0 and 1, and the construction must be cut off for increasing particle number in the same way as at $|0\rangle$ is:

$$0 = \hat{c}^\dagger|1\rangle = \hat{c}^\dagger\hat{c}^\dagger|0\rangle \quad . \tag{3.15}$$

Clearly, any power of $\hat{c}^\dagger$ higher than 1 must vanish, which can be fulfilled by requiring $\hat{c}^{\dagger 2} = 0$. Since there can be at most one particle in any state, it must also be true that $\hat{c}^2 = 0$. Also for particles in different single-particle wave functions the wave function should change sign under interchange of the particles, so that we expect that a product of two operators changes sign if they are interchanged: $\hat{c}_i^\dagger\hat{c}_j^\dagger = -\hat{c}_j^\dagger\hat{c}_i^\dagger$. All of these conditions can be formulated in terms of *anticommutation relations* (the usual notation for the *anticommutator* $\{\hat{A},\hat{B}\} = \hat{A}\hat{B} + \hat{B}\hat{A}$ is used):

$$\{\hat{c}_i,\hat{c}_j^\dagger\} = \delta_{ij} \quad , \quad \{\hat{c}_i,\hat{c}_j\} = 0 \quad , \quad \{\hat{c}_i^\dagger,\hat{c}_j^\dagger\} = 0 \quad . \tag{3.16}$$

The first of these relations still has to be checked for its validity. For $i = j$ the only matrix elements are

$$\langle 0|\hat{c}\hat{c}^\dagger + \hat{c}^\dagger\hat{c}|0\rangle = \langle 0|\hat{c}c^\dagger|0\rangle = 1 \tag{3.17}$$

and

$$\langle 1|\hat{c}c^\dagger + \hat{c}^\dagger\hat{c}|1\rangle = \langle 1|\hat{c}^\dagger\hat{c}|1\rangle = 1 \tag{3.18}$$

because $\hat{c}|0\rangle = 0$ and $\hat{c}^\dagger|1\rangle = 0$, so that the anticommutation relation is reasonable also for this combination.

This completes the motivation for the commutation and anticommutation relations in second quantization; we will now use these as a starting point for developping the formalism of second quantization in the next section.

3.1.2 Second Quantization for Bosons

As motivated in the preceding section, two operators are defined which are Hermitian adjoints of each other and which will now be designated by their customary name and notation: the *annihilation operator* $\hat{a}$ and the *creation operator* $\hat{a}^\dagger$, describing the annihilation and creation of particles, respectively, in a given single-particle state. For bosons they are required to fulfil the commutation relations

$$[\hat{a},\hat{a}] = 0 \quad , \quad [\hat{a}^\dagger,\hat{a}^\dagger] = 0 \quad , \quad [\hat{a},\hat{a}^\dagger] = 1 \quad . \tag{3.19}$$

Now define the *particle-number operator* $\hat{n}$ as

$$\hat{n} = \hat{a}^\dagger\hat{a} \quad , \tag{3.20}$$

and study its commutation rules with $\hat{a}$ and $\hat{a}^\dagger$:

$$[\hat{a},\hat{n}] = [\hat{a},\hat{a}^\dagger\hat{a}] = \hat{a}\hat{a}^\dagger\hat{a} - \hat{a}^\dagger\hat{a}\hat{a} = [\hat{a},\hat{a}^\dagger]\hat{a} = \hat{a} \quad , \tag{3.21}$$

and

$$[\hat{a}^\dagger,\hat{n}] = [\hat{a}^\dagger,\hat{a}^\dagger\hat{a}] = \hat{a}^\dagger\hat{a}^\dagger\hat{a} - \hat{a}^\dagger\hat{a}\hat{a}^\dagger = \hat{a}^\dagger[\hat{a}^\dagger,\hat{a}] = -\hat{a}^\dagger \quad . \tag{3.22}$$

These are commutation relations of the shift-operator type as discussed for the angular-momentum operators in Sect. 2.3.2. Thus we may immediately conclude that $\hat{a}$ and $\hat{a}^\dagger$ lower and raise, respectively, the eigenvalues of $\hat{n}$ by 1. Assuming eigenstates of $\hat{n}$ in the form

$$\hat{n}|n\rangle = n|n\rangle \quad , \tag{3.23}$$

we get as in the preceding section

$$\hat{a}|n\rangle = \sqrt{n}|n-1\rangle \quad , \quad \hat{a}^\dagger|n\rangle = \sqrt{n+1}|n+1\rangle \quad . \tag{3.24}$$

What values can n have? Since

$$n = \langle n|\hat{n}|n\rangle = \langle n|\hat{a}^\dagger\hat{a}|n\rangle \tag{3.25}$$

corresponds to the square of the norm of the state $\hat{a}|n\rangle$, it must be a positive number. If we repeatedly apply $\hat{a}$ to such a state, the eigenvalue must eventually become negative, which would be inconsistent, or the construction must break off at some place. This happens for the case $n = 0$, the vacuum state, where

$$a|0\rangle = 0 \quad . \tag{3.26}$$

Equation (3.26) may be taken as defining the vacuum state.

The states n may also be generated from the vacuum by repeated application of $\hat{a}^\dagger$. With the normalization factors from (3.24) taken into account this yields

$$|n\rangle = \frac{1}{\sqrt{n!}}(\hat{a}^\dagger)^n|0\rangle \quad . \tag{3.27}$$

For the case of many single particle states with index i that are to be occupied by bosons, the operators are indexed by i to denote which state they affect, and the commutation relations become

$$\left[\hat{a}_i, \hat{a}_j\right] = 0 \quad , \quad \left[\hat{a}_i^\dagger, \hat{a}_j^\dagger\right] = 0 \quad , \quad \left[\hat{a}_i, \hat{a}_j^\dagger\right] = \delta_{ij} \quad . \tag{3.28}$$

There is now also a particle-number operator for each single-particle state,

$$\hat{n}_i = \hat{a}_i^\dagger \hat{a}_i \tag{3.29}$$

as well as one counting the total number of particles,

$$\hat{n} = \sum_i \hat{n}_i = \sum_i \hat{a}_i^\dagger \hat{a}_i \quad . \tag{3.30}$$

The states of the system are characterized by *all* the occupation numbers:

$$\hat{n}_i|n_1, n_2, \ldots, n_i, \ldots\rangle = n_i|n_1, n_2, \ldots, n_i, \ldots\rangle \tag{3.31}$$

and may be written in terms of the vacuum state as

$$|n_1, n_2, \ldots, n_i, \ldots\rangle = \prod_i \frac{(\hat{a}_i^\dagger)^{n_i}}{\sqrt{n_i!}}|0\rangle \quad , \tag{3.32}$$

where usually $|0\rangle$ is written instead of the more correct but lengthy $|0, 0, \ldots, 0, \ldots\rangle$ with one zero for each quantum number. Note that the ordering of the operators is unimportant because they can be commuted freely; this will be different for fermions. The application of the operators $\hat{a}_i$ and $\hat{a}_i^\dagger$ is simply given by

$$\begin{aligned} \hat{a}_i|n_1, n_2, \ldots, n_i, \ldots\rangle &= \sqrt{n_i}\;|n_1, n_2, \ldots, n_i - 1, \ldots\rangle \quad , \\ \hat{a}_i^\dagger|n_1, n_2, \ldots, n_i, \ldots\rangle &= \sqrt{n_i + 1}\;|n_1, n_2, \ldots, n_i + 1, \ldots\rangle \quad . \end{aligned} \tag{3.33}$$

In this book the applications of second quantization for bosons include the photons of the radiation field and the various phonons present in the collective models.

3.1.3 Second Quantization for Fermions

In the fermion case the basic *anti*commutation relations are – again first quoted for the case of a single level –

$$\{\hat{a},\hat{a}\} = 0 \quad , \quad \{\hat{a}^\dagger,\hat{a}^\dagger\} = 0 \quad , \quad \{\hat{a},\hat{a}^\dagger\} = 1 \quad . \tag{3.34}$$

Many of the developments are now formally similar to the boson case, but the small differences in mathematical detail cause tremendous differences in physics. The particle-number operator is defined in the same way,

$$\hat{n} = \hat{a}^\dagger\hat{a} \quad , \tag{3.35}$$

and the shift operator properties can be derived analogously, for example

$$\begin{aligned}[\hat{a},\hat{n}] &= [\hat{a},\hat{a}^\dagger\hat{a}] = \hat{a}\hat{a}^\dagger\hat{a} - \hat{a}^\dagger\hat{a}\hat{a} = \hat{a}\hat{a}^\dagger\hat{a} + \hat{a}^\dagger\hat{a}\hat{a}\\ &= \{\hat{a},\hat{a}^\dagger\}\hat{a} = \hat{a} \quad .\end{aligned} \tag{3.36}$$

The sign change leading to the last equality in the first line is due to the anticommutation of the two $\hat{a}$s (in fact, this product is zero, so that a vanishing term is being manipulated). Thus we can again use eigenstates of $\hat{n}$, with the eigenvalue raised and lowered by $\hat{a}^\dagger$ and $\hat{a}$, respectively. That only two values of n are possible can be inferred from the fact that $\hat{a}^\dagger\hat{a}^\dagger = \hat{a}\hat{a} = 0$, leading to

$$\begin{aligned}\hat{n}^2 &= \hat{a}^\dagger\hat{a}\hat{a}^\dagger\hat{a} = \hat{a}^\dagger\left(\{\hat{a},\hat{a}^\dagger\} - \hat{a}^\dagger\hat{a}\right)\hat{a} = \hat{a}^\dagger\hat{a} - \hat{a}^\dagger\hat{a}^\dagger\hat{a}\hat{a}\\ &= \hat{a}^\dagger\hat{a} = \hat{n} \quad ,\end{aligned} \tag{3.37}$$

so that $\hat{n}$ can have eigenvalues of 0 and 1 only. There are thus only two eigenstates, the vacuum state $|0\rangle$ and the one-particle state $|1\rangle$, and the action of the operators is given by

$$\begin{aligned}&\hat{n}|0\rangle = 0 \quad , \quad \hat{a}^\dagger|0\rangle = |1\rangle \quad , \quad \hat{a}|0\rangle = 0 \quad ,\\ &\hat{n}|1\rangle = |1\rangle \quad , \quad \hat{a}^\dagger|1\rangle = 0 \quad , \quad \hat{a}|1\rangle = |0\rangle \quad .\end{aligned} \tag{3.38}$$

The matrix elements of the operators have been calculated exactly in the same way as for bosons, but reduce to trivial factors of 1 in this case. These equations can be brought into the same form for both states as follows:

$$\hat{n}|n\rangle = n|n\rangle \quad , \quad \hat{a}^\dagger|n\rangle = (1-n)|n+1\rangle \quad , \quad a|n\rangle = n|n-1\rangle \quad . \tag{3.39}$$

For systems with many single-particle levels the procedure starts off similarly as for bosons. We now have many operators $\hat{a}_i$ and $\hat{a}_i^\dagger$, and anticommutation relations

$$\{\hat{a}_i,\hat{a}_j\} = 0 \quad , \quad \{\hat{a}_i^\dagger,\hat{a}_j^\dagger\} = 0 \quad , \quad \{\hat{a}_i,\hat{a}_j^\dagger\} = \delta_{ij} \quad , \tag{3.40}$$

and again the states are characterized by the eigenvalues of the particle-number operators,

$$\hat{n}_i = \hat{a}_i^\dagger\hat{a}_i \quad , \quad \hat{n}_i|n_1,n_2,\ldots,n_i,\ldots\rangle = n_i|n_1,n_2,\ldots,n_i,\ldots\rangle \quad , \tag{3.41}$$

albeit with all n_i restricted to values of 0 or 1. There is one new problem, however, in that the order of the operators is important for the overall sign of the wave function when it is expressed as a product of creation operators acting on the vacuum state. We use the convention

$$|n_1, n_2, \ldots, n_i, \ldots\rangle = \left(\hat{a}_1^\dagger\right)^{n_1} \left(\hat{a}_2^\dagger\right)^{n_2} \cdots \left(\hat{a}_i^\dagger\right)^{n_i} \cdots |0\rangle \quad , \tag{3.42}$$

i.e., order the operators in the same way as the quantum numbers in the ket. There is no normalization factor, because 1! and 0! both reduce to unity.

Applying the operator $\hat{a}_i$ or $\hat{a}_j$ onto a state similar to that in (3.42) will yield a sign that depends on the exact ordering of the creation operators. Regard the operator $\hat{a}_i$, for example. If the state has $n_i = 0$, then application of $\hat{a}_i$ must yield zero. But if $n_i = 1$, then the expression (3.42) contains $\hat{a}_i^\dagger$ once at a certain position. To evaluate the sign, we anticommute $\hat{a}_i$ with all operators preceding $\hat{a}_i^\dagger$ in the product:

$$\hat{a}_i \left(\hat{a}_1^\dagger\right)^{n_1} \cdots \left(\hat{a}_{i-1}^\dagger\right)^{n_{i-1}} \hat{a}_i^\dagger \cdots |0\rangle = \sigma_i \left(\hat{a}_1^\dagger\right)^{n_1} \cdots \left(\hat{a}_{i-1}^\dagger\right)^{n_{i-1}} \hat{a}_i \hat{a}_i^\dagger \cdots |0\rangle \quad . \tag{3.43}$$

The sign factor σ_i is determined by the number of operator factors preceding $\hat{a}_i^\dagger$ in the product and is thus given by

$$\sigma_i = (-1)^{\sum_{j=1}^{i-1} n_j} \quad . \tag{3.44}$$

The product $\hat{a}_i \hat{a}_i^\dagger$ can now be commuted with all other operators between its present position and the state $|0\rangle$; this is possible because all these operators anticommute with both $\hat{a}_i$ and $\hat{a}_i^\dagger$ and so commute with their product because of the double sign change. Finally, from (3.39) one has

$$\hat{a}_i \hat{a}_i^\dagger |0\rangle = |0\rangle \quad , \tag{3.45}$$

so that

$$\begin{aligned} \hat{a}_i \left(\hat{a}_1^\dagger\right)^{n_1} \cdots \left(\hat{a}_{i-1}^\dagger\right)^{n_{i-1}} \hat{a}_i^\dagger \cdots |0\rangle &= \sigma_i \left(\hat{a}_1^\dagger\right)^{n_1} \cdots \left(\hat{a}_{i-1}^\dagger\right)^{n_{i-1}} \left(\hat{a}_{i+1}^\dagger\right)^{n_{i+1}} |0\rangle \\ &= \sigma_i |n_1, n_2, \ldots, n_{i-1}, 0, n_{i+1}, \ldots\rangle \quad . \end{aligned} \tag{3.46}$$

We may summarize the results of doing the same derivations for $\hat{a}_i^\dagger$ as

$$\begin{aligned} \hat{a}_i |n_1, \ldots, n_i, \ldots\rangle &= \sigma_i (1 - n_i) |n_1, \ldots, n_i - 1, \ldots\rangle \quad , \\ \hat{a}_i^\dagger |n_1, \ldots, n_i, \ldots\rangle &= \sigma_i n_i |n_1, \ldots, n_i + 1, \ldots\rangle \quad . \end{aligned} \tag{3.47}$$

3.2 Representation of Operators

3.2.1 One-Particle Operators

Up to now we have seen how the wave functions of a noninteracting many-body system can be represented in the occupation-number representation. To complete this description, we also have to translate operators into this representation. The types of operators that will be needed are one-body operators that depend only on the coordinates of one particle, for example the kinetic energy or an external potential, and two-body operators involving coordinates of two particles, such as an interaction potential. Matrix elements of such operators have to be evaluated between Slater determinants and then an operator in second quantization has to be constructed that yields the same matrix elements in the equivalent occupation-number states. This provides a good illustration of how cumbersome it would be to work with the original representation.

As our use of Slater determinants has already hinted at, we will only consider the case of Fermions. For bosons the developments are quite similar, but only trivial instances will be needed in this book.

A one-particle operator has the general form $\hat{f}(\boldsymbol{r}_k)$, where the coordinate $\boldsymbol{r}_k$ of the kth particle also represents the momentum, spin, and any other needed degrees of freedom. Now in a system of nondistinguishable particles it makes no sense to ask for the properties of a certain particle; instead only quantities should be evaluated that are invariant under an arbitrary permutation of the particles. The reasonable definition of a one-particle operator is thus

$$\hat{f} = \sum_{k=1}^{A} \hat{f}(\boldsymbol{r}_k) \quad . \tag{3.48}$$

What are its matrix elements? Take one of the terms in the sum between two determinantal wave functions

$$\Psi(\boldsymbol{r}_1, \ldots, \boldsymbol{r}_A) = \frac{1}{\sqrt{A!}} \sum_{\pi} (-1)^{\pi} \prod_{i \in O} \psi_{i_\pi}(\boldsymbol{r}_i) \tag{3.49}$$

and

$$\Psi'(\boldsymbol{r}_1, \ldots, \boldsymbol{r}_A) = \frac{1}{\sqrt{A!}} \sum_{\pi'} (-1)^{\pi'} \prod_{i' \in O'} \psi_{i'_{\pi'}}(\boldsymbol{r}_{i'}) \quad . \tag{3.50}$$

Note that the indices i' and i are taken from sets of indices O' and O, which refer to different choices of A single-particle wave functions from a complete orthonormal set $\psi_k(\boldsymbol{r})$, $k = 1, \ldots, \infty$. The matrix element becomes

$$\begin{aligned} \langle \Psi | \hat{f} | \Psi' \rangle = & \sum_{k=1}^{A} \frac{1}{A!} \int \mathrm{d}^3 r_1 \cdots \int \mathrm{d}^3 r_A \\ & \times \sum_{\pi\pi'} (-1)^{\pi+\pi'} \Big(\prod_{i \in O} \psi^*_{i_\pi}(\boldsymbol{r}_i) \Big) \hat{f}(\boldsymbol{r}_k) \Big(\prod_{i' \in O'} \psi_{i'_{\pi'}}(\boldsymbol{r}_{i'}) \Big) \quad . \end{aligned} \tag{3.51}$$

The products can be split up into A single-particle matrix elements. One of these involves the operator $\hat{f}(\boldsymbol{r}_k)$,

$$\int \mathrm{d}^3 r_k \; \psi^*_{k_\pi}(\boldsymbol{r}_k) \hat{f}(\boldsymbol{r}_k) \, \psi_{k_{\pi'}}(\boldsymbol{r}_k) = f_{k_\pi k_{\pi'}} \quad , \tag{3.52}$$

whereas the others can be reduced because of orthonormality of the wave functions,

$$\int \mathrm{d}^3 r_k \; \psi^*_{i_\pi}(\boldsymbol{r}_i) \, \psi_{i'_{\pi'}}(\boldsymbol{r}_i) = \delta_{i_\pi i'_{\pi'}} \quad , \tag{3.53}$$

and the total matrix element may now be written as

$$\langle \Psi | \hat{f} | \Psi' \rangle = \sum_k \frac{1}{A!} \sum_{\pi\pi'} (-1)^{\pi+\pi'} f_{k_\pi k_{\pi'}} \prod_{\substack{i\in O, i'\in O' \\ i\neq k, i'\neq k}}^{A} \delta_{i_\pi i'_{\pi'}} \quad . \tag{3.54}$$

Before proceeding further it is worthwhile to state a few consequences of this formula.

1. The Kronecker symbols require that for a nonvanishing matrix element the same states must be occupied in Ψ as in Ψ' with a single exception. Thus a single-particle operator changes the state of a single particle only.
2. To obtain a nonvanishing matrix element, for a fixed permutation π there is one and only one permutation π' which correctly pairs the single-particle wave functions in Ψ' with those in Ψ. Instead of a double sum over permutations, a single sum then suffices; we only have to keep track of the sign. As π and π' are the permutations needed to bring the states numbered by i and i' from their original ordering into the same order (with the indices k_π and $k_{\pi'}$ also in the same position), the factor $\sigma = (-1)^{\pi+\pi'}$ tells whether an even or odd permutation is needed to transform these original orderings into each other and it does not depend on π or π'.
3. The sum over permutations π then effectively runs only over the various ways to number the $A-1$ states occupied in both Ψ and Ψ'.

The matrix element thus is totally independent of the permutation. It only contains the factor σ and the matrix element $f_{k_\pi k_{\pi'}}$ always obtains the same indices: those of the two single-particle states which differ between Ψ and Ψ'; let us simply call them j and j'. The matrix element is now

$$\langle \Psi | \hat{f} \ | \Psi' \rangle = \sum_{k=1}^{A} \frac{1}{A!} \, \sigma \, f_{jj'} \sum_{\pi, j \text{ fixed}} 1 = \frac{1}{A!} \, \sigma \, A \, f_{jj'} \, (A-1)! = \sigma \, f_{jj'} \quad . \tag{3.55}$$

What should the equivalent operator in second quantization look like? It must remove one particle from state j' and put it into the state j while not doing anything to the other states; the resulting matrix element is $f_{jj'}$. It seems natural to try

$$\hat{f} = \sum_{jj'} f_{jj'} \, \hat{a}_j^\dagger \hat{a}_{j'} \quad . \tag{3.56}$$

The sum is included because it should work for all possible states j and j'. It only remains for us to check whether the sign comes out alright. In second quantization the states are

$$|\Psi\rangle = \hat{a}_{i_1}^\dagger \cdots \hat{a}_{i_A}^\dagger |0\rangle \quad , \quad |\Psi'\rangle = \hat{a}_{i'_1}^\dagger \cdots \hat{a}_{i'_A}^\dagger |0\rangle \quad , \tag{3.57}$$

and the matrix element is

$$\langle \Psi | \hat{f} | \Psi' \rangle = \sum_{jj'} f_{jj'} \langle 0 | \hat{a}_{i_A} \cdots \hat{a}_{i_1} \hat{a}_j^\dagger \hat{a}_{j'} \hat{a}_{i'_1}^\dagger \cdots \hat{a}_{i'_A}^\dagger |0\rangle \quad . \tag{3.58}$$

It is clear that again the indices i must denote the same states as the i', except that j replaces j'. Permute the i' in such a way that they are in the same order as the i

(and with j' at the same place as j), and this will yield the same sign factor σ as defined above. Permuting the operator combination $\hat{a}_j^\dagger \hat{a}_{j'}$ in front of $\hat{a}_{j'}^\dagger$ does not change the sign, and we get

$$\langle \Psi|\hat{f}|\Psi'\rangle = \sum_{jj'} f_{jj'} \langle 0|\hat{a}_{i_A} \cdots \hat{a}_{i_1} \hat{a}_{i_1}^\dagger \cdots \hat{a}_j^\dagger \hat{a}_{j'} \hat{a}_{j'}^\dagger \cdots \hat{a}_{i_A}^\dagger |0\rangle \quad . \tag{3.59}$$

In this expression the operator combinations $\hat{a}_i^\dagger \hat{a}_{i'}$ yield factors of 1, so that we end up with

$$\langle \Psi|\hat{a}|\Psi'\rangle = \sigma f_{jj'} \quad , \tag{3.60}$$

and this agrees with the previous result.

The rule for transcribing a single-particle operator into second-quantized form is thus

$$\hat{f} = \sum_{k=1}^{A} \hat{f}(\boldsymbol{r}_k) \quad \rightarrow \quad \hat{f} = \sum_{jj'} f_{jj'} \hat{a}_j^\dagger \hat{a}_{j'} \quad , \tag{3.61}$$

with the single-particle matrix elements $f_{jj'}$ given by (3.52).

3.2.2 Two-Particle Operators

The analogous result for two-particle operators such as the potential energy

$$\hat{V} = \tfrac{1}{2} \sum_{k \neq k'} \hat{v}(\boldsymbol{r}_k, \boldsymbol{r}_{k'}) \tag{3.62}$$

can be derived similarly, but naturally more laboriously. We prefer to give only the result, providing a check in Exercise 3.1 for a simple special case.

The second quantized operator is

$$\hat{V} = \tfrac{1}{2} \sum_{ijkl} v_{ijkl} \, \hat{a}_i^\dagger \hat{a}_j^\dagger \hat{a}_l \hat{a}_k \quad , \tag{3.63}$$

with the two-particle matrix element defined by

$$v_{ijkl} = \int \mathrm{d}^3 r \int \mathrm{d}^3 r' \; \psi_i^*(\boldsymbol{r}) \, \psi_j^*(\boldsymbol{r}') \, v(\boldsymbol{r}, \boldsymbol{r}') \, \psi_k(\boldsymbol{r}) \, \psi_l(\boldsymbol{r}') \quad . \tag{3.64}$$

Note that the operator can, as expected, change two single-particle states simultaneously, and that the index order in the operator products has the last two indices interchanged relative to the ordering in the matrix element.

In many calculations the evaluation of the matrix elements leads to an antisymmetric combination, which is therefore given a special abbreviation:

$$\bar{v}_{ijkl} = v_{ijkl} - v_{ijlk} \quad . \tag{3.65}$$

Note also the symmetry of the matrix element under the interchange of the two pairs of single-particle wave functions:

$$v_{ijkl} = v_{klij} \quad . \tag{3.66}$$

EXERCISE

3.1 Two-Body Operators in Second Quantization

Problem. Derive the transcription of two-body operators into second quantization for the special case of two-particle wave functions.

Solution. Written as Slater determinants, the two-particle wave functions are

$$\Psi_{ij}(\boldsymbol{r},\boldsymbol{r}') = \frac{1}{\sqrt{2}}\left(\psi_i(\boldsymbol{r})\,\psi_j(\boldsymbol{r}') - \psi_i(\boldsymbol{r}')\,\psi_j(\boldsymbol{r})\right) \tag{1}$$

and

$$\Psi_{kl}(\boldsymbol{r},\boldsymbol{r}') = \frac{1}{\sqrt{2}}\left(\psi_k(\boldsymbol{r})\,\psi_l(\boldsymbol{r}') - \psi_k(\boldsymbol{r}')\,\psi_l(\boldsymbol{r})\right) \quad , \tag{2}$$

so that the matrix element becomes

$$\begin{aligned}\langle\Psi_{ij}|\hat{V}|\Psi_{kl}\rangle = {} & \tfrac{1}{2}\int d^3r \int d^3r'\; \psi_i^*(\boldsymbol{r})\,\psi_j^*(\boldsymbol{r}')\,\hat{v}(\boldsymbol{r},\boldsymbol{r}')\,\psi_k(\boldsymbol{r})\,\psi_l(\boldsymbol{r}') \\ & - \tfrac{1}{2}\int d^3r \int d^3r'\; \psi_i^*(\boldsymbol{r}')\,\psi_j^*(\boldsymbol{r})\,\hat{v}(\boldsymbol{r},\boldsymbol{r}')\,\psi_k(\boldsymbol{r})\,\psi_l(\boldsymbol{r}') \\ & - \tfrac{1}{2}\int d^3r \int d^3r'\; \psi_i^*(\boldsymbol{r})\,\psi_j^*(\boldsymbol{r}')\,\hat{v}(\boldsymbol{r},\boldsymbol{r}')\,\psi_k(\boldsymbol{r}')\,\psi_l(\boldsymbol{r}) \\ & + \tfrac{1}{2}\int d^3r \int d^3r'\; \psi_i^*(\boldsymbol{r}')\,\psi_j^*(\boldsymbol{r})\,\hat{v}(\boldsymbol{r},\boldsymbol{r}')\,\psi_k(\boldsymbol{r}')\,\psi_l(\boldsymbol{r}) \quad . \end{aligned} \tag{3}$$

Using the definition (3.64) and the symmetry of $\hat{v}(\boldsymbol{r},\boldsymbol{r}') = \hat{v}(\boldsymbol{r}',\boldsymbol{r})$ this may be rewritten as

$$\begin{aligned}\langle\Psi_{ij}|\hat{V}|\Psi_{kl}\rangle &= \tfrac{1}{2}(v_{ijkl} - v_{ijlk} - v_{ijlk} + v_{ijkl}) \\ &= v_{ijkl} - v_{ijlk} \\ &= \bar{v}_{ijkl} \quad . \end{aligned} \tag{4}$$

Now comes the same calculation in second quantization. The two states are

$$|\Psi_{ij}\rangle = \hat{a}_i^\dagger \hat{a}_j^\dagger |0\rangle \tag{5}$$

and

$$|\Psi_{kl}\rangle = \hat{a}_k^\dagger \hat{a}_l^\dagger |0\rangle \quad , \tag{6}$$

and the matrix element becomes

$$\langle\Psi_{ij}|\hat{V}|\Psi_{kl}\rangle = \tfrac{1}{2}\sum_{i'j'k'l'} v_{i'j'k'l'}\langle 0|\hat{a}_j\hat{a}_i\hat{a}_{i'}^\dagger\hat{a}_{j'}^\dagger\hat{a}_{l'}\hat{a}_{k'}\hat{a}_k^\dagger\hat{a}_l^\dagger|0\rangle \quad . \tag{7}$$

Evaluating it with the methods presented in Sect. 3.3 leads to

Exercise 3.1

$$\begin{aligned}\langle\Psi_{ij}|\hat{V}|\Psi_{kl}\rangle &= \frac{1}{2}\sum_{i'j'k'l'} v_{i'j'k'l'}(\delta_{i'i}\,\delta_{j'j}\,\delta_{k'k}\,\delta_{l'l} - \delta_{i'j}\,\delta_{j'i}\,\delta_{k'k}\,\delta_{l'l} \\ &\quad - \delta_{i'i}\,\delta_{j'j}\,\delta_{k'l}\,\delta_{l'k} + \delta_{i'j}\,\delta_{j'j}\,\delta_{k'l}\,\delta_{l'k}) \\ &= \frac{1}{2}(v_{ijkl} - v_{jikl} - v_{ijlk} + v_{jilk}) \\ &= v_{ijkl} - v_{ijlk} \\ &= \bar{v}_{ijkl} \quad . \end{aligned} \tag{8}$$

In the last steps the symmetry of the matrix elements was exploited.

This derivation makes it quite apparent what causes the difference in the order of indices in the matrix element and the operator product: it has to make up for the interchange of operators when changing the wave function from a ket to a bra vector.

3.3 Evaluation of Matrix Elements for Fermions

The most frequent type of calculation required for microscopic nuclear models is that of matrix elements of a product of fermion creation and annihilation operators. The simplest case is an expectation value of such an operator in the vacuum, and we will illustrate the general procedure for the example

$$M = \langle 0|\hat{a}_i\hat{a}_j\hat{a}_k^\dagger\hat{a}_l\hat{a}_m^\dagger\hat{a}_n^\dagger|0\rangle \quad . \tag{3.67}$$

Any nonvanishing matrix element must have equal numbers of creation and annihilation operators, because applying the operator product to the vacuum state on the right must lead back to the vacuum: all created particles must be destroyed again. If there is a surplus of annihilation operators, their action on the vacuum produces zero in any case.

Note that if there were an annihilation operator to the right of the product, or a creation operator to the left, the matrix element would vanish because of

$$\hat{a}_i|0\rangle = 0 \quad , \quad \langle 0|\hat{a}_i^\dagger = 0 \quad . \tag{3.68}$$

The strategy will now be to exploit this very fact and to commute the annihilation operators to the right, which, of course, also commutes the creation operators to the left.

For this purpose, the anticommutation rules can be stated as the following two simple rules.

1. The commutation of two operators of the same type (both creation or both annihilation operators) only inverts the sign.
2. Two operators of opposite type yield an additional term with a Kronecker symbol replacing both operators:

$$\hat{a}_i^\dagger\hat{a}_j = -\hat{a}_j\hat{a}_i^\dagger + \delta_{ij} \quad , \quad \hat{a}_i\hat{a}_j^\dagger = -\hat{a}_j^\dagger\hat{a}_i + \delta_{ij} \quad . \tag{3.69}$$

Thus in our procedure each commutation either only changes the sign or adds a term with a Kronecker symbol replacing the two operators. If the process is continued finally only terms consisting entirely of Kronecker symbols will remain, because the operators have been commuted far enough to yield zero.

Let us apply this to the example matrix element. One possibility for starting is to commute $\hat{a}_l$ to the right. After the first step, commuting with $\hat{a}_m^\dagger$, this yields

$$M = -\langle 0|\hat{a}_i\hat{a}_j\hat{a}_k^\dagger\hat{a}_m^\dagger\hat{a}_l\hat{a}_n^\dagger|0\rangle + \delta_{lm}\langle 0|\hat{a}_i\hat{a}_j\hat{a}_k^\dagger\hat{a}_n^\dagger|0\rangle \quad . \tag{3.70}$$

In the first matrix element on the right-hand side, the process has to be continued by commuting with $\hat{a}_n^\dagger$, whereas in the second matrix element we can start commuting $\hat{a}_j$ to the right. The result is

$$\begin{aligned} M &= \langle 0|\hat{a}_i\hat{a}_j\hat{a}_k^\dagger\hat{a}_m^\dagger\hat{a}_n^\dagger\hat{a}_l|0\rangle - \delta_{ln}\langle 0|\hat{a}_i\hat{a}_j\hat{a}_k^\dagger\hat{a}_m^\dagger|0\rangle \\ &\quad - \delta_{lm}\langle 0|\hat{a}_i\hat{a}_k^\dagger\hat{a}_j\hat{a}_n^\dagger|0\rangle + \delta_{lm}\delta_{jk}\langle 0|\hat{a}_i\hat{a}_n^\dagger|0\rangle \quad . \end{aligned} \tag{3.71}$$

The first matrix element on the right vanishes, because $\hat{a}_l$ acts directly on the vacuum. In the second and third matrix elements commutation of $\hat{a}_j$ to the right continues, whereas the last one needs one final commutation. This leaves

$$\begin{aligned} M &= \delta_{ln}\langle 0|\hat{a}_i\hat{a}_k^\dagger\hat{a}_j\hat{a}_m^\dagger|0\rangle - \delta_{ln}\delta_{jk}\langle 0|\hat{a}_i\hat{a}_m^\dagger|0\rangle \\ &\quad - \delta_{lm}\delta_{jn}\langle 0|\hat{a}_i\hat{a}_k^\dagger|0\rangle + \delta_{lm}\delta_{jk}\delta_{in} \quad . \end{aligned} \tag{3.72}$$

It is worth while to memorize as a shortcut that

$$\langle 0|\hat{a}_i\hat{a}_j^\dagger|0\rangle = \delta_{ij} \quad . \tag{3.73}$$

This means that creating a particle from the vacuum and annihilating it immediately just provides a factor of 1. This fact already makes two of the three remaining matrix elements trivial, and in the other one we continue commuting:

$$\begin{aligned} M &= \delta_{ln}\delta_{jm}\langle 0|\hat{a}_i\hat{a}_k^\dagger|0\rangle - \delta_{ln}\delta_{jk}\delta_{im} - \delta_{lm}\delta_{jn}\delta_{ik} + \delta_{lm}\delta_{jk}\delta_{in} \\ &= \delta_{ln}\delta_{jm}\delta_{ik} - \delta_{ln}\delta_{jk}\delta_{im} - \delta_{lm}\delta_{jn}\delta_{ik} + \delta_{lm}\delta_{jk}\delta_{in} \quad . \end{aligned} \tag{3.74}$$

Future calculations of this type can be done much more rapidly by noting the shape of this final result while keeping in mind how it was reached.

- The matrix element is reduced to combinations of Kronecker symbols with different signs taking care of the antisymmetry.
- The index combinations in the Kronecker symbols show all possible combinations of one annihilation and one creation operator, *where the annihilation operator preceded the creation operator in the original matrix element.* These are exactly the cases where the commutation process will leave a Kronecker symbol.
- The sign of a term in this final result can be obtained from the following consideration: reorder the operators in the original matrix element such as to produce the combinations appearing in the Kronecker symbols with annihilator preceding creator. For this reordering only the sign changes are considered, not the commutation remainders, and the resulting sign is the desired one. It does not matter whether the other operators are in front of or behind the pair, as

the sign of the matrix element does not depend on this (commuting with a pair always produces a plus sign).

Going back to the example original matrix element, we can generate the final result more simply by noting that the indices to be combined into Kronecker symbols are i and j with k, m, and n, respectively, while l combines only with m and n, because $\hat{a}_l$ is to the right of $\hat{a}_k^\dagger$. This yields the combinations $\delta_{ln}\delta_{jm}\delta_{ik}$, $\delta_{ln}\delta_{jk}\delta_{im}$, $\delta_{lm}\delta_{jn}\delta_{ik}$, and $\delta_{lm}\delta_{jk}\delta_{in}$, exactly as seen in the matrix element. To understand the sign, for example, of the second term note that getting from $\hat{a}_i\hat{a}_j\hat{a}_k^\dagger\hat{a}_l\hat{a}_m^\dagger\hat{a}_n^\dagger$ to $\hat{a}_j\hat{a}_k^\dagger\hat{a}_i\hat{a}_m^\dagger\hat{a}_l\hat{a}_n^\dagger$ requires three commutations, so that the sign must be negative.

Actually the method derived here constitutes a simple case of *Wick's Theorem*, which also deals with more general matrix elements. However, for the purposes of this book the present results are quite sufficient.

The method only needs to be modified slightly if the expectation value is taken in a more complicated state; this is the subject of the following section.

3.4 The Particle-Hole Picture

Expectation values of operators in the vacuum do not really occur very often in theoretical nuclear physics. The more frequent case will be matrix elements in the nuclear ground state, which in the single-particle model is given by A nucleons occupying the lowest-available single-particle states. If we arrange the indices in order of increasing single-particle energy,

$$\varepsilon_1 < \varepsilon_2 < \cdots \varepsilon_A < \varepsilon_{A+1} < \cdots \quad , \tag{3.75}$$

the lowest state of the A-nucleon system is

$$|\Psi_0\rangle = \prod_{i=1}^{A} \hat{a}_i^\dagger |0\rangle \tag{3.76}$$

with an energy

$$E_0 = \sum_{i=1}^{A} \varepsilon_i \quad . \tag{3.77}$$

The highest occupied state with energy ε_A is the Fermi level. The expectation value of an operator $\hat{O}$ in the ground state may then be written as

$$\langle\Psi_0|\hat{O}|\Psi_0\rangle = \langle 0|\hat{a}_A\cdots\hat{a}_1\hat{O}\hat{a}_1^\dagger\cdots\hat{a}_A^\dagger|0\rangle \quad , \tag{3.78}$$

and this is just the vacuum expectation of the much more complicated operator contained in the matrix element on the right, so that in this sense such expectation values can always be rewritten as vacuum expectation values. They could in principle be evaluated with the method discussed above, but they of course involve an impractically large number of operators. Some simplification is possible, however, if we remember that mathematically the ground state has a similar property to the vacuum. While for the latter we have

$$\hat{a}_i|0\rangle = 0 \quad \text{for all } i \quad , \tag{3.79}$$

the ground state fulfills

$$\begin{aligned} \hat{a}_i|0\rangle &= 0 \quad , \quad i > A \\ \hat{a}_i^\dagger|0\rangle &= 0 \quad , \quad i \le A \quad . \end{aligned} \tag{3.80}$$

This implies that similar methods may be used *if indices above and below A are treated differently*. There are two ways of handling this formally:

- keep the present notation and explicitly note in the formulas which index in which term takes what range of values, or,
- redefine the operators such that the ground state takes over the role of the vacuum.

Let us exemplify this by looking at matrix elements and at the excited states of the system in both notations.

The simplest excited states will have one particle lifted from an occupied state into an unoccupied one. They can be written as

$$|\Psi_{mi}\rangle = \hat{a}_m^\dagger \hat{a}_i |\Psi_0\rangle \quad , \quad m > A \quad , \quad i \le A \quad , \tag{3.81}$$

and the associated excitation energy is

$$E_{mi} - E_0 = \varepsilon_m - \varepsilon_i \quad . \tag{3.82}$$

The state $|\Psi_{mi}\rangle$ has an unoccupied level i, a *hole* below the Fermi energy, and a particle in state m above the Fermi energy. For that reason it is called a *one-particle/one-hole state* or 1p1h state. The next more complicated type of excitation is a *two-particle/two-hole* (2p2h) *state* like

$$|\Psi_{mnij}\rangle = \hat{a}_m^\dagger \hat{a}_n^\dagger \hat{a}_i \hat{a}_j |\Psi_0\rangle \tag{3.83}$$

with excitation energy

$$E_{mnij} = \varepsilon_m + \varepsilon_n - \varepsilon_i - \varepsilon_j \quad . \tag{3.84}$$

Looking at the expectation value of a one-particle operator like

$$\langle\Psi_0| \sum_{ij=1}^{\infty} t_{ij}\ \hat{a}_i^\dagger \hat{a}_j |\Psi_0\rangle = \sum_{ij=1}^{\infty} t_{ij}\ \langle\Psi_0|\hat{a}_i^\dagger \hat{a}_j|\Psi_0\rangle \quad , \tag{3.85}$$

it is clear that for $j > A$ the contributions will vanish. For $j \le A$ the operators can be commuted to yield

$$\langle\Psi_0|\hat{a}_i^\dagger \hat{a}_j|\Psi_0\rangle = \delta_{ij} - \langle\Psi_0|\hat{a}_j \hat{a}_i^\dagger|\Psi_0\rangle \quad , \tag{3.86}$$

and the second term on the right-hand side now vanishes. A convenient notation for this result is

$$\langle\Psi_0|\hat{a}_i^\dagger \hat{a}_j|\Psi_0\rangle = \delta_{ij}\,\theta_{iA} \quad . \tag{3.87}$$

The symbol θ_{kl} is an adapted analog to the θ function and is defined by

$$\theta_{iA} = \begin{cases} 1 & \text{if } i \le A \quad , \\ 0 & \text{if } i > A \quad . \end{cases} \tag{3.88}$$

Its function is to restrict summations over the index i to the range $1, \ldots, A$. The full matrix element now becomes

$$\langle \Psi_0 | \sum_{ij=1}^{\infty} t_{ij} \, \hat{a}_i^\dagger \hat{a}_j | \Psi_0 \rangle = \sum_{ij=1}^{\infty} t_{ij} \, \delta_{ij} \, \theta_{iA} = \sum_{i=1}^{A} t_{ii} \quad , \tag{3.89}$$

i.e., simply a sum of the diagonal matrix elements over all occupied states.

In this book this method of treating the ground state will usually be used. An alternative method consists in redefining the operators in the following way. We introduce new creation and annihilation operators $\hat{\beta}_i^\dagger$ and $\hat{\beta}_i$ related to the usual ones via

$$\hat{\beta}_i^\dagger = \begin{cases} \hat{a}_i^\dagger, & i > A \quad , \\ \hat{a}_i, & i \le A \quad , \end{cases} \qquad \hat{\beta}_i = \begin{cases} \hat{a}_i, & i > A \quad , \\ \hat{a}_i^\dagger, & i \le A \quad . \end{cases} \tag{3.90}$$

For these operators the nuclear ground state is the vacuum state:

$$\hat{\beta}_i | \Psi_o \rangle = 0 \quad \text{for all } i \quad , \tag{3.91}$$

and the operators $\hat{\beta}_i^\dagger$, $i \le A$, describe the *creation of holes* through the destruction of the particles in the corresponding states.

In this framework the 1p1h states are given by

$$|\Psi_{mi}\rangle = \hat{\beta}_m^\dagger \hat{\beta}_i^\dagger |\Psi_0\rangle \quad , \quad m > A \quad , \quad i \le A \quad , \tag{3.92}$$

and the 2p2h states are

$$|\Psi_{mnij}\rangle = \hat{\beta}_m^\dagger \hat{\beta}_n^\dagger \hat{\beta}_i^\dagger \hat{\beta}_j^\dagger |\Psi_0\rangle \quad , \quad m,n > A \quad , \quad i,j \le A \quad . \tag{3.93}$$

The matrix element discussed above can now also be calculated in this notation after rewriting the operators:

$$\begin{aligned} \sum_{i,j=1}^{\infty} t_{ij} \, \hat{a}_i^\dagger \hat{a}_j = & \sum_{i=1}^{A} \sum_{j=1}^{A} t_{ij} \, \hat{\beta}_i \hat{\beta}_j^\dagger + \sum_{i=A+1}^{\infty} \sum_{j=1}^{A} t_{ij} \, \hat{\beta}_i^\dagger \hat{\beta}_j^\dagger \\ & + \sum_{i=1}^{A} \sum_{j=A+1}^{\infty} t_{ij} \, \hat{\beta}_i \hat{\beta}_j + \sum_{i=A+1}^{\infty} \sum_{j=A+1}^{\infty} t_{ij} \, \hat{\beta}_i^\dagger \hat{\beta}_j \quad . \end{aligned} \tag{3.94}$$

Remembering that for these operators $|\Psi_0\rangle$ is the vacuum, we see immediately that the second through last terms will yield a zero expectation value, whereas the first one gives the result calculated above.

It is clear that this second method has no particular advantage for the type of problem discussed here; in both cases the index range has to be divided up into occupied and unoccupied states. Its true power becomes apparent in situations where holes can be treated analogously to particles; for example, a hole inside a closed shell can be treated quite similarly to a single particle in the empty shell (for examples see the section on pairing, Sect. 7.5.5).

Let us just briefly mention how the quantum numbers of holes should be related to those of a particle in the same state. Since a hole is created by destroying a particle, all its additive quantum numbers are opposite in sign to those of the particle destroyed. This requires a modification of (3.90) in the case of such quantum numbers. For example, if the states have good angular momentum there are multiplets of operators $\hat{a}^\dagger_{jm}$, $m = -j, \ldots, j$. In this case the hole-creation operator should be defined as

$$\hat{\beta}^\dagger_{jm} = (-1)^m \hat{a}_{j-m} \quad , \tag{3.95}$$

with the phase factor added to fulfil the Condon–Shortley phase convention (see Sect. 2.3.2).

4. Group Theory in Nuclear Physics

4.1 Lie Groups and Lie Algebras

In Sect. 2.1 we discussed the concept of *groups*, which was then applied principally to the study of rotations. In this chapter the treatment of group theory will be expanded sufficiently to understand the mathematics, for example, of the interacting boson model.

A *continous group* is a group whose elements g depend continously on r parameters α_i ($i = 1, 2, \ldots, r$), so that they can be written as functions $g(\alpha_1, \alpha_2, \ldots, \alpha_r)$. To shorten the notation we also denote the r tuplet of numbers α_i by the vector $\boldsymbol{\alpha}$, so that group elements are written as $g(\boldsymbol{\alpha})$ (to avoid confusion, we will always use Greek letters for group parameters). r is the *order* of the continous group G. If the parameter range spanned by $(\alpha_1, \alpha_2, \ldots, \alpha_r)$ in an r-dimensional space is bounded and closed (i.e., compact), we call G a compact group.

A *Lie group* is a continuous group fulfilling continuity requirements for all group operations; since this is usually fulfilled for the applications in physics, we will always deal with Lie groups and assume that all mathematical functions describing the group are arbitrarily differentiable.

Usually one considers continous groups with elements that act as transformations on a multi-dimensional vector space. Consider, for example, the three-dimensional rotation group SO(3) whose elements rotate the vectors of the usual three-dimensional space. Another example is the group of unitary transformations in two dimensions, SU(2), which mixes the two isospin projections in a wavefunction.

The application of a group element $g(\boldsymbol{\alpha})$ on a vector $\boldsymbol{x} = (x_1, x_2, \ldots, x_N)$ in an N-dimensional space can be written as

$$\boldsymbol{x}' = g(\boldsymbol{\alpha})\boldsymbol{x} = \boldsymbol{f}(\boldsymbol{x}, \boldsymbol{\alpha}) \tag{4.1}$$

with a vector function $\boldsymbol{f}(\boldsymbol{x}, \boldsymbol{\alpha})$ still to be determined. Within this formalism we write for the group multiplication

$$\left[g(\boldsymbol{\alpha}) \cdot g(\boldsymbol{\beta})\right]\boldsymbol{x} = \boldsymbol{f}\big(\boldsymbol{f}(\boldsymbol{x}, \boldsymbol{\beta}), \boldsymbol{\alpha}\big) = \boldsymbol{f}(\boldsymbol{x}, \boldsymbol{\gamma}) \quad , \tag{4.2}$$

if $\boldsymbol{\gamma}$ is such that $g(\boldsymbol{\alpha}) \cdot g(\boldsymbol{\beta}) = g(\boldsymbol{\gamma})$ holds. This expresses the law of multiplication in the group in terms of the parameters and can be quite complicated; it is often not expressible analytically.

Now consider as an example the two-dimensional translation group T_2 acting on two-dimensional space according to

$$g(\alpha,\beta)\boldsymbol{x} = g(\alpha,\beta)(x,y) = (x+\alpha, y+\beta) \quad , \tag{4.3}$$

where α and β, the displacements, are real numbers. Note that T_2 is a noncompact group since the parameter range of (α,β) is noncompact (both have infinite range) and that both the order and the dimension of this group equal 2. In the following we check the four fundamental group relations.

- Closure:

$$\begin{aligned}\left[g(\alpha,\beta)\cdot g(\gamma,\delta)\right](x,y) &= g(\alpha,\beta)(x+\gamma, y+\delta)\\ &= (x+\alpha+\gamma, y+\beta+\delta)\\ &= g(\alpha+\gamma,\beta+\delta)(x,y) \quad ,\end{aligned} \tag{4.4}$$

  which shows that the multiplication rule in this group is expressed by addition of the parameters.

- Associativity:

$$\begin{aligned}&\left[g(\alpha,\beta)\cdot\big(g(\gamma,\delta)\cdot g(\varepsilon,\zeta)\big)\right](x,y)\\ &\quad= \big(x+\alpha+(\gamma+\varepsilon), y+\beta+(\delta+\zeta)\big)\\ &\quad= \big(x+(\alpha+\gamma)+\varepsilon, y+(\beta+\delta)+\zeta\big)\\ &\quad= \left[\big(g(\alpha,\beta)\cdot g(\gamma,\delta)\big)\cdot g(\varepsilon,\zeta)\right](x,y) \quad .\end{aligned} \tag{4.5}$$

- Neutral element: this is clearly given by $g(0,0)$.
- Inverse element:

$$g^{-1}(\alpha,\beta) = g(-\alpha,-\beta) \quad . \tag{4.6}$$

Generalizing the discussion in Sect. 2.2.1, we now introduce the *generators* of a continous group, which are determined by the group elements near the identity element, which is conventionally assumed to correspond to parameter values of zero, $\boldsymbol{\alpha} = \boldsymbol{0}$. Consider the transformation

$$\boldsymbol{x} \to \boldsymbol{x}' = \boldsymbol{f}(\boldsymbol{x},\boldsymbol{\alpha}) \tag{4.7}$$

and expand $\boldsymbol{f}(\boldsymbol{x},\boldsymbol{\alpha})$ in a Taylor series for small values of $\boldsymbol{\alpha}$:

$$\boldsymbol{f}(\boldsymbol{x},\boldsymbol{\alpha}) = \boldsymbol{f}(\boldsymbol{x},\boldsymbol{0}) + \boldsymbol{\alpha}\cdot\nabla_\alpha \boldsymbol{f}(\boldsymbol{x},\boldsymbol{\alpha})\Big|_{\alpha=0} + \cdots \tag{4.8}$$

with ∇_α the gradient operator in the r-dimensional parameter space. Setting

$$\boldsymbol{x}' = \boldsymbol{x} + \mathrm{d}\boldsymbol{x} \quad , \tag{4.9}$$

we obtain for small values of $\boldsymbol{\alpha}$, i.e., $\boldsymbol{\alpha} \to \mathrm{d}\boldsymbol{\alpha}$,

$$\mathrm{d}\boldsymbol{x} = \mathrm{d}\boldsymbol{\alpha}\cdot\nabla_\alpha \boldsymbol{f}(\boldsymbol{x},\boldsymbol{\alpha})\Big|_{\alpha=0} = \sum_{i=1}^{r} \mathrm{d}\alpha_i \frac{\partial}{\partial\alpha_i}\boldsymbol{f}(\boldsymbol{x},\boldsymbol{\alpha})\Big|_{\alpha=0} \quad . \tag{4.10}$$

In terms of components we can write

$$\mathrm{d}x_k = \sum_{i=1}^{r} \mathrm{d}\alpha_i\, U_i^k \quad , \tag{4.11}$$

with the definition

$$U_i^k = \frac{\partial}{\partial \alpha_i} f_k(\boldsymbol{x}, \boldsymbol{\alpha})\bigg|_{\alpha=0} \quad . \tag{4.12}$$

In the following we consider the change of an arbitrary scalar function $F(\boldsymbol{x})$ if an infinitesimal group transformation, i.e., a group element close to the identity element, is applied to its argument $\boldsymbol{x}$. The total differential of $F(\boldsymbol{x})$ reads

$$\begin{aligned} \mathrm{d}F(\boldsymbol{x}) &= \sum_{k=1}^{N} \frac{\partial F(x_1, \ldots, x_N)}{\partial x_k} \mathrm{d}x^k \\ &= \sum_{k=1}^{N} \frac{\partial F(\boldsymbol{x})}{\partial x_k} \sum_{i=1}^{r} \mathrm{d}\alpha_i \, U_i^k = \sum_{i=1}^{r} \mathrm{d}\alpha_i \, \hat{G}_i F(\boldsymbol{x}) \quad , \end{aligned} \tag{4.13}$$

with the definition

$$\hat{G}_i = \sum_{k=1}^{N} U_i^k \frac{\partial}{\partial x_k} \quad . \tag{4.14}$$

The quantities $\hat{G}_i$ are called the *generators* of the group G, since they induce the infinitesimal changes of an arbitrary function $F(\boldsymbol{x})$ due to the transformation.

As an example we discuss the SO(2), the group of orthogonal transformations in two dimensions, i.e.,

$$\begin{aligned} x' &\equiv f_x(x,y) = x \cos\phi - y \sin\phi \quad , \\ y' &\equiv f_y(x,y) = x \sin\phi + y \cos\phi \quad . \end{aligned} \tag{4.15}$$

Obviously SO(2) is of order 1, since it has only one parameter, the rotation angle ϕ. For the functions U_i^k when we substitute the notation $i \equiv \phi$, $k \equiv x, y$ we get

$$U_\phi^x = \frac{\partial f_x}{\partial \phi}\bigg|_{\phi=0} = -y \quad , \quad U_\phi^y = \frac{\partial f_y}{\partial \phi}\bigg|_{\phi=0} = +x \quad , \tag{4.16}$$

and for the generator $\hat{G}_\phi$,

$$\hat{G}_\phi = U_\phi^x \frac{\partial}{\partial x} + U_\phi^y \frac{\partial}{\partial y} = x\frac{\partial}{\partial y} - y\frac{\partial}{\partial x} = -\tfrac{\mathrm{i}}{\hbar}\hat{L}_z \quad , \tag{4.17}$$

which equals the z component of the angular-momentum operator.

A *Lie algebra* is a vector space with an additional antisymmetric multiplication given by the commutator. The generators of a Lie group form a Lie algebra, because they can be added and multiplied by scalars, and it can be shown that they also form a closed set under commutation, i.e.,

$$[\hat{G}_\mu, \hat{G}_\nu] = \sum_{\lambda=1}^{r} C_{\mu\nu}^{\lambda} \hat{G}_\lambda \quad , \tag{4.18}$$

where the $C_{\mu\nu}$ are the *structure constants*. The properties fulfilled by the commutator are as follows.

- Definition:

$$[\hat{G}_\mu, \hat{G}_\nu] = \hat{G}_\mu \hat{G}_\nu - \hat{G}_\nu \hat{G}_\mu \quad . \tag{4.19}$$

- Antisymmetry:

$$[\hat{G}_\mu, \hat{G}_\nu] = -[\hat{G}_\nu, \hat{G}_\mu] \quad . \tag{4.20}$$

- Jacobi identity:

$$\left[[\hat{G}_\mu, \hat{G}_\nu], \hat{G}_\tau\right] = \left[[\hat{G}_\tau, \hat{G}_\mu], \hat{G}_\nu\right] = \left[[\hat{G}_\nu, \hat{G}_\tau], \hat{G}_\mu\right] \quad . \tag{4.21}$$

As a direct consequence of these conditions for the generators of a Lie algebra the structure constants must themselves be antisymmetric, $C^\lambda_{\mu\nu} = -C^\lambda_{\nu\mu}$, and fulfil the Jacobi identity,

$$\sum_\sigma \left(C^\sigma_{\alpha\beta} C^\rho_{\sigma\gamma} + C^\sigma_{\beta\gamma} C^\rho_{\sigma\alpha} + C^\sigma_{\gamma\alpha} C^\rho_{\sigma\beta}\right) = 0 \quad . \tag{4.22}$$

Conversely it can be shown that if the structure constants have these two properties, the corresponding generators form a Lie algebra with the properties given.

EXERCISE

4.1 The Lie Algebra of Angular-Momentum Operators

Problem. Show that the angular-momentum operators form a Lie algebra.

Solution. In Chap. 2 the angular-momentum operators $\hat{L}_i$ were constructed as generators for rotations about the three coordinate axes (note that the parametrization in terms of Euler angles cannot be used for this purpose, as they are not independent for small rotations: in fact, both θ_1 and θ_3 have $\hat{L}_z$ as generator). They obey the commutation relations

$$[\hat{L}_i, \hat{L}_k] = \mathrm{i}\hbar \varepsilon_{ijk} \hat{L}_k \quad . \tag{1}$$

Hence the structure constants of the algebra associated with $SO(3)$ are given by the antisymmetric tensor ε_{ijk}. The antisymmetry is obviously fulfilled since $\varepsilon_{jik} = -\varepsilon_{ijk}$, and the Jacobi identity can be seen from

$$\begin{aligned}
&\sum_\sigma (\varepsilon_{\alpha\beta\sigma}\varepsilon_{\sigma\gamma\rho} + \varepsilon_{\beta\gamma\sigma}\varepsilon_{\sigma\alpha\rho} + \varepsilon_{\gamma\alpha\sigma}\varepsilon_{\sigma\beta\rho}) \\
&\qquad = \sum_\sigma (\varepsilon_{\alpha\beta\sigma}\varepsilon_{\gamma\rho\sigma} + \varepsilon_{\beta\gamma\sigma}\varepsilon_{\alpha\rho\sigma} + \varepsilon_{\gamma\alpha\sigma}\varepsilon_{\beta\rho\sigma}) \\
&\qquad = \delta_{\alpha\gamma}\delta_{\beta\rho} - \delta_{\alpha\rho}\delta_{\beta\gamma} + \delta_{\beta\alpha}\delta_{\gamma\rho} - \delta_{\beta\rho}\delta_{\gamma\alpha} + \delta_{\gamma\beta}\delta_{\alpha\rho} - \delta_{\gamma\rho}\delta_{\alpha\beta} \\
&\qquad = 0 \quad .
\end{aligned} \tag{2}$$

Of central importance for the application of Lie groups to physics are the *Casimir operators* $\hat{C}_\lambda$, which are defined as operators which commute with all generators $\hat{G}_i$ of the algebra:

$$[\hat{C}_\lambda, \hat{G}_i] = 0 \quad . \tag{4.23}$$

For a given group there exist in general several Casimir operators and their explicit form is not unique: if $\hat{C}_\lambda$ and $\hat{C}_\sigma$ are Casimir operators, any combination of them is, too; e.g., $\hat{C}_\lambda - \hat{C}_\sigma$, $\hat{C}_\lambda + \hat{C}_\sigma$, $\hat{C}_\lambda \hat{C}_\sigma$, and so on.

In general it is very difficult to determine all the Casimir operators for a given algebra, but there is a simple rule for obtaining at least one nontrivial Casimir operator from the generators, namely

$$\hat{C}_\lambda = \sum_{\mu,\nu} g_{\mu\nu} \hat{G}_\mu \hat{G}_\nu \quad . \tag{4.24}$$

The tensor $g_{\mu\nu}$ is called the *metric* of the algebra and is defined as

$$g_{\mu\nu} = \sum_{\tau\rho} C^\rho_{\mu\tau} C^\tau_{\nu\rho} \quad . \tag{4.25}$$

EXERCISE

4.2 The Casimir Operator of the Angular-Momentum Algebra

Problem. Construct the Casimir operator for the angular-momentum algebra using (4.24).

Solution. Using the structure constants $C^k_{ij} = \varepsilon_{ijk}$, one obtains for the metric

$$\begin{aligned} g_{\mu\nu} &= \sum_{\tau\rho} C^\rho_{\mu\tau} C^\tau_{\nu\rho} = \sum_{\tau\sigma} \varepsilon_{\mu\tau\rho} \varepsilon_{\nu\rho\tau} \\ &= -\sum_{\tau\sigma} \varepsilon_{\mu\tau\rho} \varepsilon_{\nu\tau\rho} = -2\delta_{\mu\nu} \quad , \end{aligned} \tag{1}$$

and the Casimir operator becomes

$$\begin{aligned} \hat{C}_{\mathrm{SO(3)}} &= \sum_{\mu\nu} g_{\mu\nu} \hat{L}_\mu \hat{L}_\nu = -2 \sum_{\mu\nu} \delta_{\mu\nu} \hat{L}_\mu \hat{L}_\nu \\ &= -2\hat{J}^2 . \end{aligned} \tag{2}$$

This simply reproduces the well-known fact that the $\hat{J}^2$ operator is the Casimir operator of the angular-momentum algebra, commuting with all angular-momentum operators $\hat{J}_i$. It is also the only Casimir operator for this algebra.

Finally we want to make the concept of *representations* more precise. Consider a group G with elements g_m. If a mapping from the elements g_m to matrices $D(g_m)$ exists, so that $D(g_1) \cdot D(g_2) = D(g_3)$ holds whenever $g_1 \cdot g_2 = g_3$, we call the matrices $D(g_m)$ a *representation* of the group G. An example is provided by the rotation matrices given in Sect. 2.3.3.

In the general case we obtain a representation of a group by considering the n-dimensional vector space the group elements are applied to. If $|\phi_i\rangle$ $(i = 1, \ldots, n)$ is a basis for this vector space, the effect of a group element on a basic vector can be expanded as

$$g_m|\phi_j\rangle = \sum_{i=1}^{n} D_{ji}^{m}|\phi_i\rangle \quad , \tag{4.26}$$

with

$$D(g_m)_{ji} = \langle\phi_i|g_m|\phi_j\rangle \quad . \tag{4.27}$$

The c numbers $D(g_m)_{ji}$ are to be interpreted as the matrix elements of the representation matrix $D(g_m)$. The $D(g)$ form a matrix group with the required representation property, which may be seen by calculating

$$\begin{aligned}\big(D(g_1) \cdot D(g_2)\big)_{ik} &= \sum_{j=1}^{n} D(g_1)_{ij}\, D(g_2)_{jk} \\ &= \sum_{j=1}^{n} \langle\phi_i|g_1|\phi_j\rangle\langle\phi_j|g_2|\phi_k\rangle \\ &= \langle i|g_1 g_2|k\rangle = D(g_1 g_2)_{ik} \quad .\end{aligned} \tag{4.28}$$

Thus it is easy to obtain a special representation; usually the important and much more difficult task is to systematically determine all possible representation by constructing all irreducible representations as building blocks, as was done for the rotation group in Sect. 2.3.2.

EXERCISE

4.3 The Lie Algebra of SO(n)

Problem. Derive the Lie algebra for the group SO(n), i.e., the group of orthogonal transformations of dimension n with determinant 1, starting from its matrix representation.

Solution. It has already been mentioned, that the algebra of a Lie group is completely determined by the properties of the group elements near the identity, where all group parameters are infinitesimally small. We thus need the elements of SO(n) in matrix form differing infinitesimally from the unit matrix. The defining property of orthogonal transformations in n dimensions is that they leave the vector norm

Exercise 4.3

invariant. This means that if the transformed vector is defined as $\boldsymbol{x}' = A\boldsymbol{x}$ with $A \in \mathrm{SO}(n)$, the equation

$$(\boldsymbol{x}')^{\mathrm{T}}\boldsymbol{x}' = \boldsymbol{x}^{\mathrm{T}}\boldsymbol{x} \tag{1}$$

must hold. Next we consider infinitesimal transformations of the form

$$\boldsymbol{x}' = (\mathbb{1} + \delta A)\boldsymbol{x} \quad , \tag{2}$$

where $\mathbb{1}$ stands for the identity matrix. To first order in δA this implies

$$\begin{aligned}(\boldsymbol{x}')^{\mathrm{T}}\boldsymbol{x}' &= (\boldsymbol{x})^{\mathrm{T}}\left(\mathbb{1} + \delta A^{\mathrm{T}}\right)(\mathbb{1} + \delta A)\boldsymbol{x} \\ &\approx (\boldsymbol{x})^{\mathrm{T}}\left(\mathbb{1} + \delta A + \delta A^{\mathrm{T}}\right)\boldsymbol{x} \quad ,\end{aligned} \tag{3}$$

so that the matrices A must be antisymmetric:

$$\delta A^{\mathrm{T}} = -\delta A \quad . \tag{4}$$

As parameters of the group one may thus choose the independent matrix elements of an antisymmetric matrix. Their number for dimension n is $n(n-1)/2$ (this is the number of matrix elements above the diagonal, or one half of the total number excluding the diagonal, $n^2 - n$), so the order of the group $\mathrm{SO}(n)$ is $n(n-1)/2$. Each of these can be labelled by an index pair (i,j) with $i < j$ and we may call the group parameters r_{ij}. An arbitrary infinitesimal transformation is then given by

$$\delta A(r_{ij}) = r_{ij}\left(E^{(ij)} - E^{(ji)}\right) \quad , \tag{5}$$

where $E^{(ij)}$ is a matrix with a 1 in the ith row and the jth column and all other elements equal to zero. The matrix form of the generators can be read off simply as

$$G_{ij} = E^{(ij)} - E^{(ji)} \quad . \tag{6}$$

We can now easily derive a realization for the generators $\hat{G}_{ij}$ of $\mathrm{SO}(n)$ using the definitions in (4.1–2):

$$\begin{aligned}\hat{G}_{ij} &= \sum_k U_{ij}^k \frac{\partial}{\partial x_j} \\ &= \sum_k \frac{\partial}{\partial r_{ij}}\Big(\sum_l \delta A(r_{ij})_{kl}\, x_l\Big)\frac{\partial}{\partial x_k} \\ &= \sum_{kl}\left(\delta_{ik}\delta_{jl} - \delta_{il}\delta_{jk}\right) x_l \frac{\partial}{\partial x_k} \\ &= x_i \frac{\partial}{\partial x_j} - x_j \frac{\partial}{\partial x_i} \quad .\end{aligned} \tag{7}$$

Exercise 4.3

The commutation relations for the generators, i.e., the Lie algebra of SO(n), now follow as

$$\begin{aligned}[\hat{G}_{ij}, \hat{G}_{km}] &= \left(x_i \frac{\partial}{\partial x_j} - x_j \frac{\partial}{\partial x_i}\right)\left(x_k \frac{\partial}{\partial x_m} - x_m \frac{\partial}{\partial x_k}\right) \\ &\quad - \left(x_k \frac{\partial}{\partial x_m} - x_m \frac{\partial}{\partial x_k}\right)\left(x_i \frac{\partial}{\partial x_j} - x_j \frac{\partial}{\partial x_i}\right) \\ &= \delta_{jk}\hat{G}_{im} + \delta_{im}\hat{G}_{jk} + \delta_{jm}\hat{G}_{ki} + \delta_{ki}\hat{G}_{mj} \quad . \end{aligned} \tag{8}$$

In the case of $n = 3$ these equations reduce to the angular-momentum commutation relations if one identifies

$$\hat{L}_x = \mathrm{i}\hat{G}_{23} \quad , \quad \hat{L}_y = \mathrm{i}\hat{G}_{31} \quad , \quad \hat{L}_z = \mathrm{i}\hat{G}_{12} \quad . \tag{9}$$

4.2 Group Chains

A method that is often used for constructing the representations of the more complicated groups is based on *chains of subgroups* and their Casimir operators. It will be employed extensively in the IBA model (Sect. 6.8), but the simplest example is the derivation of the representations of the rotation group as given in Sect. 2.3.2, and we will use this example for an explanation of the ideas involved.

The operators employed in Sect. 2.3.2 were $\hat{J}^2$, $\hat{J}_\pm$, and $\hat{J}_z$. Of these the group-theoretic role is clear only for $\hat{J}^2$, which is simply the Casimir operator of the group SU(2) (or SO(3)) and thus distinguishes the different irreducible representations according to Schur's lemma. What is the significance of $\hat{J}_z$? It is the generator of the group of rotations about the z axis, which is obviously a subgroup of the full rotation group.

The two-dimensional rotation group SO(2) consists of matrices of the type

$$\begin{pmatrix} \cos\theta & -\sin\theta \\ \sin\theta & \cos\theta \end{pmatrix} \quad , \tag{4.29}$$

which are in one-to-one correspondence with complex phases factors $\mathrm{e}^{\mathrm{i}\theta}$, so that the group SO(2) is isomorphic to the group U(1). Since there is only one group parameter θ, there is only one generator $\hat{J}_z$, which is also trivially a Casimir operator of the group. The irreducible repesentations are thus all one-dimensional and are labelled by the eigenvalue $\hbar m$ of $\hat{J}_z$.

What happened in Sect. 2.3.2 now becomes clear on a more general level: the representations of SU(2) or SO(3) were decomposed into representations of the subgroup U(1) or SO(2). The subspaces of fixed projection m are invariant only under SO(2), not under the full rotation group, while the additional generators $\hat{J}_\pm$ or $\hat{J}_x$ and $\hat{J}_y$ link the different projections. Formally we write the relation of groups, Casimir operators, and eigenvalues as

$$\begin{array}{ccc} \mathrm{SO}(3) & \supset & \mathrm{SO}(2) \\ \hat{J}^2 & & \hat{J}_z \\ \hbar^2 j(j+1) & & \hbar m \end{array} \quad . \tag{4.30}$$

Of course this is an extremely simple example. In the IBA the group chain will be, for example,

$$\mathrm{U}(6) \supset \mathrm{U}(5) \supset \mathrm{O}(5) \supset \mathrm{O}(3) \supset \mathrm{O}(2) \quad . \tag{4.31}$$

It is always desirable to have a chain that concludes with O(3) and O(2), because then the angular-momentum quantum numbers will appear and the representations become more useful for physics.

Each representation is characterized by the eigenvalue of the Casimir operator(s) of the first group in the chain, and the representation space is then decomposed into subspaces invariant under the next group and characterized in turn by its Casimir operator(s), and so on. This yields a set of quantum numbers and a full solution of the problem if the Hamiltonian can be expressed in terms of the Casimir operators.

If the Hamiltonian is invariant under the group, it must commute with all generators of the group and is itself a Casimir operator. An example is provided by the Hamiltonian of a spherically symmetric system in ordinary space, whose dependence on the rotation angles is contained purely in the centrifugal term proportional to $\hat{J}^2$. In this case the angular dependence of the wave functions and the associated quantum numbers can be obtained purely by group-theoretical methods; the radial dependence requires a more complicated treatment and it is often the case that a more general symmetry group encompassing all degress of freedom of the system can be found and the full solution be constructed group-theoretically.

4.3 Lie Algebras in Second Quantization

The generators of Lie algebras often appear in the guise of second quantization. The mathematical reason for this is quite simple: if a particle is created in a space of n single-particle wave functions, it can be described by some linear combination of these wave functions. A transformation that changes its wave function to another such linear combination corresponds to an element of the group U(n). Alternatively, this redistribution may be achieved by some combination $\hat{a}_j^\dagger \hat{a}_k$ of creation and annihilation operators, which remove the particle from the original states and put it into the new ones, instead. The simplest example is again provided by the ubiquitous rotation group, in this case in its incarnation as SU(2).

Regard a fermion that can exist in two different states. Thinking of isospin as the simplest example, call them $|\mathrm{p}\rangle$ and $|\mathrm{n}\rangle$; they will be created using the operators $\hat{a}_\mathrm{p}^\dagger$ and $\hat{a}_\mathrm{n}^\dagger$, so that

$$|\mathrm{p}\rangle = \hat{a}_\mathrm{p}^\dagger|0\rangle \quad , \quad |\mathrm{n}\rangle = \hat{a}_\mathrm{n}^\dagger|0\rangle \quad . \tag{4.32}$$

A change in the state of the particle can then be mediated by the operators $\hat{a}_\mathrm{n}^\dagger \hat{a}_\mathrm{p}$ and $\hat{a}_\mathrm{p}^\dagger \hat{a}_\mathrm{n}$, respectively. These might thus correspond to the shift operators of angular-momentum theory. If we assign a projection of $\frac{1}{2}$ to the state $|\mathrm{p}\rangle$ and $-\frac{1}{2}$ to $|\mathrm{n}\rangle$, we may define "angular-momentum operators"

$$\hat{J}_+ = \hat{a}_\mathrm{p}^\dagger \hat{a}_\mathrm{n} \quad , \quad \hat{J}_- = \hat{a}_\mathrm{n}^\dagger \hat{a}_\mathrm{p} \quad . \tag{4.33}$$

In addition the z component of the angular-momentum operator may be conjectured to be

$$\hat{J}_z = \tfrac{1}{2}\left(\hat{a}_\mathrm{p}^\dagger \hat{a}_\mathrm{p} - \hat{a}_\mathrm{n}^\dagger \hat{a}_\mathrm{n}\right) \quad , \tag{4.34}$$

which clearly the yields desired eigenvalues $\pm\frac{1}{2}$ for the two states.

Whether this indeed yields the Lie algebra can be checked by evaluating the commutation relations. Since the operator $\hat{a}_\mathrm{n}$ can be anticommuted to the back, we have

$$\left[\hat{a}_\mathrm{p}^\dagger \hat{a}_\mathrm{p}, \hat{a}_\mathrm{p}^\dagger \hat{a}_\mathrm{n}\right] = \left[\hat{a}_\mathrm{p}^\dagger \hat{a}_\mathrm{p}, \hat{a}_\mathrm{p}^\dagger\right]\hat{a}_\mathrm{n} = \hat{a}_\mathrm{p}^\dagger \hat{a}_\mathrm{n} \tag{4.35}$$

(the latter commutator is one of the "shift-operator" relations of Sect. 3.1.2), and using the similar result for the $\hat{a}_\mathrm{n}^\dagger \hat{a}_\mathrm{n}$ operator, we get

$$\left[\tfrac{1}{2}\left(\hat{a}_\mathrm{p}^\dagger \hat{a}_\mathrm{p} - \hat{a}_\mathrm{n}^\dagger \hat{a}_\mathrm{n}\right), \hat{a}_\mathrm{p}^\dagger \hat{a}_\mathrm{m}\right] = \hat{a}_\mathrm{p}^\dagger \hat{a}_\mathrm{n} \tag{4.36}$$

or

$$\left[\hat{J}_z, \hat{J}_+\right] = \hat{J}_+ \quad , \tag{4.37}$$

exactly as it should be for the angular-momentum shift operators. Similarly all the other angular-momentum commutation rules can be confirmed. Thus we have the full Lie algebra of the group SU(2) realized in terms of the second-quantization operators.

Similarly the generators of SO(n) can be seen simply from the matrix representation of (6) in Exercise 4.3. Each matrix $E^{(ij)}$ corresponds to an operator pair $\hat{a}_i^\dagger \hat{a}_j$, so that the second-quantization form becomes

$$\hat{G}_{ij} = \hat{a}_i^\dagger \hat{a}_j - \hat{a}_j^\dagger \hat{a}_i \quad . \tag{4.38}$$

This methodology will be applied to the much more complicated case of a six-dimensional space in the context of the IBA (Sect. 6.8). There is, however, also an example of an application of the angular-momentum algebra in a situation that is not so trivially connected with the group SU(2): this is the quasispin model explained in Sect. 7.5.3.

5. Electromagnetic Moments and Transitions

5.1 Introduction

The measurement of the interaction of the electromagnetic field with the nucleus provides the most important source of experimental information. Gamma radiation can be both absorbed and emitted by the nucleus, and carries information in its angular momentum and energy as well as the associated transition probabilities. These allow direct conclusions about the angular momentum, parity, excitation energies, and transition matrix elements of the stationary states of the nucleus. In some cases the information may be summarized simply in terms of the electromagnetic multipole moments such as the quadrupole moment, which is more directly related to the shape of the nuclear charge distribution.

The purpose of this chapter is to derive the necessary formulas allowing the calculation of the electromagnetic transition probabilities and moments from nuclear models. The electromagnetic field first has to be quantized and decomposed into fields of definite parity and multipolarity. Then the interaction of these fields with the charge and current distributions within the nucleus must be formulated in such a way that nuclear model distributions can be inserted. These steps will be the subject matter of the following sections.

5.2 The Quantized Electromagnetic Field

The quantization of the electromagnetic field will be shown here only in its final form, without many of the subtleties associated with field theory, because we are essentially only interested in it as a tool for studying nuclei. Regard the vector potential $\boldsymbol{A}(\boldsymbol{r}, t)$ as the principal dynamic field and use the Coulomb (or *transverse*) gauge. The field then fulfils the gauge condition

$$\nabla \cdot \boldsymbol{A} = 0 \tag{5.1}$$

and, in the absence of charges or currents, the wave equation

$$\left(\frac{1}{c^2} \frac{\partial^2}{\partial t^2} - \nabla^2 \right) \boldsymbol{A} = 0 \quad . \tag{5.2}$$

The electromagnetic fields are given in terms of $\boldsymbol{A}$ by

$$\boldsymbol{E} = -\frac{1}{c} \frac{\partial}{\partial t} \boldsymbol{A} \quad , \quad \boldsymbol{H} = \nabla \times \boldsymbol{A} \quad , \tag{5.3}$$

and the energy density of the field is

$$\varepsilon_{\mathrm{em}} = \frac{1}{8\pi}\left(|E|^2 + |H|^2\right) \quad . \tag{5.4}$$

With a plane wave for the vector potential,

$$\boldsymbol{A}_{\boldsymbol{k}}(\boldsymbol{r},t) = \boldsymbol{A}_0 \cos(\boldsymbol{k}\cdot\boldsymbol{r} - \omega t) \quad , \tag{5.5}$$

these expressions become (using $\omega = ck = c|\boldsymbol{k}|$, a direct consequence of the wave equation)

$$\boldsymbol{E}_{\boldsymbol{k}} = -k\boldsymbol{A}_0 \sin(\boldsymbol{k}\cdot\boldsymbol{r} - \omega t) \quad , \quad \boldsymbol{H}_{\boldsymbol{k}} = -\boldsymbol{k}\times\boldsymbol{A}_0 \sin(\boldsymbol{k}\cdot\boldsymbol{r} - \omega t) \quad , \tag{5.6}$$

and the energy density will be

$$\varepsilon_{\mathrm{em}} = \frac{1}{8\pi} k^2 |A_0|^2 \quad . \tag{5.7}$$

For a field corresponding to one photon, this should be equal to $\hbar\omega/V$, with V the system volume, so that the appropriate amplitude becomes

$$A_0 = \sqrt{\frac{8\pi\hbar\omega}{k^2 V}} = \sqrt{\frac{8\pi\hbar c^2}{\omega V}} \quad . \tag{5.8}$$

The corresponding complex expression yielding the same average energy density is

$$\boldsymbol{A}_{\mathrm{em}}(\boldsymbol{r},t) = \boldsymbol{\varepsilon}\sqrt{\frac{2\pi\hbar c^2}{\omega V}}\left(a_0 \mathrm{e}^{\mathrm{i}\boldsymbol{k}\cdot\boldsymbol{r}-\mathrm{i}\omega t} + a_0^* \mathrm{e}^{-\mathrm{i}\boldsymbol{k}\cdot\boldsymbol{r}+\mathrm{i}\omega t}\right) \quad , \tag{5.9}$$

where a_0 is a complex number with unit modulus, which determines the phase of the wave, and $\boldsymbol{\varepsilon}$ is the unit vector indicating the polarization. Because of the transverse nature of electromagnetic waves, there are two independent polarization directions $\boldsymbol{\varepsilon}_l$, $l = 1, 2$, both fulfilling $\boldsymbol{\varepsilon}_l \cdot \boldsymbol{k} = 0$.

We will now give the motive for the quantized form of this vector potential. The interaction Hamiltonian density with matter will be the quantized version of the classical

$$\varepsilon_{\mathrm{int}} = -\frac{1}{c}\,\boldsymbol{j}(\boldsymbol{r})\cdot\boldsymbol{A}(\boldsymbol{r}) \quad , \tag{5.10}$$

with $\boldsymbol{j}$ the current distribution. This should describe the emission and absorption of photons, and the typical transition matrix element between nuclear states Ψ_i and Ψ_f should be something like

$$\int \mathrm{d}^3 r \, \langle \Psi_f, \text{one photon}| -\tfrac{1}{c}\,\boldsymbol{j}\cdot\boldsymbol{A}\,|\Psi_i, \text{no photon}\rangle \tag{5.11}$$

for the emission of a photon, and the Hermitian conjugate for the absorption of a photon. The field operator should thus contain two Hermitian conjugate parts involving creation and annihilation operators for photons, and it appears natural to just replace the amplitudes a_0 and a_0^* by $\hat{\beta}^\dagger$ and $\hat{\beta}$, the creation and annihilation operators for photons. It only remains for us to settle which of the operators to use for which of the two terms, but this is clear from the energy balance. In the above matrix element for emission the total time-dependent phase is

$$\frac{\mathrm{i}}{\hbar}(E_f - E_i)t + \mathrm{i}\phi \tag{5.12}$$

with ϕ the unknown phase accompanying the creation operator in the radiation field. Total energy conservation implies $E_f = E_i - \hbar\omega$, so that we must have $\phi = \omega t$. It is also consistent that the creation and annihilation operators are associated with plane waves having momenta in opposite directions, since momentum conservation requires the appropriate change in total momentum for creation or destruction of a photon. The normalization factor ensures that the energy of the field will be given correctly in terms of the photon number; to see this just compute the field energy as above and remove the zero point energy as for the harmonic oscillator.

For the final form of the radiation field the operators need to acquire indices indicating which mode in the field (determined by wave vector $\boldsymbol{k}$ and polarization index μ) they act on. In the following the symbol ω will always stand for ck. Thus the resultant field operator summed up over all modes $(\boldsymbol{k}, \mu)$ is

$$\hat{\boldsymbol{A}}(\boldsymbol{r}, t) = \sum_{\boldsymbol{k}\mu} \sqrt{\frac{2\pi\hbar c^2}{\omega V}} \left(\hat{\beta}_{\boldsymbol{k}\mu}\boldsymbol{\varepsilon}^*_\mu \mathrm{e}^{\mathrm{i}\boldsymbol{k}\cdot\boldsymbol{r}-\mathrm{i}\omega t} + \hat{\beta}^\dagger_{\boldsymbol{k}\mu}\boldsymbol{\varepsilon}_\mu \mathrm{e}^{-\mathrm{i}\boldsymbol{k}\cdot\boldsymbol{r}+\mathrm{i}\omega t}\right) \tag{5.13}$$

(The choice of ε_μ or its complex conjugate took into account that the spherical basis vector ε_μ with angular momentum projection μ should accompany the *creation operator* for that projection.)

The operator for the total energy of the field, integrated over the volume V, is

$$\hat{H} = \sum_{\boldsymbol{k}\mu} \hbar\omega_k \left(\hat{\beta}^\dagger_{\boldsymbol{k}\mu}\hat{\beta}_{\boldsymbol{k}\mu} + \tfrac{1}{2}\right) \quad , \tag{5.14}$$

and the total momentum is given by

$$\hat{\boldsymbol{P}} = \sum_{\boldsymbol{k}\mu} \hbar\boldsymbol{k}\hat{\beta}^\dagger_{\boldsymbol{k}\mu}\hat{\beta}_{\boldsymbol{k}\mu} \quad . \tag{5.15}$$

The photon states treated here are eigenstates of the momentum operator. For their interaction with nuclei, however, it is much more convenient to use angular-momentum eigenstates. Their construction is carried out in the next section.

5.3 Radiation Fields of Good Angular Momentum

5.3.1 Solutions of the Scalar Helmholtz Equation

As was shown in Sect. 2.3.6, the vector field $\boldsymbol{A}$ describing the photon has an intrinsic angular momentum of 1 with basis states given by the spherical unit vectors $\boldsymbol{e}_\mu$, $\mu = -1, 0, 1$. To construct states of good total angular momentum these have to be coupled with scalar functions $\Phi_{\lambda\mu}(\boldsymbol{r})$ of good orbital angular momentum, and this is completed by the usual harmonic time dependence:

$$\boldsymbol{A}_{lm,\lambda}(\boldsymbol{r},t) = \sum_{\mu\mu'} (\lambda 1 l | \mu\mu' m)\, \Phi_{\lambda\mu}(\boldsymbol{r})\, \boldsymbol{e}_{\mu'}\, \mathrm{e}^{-\mathrm{i}\omega t} \quad . \tag{5.16}$$

Note that λ remains a good quantum number also for the coupled field. With $\omega = ck$, the wave equation reduces to the Helmholtz equation for the $\Phi_{\lambda\mu}$:

$$(\Delta + k^2)\, \Phi_{\lambda\mu}(\boldsymbol{r}) = 0 \quad . \tag{5.17}$$

The boundary condition is that the function vanish at infinity. Solutions to this problem are familiar from elementary scattering theory:

$$\Phi_{\lambda\mu}(\boldsymbol{r}) = j_\lambda(kr)\, Y_{\lambda\mu}(\Omega) \quad , \tag{5.18}$$

where $j_\lambda(kr)$ denotes the spherical Bessel function.

At this place it is helpful to summarize a few properties of the spherical Bessel functions, which will be needed in this chapter.

1. Definition in terms of Bessel functions of the first kind:

$$j_l(x) = \sqrt{\frac{\pi}{2x}} J_{l+1/2}(x) \quad , \quad l = 0, \pm 1, \pm 2, \ldots \tag{5.19}$$

2. Approximation for small argument values:

$$j_l(x) \approx \frac{x^l}{(2l+1)!!} \quad \text{for } x \to 0 \quad . \tag{5.20}$$

3. Approximation for large argument values:

$$j_l(x) \approx \frac{1}{x} \sin\left(x - \tfrac{1}{2} l \pi\right) \quad \text{for } x \to \infty \quad . \tag{5.21}$$

4. Differential equation:

$$\left(\frac{\mathrm{d}^2}{\mathrm{d}x^2} + \frac{2}{x}\frac{\mathrm{d}}{\mathrm{d}x} + 1 - \frac{l(l+1)}{x^2} \right) j_l(x) = 0 \quad . \tag{5.22}$$

The solutions $\Phi_{\lambda\mu}(\boldsymbol{r})$ obviously have angular momentum λ and projection μ. Their parity is determined by that of the spherical harmonics as $(-1)^\lambda$.

5.3.2 Solutions of the Vector Helmholtz Equation

Vector solutions of the Helmholtz equation with good angular momentum can now be constructed by angular-momentum coupling as outlined. They are

$$\boldsymbol{A}_{lm,\lambda}(\boldsymbol{r}) = \sum_{\mu\mu'} (\lambda 1 l | \mu\mu' m)\, j_\lambda(kr)\, Y_{\lambda\mu}(\Omega)\, \boldsymbol{e}_{\mu'} \quad . \tag{5.23}$$

The basic vectors $\boldsymbol{e}_\mu$ are customarily chosen such that the wave propagates in the positive z direction, i.e., the directions of $\boldsymbol{e}_0$ and $\boldsymbol{k}$ agree.

For the angular-momentum coupling the radial function appears only as a trivial factor, so that it is useful to define the *vector spherical harmonics* as the angular and vector part of this expression:

$$\boldsymbol{Y}_{lm,\lambda}(\Omega) = \sum_{\mu\mu'} (\lambda 1 l|\mu\mu' m)\, Y_{\lambda\mu}(\Omega)\, \boldsymbol{e}_{\mu'} \quad . \tag{5.24}$$

$\boldsymbol{Y}_{lm,\lambda}$ keeps the index λ because λ remains a good quantum number in the coupled basis. The other angular momentum involved is the constant 1, so that there is no need to keep it in the notation (although in other notations it is used as an index to distinguish these functions, for example, from spin spherical harmonics, in which the $Y_{\lambda\mu}$ are coupled with spinors)

Two useful properties of the vector spherical harmonics are given here

1. Complex conjugation:

$$\boldsymbol{Y}^*_{lm,\lambda}(\Omega) = (-1)^{m+l+\lambda+1}\boldsymbol{Y}_{l-m,\lambda}(\Omega) \quad . \tag{5.25}$$

2. Orthonormality:

$$\int \mathrm{d}\Omega\, \boldsymbol{Y}^*_{l'm',\lambda'}(\Omega) \cdot \boldsymbol{Y}_{lm,\lambda}(\Omega) = \delta_{ll'}\, \delta_{\lambda\lambda'}\, \delta_{mm'} \quad . \tag{5.26}$$

Note that this involves both the scalar product in the function space (the integration over Ω) as well as a scalar product of the two vectors.

EXERCISE

5.1 The Vector Spherical Harmonics

Problem. Calculate the vector spherical harmonic $\boldsymbol{Y}_{00,1}(\Omega)$. Does it show the symmetry associated with its total angular momentum of zero?

Solution. According to the definition,

$$\boldsymbol{Y}_{00,1}(\Omega) = \sum_{\mu} (110|\mu - \mu 0)\, Y_{1\mu}(\Omega)\, \boldsymbol{e}_{-\mu} \quad . \tag{1}$$

Inserting the Clebsch–Gordan coefficient yields

$$\boldsymbol{Y}_{00,1}(\Omega) = \frac{1}{\sqrt{3}} \sum_{\mu} (-1)^{1-\mu} Y_{1\mu}(\Omega)\, \boldsymbol{e}_{-\mu} \quad , \tag{2}$$

and a further reduction can only be made through explicit use of the spherical harmonics:

$$\begin{aligned}\boldsymbol{Y}_{00,1}(\Omega) &= \frac{1}{\sqrt{3}}\left[-\sqrt{\tfrac{3}{8\pi}} \sin\theta\, \mathrm{e}^{\mathrm{i}\phi} \tfrac{1}{\sqrt{2}}(\boldsymbol{e}_x - \mathrm{i}\boldsymbol{e}_y) - \sqrt{\tfrac{3}{4\pi}} \cos\theta\, \boldsymbol{e}_z \right.\\ &\qquad \left. - \sqrt{\tfrac{3}{8\pi}} \sin\theta\, \mathrm{e}^{-\mathrm{i}\phi} \tfrac{1}{\sqrt{2}}(\boldsymbol{e}_x + \mathrm{i}\boldsymbol{e}_y)\right]\\ &= -\frac{1}{\sqrt{4\pi}}(\cos\theta\, \boldsymbol{e}_z + \sin\theta\cos\phi\, \boldsymbol{e}_x + \sin\theta\sin\phi\, \boldsymbol{e}_y)\\ &= -\frac{1}{\sqrt{4\pi}}\boldsymbol{e}_r \quad . \end{aligned} \tag{3}$$

Exercise 5.1

So the result is the spherically symmetric vector field proportional to the unit vector in the radial direction at each point. It can be combined with an arbitrary function of r without destroying its scalar character; for example, we have

$$\boldsymbol{r} = -r\sqrt{4\pi}\,\boldsymbol{Y}_{00,1}(\Omega) \quad . \tag{4}$$

It is also interesting to note its parity. The components (x, y, z) of the vector change their sign under a spatial inversion, but regarded as a vector field the vector has in addition to be inverted at each point, so that the total parity is positive. This accords with the geometric picture.

An important application of the vector spherical harmonics is the *gradient formula*, which we will state here without proof. It allows the expression of the gradient of a product of a spherical harmonic and a radial function in terms of vector spherical harmonics:

$$\begin{aligned}\nabla f(r) Y_{lm}(\Omega) = & \sqrt{\frac{l}{2l+1}}\left(\frac{\mathrm{d}f}{\mathrm{d}r} + \frac{l+1}{r} f\right) \boldsymbol{Y}_{lm,l-1}(\Omega) \\ & - \sqrt{\frac{l+1}{2l+1}}\left(\frac{\mathrm{d}f}{\mathrm{d}r} - \frac{l}{r} f\right) \boldsymbol{Y}_{lm,l+1}(\Omega) \quad .\end{aligned} \tag{5.27}$$

Now it is time to go back to the construction of the vector solutions of the Helmholtz equation. They are obtained by coupling the scalar solution $\Phi_{\lambda\mu}(\Omega)$ with the polarization vectors $\boldsymbol{e}_\mu$. For a given total angular momentum l the different choices for the angular-momentum coupling, i.e., $\lambda = l, l \pm 1$, yield three independent vector fields

$$j_l(kr)\boldsymbol{Y}_{lm,l}(\Omega) \quad , \quad j_{l-1}(kr)\boldsymbol{Y}_{lm,l-1}(\Omega) \quad , \text{ and } \quad j_{l+1}(kr)\boldsymbol{Y}_{lm,l+1}(\Omega) \; . \tag{5.28}$$

This corresponds to the three-dimensional basis of the $\boldsymbol{e}_\mu$. But $\boldsymbol{e}_0$ does not describe a physical solution, because the field must be transverse, i.e., $\nabla \cdot \boldsymbol{A} = 0$ or for plane waves $\boldsymbol{k} \cdot \boldsymbol{A} = 0$. This condition leaves two independent solutions, and one may further narrow down their choice by requiring good parity.

Keeping in mind the discussion of the parity of vector fields given in Sect. 2.5.2 and using the special definition presented there, we find that the parity of the vector spherical harmonic $\boldsymbol{Y}_{lm,\lambda}$ is a product of the orbital parity, which results from the spherical harmonic $Y_{\lambda\mu}$ and is consequently $(-1)^\lambda$, and a -1 for the polar vectors $\boldsymbol{e}_\mu$, so that the total parity is $(-1)^{\lambda+1}$. The fields of good parity must be constructed like

$$\begin{aligned} \boldsymbol{A}_{lm}(\boldsymbol{r}; \mathrm{M}) &= j_l(kr)\,\boldsymbol{Y}_{lm,l}(\Omega) \quad , \\ & \qquad \text{Parity: } (-1)^{l+1} \quad , \\ \boldsymbol{A}_{lm}(\boldsymbol{r}; \mathrm{E}) &= c_{l-1} j_{l-1}(kr)\,\boldsymbol{Y}_{lm,l-1}(\Omega) + c_{l+1} j_{l+1}(kr)\,\boldsymbol{Y}_{lm,l+1}(\Omega) \quad , \\ & \qquad \text{Parity: } (-1)^l \quad . \end{aligned} \tag{5.29}$$

The variables M and E indicate the *magnetic* and *electric* multipole fields.

Since $\boldsymbol{A}_{lm}(\boldsymbol{r}; \mathrm{M})$ is the only field for that parity, it should automatically be transverse; this can be confirmed by direct calculation or seen directly from the alternative form of this field given below. For the electric multipole the condition of transversality determines the ratio of the coefficients c_{l+1} and c_{l-1}. There is a second linearly independent field of this type, which should then correspond to the longitudinal field with $\nabla \times \boldsymbol{A} = 0$.

The determination of the coefficients from the transversality condition, $\nabla \cdot \boldsymbol{A}_{lm}(\boldsymbol{r}; \mathrm{E}) = 0$, is straightforward but laborious: insert the definition of the vector spherical harmonics, use the gradient formula and recursion relations for the derivative of the spherical Bessel functions, and then again insert the definition of the vector spherical harmonics to reduce everything to a sum of radial functions times spherical harmonics. This is then seen to vanish for the choice

$$c_{l-1} = \sqrt{\frac{l+1}{2l+1}} \quad , \quad c_{l+1} = -\sqrt{\frac{l}{2l+1}} \quad . \tag{5.30}$$

Similarly the longitudinal multipole is given by

$$c_{l-1} = \sqrt{\frac{l}{2l+1}} \quad , \quad c_{l+1} = \sqrt{\frac{l+1}{2l+1}} \quad . \tag{5.31}$$

5.3.3 Properties of the Multipole Fields

We sum up these developments by noting the expressions for the different multipole fields together with alternative formulations, which are based on vector differentiations of the scalar solutions to the Helmholtz equation and are more useful in many manipulations. All of these can be checked using the gradient formula.

- The magnetic multipole field:

$$\begin{aligned} \boldsymbol{A}_{lm}(\boldsymbol{r}; \mathrm{M}) &= j_l(kr)\, \boldsymbol{Y}_{lm,l}(\Omega) \\ &= \frac{1}{\hbar\sqrt{l(l+1)}} \hat{\boldsymbol{L}}\, j_l(kr)\, Y_{lm}(\Omega) \quad . \end{aligned} \tag{5.32}$$

  The latter form immediately shows that the field is in fact transverse, as $\nabla \cdot \hat{\boldsymbol{L}} = 0$.

- The electric multipole field:

$$\begin{aligned} \boldsymbol{A}_{lm}(\boldsymbol{r}; \mathrm{E}) &= \sqrt{\frac{l+1}{2l+1}} j_{l-1}(kr)\, \boldsymbol{Y}_{lm,l-1}(\Omega) \\ &\quad - \sqrt{\frac{l}{2l+1}} j_{l+1}(kr)\, \boldsymbol{Y}_{lm,l+1}(\Omega) \\ &= \frac{-\mathrm{i}}{\hbar k \sqrt{l(l+1)}} \nabla \times \left(\hat{\boldsymbol{L}}\, j_l(kr)\, Y_{lm}(\Omega)\right) \quad , \end{aligned} \tag{5.33}$$

  again with an obviously vanishing divergence.

- The longitudinal multipole field:

$$\begin{aligned} \boldsymbol{A}_{lm}(\boldsymbol{r}; \mathrm{L}) &= \sqrt{\frac{l}{2l+1}} j_{l-1}(kr)\, \boldsymbol{Y}_{lm,l-1}(\Omega) \\ &\quad + \sqrt{\frac{l+1}{2l+1}} j_{l+1}(kr)\, \boldsymbol{Y}_{lm,l+1}(\Omega) \\ &= \frac{1}{k} \nabla j_l(kr)\, Y_{lm}(\Omega) \quad . \end{aligned} \tag{5.34}$$

The last formula makes it immediately apparent that $\nabla \times \boldsymbol{A}_{lm}(\boldsymbol{r};\mathrm{L}) = 0$.

Some additional properties of these fields may also be summarized here.

1. All of them are solutions of the Helmholtz equation:

$$\left(\nabla^2 + k^2\right)\boldsymbol{A}_{lm}(\boldsymbol{r};R) = 0 \quad \text{for} \quad R = \mathrm{E},\mathrm{M},\mathrm{L} \quad . \tag{5.35}$$

2. Transversality and longitudinality properties:

$$\nabla \cdot \boldsymbol{A}_{lm}(\boldsymbol{r};\mathrm{E}) = \nabla \cdot \boldsymbol{A}_{lm}(\boldsymbol{r};\mathrm{M}) = 0 \quad , \quad \nabla \times \boldsymbol{A}_{lm}(\boldsymbol{r};\mathrm{L}) = 0 \quad . \tag{5.36}$$

3. The parity is $(-1)^l$ for E and L, $(-1)^{l+1}$ for M multipole fields.
4. Interrelation: direct manipulation of the definitions shows that

$$\begin{aligned} \nabla \times \boldsymbol{A}_{lm}(\boldsymbol{r};\mathrm{E}) &= -\mathrm{i}k\,\boldsymbol{A}_{lm}(\boldsymbol{r};\mathrm{M}) \quad , \\ \nabla \times \boldsymbol{A}_{lm}(\boldsymbol{r};\mathrm{M}) &= \mathrm{i}k\,\boldsymbol{A}_{lm}(\boldsymbol{r};\mathrm{E}) \quad . \end{aligned} \tag{5.37}$$

5. Name: the terminology is obvious for the longitudinal multipole. The other names refer to the properties of the radiation field near the radiating source. Using the above relations between the fields, we see that the electric and magnetic multipole fields themselves are given by

$$\begin{aligned} \boldsymbol{E}(\boldsymbol{r};\mathrm{E}) &= -\frac{1}{c}\frac{\partial}{\partial t}\boldsymbol{A}(\boldsymbol{r};\mathrm{E}) = \mathrm{i}k\boldsymbol{A}(\boldsymbol{r};\mathrm{E}) \quad , \\ \boldsymbol{E}(\boldsymbol{r};\mathrm{M}) &= -\frac{1}{c}\frac{\partial}{\partial t}\boldsymbol{A}(\boldsymbol{r};\mathrm{M}) = \mathrm{i}k\boldsymbol{A}(\boldsymbol{r};\mathrm{M}) \quad , \\ \boldsymbol{H}(\boldsymbol{r};\mathrm{E}) &= \nabla \times \boldsymbol{A}(\boldsymbol{r};\mathrm{E}) = -\mathrm{i}k\boldsymbol{A}(\boldsymbol{r};\mathrm{M}) \quad , \\ \boldsymbol{H}(\boldsymbol{r};\mathrm{M}) &= \nabla \times \boldsymbol{A}(\boldsymbol{r};\mathrm{M}) = \mathrm{i}k\boldsymbol{A}(\boldsymbol{r};\mathrm{E}) \quad . \end{aligned} \tag{5.38}$$

Near the source, where $kr \ll 1$, the spherical Bessel functions $j_l(kr)$ can be approximated by $(kr)^l/(2l+1)!!$, so that the field $\boldsymbol{A}(\boldsymbol{r};\mathrm{E})$, which contains a spherical Bessel function with an index lower by 1, will dominate over the magnetic multipole field. The above relations then show that in this region

$$|\boldsymbol{E}(\boldsymbol{r};\mathrm{E})| \gg |\boldsymbol{H}(\boldsymbol{r};\mathrm{E})| \quad , \quad |\boldsymbol{H}(\boldsymbol{r};\mathrm{M})| \gg |\boldsymbol{E}(\boldsymbol{r};\mathrm{M})| \tag{5.39}$$

and justify the nomenclature.

5.3.4 Multipole Expansion of Plane Waves

In experiments the incoming or outgoing photon is usually in a plane-wave state with fixed momentum. The connection to the states of good angular momentum must thus be found by expanding the plane wave in this basis. Because the mathematics is simplest in this case, we will usually deal with waves having their wave vector $\boldsymbol{k}$ into the z direction. In this case the polarization vectors can be chosen to be the spherical unit vectors $\boldsymbol{e}_\mu$, $\mu = \pm 1$.

Define the expansion coefficients for a plane wave of unit amplitude propagating into the z direction via

$$e_\mu \mathrm{e}^{\mathrm{i}kz} = \sum_{lm} \left(c_{lm} \boldsymbol{A}_{lm}(\boldsymbol{r}; \mathrm{E}) + d_{lm} \boldsymbol{A}_{lm}(\boldsymbol{r}; \mathrm{M}) \right) \quad . \tag{5.40}$$

To evaluate the coefficients, first take the curl on both sides. On the left we get

$$\nabla \times \boldsymbol{e}_\mu \mathrm{e}^{\mathrm{i}kz} = \mathrm{i}k\, \boldsymbol{e}_0 \times \boldsymbol{e}_\mu\, \mathrm{e}^{\mathrm{i}kz} = \mu k\, \boldsymbol{e}_\mu\, \mathrm{e}^{\mathrm{i}kz} \quad , \tag{5.41}$$

for which $\boldsymbol{e}_0 \times \boldsymbol{e}_\mu = -\mathrm{i}\mu \boldsymbol{e}_\mu$ (easily checked using the definition of the spherical basis vectors) was used. On the right-hand side properties (4) of the multipole fields may be employed to yield

$$\begin{aligned} &\nabla \times \sum_{lm} \left(c_{lm} \boldsymbol{A}_{lm}(\boldsymbol{r}; \mathrm{E}) + d_{lm} \boldsymbol{A}_{lm}(\boldsymbol{r}; \mathrm{M}) \right) \\ &\quad = \sum_{lm} \left(-\mathrm{i}k c_{lm} \boldsymbol{A}_{lm}(\boldsymbol{r}; \mathrm{M}) + \mathrm{i}k d_{lm} \boldsymbol{A}_{lm}(\boldsymbol{r}; \mathrm{E}) \right) \quad , \end{aligned} \tag{5.42}$$

and a comparison of the two expressions leads to $c_{lm} = \mathrm{i}\mu d_{lm}$. The expansion now reads

$$\boldsymbol{e}_\mu \mathrm{e}^{\mathrm{i}kz} = \sum_{lm} d_{lm} \left(\boldsymbol{A}_{lm}(\boldsymbol{r}; \mathrm{M}) + \mathrm{i}\mu \boldsymbol{A}_{lm}(\boldsymbol{r}; \mathrm{E}) \right) \quad . \tag{5.43}$$

Finding the remaining coefficients d_{lm} requires a lengthier calculation. Probably the easiest way is to use the expansion of the scalar plane wave

$$\begin{aligned} \mathrm{e}^{\mathrm{i}kz} &= \sum_l (2l+1)\, \mathrm{i}^l j_l(kr)\, P_l(\cos\theta) \\ &= \sqrt{4\pi} \sum_l \sqrt{2l+1}\; \mathrm{i}^l j_l(kr)\, Y_{l0}(\theta, \phi) \quad . \end{aligned} \tag{5.44}$$

Here θ denotes the polar angle in spherical coordinates. The vector field of (5.43) can be turned into a scalar field by letting the angular momentum operator act on it with a scalar product (the simpler divergence operator cannot be used because of the transversality). We first rewrite the left-hand side of (5.43), then insert the expansion of (5.44) and the definition of the magnetic multipole field of (5.32) to get

$$\hat{\boldsymbol{L}} \cdot \boldsymbol{e}_\mu \mathrm{e}^{\mathrm{i}kz} = \sqrt{4\pi} \sum_l \hbar \sqrt{l(l+1)(2l+1)}\; \mathrm{i}^l\, \boldsymbol{e}_\mu \cdot \boldsymbol{A}_{l0}(\boldsymbol{r}; \mathrm{M}) \quad . \tag{5.45}$$

On the right-hand side of (5.43) the magnetic multipole is again replaced by its definition, so that the two $\hat{\boldsymbol{L}}$ operators combine into an $\hat{L}^2$, while in the second term the electric multipole is expressed in terms of the curl of the magnetic one:

$$\begin{aligned} \hat{\boldsymbol{L}} \cdot \boldsymbol{e}_\mu \mathrm{e}^{\mathrm{i}kz} = \sum_{lm} d_{lm} \Bigg[&\frac{\hat{L}^2}{\hbar\sqrt{l(l+1)}}\, j_l(kr)\, Y_{lm}(\Omega) \\ &+ \frac{\mu}{k} \hat{\boldsymbol{L}} \cdot \left(\nabla \times \boldsymbol{A}_{lm}(\boldsymbol{r}; \mathrm{M}) \right) \Bigg] \quad . \end{aligned} \tag{5.46}$$

In the second term on the right the vector products can be manipulated as follows:

$$\begin{aligned}\hat{\boldsymbol{L}} \cdot \big(\nabla \times \boldsymbol{A}_{lm}(\boldsymbol{r};\mathrm{M})\big) &= -\mathrm{i}\hbar(\boldsymbol{r} \times \nabla) \cdot \big(\nabla \times \boldsymbol{A}_{lm}(\boldsymbol{r};\mathrm{M})\big) \\ &= -\mathrm{i}\hbar\,\boldsymbol{r} \cdot \big[\nabla \times \big(\nabla \times \boldsymbol{A}_{lm}(\boldsymbol{r};\mathrm{M})\big)\big] \\ &= -\mathrm{i}\hbar\,\boldsymbol{r} \cdot \big[\nabla\big(\nabla \cdot \boldsymbol{A}_{lm}(\boldsymbol{r};\mathrm{M})\big) - \nabla^2 \boldsymbol{A}_{lm}(\boldsymbol{r};\mathrm{M})\big] \\ &= -\mathrm{i}\hbar k^2 \boldsymbol{r} \cdot \boldsymbol{A}_{lm}(\boldsymbol{r};\mathrm{M}) \quad . \end{aligned} \tag{5.47}$$

In the next-to-last step the transversality of the field and the Helmholtz equation were used. The final result is zero, however, as can be seen from the definition in (5.32).

Combining the two intermediate results now yields

$$\sqrt{4\pi}\sum_l \hbar\sqrt{l(l+1)(2l+1)}\,\mathrm{i}^l \boldsymbol{e}_\mu \cdot \boldsymbol{A}_{l0}(\boldsymbol{r};\mathrm{M}) = \sum_{lm} d_{lm} \frac{\hat{L}^2}{\hbar\sqrt{l(l+1)}} j_l(kr)\, Y_{lm}(\Omega) \quad , \tag{5.48}$$

and determines the coefficients d_{lm} still requires expressing the sums on both sides in the same orthonormal set of functions. On the left the magnetic multipole may be expressed in terms of the spherical harmonics:

$$\begin{aligned}\boldsymbol{e}_\mu \cdot \boldsymbol{A}_{l0}(\boldsymbol{r};\mathrm{M}) &= j_l(kr)\sum_{\mu'}(l1l|\mu' - \mu'0)\, Y_{l\mu'}(\Omega)\, \boldsymbol{e}_\mu \cdot \boldsymbol{e}_{-\mu'} \\ &= -j_l(kr)\sum_{\mu}(l1l|\mu{-}\mu 0)\, Y_{l\mu}(\Omega) \quad . \end{aligned} \tag{5.49}$$

Since for the transverse fields only $\mu = \pm 1$ appears, the properties of the spherical basis vectors (see Sect. 2.3.6) lead to

$$\boldsymbol{e}_\mu \cdot \boldsymbol{e}_{-\mu'} = -\boldsymbol{e}_\mu \cdot \boldsymbol{e}^*_{\mu'} = -\delta_{\mu\mu'} \quad \text{for} \quad \mu = \pm 1 \quad , \tag{5.50}$$

which was used above. Comparing coefficients on both sides of (5.48) after replacing $\hat{\boldsymbol{L}}^2$ by its eigenvalue $\hbar^2 l(l+1)$ shows that

$$-\sqrt{4\pi}\hbar\sqrt{l(l+1)(2l+1)}\,\mathrm{i}^l\,(l1l|\mu{-}\mu 0)\,\delta_{m\mu} = d_{lm}\hbar\sqrt{l(l+1)} \quad , \tag{5.51}$$

so that

$$d_{lm} = -\sqrt{4\pi(2l+1)}\,\mathrm{i}^l\,(l1l|\mu - \mu 0)\,\delta_{m\mu} \quad , \tag{5.52}$$

and inserting the Clebsch–Gordan coefficient from the Appendix:[1]

$$d_{lm} = -\mu\sqrt{2\pi(2l+1)}\,\mathrm{i}^l\,\delta_{m\mu} \quad . \tag{5.53}$$

This completes the final result for the multipole expansion of the plane wave:

$$\boldsymbol{e}_\mu \mathrm{e}^{\mathrm{i}kz} = -\mu\sqrt{2\pi}\sum_l \sqrt{2l+1}\,\mathrm{i}^l\big(\boldsymbol{A}_{l\mu}(\boldsymbol{r};\mathrm{M}) + \mathrm{i}\mu\boldsymbol{A}_{l\mu}(\boldsymbol{r};\mathrm{E})\big) \quad . \tag{5.54}$$

[1] Note that the expressions there can, for the special case $M = 0$, be combined into $(l1l|m - m0) = -m/\sqrt{2}$

This formula can be generalized to an arbitrary direction of the wave vector $\boldsymbol{k}$ by the application of the appropriate rotation matrix, but this will be needed later only for an intermediate manipulation, so we skip the details.

5.4 Coupling of Radiation and Matter

5.4.1 Basic Matrix Elements

The rate for transitions between the initial state $|i\rangle$ and the final state $|f\rangle$ under the influence of a perturbation Hamiltonian $\hat{H}_{\rm int}$ is given by *Fermi's golden rule*

$$w_{f\leftarrow i} = \frac{2\pi}{\hbar}\left|\langle f|\hat{H}_{\rm int}|i\rangle\right|^2 \rho_f \quad , \tag{5.55}$$

where ρ_f is the density of final states. In this chapter we will have to supply detailed expressions for all the terms in this expression in the special case of a nucleus interacting with the electromagnetic field.

The initial and final states in our case are made up of nuclear states coupled to states of the radiation field. The nuclear states will be denoted in the following by $|\alpha(t)\rangle$, $|\beta(t)\rangle$, where details have to be supplied by the nuclear model, while the radiation field will be either in the vacuum, $|0\rangle$, or the one-phonon state determined by the wave vector and polarization, $|\boldsymbol{k}\mu\rangle$. Note that for the radiation field we have assumed the Heisenberg picture with the time dependence in the operators, whereas the nucleus is treated in the Schrödinger picture and has time-dependent wave functions. This was indicated by writing $\alpha(t)$ in the state vectors. For emission of radiation we will thus have

$$|i\rangle = |\alpha(t), 0\rangle = |\alpha(t)\rangle|0\rangle \quad , \quad |f\rangle = |\beta(t), \boldsymbol{k}\mu\rangle = |\beta(t)\rangle|\boldsymbol{k}\mu\rangle \quad , \tag{5.56}$$

and for absorption the roles of the initial and final states are reversed.

For the Hamiltonian of interaction between radiation and matter we will take the usual expression

$$\hat{H}_{\rm int} = -\frac{1}{c}\int {\rm d}^3r\, \hat{\jmath}(\boldsymbol{r})\cdot\hat{\boldsymbol{A}}(\boldsymbol{r},t) \quad . \tag{5.57}$$

The field operator $\hat{\boldsymbol{A}}$ is the Fock-space operator as constructed in Sect. 5.2. The matrix elements of the operators $\hat{\beta}^\dagger_{\boldsymbol{k}\mu}$ and $\hat{\beta}_{\boldsymbol{k}\mu}$ between the one-photon and vacuum states are equal to 1, so that the combined matrix element becomes

$$\langle\beta(t), \boldsymbol{k}\mu|\hat{H}_{\rm int}|\alpha(t), 0\rangle = \int {\rm d}^3r\sqrt{\frac{2\pi\hbar c^2}{\omega V}}\varepsilon^*_{\boldsymbol{k}\mu}{\rm e}^{-{\rm i}\boldsymbol{k}\cdot\boldsymbol{r}+{\rm i}\omega t}\cdot\langle\beta(t)|\hat{\jmath}(\boldsymbol{r})|\alpha(t)\rangle \tag{5.58}$$

for the emission of photons, and

$$\langle\beta(t), 0|\hat{H}_{\rm int}|\alpha(t), \boldsymbol{k}\mu\rangle = \int {\rm d}^3r\sqrt{\frac{2\pi\hbar c^2}{\omega V}}\varepsilon_{\boldsymbol{k}\mu}{\rm e}^{{\rm i}\boldsymbol{k}\cdot\boldsymbol{r}-{\rm i}\omega t}\cdot\langle\beta(t)|\hat{\jmath}(\boldsymbol{r})|\alpha(t)\rangle \tag{5.59}$$

for absorption. Note that the two expressions differ only by the signs in the exponential function; it therefore suffices to do the following manipulations for the case of emission only and then obtain the corresponding absorption result by simple sign changes.

The matrix element of the operator $\hat{\jmath}(\boldsymbol{r})$, which describes the current density in the nucleus, obtains its time dependence through that of the wave functions. Since these are stationary states, we have

$$|\alpha(t)\rangle = |\alpha\rangle \exp\left(-\mathrm{i}E_\alpha t/\hbar\right) \quad , \quad |\beta(t)\rangle = |\beta\rangle \exp\left(-\mathrm{i}E_\beta t/\hbar\right) \quad , \tag{5.60}$$

and

$$\langle\beta(t)|\hat{\jmath}(\boldsymbol{r})|\alpha(t)\rangle = \langle\beta|\hat{\jmath}(\boldsymbol{r})|\alpha\rangle \, \exp\left(\mathrm{i}(E_\beta - E_\alpha)t/\hbar\right) \quad . \tag{5.61}$$

As usual for Fermi's golden rule, the time-dependent phase of the matrix element has to vanish to ensure energy conservation; combining the phases of the nuclear states with that of the photon operator yields the condition $E_\beta = E_\alpha + \hbar\omega$ for absorption and $E_\beta = E_\alpha - \hbar\omega$ for emission.

In special cases to be treated later the operator $\hat{\jmath}(\boldsymbol{r})$ can be replaced by the charge-density operator $\hat{\rho}(\boldsymbol{r})$ in a simple way: the continuity equation must be valid also for the operators in the Heisenberg picture,

$$\frac{\partial}{\partial t}\hat{\rho}(\boldsymbol{r},t) + \nabla \cdot \hat{\jmath}(\boldsymbol{r},t) = 0 \quad . \tag{5.62}$$

Taking matrix elements of this equation yields

$$\frac{\partial}{\partial t}\langle\beta|\hat{\rho}(\boldsymbol{r},t)|\alpha\rangle + \nabla \cdot \langle\beta|\hat{\jmath}(\boldsymbol{r},t)|\alpha\rangle = 0 \quad , \tag{5.63}$$

because the wave functions in $|\alpha\rangle$ and $|\beta\rangle$ depend neither on $\boldsymbol{r}$, the fixed position in space where the density and current density are being evaluated, nor on time. But this may be transformed easily into the Schrödinger picture:

$$\frac{\partial}{\partial t}\langle\beta(t)|\hat{\rho}(\boldsymbol{r})|\alpha(t)\rangle + \nabla \cdot \langle\beta(t)|\hat{\jmath}(\boldsymbol{r})|\alpha(t)\rangle = 0 \quad , \tag{5.64}$$

and now the explicit time dependence of the states can be inserted,

$$-\tfrac{\mathrm{i}}{\hbar}(E_\alpha - E_\beta)\langle\beta|\hat{\rho}(\boldsymbol{r})|\alpha\rangle + \nabla \cdot \langle\beta|\hat{\jmath}(\boldsymbol{r})|\alpha\rangle = 0 \quad . \tag{5.65}$$

Thus all matrix elements involving the divergence of $\hat{\jmath}(\boldsymbol{r})$ can be simplified to matrix elements of the density operator, which is usually much more easily handled, as is already plausible because of the scalar nature of the density.

The remaining ingredient in the formula for the transition rate is the density of states. Again we have to distinguish the cases of emission and absorption.

- *Absorption:* the cases of interest to this book are the transition to an isolated energy level or a narrow resonance of the nucleus, where in both cases one is interested only in the total transition rate integrated over the extent of the line or resonance and may assume that the matrix element varies little as a function of the energy within the resonance. In both cases

$$\int_{\text{line}} \mathrm{d}E \, \frac{2\pi}{\hbar}\left|\langle f|\hat{H}_{\text{int}}|i\rangle\right|^2 \rho(E) \approx \frac{2\pi}{\hbar}\left|\langle f|\hat{H}_{\text{int}}|i\rangle\right|^2_{\text{avg}} \int \mathrm{d}E \; \rho(E) \quad , \tag{5.66}$$

and the remaining integral over the width of the line is unity. The absorption cross section is defined as the absorption rate divided by the incoming photon flux, which in our case must correspond to the flux due to the one-photon field. This can be calculated from the Poynting vector of the field, but may also be obtained more easily by observing that we have assumed that there is only one photon in the volume V, i.e., a photon density of $1/V$, and this produces a photon-number flux density of c/V. The cross section is thus

$$\sigma_{\text{absorption}} = \frac{2\pi V}{\hbar c} \left| \langle f | \hat{H}_{\text{int}} | i \rangle \right|^2 \quad , \tag{5.67}$$

and inserting the expression of (5.58) we have

$$\sigma_{\text{absorption}} = \frac{(2\pi)^2 c}{\omega} \left| \int \mathrm{d}^3 r \, \langle \beta | \hat{\jmath}(\boldsymbol{r}) | \alpha \rangle \cdot \varepsilon_{\boldsymbol{k}\mu} \, \mathrm{e}^{\mathrm{i}\boldsymbol{k}\cdot\boldsymbol{r}} \right|^2 \quad . \tag{5.68}$$

- *Emission:* in this case the final state will be a discrete state or narrow resonance of the nucleus combined with a photon state of given energy but arbitrary direction, so that the density of available states is determined by the photon. To count the number of photon states take the volume V as a cube with side length L. Imposing periodic boundary conditions on the photon field along each cartesian direction in the cube restricts the wave number in each direction to multiples of $2\pi/L$, so that, conversely, the number of photon states is given by $L/2\pi$ times the range of wave numbers. Translating this into spherical coordinates implies that

$$\mathrm{d}^3 n = \left(\frac{L}{2\pi} \right)^3 \mathrm{d}^3 k = \left(\frac{L}{2\pi} \right)^3 k^2 \, \mathrm{d}k \, \mathrm{d}\Omega_k \quad , \tag{5.69}$$

where Ω_k is the solid angle of emission. The density of states can now be written as

$$\rho(E) = \frac{\mathrm{d}^3 n}{\mathrm{d}E} = \left(\frac{L}{2\pi} \right)^3 \frac{k^2}{\hbar c} \, \mathrm{d}\Omega_k \quad , \tag{5.70}$$

where $E = \hbar\omega = \hbar c k$ was used. The transition rate is then

$$w_{f \leftarrow i} = \frac{V}{(2\pi)^2} \frac{k^2}{\hbar^2 c} \left| \langle f | \hat{H}_{\text{int}} | i \rangle \right|^2 \mathrm{d}\Omega_k \tag{5.71}$$

and after insertion of (5.59)

$$w_{f \leftarrow i} = \frac{k}{2\pi \hbar c^2} \left| \int \mathrm{d}^3 r \, \langle \beta | \hat{\jmath}(\boldsymbol{r}) | \alpha \rangle \cdot \varepsilon_{\boldsymbol{k}\mu} \mathrm{e}^{-\mathrm{i}\boldsymbol{k}\cdot\boldsymbol{r}} \right|^2 \mathrm{d}\Omega_k \quad . \tag{5.72}$$

5.4.2 Multipole Expansion of the Matrix Elements and Selection Rules

The crucial matrix element containing the nuclear model for both absorption and emission has the form

$$M_{\beta\alpha}(k\mu) = \int \mathrm{d}^3r \, \langle\beta|\hat{\jmath}(\boldsymbol{r})|\alpha\rangle \cdot \boldsymbol{e}_\mu \mathrm{e}^{ikr} \quad , \tag{5.73}$$

where to utilize the results of Sect. 5.3.4 it was assumed that the wave runs into the z direction and its polarization is given by one of the spherical unit vectors $\boldsymbol{e}_{\pm 1}$. Inserting the expansion of the plane wave into fields of good angular momentum we get

$$\begin{aligned} M_{\beta\alpha}(k\mu) = & -\mu\sqrt{2\pi}\sum_l \sqrt{2l+1}\,\mathrm{i}^l \\ & \times \int \mathrm{d}^3r \, \langle\beta|\hat{\jmath}(\boldsymbol{r})|\alpha\rangle \cdot \left(\boldsymbol{A}_{l\mu}(\boldsymbol{r};\mathrm{M}) + \mathrm{i}\mu\boldsymbol{A}_{l\mu}(\boldsymbol{r};\mathrm{E})\right) \quad . \end{aligned} \tag{5.74}$$

This result is sufficient already for discussing the selection rules for angular momentum and parity.

- *Angular momentum:* the integral in (5.74) is nonzero only if the various terms can be coupled to a total angular momentum of zero (see this by inserting the constant function $Y_{00}(\Omega)$, which is orthogonal to any function with nonzero angular momentum). The vector character of the operator $\hat{\jmath}$ and the fields is unimportant for the integration and they are anyway coupled to zero by the scalar product. This requires that J_α, J_β, and l obey the triangular rule:

$$|J_\alpha - l| \leq J_\beta \leq J_\alpha + l \quad . \tag{5.75}$$

- *Parity:* the total parity of the fields contained in the integral must be positive, as otherwise the contributions from $\boldsymbol{r}$ and $-\boldsymbol{r}$ would cancel. The multipole field has a parity of $(-1)^l$ and $(-1)^{l+1}$ for the electric and magnetic cases, respectively, while the vector field $\langle\beta|\hat{\jmath}(\boldsymbol{r})|\alpha\rangle$ has a parity given by the two nuclear states involved: $\pi_\alpha\pi_\beta$. So the selection rule is

$$\pi_\alpha\pi_\beta = \begin{cases} (-1)^l & \text{for electric transitions} \quad , \\ (-1)^{l+1} & \text{for magnetic transitions} \quad . \end{cases} \tag{5.76}$$

Why does the vector character of $\hat{\jmath}$ not contribute an additional minus sign to the total parity of the matrix element? Actually the parity of the nuclear matrix element is a bit more involved, as is usual for questions about vector field parity. To understand the result given, take as a simple example the probability current of one particle, which is given by $\boldsymbol{j}(\boldsymbol{r}) = \hbar[\psi^*(\boldsymbol{r})\nabla\psi(\boldsymbol{r}) - \psi(\boldsymbol{r})\nabla\psi^*(\boldsymbol{r})]/2m\mathrm{i}$. This can be translated into an operator as

$$\hat{\jmath}(\boldsymbol{r}) = \frac{\hbar}{2m\mathrm{i}}\left[\delta(\boldsymbol{r}-\boldsymbol{r}')\nabla' - \left(\delta(\boldsymbol{r}-\boldsymbol{r}')\nabla'\right)^\dagger\right] \quad , \tag{5.77}$$

where $\boldsymbol{r}'$ is the variable to be integrated over in $\langle\psi|\hat{\jmath}(\boldsymbol{r})|\psi\rangle$. This expression works because

$$
\begin{aligned}
\langle\psi|\hat{\jmath}(\boldsymbol{r})|\psi\rangle &= \int \mathrm{d}^3r'\,\psi^*(\boldsymbol{r}')\hat{\jmath}(\boldsymbol{r})\psi(\boldsymbol{r}') \\
&= \frac{\hbar}{2m\mathrm{i}}\int \mathrm{d}^3r'\,\big(\psi^*(\boldsymbol{r}')\delta(\boldsymbol{r}'-\boldsymbol{r}) \\
&\quad\times \nabla'\psi(\boldsymbol{r}') - \nabla'\psi^*(\boldsymbol{r}')\delta(\boldsymbol{r}'-\boldsymbol{r})\psi(\boldsymbol{r}')\big) \\
&= \frac{\hbar}{2m\mathrm{i}}\big(\psi^*(\boldsymbol{r})\nabla\psi(\boldsymbol{r}) - \nabla\psi^*(\boldsymbol{r})\psi(\boldsymbol{r})\big) \quad . \qquad (5.78)
\end{aligned}
$$

Substituting $\boldsymbol{r} \longrightarrow -\boldsymbol{r}$ yields

$$
\begin{aligned}
\langle\psi|\hat{\jmath}(-\boldsymbol{r})|\psi\rangle &= \frac{\hbar}{2m\mathrm{i}}\big(-\pi'\pi\psi^*(\boldsymbol{r})\nabla\psi(\boldsymbol{r}) + \pi'\pi\nabla\psi^*(\boldsymbol{r})\psi(\boldsymbol{r})\big) \\
&= -\frac{\hbar}{2m\mathrm{i}}\pi\pi'\langle\psi|\hat{\jmath}(\boldsymbol{r})|\psi\rangle \qquad (5.79)
\end{aligned}
$$

with π and π' the parities of the wave functions. The minus sign comes from

$$
\left.\frac{\mathrm{d}f(x)}{\mathrm{d}x}\right|_{x\longrightarrow -x} = \frac{\mathrm{d}f(-x)}{\mathrm{d}(-x)} = -\frac{\mathrm{d}f(-x)}{\mathrm{d}x} = -\pi\frac{\mathrm{d}f(x)}{\mathrm{d}x} \qquad (5.80)
$$

and is clearly associated with the *orbital* parity of the operator. Then an additional sign is added by the vector character of the operator ∇. Compare the scalar field $\frac{1}{2}r^2$, which has positive parity, and whose derivative $\nabla\frac{1}{2}r^2 = \boldsymbol{r}$ is a vector field of positive parity also.

Clearly the different selection rules for electric and magnetic multipoles make it advantageous to separate their contributions and to assign an appropriate nomenclature:

- El transitions: electric multipole transitions with angular momentum l. Parity selection rule: $\pi_\alpha\pi_\beta = (-1)^l$. Matrix element:

$$
M_{\beta\alpha}(k\mu;\mathrm{E}l) = -\sqrt{2\pi}\,\mathrm{i}^{l+1}\sqrt{2l+1}\int \mathrm{d}^3r\,\langle\beta|\hat{\jmath}(\boldsymbol{r})|\alpha\rangle\cdot\boldsymbol{A}_{l\mu}(\boldsymbol{r};\mathrm{E}) \quad . \qquad (5.81)
$$

- Ml transitions: magnetic multipole transitions with angular momentum l. Parity selection rule: $\pi_\alpha\pi_\beta = (-1)^{l+1}$. Matrix element:

$$
M_{\beta\alpha}(k\mu;\mathrm{M}l) = -\mu\sqrt{2\pi}\,\mathrm{i}^{l}\sqrt{2l+1}\int \mathrm{d}^3r\,\langle\beta|\hat{\jmath}(\boldsymbol{r})|\alpha\rangle\cdot\boldsymbol{A}_{l\mu}(\boldsymbol{r};\mathrm{M}) \quad . \qquad (5.82)
$$

The angular momentum selection rule is the triangle rule for (J_α, J_β, l) in both cases.

For two given nuclear states the parity selection rules allow transitions of both types but with alternating angular momenta. For example, transitions between a 4^+ and a 6^+ state can be of type E2, M3, E4, M5, E6, M7, E8, M9, E10. The upper and lower cutoffs in angular momenta are caused by the triangle rule. For a transition between a 4^- and a 6^+ the same series with M and E interchanged is allowed. We will see later that the lowest possible angular momenta dominate in each case. The upper cutoff restricts the available transitions severely if one of the angular momenta is small; for example a $0^+ \longrightarrow J^+$ transition allows only EJ transitions for J even and MJ for J odd.

5.4.3 Siegert's Theorem

The general matrix elements derived above have somewhat simpler forms in the low-energy limit, the most important of which is the application of Siegert's theorem for the case of electric radiation. "Low energy" refers to the energy of the radiation, i.e., the wave length is assumed to be small compared to the typical length scale in the nucleus, the nuclear radius R:

$$kR \ll 1 \quad . \tag{5.83}$$

As the nuclear radii are smaller than 10 fm, this corresponds to energies

$$E = \hbar ck \approx (200\ \mathrm{MeV\,fm})\,k \ll \frac{200\ \mathrm{MeV\,fm}}{10\ \mathrm{fm}} = 20\ \mathrm{MeV} \quad , \tag{5.84}$$

which is practically fulfilled for all transitions of interest in this book with the exception of the giant resonances.

As the current distribution is restricted to the nuclear volume, the assumption implies that in the integrals always $kr \ll 1$, and this may be used to simplify the radiation fields. Recalling the definitions of the electric and longitudinal multipole fields,

$$\begin{aligned} \boldsymbol{A}_{lm}(\boldsymbol{r};\mathrm{E}) &= \sqrt{\frac{l+1}{2l+1}}\, j_{l-1}(kr)\boldsymbol{Y}_{lm,l-1}(\Omega) \\ &\quad - \sqrt{\frac{l}{2l+1}}\, j_{l+1}(kr)\boldsymbol{Y}_{lm,l+1}(\Omega) \quad , \\ \boldsymbol{A}_{lm}(\boldsymbol{r};\mathrm{L}) &= \sqrt{\frac{l}{2l+1}}\, j_{l-1}(kr)\boldsymbol{Y}_{lm,l-1}(\Omega) \\ &\quad + \sqrt{\frac{l+1}{2l+1}}\, j_{l+1}(kr)\boldsymbol{Y}_{lm,l+1}(\Omega) \quad , \end{aligned} \tag{5.85}$$

one notes that aside from a trivial factor they are identical if the second term could be neglected in each case. This is in fact the case in the low-energy limit, because

$$j_l(kr) \approx \frac{(kr)^l}{(2l+1)!!} \quad \rightarrow \quad \frac{j_{l+1}(kr)}{j_{l-1}(kr)} \approx \frac{(kr)^2}{(2l+1)(2l+3)} \ll 1 \quad . \tag{5.86}$$

Thus we can write

$$\boldsymbol{A}_{lm}(\boldsymbol{r};\mathrm{E}) \approx \sqrt{\frac{l+1}{l}}\boldsymbol{A}_{lm}(\boldsymbol{r};\mathrm{L}) = \sqrt{\frac{l+1}{l}}\,\frac{1}{k}\nabla\big(j_l(kr)\,Y_{lm}(\Omega)\big) \quad . \tag{5.87}$$

The advantage gained is that we can express this field as the gradient of a scalar, which enables us to perform a partial integration:

$$\begin{aligned} M_{\beta\alpha}(\boldsymbol{k}\mu;\mathrm{E}l) &\approx -\frac{\sqrt{2\pi}\,\mathrm{i}^{l+1}}{k}\sqrt{\frac{(2l+1)(l+1)}{l}} \\ &\quad \times \int \mathrm{d}^3r\,\langle\beta|\hat{\boldsymbol{\jmath}}(\boldsymbol{r})|\alpha\rangle\cdot\nabla\big(j_l(kr)\,Y_{l\mu}(\Omega)\big) \\ &= \frac{\sqrt{2\pi}\,\mathrm{i}^{l+1}}{k}\sqrt{\frac{(2l+1)(l+1)}{l}} \\ &\quad \times \int \mathrm{d}^3r\,\nabla\cdot\langle\beta|\hat{\boldsymbol{\jmath}}(\boldsymbol{r})|\alpha\rangle\, j_l(kr)\,Y_{l\mu}(\Omega) \quad . \end{aligned} \tag{5.88}$$

The surface terms vanish because of the finite extent of the nuclear current distribution.

Now the continuity equation for the current can be applied according to (5.65),

$$\begin{aligned} M_{\beta\alpha}(\boldsymbol{k}\mu; \mathrm{E}l) &\approx \frac{\sqrt{2\pi}\,\mathrm{i}^{l+1}}{k}\sqrt{\frac{(2l+1)(l+1)}{l}} \\ &\quad \times \int \mathrm{d}^3r\, \tfrac{\mathrm{i}}{\hbar}(E_\alpha - E_\beta)\langle\beta|\hat{\rho}(\boldsymbol{r})|\alpha\rangle\, j_l(kr)\, Y_{l\mu}(\Omega) \\ &= \mp\sqrt{2\pi}\,\mathrm{i}^l\, c\sqrt{\frac{(2l+1)(l+1)}{l}} \\ &\quad \times \int \mathrm{d}^3r\, \langle\beta|\hat{\rho}(\boldsymbol{r})|\alpha\rangle\, j_l(kr)\, Y_{l\mu}(\Omega) \quad , \end{aligned} \tag{5.89}$$

where the difference in energy was replaced by the photon energy

$$E_\alpha - E_\beta = \begin{cases} \hbar ck & \text{for emission} \quad , \\ -\hbar ck & \text{for absorption} \quad . \end{cases} \tag{5.90}$$

So the upper sign in (5.89) is correct for emission and the lower one for absorption.

Siegert's theorem essentially consists in the replacement of $\hat{\jmath}(\boldsymbol{r})$ by $\hat{\rho}(\boldsymbol{r})$ in the matrix element. The density is usually much easier to calculate and less sensitive to dynamic currents such as pion exchanges in the nucleus.

5.4.4 Matrix Elements for Emission in the Long-Wavelength Limit

For the emission of photons we collect here the results obtained up to now and then separate the complicated phase space and angular momentum factors from the basic nuclear matrix elements. The transition rate calculated in Sect. 5.4.1 was

$$w_{f\leftarrow i} = \frac{k}{2\pi\hbar c^2}\left|\int \mathrm{d}^3r\, \langle\beta|\hat{\jmath}(\boldsymbol{r})|\alpha\rangle \cdot \boldsymbol{e}_\mu \mathrm{e}^{-\mathrm{i}kz}\right|^2 \mathrm{d}\Omega_k \quad . \tag{5.91}$$

Inserting the multipole expansion developed in Sect. 5.4.2 and renaming the transition rate for a given mode $T_{\beta\alpha}(l\mu; R)$ for type Rl, where R is either E or M depending on the electric or magnetic character of the radiation, we can rewrite the rate as

$$T_{\beta\alpha}(l\mu; R) = \frac{k}{2\pi\hbar c^2}\left|M_{\beta\alpha}(k\mu, Rl)\right|^2 \mathrm{d}\Omega_k \tag{5.92}$$

with the matrix elements

$$M_{\beta\alpha}(k\mu, El) = -\mathrm{i}\sqrt{2\pi(2l+1)}\,\mathrm{i}^l \int \mathrm{d}^3r\, \langle\beta|\hat{\rho}(\boldsymbol{r})|\alpha\rangle \cdot \boldsymbol{A}_{l\mu}(\boldsymbol{r}; \mathrm{E}) \tag{5.93}$$

and

$$M_{\beta\alpha}(k\mu, Ml) = -\mu\sqrt{2\pi(2l+1)}\,\mathrm{i}^l \int \mathrm{d}^3r\, \langle\beta|\hat{\jmath}(\boldsymbol{r})|\alpha\rangle \cdot \boldsymbol{A}_{l\mu}(\boldsymbol{r}; \mathrm{M}) \quad . \tag{5.94}$$

Now for the limit of long wavelengths we will approximate the spherical Bessel function in both cases with the by now familiar formula $j_l(kr) \approx (kr)^l/(2l+1)!!$.

For electric transitions we may use Siegert's theorem in addition, yielding

$$M_{\beta\alpha}(k\mu;\mathrm{E}l) = -\sqrt{2\pi}\,\mathrm{i}^l c\sqrt{\frac{(2l+1)(l+1)}{l}} \times \int \mathrm{d}^3r\,\langle\beta|\hat{\jmath}(\boldsymbol{r})|\alpha\rangle \frac{(kr)^l}{(2l+1)!!} Y_{l\mu}(\Omega) \quad , \tag{5.95}$$

and inserting this into the transition rate leads to

$$T_{\beta\alpha}(l\mu;\mathrm{E}) = \frac{(2l+1)(l+1)}{l[(2l+1)!!]^2}\frac{k^{2l+1}}{\hbar}\left|\langle\beta|\hat{\Omega}_{l\mu}(E)|\alpha\rangle\right|^2 \mathrm{d}\Omega_k \quad . \tag{5.96}$$

The relatively simple part, dependent on the nuclear model, has been defined as

$$\langle\beta|\hat{\Omega}_{l\mu}(E)|\alpha\rangle = \int \mathrm{d}^3r\,\langle\beta|\hat{\rho}(\boldsymbol{r})|\alpha\rangle r^l Y_{l\mu}(\Omega) \quad . \tag{5.97}$$

For magnetic transitions, the expansion of the multipole field is

$$\begin{aligned} \boldsymbol{A}_{l\mu}(\boldsymbol{r};\mathrm{M}) &= \frac{1}{\sqrt{l(l+1)}}\hat{\boldsymbol{L}} j_l(kr)\,Y_{l\mu}(\Omega) \\ &\approx \frac{1}{\sqrt{l(l+1)}}\hat{\boldsymbol{L}}\frac{(kr)^l}{(2l+1)!!} Y_{l\mu(\Omega)} \quad . \end{aligned} \tag{5.98}$$

After inserting this into the transition rate, we separate off the same factors as for the electric case to get

$$T_{\beta\alpha}(l\mu;\mathrm{M}) = \frac{(2l+1)(l+1)}{l[(2l+1)!!]^2}\frac{k^{2l+1}}{\hbar}\left|\langle\beta|\hat{\Omega}_{l\mu}(M)|\alpha\rangle\right|^2 \mathrm{d}\Omega_k \tag{5.99}$$

with

$$\langle\beta|\hat{\Omega}_{l\mu}(M)|\alpha\rangle = \frac{-1}{c(l+1)}\int \mathrm{d}^3r\,\langle\beta|\hat{\jmath}(\boldsymbol{r})|\alpha\rangle\cdot\hat{\boldsymbol{L}}\left(r^l Y_{l\mu}(\Omega)\right) \quad . \tag{5.100}$$

These results already show a separation into the crucial nuclear matrix elements, which contain the nuclear model information, and the kinematic and radiation field factors. Experimentally, however, one is often interested only in averaged transition rates. For example, the direction of the outgoing photon is often unimportant. In the formulas obtained we have assumed the photon is emitted in the z direction exclusively. Other directions can be calculated by rotating the expansion of the plane wave, which in turn corresponds to rotating the multipole fields. Since these are spherical tensors, it can be achieved through the rotation matrices, with $\theta_3 = 0$ since two Euler angles suffice for selecting a direction:

$$\boldsymbol{A}_{l\mu}(\boldsymbol{r};R) \to \sum_{\mu'} \mathcal{D}^{(l)}_{\mu\mu'}(\theta_1,\theta_2,0)\,\boldsymbol{A}_{l\mu'}(\boldsymbol{r};R) \quad . \tag{5.101}$$

There is no need to worry about what direction this actually corresponds to, since we will sum over those directions in a moment anyway. Note, however, that the Euler angle θ_1 corresponds to an azimuthal angle, as it rotates about the z axis, while θ_2 takes over the polar rotation role.

The rotation does not interfere with any of the manipulations carried out, so that we may simply rotate the final result for the matrix elements,

$$\langle\beta|\hat{\Omega}_{l\mu}(R)|\alpha\rangle \rightarrow \sum_{\mu'} \mathcal{D}^{(l)}_{\mu\mu'}(\phi,\theta,0)\,\langle\beta|\hat{\Omega}_{l\mu'}(R)|\alpha\rangle \quad , \tag{5.102}$$

but we have to sum the transition rates and not the matrix elements. This requires forming the absolute value squared and then integrating over the Euler angles. Since only the rotation matrices participate in this, it boils down to evaluating the factor

$$\int \mathrm{d}\Omega\, \mathcal{D}^{(l)\,*}_{mm'}(\phi,\theta,0)\,\mathcal{D}^{(l)}_{\mu\mu'}(\phi,\theta,0) = \frac{4\pi}{2l+1}\,\delta_{m\mu}\,\delta_{m'\mu'} \quad , \tag{5.103}$$

which is an orthogonality relation for the rotation matrices.

Summing also over the two polarization states of the photon produces an additional factor of 2, and we get

$$T_{\beta\alpha}(l;R) = \frac{8\pi(l+1)}{l[(2l+1)!!]^2}\,\frac{k^{2l+1}}{\hbar}\left|\langle\beta|\hat{\Omega}_{l\mu}(R)|\alpha\rangle\right|^2 \quad , \tag{5.104}$$

where, as before, R stands for either E or M in formulas like these that have the same form for electric and magnetic radiation.

Finally the projections M_β and M_α of the nuclear angular momenta are not needed unless experiments with polarized beams and/or targets are considered. The transition rate in this case has to be *averaged* over the initial projection M_α and summed over the final one, M_β. This can be done in closed form by using the reduced matrix element. The reduced transition probability is defined as

$$\begin{aligned} B(Rl, J_\alpha \rightarrow J_\beta) &= \frac{1}{2J_\alpha+1}\sum_{M_\alpha M_\beta}\left|\langle\beta|\hat{\Omega}_{l\mu}(R)|\alpha\rangle\right|^2 \\ &= \frac{1}{2J_\alpha+1}\left|\langle\beta||\hat{\Omega}_l(R)||\alpha\rangle\right|^2 \sum_{M_\alpha M_\beta}\left(J_\alpha l J_\beta|M_\alpha \mu M_\beta\right)^2 \quad . \end{aligned} \tag{5.105}$$

Evaluating the sum over Clebsch-Gordan coefficients is not too difficult. The fact that the uncoupled basis $|J_\alpha M_\alpha J_\beta M_\beta\rangle$ and the coupled basis $|l\mu J_\alpha J_\beta\rangle$ both form orthonormal sets requires that

$$\begin{aligned} &\sum_{M_\alpha M_\beta}\langle l\mu J_\alpha J_\beta|J_\alpha J_\beta M_\alpha M_\beta\rangle\langle J_\alpha J_\beta M_\alpha M_\beta|l'\mu' J_\alpha J_\beta\rangle \\ &\quad = \sum_{M_\alpha M_\beta}\left(J_\alpha J_\beta l|M_\alpha M_\beta \mu\right)\left(J_\alpha J_\beta l'|M_\alpha M_\beta \mu'\right) \\ &\quad = \delta_{ll'}\,\delta_{\mu\mu'} \quad . \end{aligned} \tag{5.106}$$

For the diagonal case this reads

$$\sum_{M_\alpha M_\beta}\left(J_\alpha J_\beta l|M_\alpha M_\beta \mu\right)^2 = 1 \quad , \tag{5.107}$$

and one may convert it into the desired result by using the symmetry property of the Clebsch–Gordan coefficients

$$\left(j_1 j_2 J \,|m_1 m_2 m\right) = (-1)^{j_1 - m_1} \sqrt{\frac{2J+1}{2j_2+1}} \left(j_1 J j_2 | m_1 - m - m_2\right) \tag{5.108}$$

reaching

$$\sum_{M_\alpha M_\beta} \left(J_\alpha l J_\beta | M_\alpha \mu M_\beta\right)^2 = \frac{2J_\beta + 1}{2l+1} \quad . \tag{5.109}$$

Note that there is one crucial difference between what we needed and this derivation. The desired sum runs over all possible values of M_α and M_β with μ varying as necessary to fulfil the selection rule $M_\alpha + \mu = M_\beta$ for the projections. In the Clebsch–Gordan sum, however, μ is implicitly assumed to be fixed, so that only one of the projections can vary freely. We have to make up for this by summing over the values of μ also, which adds a factor of $2l + 1$, so that the denominator drops out and the result for the reduced transition probability becomes

$$B\left(Rl, J_\alpha \to J_\beta\right) = \frac{2J_\beta + 1}{2J_\alpha + 1} \left|\langle \beta || \hat{\Omega}_l(R) || \alpha \rangle\right|^2 \quad , \tag{5.110}$$

which is the final version.

To sum up the principal results of these lengthy manipulations we repeat the three crucial formulas, replacing the letters α and β by i and f to make it obvious which is the initial and which the final state of the nucleus.

- The transition rate for the emission of a photon:

$$T_{fi}(l;R) = \frac{8\pi(l+1)}{l[(2l+1)!!]^2} \frac{k^{2l+1}}{\hbar} B\left(Rl, J_i \to J_f\right) \quad , \tag{5.111}$$

  where R stands for E or M.
- The reduced transition probability for electric multipole radiation:

$$B\left(\mathrm{E}l, J_i \to J_f\right) = \frac{2J_f + 1}{2J_i + 1} \left| \int \mathrm{d}^3 r \, \langle f || \hat{\rho}(\boldsymbol{r}) r^l Y_l(\Omega) || i \rangle \right|^2 \quad . \tag{5.112}$$

- The reduced transition probability for magnetic multipole radiation:

$$\begin{aligned} B\left(\mathrm{M}l, J_i \to J_f\right) = & \frac{2J_f + 1}{2J_i + 1} \frac{1}{c^2(l+1)^2} \\ & \times \left| \int \mathrm{d}^3 r \, \langle f || \hat{\boldsymbol{\jmath}}(\boldsymbol{r}) \cdot \hat{\boldsymbol{L}} r^l Y_l(\Omega) || i \rangle \right|^2 \quad . \end{aligned} \tag{5.113}$$

These values are usually referred to simply as $B(\mathrm{E}l)$ or $B(\mathrm{M}l)$ values, respectively.

5.4.5 Relative Importance of Transitions and Weisskopf Estimates

Even if we take into account the selection rules, often several transition types will be allowed between two nuclear states, and it is important to find some guidance as to which of the allowed transitions will be strongest. First examine the relative magnitude of two electric transitions. The $B(\mathrm{E}l)$ value scales roughly as R^{2l}, where R is the nuclear radius, so that the transition rate will depend on l as $(kR)^{2l}$. Since it was assumed that $kr \ll 1$, this means that the lowest multipole will be dominant. The same argument applies to magnetic transitions.

Estimating the relative strength of electric and magnetic transitions is somewhat more involved, because one has to specify in detail how the currents inside the nucleus are generated. Instead of doing this, we just cite the results of the simple *Weisskopf estimates*, which are useful in their own right and give an indication of the relative strengths; this is also applicable in the general case.

The Weisskopf estimates are based on electromagnetic transitions due to a single nucleon, which is described by a schematic wave function that has a constant radial wave function within $r < R$ as well as an angular-momentum part given by a spherical harmonic and a spinor, but carries angular-momentum properties. Assuming a transition between the state with $l = 0$ and a general value of l with the same spin state produces the transition probabilities

$$T(l;\mathrm{E}) = \frac{2(l+1)}{l[(2l+1)!!]^2}\left(\frac{3}{l+3}\right)^2 \frac{e^2}{\hbar c}(kR)^{2l}\omega \tag{5.114}$$

and

$$T(l;\mathrm{M}) = \frac{20(l+1)}{l[(2l+1)!!]^2}\left(\frac{3}{l+3}\right)^2 \left(\frac{\hbar}{mcR}\right)^2 \frac{e^2}{\hbar c}(kR)^{2l}\omega \quad . \tag{5.115}$$

The factor $\hbar/mcR$ arises from the nuclear magneton present in the magnetic moments of the nucleons.

Comparing the two formulas, it is apparent that the ratio between magnetic and electric transitions of the same angular momentum is

$$\frac{T(l;\mathrm{M})}{T(l;\mathrm{E})} = 10\left(\frac{\hbar}{mcR}\right)^2 \approx 0.3\,A^{-2/3} \quad , \tag{5.116}$$

where for the last estimate standard values for the nucleon mass $mc^2 \approx 938$ MeV and for the nuclear radius $R \approx 1.2\ A^{1/3}$ were used. Thus at the same angular momentum the electric transitions are somewhat stronger. This is, however, not usually a meaningful comparison, since the parity selection rule forbids us from having both transitions between the same levels. The case of practical interest is that of an $\mathrm{E}(l+1)$ transition mixing with an $\mathrm{M}l$ transition. For this case we get

$$\frac{T(l+1;\mathrm{E})}{T(l;\mathrm{M})} = \frac{(kR)^2}{10}\left(\frac{\hbar}{mcR}\right)^{-2} l(l+2)\left(\frac{l+3}{(2l+3)(l+1)(l+4)}\right)^2 \quad . \tag{5.117}$$

The factors containing the angular momentum produce something close to unity; inserting numerical values in the rest with k corresponding to a 1 MeV transition yields

$$\frac{T(l+1;\mathrm{E})}{T(l;\mathrm{M})} \approx 1.2 \times 10^{-4} A^{4/3} \left(\hbar\omega \ [\mathrm{MeV}]\right)^2 \quad . \tag{5.118}$$

For masses in, for example, the rare-earth region ($A \approx 150$) the factor is about 0.1, showing that magnetic transitions should dominate appreciably under these conditions.

Of course it should be borne in mind that all of these estimates ignore nuclear structure effects, which may change the matrix elements very strongly; also they are strictly valid only in the long wavelength limit.

There is an additional value in knowing the Weisskopf estimates: since they assume that only one nucleon takes part in the transition, experimental values much larger than these estimates point to a collective transition, in which a large fraction of the nucleons in the nucleus participate in the excitation. Such experimental observations gave one of the crucial hints for collective motion in nuclei.

EXERCISE

5.2 The Weisskopf Estimates for Electric Transitions

Problem. Derive the Weisskopf estimates for the electric transitions.

Solution. As mentioned in the text, the wave functions assumed for the Weisskopf estimates consist of a constant radial part for $r < R$, a spherical harmonic, and a spinor:

$$\psi(\boldsymbol{r}) = N Y_{lm}(\Omega)\chi \quad . \tag{1}$$

The constant N can be determined from the normalization condition

$$1 = N^2 \int_0^R \mathrm{d}r\, r^2 = N^2 \frac{R^3}{3} \tag{2}$$

to be $N = \sqrt{3/R^3}$.

Now for a single point particle with charge e the electric density operator is

$$\hat{\rho}(\boldsymbol{r}) = e\delta(\boldsymbol{r} - \boldsymbol{r}') \quad , \tag{3}$$

where $\boldsymbol{r}'$ is the particle coordinate. The integration over $\boldsymbol{r}$ in the $B(\mathrm{E}l)$ thus drops out and the integration inherent in the matrix element is left. For a transition from $l = 0$ to l it can be split into an r integral and a reduced matrix element according to

$$\langle l || \hat{\rho} r^l Y_l || 0 \rangle = e^2 \frac{3}{R^3} \int_0^R \mathrm{d}r\, r^{l+2} \times \langle l || Y_l || 0 \rangle \quad . \tag{4}$$

Using the general formula for reduced matrix elements of spherical harmonics from the Appendix, we get

Exercise 5.2

$$\langle l||Y_l||0\rangle = \tfrac{1}{\sqrt{4\pi}}(0ll|000) = \tfrac{1}{\sqrt{4\pi}} \quad , \tag{5}$$

while the radial integral is

$$\frac{3}{R^3}\int_0^R \mathrm{d}r\, r^{l+2} = \frac{3R^l}{l+3} \quad , \tag{6}$$

and the full $B(\mathrm{E}l)$ becomes

$$B(El; 0 \to l) = \frac{2l+1}{4\pi}\left(\frac{3R^l}{l+3}\right)^2 \quad . \tag{7}$$

Inserting all of these results into the formula for the transition rate yields the final result given in the text, except for an additional factor of $2l + 1$, which is missing there because the Weisskopf estimates pertain to specific polarization of the radiation.

5.4.6 Electric Multipole Moments

The operator appearing in the electric transition rate of (5.112) is also used for determining electric multipole *moments*. We define the *electric multipole operator* as

$$\hat{Q}_{lm} = \int \mathrm{d}^3r\, r^l Y_{lm}(\Omega)\hat{\rho}(\boldsymbol{r}) \quad . \tag{5.119}$$

Its diagonal matrix elements then yield the multipole moments, but are modified by conventional factors, which can be traced back to original definitions in cartesian coordinates. So, for example, the quadrupole moment is defined as

$$\begin{aligned} Q &= \int \mathrm{d}^3r\, \langle\Psi|\left(3z^2 - r^2\right)\hat{\rho}(\boldsymbol{r})|\Psi\rangle \\ &= \sqrt{\frac{16\pi}{5}}\int \mathrm{d}^3r\, \langle\Psi|r^2 Y_{20}(\Omega)\hat{\rho}(\boldsymbol{r})|\Psi\rangle \quad , \end{aligned} \tag{5.120}$$

and customarily this matrix element is evaluated for the substate with the highest angular momentum projection, i.e., $m = j$.

5.4.7 Effective Charges

Naively one would expect that in the description of electromagnetic transitions only the protons should appear, since they alone carry charge inside the nucleus. This is, however, not quite the full story. On the one hand, the interactions between the nucleons may exchange charge and thus contribute to the current, and on the other hand, neutrons and protons are coupled by *center-of-mass conservation*. The former effect is beyond the scope of this book, but the latter will be discussed now.

Assume, for example, that a neutron is excited in its orbital motion. By itself it carries no charge, but because the total center of mass must be stationary, the rest of the nucleus will move to balance that and produce a displacement of charge also. The formal application is simple in the case of the dipole moment, which is given by the vector

$$\boldsymbol{D} = \sum_{i=1}^{A} q_i(\boldsymbol{r}_i - \boldsymbol{R}) \quad , \tag{5.121}$$

where q_i is the charge of the ith nucleon and $\boldsymbol{R}$ is the center-of-mass vector. Now focus attention on one particular nucleon, for example, that with $i = 1$, which is excited in a single-particle mode, and assume that the remaining $A - 1$ particles are correspondingly shifted as a group to a position $\boldsymbol{R}'$. We then have

$$\boldsymbol{D} = q_1(\boldsymbol{r}_1 - \boldsymbol{R}) + (Ze - q_1)(\boldsymbol{R}' - \boldsymbol{R}) \quad . \tag{5.122}$$

The center of mass is

$$\boldsymbol{R} = \frac{\boldsymbol{r}_1 + (A-1)\boldsymbol{R}'}{A} \quad , \tag{5.123}$$

and insertion into (5.122) yields

$$\boldsymbol{D} = q_1^{\text{eff}} \boldsymbol{r}_1' \quad , \tag{5.124}$$

where $\boldsymbol{r}_1' = \boldsymbol{r}_1 - \boldsymbol{R}'$ is the relative separation between the single particle and the remainder, and

$$q^{\text{eff}} = \frac{A-1}{A} q - \frac{Ze - q}{A} \tag{5.125}$$

is the *effective charge* of the nucleon. Often this expression is simplified by setting $A - 1 \approx A$ and $Ze - q \approx Ze$, so that

$$q^{\text{eff}} = q - \frac{Ze}{A} \tag{5.126}$$

is used for the effective charge. Note that this implies a reduced charge for the protons while the neutrons acquire a negative charge. For $Z = A$ the charges of the nucleons are $\pm\frac{1}{2}e$, showing that the correction can be quite essential.

For higher angular momenta the formulas are not as simple and the corrections become much smaller, so that usually this effect is ignored. In this book the effective charges will be used explicitly during the discussion of giant dipole resonances, see Sect. 6.9.

6. Collective Models

6.1 Nuclear Matter

6.1.1 Mass Formulas

In many nuclear models the properties of *nuclear matter* enter as limiting conditions. Nuclear matter is a fictitious concept: it should be thought of as the extrapolation of the almost homogeneous conditions in the center of heavy nuclei to an infinite geometry. "Fictitious" means that it is unrealistic for the following reasons.

- A fixed ratio of protons to neutrons close to 1 is assumed; this would imply an infinite Coulomb energy, so that the Coulomb interaction is disregarded. In reality beta decay would produce an equilibrium ratio. Note that in the related case of *neutron star matter* the number of protons is very small, and the study of nuclear matter is certainly of interest to the theory of neutron stars.
- For the same reason the density inside heavy nuclei is actually depressed, so that an extrapolation has to eliminate the effects of the Coulomb interaction.
- A homogeneous system of protons and neutrons is not always the lowest state. At lower densities it will break up into smaller fragments, whereas at a very high density the nucleons may dissolve and form a quark–gluon plasma instead.

Although nuclear matter is therefore not a physically measurable system, it is extremely useful in providing a simple theoretical limiting case for which theories may be tested and interesting parameters be defined.

The reader should be warned that for reasons of space we do omit the very important and interesting field of *nuclear matter theory*, which tries to derive the properties of nuclear matter from the nucleon–nucleon interaction. This theory is treated in many of the larger treatises on theoretical nuclear physics, such as, [Ei72].

Basic information about nuclear matter comes from semiempirical mass formulas, which determine the binding energies of nuclei in terms of a nuclear-matter contribution and various corrections for finite nuclei. The oldest, most well-known, and still useful one of these is the *Bethe–Weizsäcker* formula [We35, Be36]

$$\begin{aligned} B(A) &= a_{\text{vol}}A + a_{\text{surf}}A^{2/3} + a_{\text{coul}}Z^2A^{-1/3} + a_{\text{sym}}\frac{(N-Z)^2}{A} \quad , \\ a_{\text{vol}} &\approx -16\ \text{MeV} \quad , \quad a_{\text{surf}} \approx 20\ \text{MeV} \quad , \\ a_{\text{coul}} &\approx 0.751\ \text{MeV} \quad , \quad a_{\text{sym}} \approx 21.4\ \text{MeV} \quad . \end{aligned} \tag{6.1}$$

Here the first term is the *volume term*, which indicates the constant binding energy per nucleon at equal density of protons and neutrons and so provides one of the important parameters of nuclear matter. The second term is proportional to the square of the nuclear radius, so that it describes the reduction of binding due to the nucleons on the surface – the *surface energy*. The third term describes the Coulomb energy, which for a homogeneously charged sphere is proportional to Z^2/R and so to $Z^2/A^{1/3}$. Finally the last term is the *symmetry energy* showing the decrease in binding for unequal numbers of protons and neutrons.

The surface and Coulomb energies will be used for developing the dependence of the energy on surface shape, while the volume and symmetry energies are directly related to the properties of nuclear matter.

A more modern version of the mass formula was given by Seeger [Se68]. It is defined by

$$\begin{aligned} B(A) = {} & W_0 A - \gamma A^{2/3} \\ & - 0.86\, r_0^{-1} Z^2 A^{-1/3} \left(1 - 0.76361\, Z^{-1/3} - 2.543\, r_0^{-2} A^{-2/3}\right) \\ & - \left(\eta A^{-4/3} - \beta A^{-1}\right)(N - Z)^2 + \delta A^{-1/2}\,(0, \pm 1) \\ & + 7\mathrm{e}^{-6\,|N-Z|/A} + 14.33 \times 10^{-6} Z^{2.39} \quad . \end{aligned} \tag{6.2}$$

Here the first term again describes the volume contribution and the second one the surface energy, while the third one corrects the Coulomb energy by an exchange and a surface difuseness correction (second and third term in parentheses). The fourth term is the symmetry energy now with an A dependence due to surface effects. In addition there is a pairing term (proportional to δ), for which the expression in parentheses is zero for odd–even, positive for even–even, and negative for odd–odd nuclei. This term represents the fact that pairs of nucleons are more tightly bound; see Sect. 7.5 for details. The final two terms have a less obvious physical significance but increase the accuracy of the mass predictions. The standard values for the coefficients not explicitly given are $W_0 = 15.645$ MeV, $\gamma = 19.655$ MeV, $\beta = 30.586$ MeV, $\eta = 53.767$ MeV, $\delta = 10.609$ MeV, and $r_0 = 1.2025$ fm.

This type of mass formula is sufficent to describe the bulk properties of nuclei throughout the periodic table to within about ± 10 MeV in the binding energies, including the position of the *valley of β stability*. In addition, because of the physical motivation of the major terms, it is useful for deriving deformation properties up to and including the fission barrier. Although for the quantitative analysis of fission and for the description of deformed ground states *shell effects* (see Sect. 9.2) are necessary, it should be kept in mind that these are small perturbations on a large binding energy which is described quite well by such a simple formula.

There are more recent enhanced mass formulas, amongst which the *droplet model* [My76] is probably the most important one. It contains a larger number of terms to describe the surface dependence of all the physical contributions in more detail – similar in spirit to the density-dependent formulas such as given in Sect. 6.1.3.

6.1.2 The Fermi-Gas Model

Another model that is very simple yet gives insight into a number of important properties of nuclear matter is the *Fermi-gas model.* In contrast to the liquid-drop simile underlying the previous chapter, it focuses on single-particle properties by treating the nucleus as a gas of noninteracting fermions with prescribed densities of protons and neutrons.

Free particles in an infinite volume are described by wave functions and eigenenergies given by

$$\psi_{\boldsymbol{k}}(\boldsymbol{r}) = \sin(\boldsymbol{k}\cdot\boldsymbol{r}) \quad , \quad \varepsilon_{\boldsymbol{k}} = \frac{\hbar^2 k^2}{2m} \quad . \tag{6.3}$$

To discretize the continuum of states we restrict the wave functions to a cube with side length a; $x, y, z = 0, \ldots, a$. The requirement that the wave functions vanish at both ends of the interval immediately leads to

$$k_x = \frac{\pi}{a} n_x \quad , \quad k_y = \frac{\pi}{a} n_y \quad , \quad k_z = \frac{\pi}{a} n_z \quad , \\ n_x,\, n_y,\, n_z = 1, 2, \ldots \tag{6.4}$$

and the eigenenergies are then

$$\varepsilon_{n_x n_y n_z} = \frac{\hbar^2}{2m}\frac{\pi^2}{a^2}\left(n_x^2 + n_y^2 + n_z^2\right) \quad . \tag{6.5}$$

The next argument is quite similar to that given for photons at the end of Sect. 6.4.1: in the limit of large a the number N of triples (n_x, n_y, n_z) can be determined by going over to spherical coordinates (n, Ω) in the differential:

$$\mathrm{d}N = \mathrm{d}n_x\, \mathrm{d}n_y\, \mathrm{d}n_z = n^2 \mathrm{d}n\, \mathrm{d}\Omega \quad . \tag{6.6}$$

Using spherical symmetry we can replace the integration over Ω by a factor of $4\pi/8$ (the 8 because only positive values of the n are allowed, corresponding to one eighth of angle space), and n is then the only independent quantity. On the other hand, because of (6.4) we have

$$\varepsilon_n = \frac{\hbar^2}{2m}\frac{\pi^2}{a^2} n^2 \quad , \tag{6.7}$$

which can be used to replace n on the right of (6.6)

$$\mathrm{d}N = \frac{a^3}{\pi^2\sqrt{2}}\left(\frac{m}{\hbar^2}\right)^{3/2} \sqrt{\varepsilon_n}\, \mathrm{d}\varepsilon_n \quad . \tag{6.8}$$

In the derivation is was assumed that all states in the spherical region corresponding to an energy $\varepsilon < \varepsilon_n$ are filled. The density of states has to be integrated up to the Fermi energy ε_{F}, and dividing by the volume a^3 yields the density of particles ϱ:

$$\varrho = \frac{N}{a^3} = \frac{1}{\pi^2\sqrt{2}}\left(\frac{m}{\hbar^2}\right)^{3/2} \int\limits_0^{\varepsilon_{\mathrm{F}}} \mathrm{d}\varepsilon\, \sqrt{\varepsilon} \quad . \tag{6.9}$$

The trivial calculation of the integral produces the desired result

$$\varepsilon_{\mathrm{F}} = \frac{\hbar^2}{2m}\sqrt{6\pi^2\varrho} \quad . \tag{6.10}$$

Before applying this result, we have to add one crucial ingredient: degeneracy. All the states constructed can actually be filled twice, so that the same number of states and the same Fermi energy correspond to twice the number of protons or neutrons. One might also deal with protons and neutrons together, so that an additional occupancy factor of 2 comes in; in general, therefore, a *degeneracy factor* g is needed. It is easy to insert this factor into the derivation; we only note that in the final result the density has to be divided by that factor, so that we get

$$\varepsilon_{\mathrm{F}} = \frac{\hbar^2}{2m}\left(\frac{3\pi^2}{\varrho/g}\right)^{3/2} \quad . \tag{6.11}$$

A related and important additional quantity is the *Fermi momentum*, which is often given in nuclear theory instead of the density,

$$k_{\mathrm{F}} = \sqrt[3]{3\pi^2\varrho/g} \quad . \tag{6.12}$$

The *total kinetic energy density* is the sum of the single-particle energies over occupied states,

$$\begin{aligned} E_{\mathrm{kin}} &= g\frac{1}{\pi^2\sqrt{2}}\left(\frac{m}{\hbar^2}\right)^{3/2}\int_0^{\varepsilon_{\mathrm{F}}} \mathrm{d}\varepsilon\, \varepsilon\sqrt{\varepsilon} \\ &= \frac{1}{5}\frac{g\,m^{3/2}\sqrt{2}}{\pi^2\hbar^3}\,\varepsilon_{\mathrm{F}}^{5/2} \quad . \end{aligned} \tag{6.13}$$

It was called *kinetic* energy density, because it does not include the binding potentials in any way. In this model the nucleons move like free particles, but should actually be bound in a very large homogeneous potential well that contributes a negative potential energy, which is not considered. The *average energy per particle* is given by the simple relation

$$\bar{\varepsilon} = \frac{E_{\mathrm{kin}}}{\varrho} = \tfrac{3}{5}\varepsilon_{\mathrm{F}} \quad . \tag{6.14}$$

It increases with $\varrho^{2/3}$ and thus is a density-dependent term in the nuclear matter energy formula.

Note that the situation considered was the lowest state of the Fermi gas only, corresponding to nuclear matter in the ground state or, in the statistical limit, at zero temperature. The energy arises only from the Pauli exclusion principle, which forced us to occupy higher and higher levels with increasing density. This is the case of a degenerate gas.

One important consequence of the Fermi-gas model is an explanation of the origin of the nuclear symmetry energy. For different proton number Z and neutron number N, $Z + N = A$, in the volume considered, the densities of protons and neutrons, ϱ_{p} and ϱ_{n} can be different but should still add up to the total density,

$$\varrho_{\mathrm{p}} + \varrho_{\mathrm{n}} = \varrho_0 \quad . \tag{6.15}$$

Consequently the Fermi energies will differ, and all the formulas above have to be adapted to such a mixture. Comparing the integrals for total kinetic energy and particle number yields the proper averaging for the average kinetic energy,

$$\bar{\varepsilon} = \frac{3}{5} \frac{N\varepsilon_{\mathrm{F}}^{\mathrm{n}} + Z\varepsilon_{\mathrm{F}}^{\mathrm{p}}}{A} \quad . \tag{6.16}$$

The Fermi energies are given in terms of the densities according to (6.11), so that only the densities need to be expressed in terms of Z, N, A, and the total density. Remembering the dependence of the symmetry energy on the expression $(Z-N)/A$, we can write

$$\begin{aligned} \varrho_{\mathrm{p}} &= \frac{Z}{A}\varrho_0 = \frac{\varrho_0}{2}\left(1 + \frac{Z-N}{A}\right) \\ \varrho_{\mathrm{n}} &= \frac{N}{A}\varrho_0 = \frac{\varrho_0}{2}\left(1 - \frac{Z-N}{A}\right) \quad . \end{aligned} \tag{6.17}$$

Inserting this into the Fermi energies, then into the average kinetic energy, and finally expanding the powers yields to lowest order in $(Z-N)/A$

$$\Delta\bar{\varepsilon} \approx \bar{t}_N \left[1 + \frac{5}{9}\frac{(Z-N)^2}{A^2}\right] \quad , \tag{6.18}$$

where the constant coefficient is given by

$$\bar{t}_N = \frac{3}{5}\frac{\hbar^2}{2m}\left(\frac{3\pi^2\varrho_0}{2}\right)^{2/3} \approx 21\,\mathrm{MeV} \quad . \tag{6.19}$$

This result shows the correct behavior and thus explains at least one simple mechanism that can lead to a symmetry energy. A quantitative comparison shows that it is too small by a factor of 2. Some arguments for correcting it by potential-energy contributions can be found in [Ei87].

6.1.3 Density-Functional Models

A natural generalization of the Fermi-gas model is provided by Thomas–Fermi theory, in which the true single-particle wave functions are replaced by plane waves locally, so that at each point there is a density of nucleons in proper relation to the local Fermi momentum. Fixing both the momentum and the position clearly violates quantum mechanics, but this method, which is also called *local density approximation (LDA)* still is quite succesful, for example, in atomic physics. For modern applications in nuclei, many corrections and higher-order terms have to be added, leading to the so-called *energy density formalism*. It can be summarized in a variational principle

$$\delta \int \mathrm{d}^3r \,\bigl(E[\varrho(\boldsymbol{r})] - \lambda\,\varrho(\boldsymbol{r})\bigr) = 0 \quad . \tag{6.20}$$

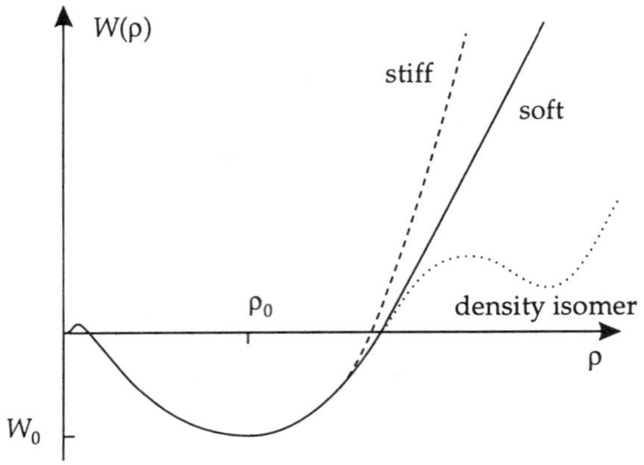

Fig. 6.1 Schematic plot of the "nuclear equation of state" $W(\varrho)$. Note the small positive region above zero density due to the Fermi-gas limit. Above equilibrium density little experimental information is available and exotic effects may be conjectured; here two equations of state with different stiffness and one with an isomeric minimum are shown.

Here $\varrho(\boldsymbol{r})$ is the density inside the nucleus, which is varied until a minimum of the integral is found, and $E[\varrho]$ is the energy functional, which expresses the energy of the nucleus in terms of the local density. The second term in the integral expresses the subsidiary condition that the total number of nucleons must be constant.

The main problem in this approach is, of course, to find a suitable functional $E(\varrho)$. In nuclear physics the development began in connection with the Bethe–Weizsäcker formula [We35, Be36], was increasingly developed in parallel with the growing understanding of the nuclear interaction [Sk56, Sk59, Wi58], and finally lead to a fully developed energy density formalism [Br68a, Be73, Lo73]. In the succeeding years many refinements were added and the method has been applied successfully to many aspects of nuclear structure and collective excitations. In this book practical examples of density functionals will appear in connection with giant resonances (Sect. 6.9) and with the Skyrme-force Hartree–Fock approach (Sect. 7.2.9). In this section we do not go into further detail (see, e.g., the review by M. Brack [Br85] and the other articles in that volume); instead we only discuss a few elementary definitions in a schematic model along the lines of [Sc68].

This density functional contains only the most basic ingredients but is already sufficient to describe the density distributions in nuclei quite well. It is given by

$$\begin{aligned} E[\varrho] = & \int d^3r\, \varrho\, W(\varrho) + \frac{e^2}{2} \int d^3r' \int d^3r\, \frac{\varrho_p(\boldsymbol{r})\varrho_p(\boldsymbol{r}')}{|\boldsymbol{r}-\boldsymbol{r}'|} \\ & + C_{ex} \int d^3r\, \varrho_p^{4/3} + C_{sym} \int d^3r\, (\varrho_p - \varrho_n)^2 \varrho^{\nu} \\ & + \frac{V_0}{4\pi} \int d^3r' \int d^3r\, \varrho(\boldsymbol{r}) \frac{e^{-|\boldsymbol{r}-\boldsymbol{r}'|/\mu}}{|\boldsymbol{r}-\boldsymbol{r}'|} \left(\varrho(\boldsymbol{r}') - \varrho(\boldsymbol{r})\right) \quad . \end{aligned} \tag{6.21}$$

Most of the terms are easily understood in the context of mass formulas. The first term is a volume energy contribution with the new twist that a density dependence for the volume energy per nucleon $W(\varrho)$ is allowed. In this form, it is often called the *equation of state of nuclear matter*, although strictly speaking the equation of state should also describe the thermal properties.

In normal nuclei, most of the matter is close to equilibrium density $\varrho_0 \approx 0.17\,\mathrm{fm}^{-3}$ with a binding energy of $W_0 = W(\varrho_0) \approx -16\,\mathrm{MeV}$. Little is known about its behavior away from equilibrium. Fig. 6.1 gives an indication of what this function might look like. There is a minimum at (ϱ_0, W_0) and a parabolic rise on both sides. At zero density the proper limit of zero energy should be reached, and for very small densities, where the nucleons do not yet feel sufficient attraction, the energy must be positive because of the Fermi-gas kinetic energy. In principle the behavior between these two points should be felt in the structure of the nuclear surface, but in this region of rapidly changing density such a simple energy functional is certainly not good enough for us to draw firm conclusions.

In recent years the question of what happens at higher densities has become the focus of attention, because in the collision of nuclei at high energies it may be possible to create nuclear matter at high density and temperature. Many exotic effects such as a secondary metastable state of nuclear matter, a *density isomer*, or the phase transition to a quark–gluon plasma have been discussed, but at present the situation is still open. Since this topic is outside the subject matter of this book, we refer the reader to reviews such as [Ma85] or the excellent articles in [Cs91] for an introduction.

An important quantity for nuclear models is the *incompressibility* of nuclear matter, which is related to the curvature of $W(\varrho)$ at the minimum and is conventionally defined as

$$K = 9\varrho_0^2 \left. \frac{\mathrm{d}W}{\mathrm{d}\varrho} \right|_{\varrho=\varrho_0} \quad . \tag{6.22}$$

K cannot be measured directly but must be deduced from density vibrations, the so-called giant monopole resonances, which unfortunately require a detailed model of the behavior of the surface. Thus there is still no unanimous consent about the proper value of K, but $K \approx 210\mathrm{MeV}$ [Bl80] seems to be widely accepted. A simple equation of state with the correct incompressibility is given by a parabolic expansion

$$W(\varrho) = \frac{K}{18\varrho_0^2}(\varrho - \varrho_0)^2 \quad , \tag{6.23}$$

but note that although it should be valid for small oscillations around equilibrium density, it does not correctly produce the vacuum limit and it is quite unclear how far the parabolic behavior may be extrapolated to higher densities. Various parametrizations of the equation of state of similar simplicity are being employed in simulations of high-energy heavy-ion reactions.

We now return to a discussion of (6.21). The second term is the familiar Coulomb energy of a charge distribution $e\varrho_\mathrm{p}(\boldsymbol{r})$ produced by the protons. Note that this implies that the proton and neutron densities are different and have to be varied independently to determine the nuclear density distribution. The third term is an approximation to the Coulomb exchange energy [Di30]. Such exchange effects will be discussed in connection with the Hartree–Fock method (Sect. 7.2).

The next term is the symmetry energy, now written as an integral over local deviations between neutron and proton densities. This is quite in the spirit of the local-density aprroximation, which produces a local Fermi-gas model at each point and thus a simple generalization of (6.18). The power ν of the additional density factor is not well known. The Fermi-gas model would suggest $\nu = -\frac{1}{3}$, but as we saw it is not quantitatively reliable, so we try to fit it to experimental data; in practice values between -1 and $\frac{1}{3}$ work well.

The last term is the density-functional version of the surface energy. It is basically the energy due to a Yukawa interaction, but the energy for a homogeneous distribution has been subtracted (that is why $\varrho(\boldsymbol{r}') - \varrho(\boldsymbol{r})$ appears), so that it contributes only on the surface.

6.2 Nuclear Surface Deformations

6.2.1 General Parametrization

The excitation spectra of even–even nuclei in the energy range of up to 2 MeV show characteristic band structures that are interpreted as vibrations and rotations of the nuclear surface in the *geometric collective model* first proposed by Bohr and Mottelson [Bo52, Bo53a, Bo54] and elaborated by Faessler and Greiner [Fa62a,b, Fa64a,b, Fa65a–c]. The underlying physical picture is that of a nucleus modeled by a classical charged liquid drop, generalizing the concept that led to the Bethe–Weizsäcker mass formula to dynamic excitation of the nucleus. For low-energy excitations the compression of nuclear matter is unimportant, and the thickness of the nuclear surface layer will also be neglected, so that we start with the model of a liquid drop of constant density and with a sharp surface. The interior structure, i.e., the existence of individual nucleons, is neglected in favor of the picture of homogeneous fluid-like nuclear matter.

The comparison of the results with experiment will show to what extent these approximations may be justified. However, it is already clear that the model of a liquid drop will be applicable only if the size of the nucleon can be neglected with respect to the size of the nucleus as a whole. This, as well as the neglect of the surface layer thickness, should work best in the case of heavy nuclei.

With these assumptions, the moving nuclear surface may be described quite generally by an expansion in spherical harmonics with time-dependent *shape parameters* as coefficients:

$$R(\theta, \phi, t) = R_0\left(1 + \sum_{\lambda=0}^{\infty} \sum_{\mu=-\lambda}^{\lambda} \alpha^*_{\lambda\mu}(t)\, Y_{\lambda\mu}(\theta, \phi)\right) \quad , \tag{6.24}$$

where $R(\theta, \phi, t)$ denotes the nuclear radius in the direction (θ, ϕ) at time t, and R_0 is the radius of the spherical nucleus, which is realized when all the $\alpha_{\lambda\mu}$ vanish.

In all the following formulas the ranges of summation $\lambda = 0, \ldots, \infty$ and $\mu = -\lambda, \ldots, \lambda$ will be implied if not otherwise indicated.

The time-dependent amplitudes $\alpha_{\lambda\mu}(t)$ describe the vibrations of the nucleus and thus serve as *collective coordinates*.

To complete the formulation of the collective model a Hamiltonian depending on the $\alpha_{\lambda\mu}$ and their suitably defined conjugate momenta has to be set up. Before doing that, however, let us examine the physical meaning of the $\alpha_{\lambda\mu}$ in more detail.

From (6.24) one may easily derive some properties of the coefficients $\alpha_{\lambda\mu}$.

1. *Complex conjugation:* the nuclear radius must be real, i.e., $R(\theta, \phi, t) = R^*(\theta, \phi, t)$. Applying this to (6.24) and using the property of the spherical harmonics that

$$Y^*_{\lambda\mu}(\theta, \phi) = (-1)^\mu Y_{\lambda-\mu}(\theta, \phi) \quad , \tag{6.25}$$

it is clear that the $\alpha_{\lambda\mu}$ have to fulfil

$$\alpha^*_{\lambda\mu} = (-1)^\mu \alpha_{\lambda-\mu} \quad . \tag{6.26}$$

2. *Spherical tensor character:* the behavior of the $\alpha_{\lambda\mu}$ under rotations follows from the invariance of the function $R(\theta, \phi)$, which must be a scalar under rotations. At first this statement may seem surprising, because the nuclear shape described obviously does not show rotational symmetry. To explain this point, therefore, let us carefully reexamine what happens as we rotate the system.

The original nuclear shape is described by the function $R(\theta, \phi)$. A rotation carries the direction (θ, ϕ) into (θ', ϕ'), and we have to describe the shape by a new function $R'(\theta', \phi')$, which fulfills (cf. the discussion at the beginning of Sect. 2.2.1)

$$R'(\theta', \phi') = R(\theta, \phi) \quad . \tag{6.27}$$

The idea now is to demand that the definition of the nuclear surface be rotationally invariant in the sense that the rotated surface $R'(\theta, \phi)$ *has the same functional form, but with rotated parameters* $\alpha'_{\lambda\mu}$. These can then be determined from

$$\sum_{\lambda\mu} \alpha'^*_{\lambda\mu} Y'_{\lambda\mu}(\theta, \phi) = \sum_{\lambda\mu} \alpha^*_{\lambda\mu} Y_{\lambda\mu}(\theta, \phi) \quad , \tag{6.28}$$

where $Y'_{\lambda\mu}$ is obtained from $Y_{\lambda\mu}$ through application of the usual rotation matrices. It is easy to see from here how the $\alpha_{\lambda\mu}$ have to transform: because of (6.26) the sum over μ can be expressed as a coupling to zero angular momentum:

$$\begin{aligned}
\sum_{\mu} \alpha^*_{\lambda\mu} Y_{\lambda\mu} &= \sum_{\mu} (-1)^\mu \alpha_{\lambda-\mu} Y_{\lambda\mu} \\
&= (-1)^\lambda \sqrt{2\lambda+1} \sum_{\mu} \frac{(-1)^{\lambda-\mu}}{\sqrt{2\lambda+1}} \alpha_{\lambda-\mu} Y_{\lambda\mu} \\
&= (-1)^\lambda \sqrt{2\lambda+1} \sum_{\mu\mu'} (\lambda\lambda 0|\mu\mu' 0)\, \alpha_{\lambda\mu'} Y_{\lambda\mu} \quad .
\end{aligned} \tag{6.29}$$

Thus the desired invariance of the definition in (6.24) is attained if the set of parameters $\alpha_{\lambda\mu}$, $\mu = -\lambda, \ldots, \lambda$ transforms like a spherical tensor with angular momentum λ; more precisely,

$$\alpha'_{\lambda\mu} = \sum_{\mu} \mathcal{D}^{(\lambda)}_{\mu\mu'} \alpha_{\lambda\mu'} \quad . \tag{6.30}$$

3. *Parity:* the same arguments hold for the parity transformation. If the spherical harmonics are reflected, the tensor $\alpha_{\lambda\mu}$ must undergo the same sign change to

preserve the invariance of the surface definition. As the spherical harmonics have a parity of $(-1)^\lambda$, so must $\alpha_{\lambda\mu}$.

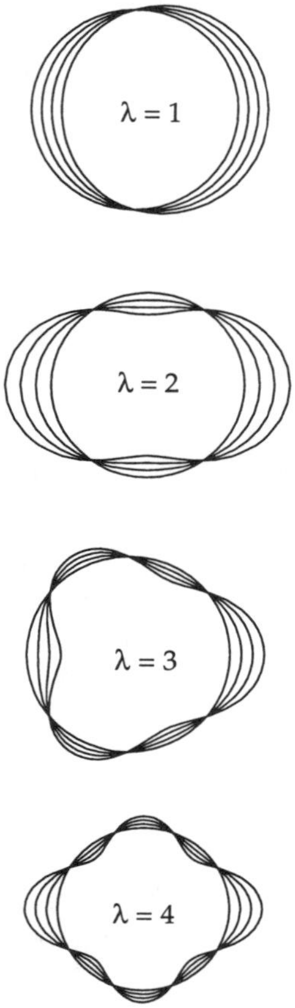

Fig. 6.2 Illustration of the multipole deformations for $\lambda = 1, \ldots, 4$. In each case the shapes are plotted for $\alpha_{\lambda 0} = n a_\lambda$, with $n = 0, \ldots, 3$ and $a_1 = 0.15$, $a_2 = 0.25$, $a_3 = a_4 = 0.15$. The figures are not scaled correctly with respect to each other and the volume is also not kept constant with increasing deformation. Note the complicated shape constant with increasing deformation. Note the complicated shapes for larger deformation.

6.2.2 Types of Multipole Deformations

The general expansion of the nuclear surface in (6.24) allows for arbitrary distortions. In this section the physical meaning of the various multipole orders and their application will be examined for increasing values of λ.

1. The monopole mode, $\lambda = 0$. The spherical harmonic $Y_{00}(\Omega)$ is constant, so that a nonvanishing value of α_{00} corresponds to a change of the radius of the sphere. The associated excitation is the so-called *breathing mode* of the nucleus. Because of the large amount of energy needed for compression of nuclear matter, however, this mode is far too high in energy to be important for the low-energy spectra discussed here.[1] The deformation parameter α_{00} can be used to cancel the overall density change present as a side effect in the other multipole deformations. Calculating the nuclear volume (see Exercise 6.1) shows that this requires

$$\alpha_{00} = -\frac{1}{\sqrt{4\pi}} \sum_{\lambda\mu} |\alpha_{\lambda\mu}|^2 \tag{6.31}$$

to second order.

2. Dipole deformations, $\lambda = 1$ to lowest order, really do not correspond to a deformation of the nucleus but rather to a shift of the center of mass (see Exercise 6.1). Thus to lowest order $\lambda = 1$ corresponds only to a translation of the nucleus, and should be disregarded for nuclear excitations.
3. Quadrupole deformations: the modes with $\lambda = 2$ turn out to be the most important collective excitations of the nucleus. Most of the coming discussion of collective models will be devoted to this case, so a more detailed treatment is given in the next section.
3. The octupole deformations, $\lambda = 3$, are the principal asymmetric modes of the nucleus associated with negative-parity bands. An octupole-deformed shape looks somewhat like a pear.
4. Hexadecupole deformations, $\lambda = 4$: this is the highest angular momentum that has been of any importance in nuclear theory. While there is no evidence for pure hexadecupole excitation in spectra, it seems to play an important role as an admixture to quadrupole excitations and for the ground-state shape of heavy nuclei.
5. Modes with higher angular momentum are of no practical importance. It should be mentioned, though, that there is also a fundamental limitation in λ, because the range of the individual "bumps" on the nuclear surface described by

[1] The compression is seen experimentally and is quite important for determining the compression properties of nuclear matter. The assumption of a sharp surface, though, is not good enough for a quantitative description as its energy depends sensitively on how the density profile on the surface changes during the vibration.

$Y_{\lambda\mu}(\theta,\phi)$ decreases with increasing λ, but it should obviously not be smaller than a nucleon diameter. For a crude estimate it suffices to notice that the number of extrema of $Y_{\lambda\mu}(\Omega)$ is roughly given by λ^2 (the number of zeros of the associated Legendre function in θ times that of $\sin\phi$ or $\cos\phi$). As the number of nucleons on the surface is $A^{2/3}$, we get a limiting value of $\lambda < A^{1/3}$. This shows that even the hexadecupole deformations become marginal below $A = 64$.

An illustration of the multipole deformations for the lowest four angular momenta is given in Fig. 6.2.

EXERCISE

6.1 Volume and Center-of-Mass Vector for a Deformed Nucleus

Problem. Calculate the volume (to second order in the $\alpha_{2\mu}$) and the center-of-mass vector (to first order) for a deformed nucleus with the surface given by (6.24).

Solution. The volume of the nucleus is given by

$$V = \int d\Omega \int_0^{R(\Omega)} dr\, r^2 \quad . \tag{1}$$

Integrating over r and inserting (6.24) yields

$$\begin{aligned} V &= \frac{1}{3} R_0^3 \int d\Omega \left(1 + \sum_{\lambda\mu} \alpha^*_{\lambda\mu} Y_{\lambda\mu}(\Omega)\right)^3 \\ &\approx \frac{1}{3} R_0^3 \int d\Omega \left(1 + 3\sum_{\lambda\mu} \alpha^*_{\lambda\mu} Y_{\lambda\mu}(\Omega) + 3\sum_{\lambda\mu}\sum_{\lambda'\mu'} \alpha^*_{\lambda\mu}\alpha^*_{\lambda'\mu'} Y_{\lambda\mu}(\Omega) Y_{\lambda'\mu'}(\Omega)\right) \\ &= \frac{1}{3} R_0^3 \left(4\pi + 3\sqrt{4\pi}\,\alpha_{00} + 3\sum_{\lambda\mu} |\alpha_{\lambda\mu}|^2\right) \quad , \end{aligned} \tag{2}$$

where terms up to second order in the deformation parameters were kept and then the orthogonality of the spherical harmonics was used. Note that the integral over such a single function can be evaluated using

$$\begin{aligned} \int d\Omega\, Y_{\lambda\mu}(\Omega) &= \sqrt{4\pi} \int d\Omega \frac{1}{\sqrt{4\pi}} Y_{\lambda\mu}(\Omega) \\ &= \sqrt{4\pi} \int d\Omega\, Y_{00}(\Omega)\, Y_{\lambda\mu}(\Omega) = \sqrt{4\pi}\, \delta_{\lambda 0}\delta_{\mu 0} \quad . \end{aligned} \tag{3}$$

The first term in the parentheses in (2) is just the volume of the undeformed nucleus. Thus the volume will be unaffected by the deformation if

$$\sqrt{4\pi}\,\alpha_{00} + \sum_{\lambda\mu} |\alpha_{\lambda\mu}|^2 = 0 \quad . \tag{4}$$

Exercise 6.1

For the center-of-mass vector we have to evaluate

$$\boldsymbol{R}_{\rm cm} = \frac{\int \mathrm{d}^3r\, \boldsymbol{r}\varrho(\boldsymbol{r})}{\int \mathrm{d}^3r\, \varrho(\boldsymbol{r})} \quad , \tag{5}$$

which, for the case of constant density within the volume given by (6.25) and in spherical notation, where $r_\mu = \sqrt{\frac{4\pi}{3}}\, r\, Y_{1\mu}$, reduces to

$$R_{\mathrm{cm},\mu} = \frac{1}{V}\sqrt{\frac{4\pi}{3}} \int \mathrm{d}\Omega \int\limits_0^{R(\Omega)} \mathrm{d}r\, r^2\, r\, Y_{1\mu} \quad . \tag{6}$$

Proceeding in the same way as for the volume we get

$$\begin{aligned} R_{\mathrm{cm},\mu} &= \frac{1}{V}\sqrt{\frac{4\pi}{3}} \int \mathrm{d}\Omega\, \frac{R_0^4}{4}\left(1 + \sum_{\lambda\mu} \alpha_{\lambda\mu} Y^*_{\lambda\mu}\right)^4 Y_{1\mu} \\ &\approx \frac{1}{V}\sqrt{\frac{4\pi}{3}} \int \mathrm{d}\Omega\, \frac{R_0^4}{4}\left(1 + 4\sum_{\lambda\mu} \alpha_{\lambda\mu} Y^*_{\lambda\mu}\right) Y_{1\mu} \\ &= \frac{1}{V}\sqrt{\frac{4\pi}{3}}\, R_0^4\, \alpha_{1\mu} \\ &= \sqrt{\frac{3}{4\pi}}\, R_0\, \alpha_{1\mu} \quad . \end{aligned} \tag{7}$$

6.2.3 Quadrupole Deformations

As mentioned above, these deformations are the most important vibrational degrees of freedom of the nucleus and many of the coming developments will be devoted to this case. In this section we will therefore take a closer look at the significance of the various parameters hidden in the quadrupole deformation tensor $\alpha_{\mu\nu}$.

For the case of pure quadrupole deformation the nuclear surface is given by

$$R(\theta, \phi) = R_0 \left(1 + \sum_\mu \alpha^*_{2\mu} Y_{2\mu}(\theta, \phi)\right) \quad . \tag{6.32}$$

Note that the volume-conserving term in α_{00} is of second order in $\alpha_{2\mu}$ and thus can be safely omitted. The parameters $\alpha_{2\mu}$ are not independent, because (6.26) implies $\alpha_{2\mu} = (-1)^\mu \alpha^*_{2-\mu}$, so that α_{20} is real and we are left with five independent real degrees of freedom: α_{00} and the real and imaginary parts of α_{21} and α_{22}.

To investigate the actual form of the nucleus, it is best to express this in cartesian coordinates by rewriting the spherical harmonics in terms of the cartesian components of the unit vector in the direction (θ, ϕ),

$$\xi = \sin\theta\cos\phi \quad , \quad \eta = \sin\theta\sin\phi \quad , \quad \zeta = \cos\theta \quad , \tag{6.33}$$

which fulfil the subsidiary condition $\xi^2 + \eta^2 + \zeta^2 = 1$:

$$
\begin{aligned}
Y_{20}(\theta,\phi) &= \sqrt{\frac{5}{16\pi}}\left(3\cos^2\theta - 1\right) = \sqrt{\frac{5}{16\pi}}\left(2\zeta^2 - \xi^2 - \eta^2\right) \quad , \\
Y_{2\pm1}(\theta,\phi) &= \mp\sqrt{\frac{15}{8\pi}}\sin\theta\cos\theta\, \mathrm{e}^{\pm \mathrm{i}\phi} = \mp\sqrt{\frac{15}{8\pi}}(\xi\zeta \pm \mathrm{i}\eta\zeta) \quad , \\
Y_{2\pm2}(\theta,\phi) &= \sqrt{\frac{15}{32\pi}}\sin^2\theta\, \mathrm{e}^{\pm 2\mathrm{i}\phi} = \sqrt{\frac{15}{32\pi}}\left(\xi^2 - \eta^2 \pm 2\mathrm{i}\xi\eta\right) \quad .
\end{aligned}
\tag{6.34}
$$

Inserting these into (6.32) yields an expression of the form

$$
\begin{aligned}
R(\xi,\eta,\zeta) = R_0\big(1 &+ \alpha_{\xi\xi}\xi^2 + \alpha_{\eta\eta}\eta^2 + \alpha_{\zeta\zeta}\zeta^2 \\
&+ 2\alpha_{\xi\eta}\xi\eta + 2\alpha_{\xi\zeta}\xi\zeta + 2\alpha_{\eta\zeta}\eta\zeta\big) \quad ,
\end{aligned}
\tag{6.35}
$$

where the cartesian components of the deformation are related to the spherical ones by

$$
\begin{aligned}
\alpha_{2\pm2} &= \frac{1}{2}\sqrt{\frac{8\pi}{15}}\left(\alpha_{\xi\xi} - \alpha_{\eta\eta} \pm 2\mathrm{i}\alpha_{\xi\eta}\right) \quad , \\
\alpha_{2\pm1} &= \sqrt{\frac{8\pi}{15}}\left(\alpha_{\xi\zeta} \pm \mathrm{i}\alpha_{\eta\zeta}\right) \quad , \\
\alpha_{20} &= \sqrt{\frac{8\pi}{15}}\frac{1}{\sqrt{6}}\left(2\alpha_{\zeta\zeta} - \alpha_{\xi\xi} - \alpha_{\eta\eta}\right) \quad .
\end{aligned}
\tag{6.36}
$$

There appear to be six independent cartesian components (all real), compared to the five degrees of freedom contained in the spherical components. However, the function $R(\theta,\phi)$ fulfills

$$
\int R(\Omega)\,\mathrm{d}\Omega = 4\pi R_0 \quad , \tag{6.37}
$$

because the integral over the $Y_{2\mu}(\Omega)$ vanishes. Doing the same integral in the cartesian form is easy: for symmetry reasons the mixed components contribute nothing, while the diagonal ones, from symmetry, should yield

$$
\int \xi^2\,\mathrm{d}\Omega = \int \eta^2\,\mathrm{d}\Omega = \int \zeta^2\,\mathrm{d}\Omega \equiv a \tag{6.38}
$$

(no need to evaluate a), and we have

$$
\int R(\Omega)\,\mathrm{d}\Omega = 4\pi R_0 + a\left(\alpha_{\xi\xi} + \alpha_{\eta\eta} + \alpha_{\zeta\zeta}\right) \quad , \tag{6.39}
$$

so that the cartesian components must fulfil the subsidiary condition

$$
\alpha_{\xi\xi} + \alpha_{\eta\eta} + \alpha_{\zeta\zeta} = 0 \quad . \tag{6.40}
$$

As the cartesian deformations are directly related to the stretching (or contraction) of the nucleus in the appropriate direction, we can read off that

- α_{20} describes a stretching of the z axis with respect to the y and x axes,
- $\alpha_{2\pm2}$ describes the relative length of the x axis compared to the y axis (real part), as well as an oblique deformation in the x-y plane, and
- $\alpha_{2\pm1}$ indicate an oblique deformation of the z axis.

One problem with these parameters is that the symmetry axes of the nucleus (if there are any) can still have an arbitrary orientation in space, so that the shape of the nucleus and its orientation are somehow mixed in the $\alpha_{2\mu}$. The geometry of the situation becomes clearer if this orientation is separated by going into the *principal axis system*. If we denote this new coordinate frame by primed quantities, the cartesian deformation tensor must then be diagonal, so that

$$R(\xi',\eta',\zeta') = R_0\left(1 + \alpha'_{\xi\xi}\xi'^2 + \alpha'_{\eta\eta}\eta'^2 + \alpha'_{\zeta\zeta}\zeta'^2\right) \quad , \tag{6.41}$$

and the condition $\alpha'_{\xi\eta} = \alpha'_{\xi\zeta} = \alpha'_{\eta\zeta} = 0$ implies for the spherical components that

$$\begin{aligned}
\alpha'_{2\pm1} &= 0 \quad , \\
\alpha'_{2\pm2} &= \sqrt{\frac{2\pi}{15}}\,(\alpha'_{\xi\xi} - \alpha'_{\eta\eta}) \equiv a_2 \quad , \\
\alpha'_{20} &= \sqrt{\frac{8\pi}{15}}\frac{1}{\sqrt{6}}\,(2\alpha'_{\zeta\zeta} - \alpha'_{\xi\xi} - \alpha'_{\eta\eta}) \equiv a_0 \quad .
\end{aligned} \tag{6.42}$$

There are still five independent real parameters, but now with an even clearer geometrical significance:

- a_0, indicating the stretching of the z' axis with respect to the x' and y' axes;
- a_2, which determines the difference in length between the x' and y' axes; and
- three Euler angles $\boldsymbol{\theta} = (\theta_1, \theta_2, \theta_3)$, which determine the orientation of the principal axis system (x', y', z') with respect to the laboratory-fixed frame (x, y, z).

The advantage of the principal axis system is that rotation and shape vibration are clearly separated; a change in the Euler angles denotes a pure rotation of the nucleus without any change in its shape, which is only determined by a_0 and a_2.[2] Note also that $a_2 = 0$ describes a shape with equal axis lengths in the x and y directions, i.e., one with axial symmetry around the z axis.

There is also another set of parameters introduced by A. Bohr [Bo52, Bo54]. It corresponds to something like polar coordinates in the space of (a_0, a_2) and is defined via

$$a_0 = \beta\cos\gamma, \qquad a_2 = \frac{1}{\sqrt{2}}\beta\sin\gamma \quad . \tag{6.43}$$

The factor $\frac{1}{\sqrt{2}}$ was chosen such that

[2] It is interesting to note that this is is exclusively possible for the quadrupole deformation, which contains just the right number of parameters to fix three axis directions and their relative lengths. In comparison, the dipole deformation determines only one axis direction and length, while the octupole fixes too many.

$$\sum_\mu |\alpha_{2\mu}|^2 = \sum_\mu |\alpha'_{2\mu}|^2 = a_0^2 + 2a_2^2 = \beta^2 \quad . \tag{6.44}$$

This particular sum over the components of $\alpha_{2\mu}$ is rotationally invariant since

$$\begin{aligned}\sum_\mu |\alpha_{2\mu}|^2 &= \sum_\mu (-1)^\mu \alpha_{2\mu}\alpha_{2-\mu} \\ &= \sqrt{5}\sum_{\mu\mu'}(220|\mu\mu'0)\ \alpha_\mu\alpha_{\mu'} = \sqrt{5}\,[\alpha_2 \times \alpha_2]^0 \quad ,\end{aligned} \tag{6.45}$$

and so has the same value in the laboratory and the principal axes systems, as already indicated in (6.44).

EXERCISE

6.2 Quadrupole Deformations

Problem. Describe the nuclear shapes in the principal axes system as a function of γ for fixed β.

Solution. To see the shape of the nucleus, calculate the cartesian components in terms of β and γ. First the condition (6.32) for the primed components yields

$$\alpha'_{\zeta\zeta} = -\alpha'_{\xi\xi} - \alpha'_{\eta\eta} \quad , \tag{1}$$

which when inserted into the definition of a_0 in (6.42), with the use of (6.43) leads to:

$$\alpha'_{\zeta\zeta} = \frac{\sqrt{6}}{3}\sqrt{\frac{15}{8\pi}}\ a_0 = \sqrt{\frac{5}{4\pi}}\ \beta\cos\gamma \quad . \tag{2}$$

Again from (6.40) we have

$$\alpha'_{\xi\xi} - \alpha'_{\eta\eta} = 2\alpha'_{\xi\xi} + \alpha'_{\zeta\zeta} \quad , \tag{3}$$

and this may be used in the definition of a_2 from (6.42) to yield

$$\alpha'_{\xi\xi} = \sqrt{\frac{15}{8\pi}}\left(a_2 - \frac{1}{\sqrt{6}}a_0\right) = \sqrt{\frac{5}{4\pi}}\ \beta\left(\tfrac{1}{2}\sqrt{3}\sin\gamma - \tfrac{1}{2}\cos\gamma\right) \quad . \tag{4}$$

The terms in parentheses can be combined using the addition theorem for cosines:

$$\alpha'_{\xi\xi} = \sqrt{\frac{5}{4\pi}}\ \beta\cos\left(\gamma - \tfrac{2\pi}{3}\right) \quad . \tag{5}$$

In a similar way

$$\alpha'_{\eta\eta} = -\sqrt{\frac{5}{4\pi}}\beta\left(\tfrac{1}{2}\sqrt{3}\sin\gamma + \tfrac{1}{2}\cos\gamma\right) = \sqrt{\frac{5}{4\pi}}\ \beta\cos(\gamma - 4\pi/3) \quad . \tag{6}$$

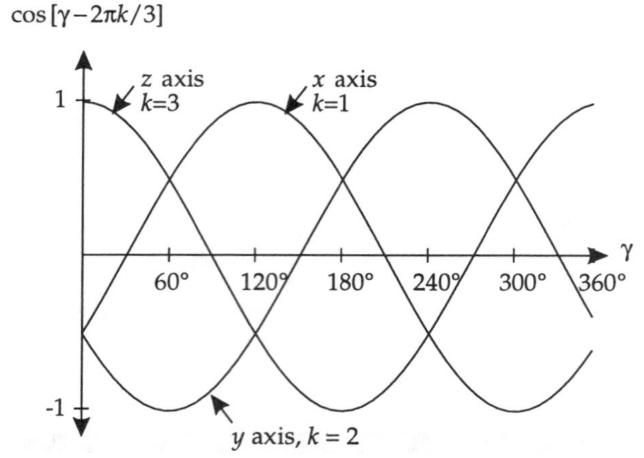

Fig. 6.3 Plot of the functions $\beta\cos(\gamma - 2\pi k/3)$ for k=1, 2, and 3, corresponding to the increase in the axis lengths in the x, y, and z directions.

The cartesian deformation components indicate the stretching of the nuclear axis in that direction. Using the new notation δR_k for these, where $k = 1, 2, 3$ corresponds to the x', y', and z' directions, respectively, one may combine these results into one equation:

$$\delta R_k = \sqrt{\frac{5}{4\pi}}\, \beta \cos\left(\gamma - \tfrac{2\pi k}{3}\right) \quad . \tag{7}$$

Figure 6.3 makes it easy to understand the variations of the three axes with γ. At $\gamma = 0°$ the nucleus is elongated along the z' axis, but the x' and y' axes are equal. This axially symmetric type of shape is somewhat reminiscent of a cigar and is called *prolate*. As we increase γ, the x' axis grows at the expense of the y' and z' axes through a region of *triaxial* shapes with three unequal axes, until axial symmetry is again reached at $\gamma = 60°$, but now with the z' and x' axes equal in length. These two axes are longer than the y' axis: the nucleus has a flat, pancake-like shape, which is called *oblate*. This pattern is repeated: every 60° axial symmetry recurs and prolate and oblate shapes alternate, but with the axes permuted in their relative length.

The results of Exercise 6.2 are again summarized in Fig. 6.4, which shows the various nuclear shapes in the (β, γ) plane and how they are repeated every 60°.

A problem that is apparent from this figure is that identical nuclear shapes occur repeatedly in the plane. For example, the oblate axially symmetric shapes at 60°, 180°, and 300° are identical: only the naming of the axes is different. Triaxial shapes, even, are represented six times in the plane. This is more clearly seen in Fig. 6.3; for example, identical shapes occur on both sides of $\gamma = 60°$ and are repeated at $\gamma = 180°$ and $\gamma = 300°$. Because the axis orientations are different, the associated Euler angles also differ. In conclusion, *the same physical shape (including its orientation in space) can be represented by different sets of deformation parameters* (β, γ) *and Euler angles.*

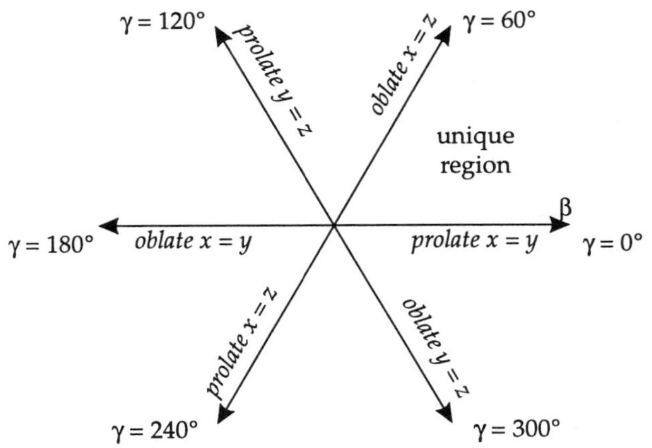

Fig. 6.4 The (β,γ) plane is divided into six equivalent parts by the symmetries; the sector between 0° and 60° contains all shapes uniquely and may be taken as the representative one. The types of shapes encountered along the axes are indicated verbally; e.g., *prolate x=y* implies prolate shapes with the z' axis as the long axis and the two other axes equal.

6.2.4 Symmetries in Collective Space

In the previous section two alternative approaches to developing a collective model were suggested: the quadrupole deformation may be described either in a laboratory-fixed reference frame through the spherical tensor $\alpha_{2\mu}$, or, alternatively, by giving the deformation of the nucleus with respect to the principal axes frame using the parameters (a_0,a_2) or (β,γ) and the Euler angles $(\theta_1,\theta_2,\theta_3)$ indicating the instantaneous orientation of that body-fixed frame. Both cases require different treatments of rotational symmetry.

In the laboratory frame the kinetic and potential energies of the nucleus must be a function of the $\alpha_{\lambda\mu}$ and the velocities $\alpha_{\lambda\mu}$. They should be rotationally invariant, and this can be fulfilled by constructing them in a manifestly invariant way. For example, a deformation-dependent potential may contain powers of $\alpha_{\lambda\mu}$ coupled to zero total angular momentum,

$$V(\alpha_{\lambda\mu}) = C\ [\alpha_2 \times \alpha_2]^0 + D\ \left[\alpha_2 \times [\alpha_2 \times \alpha_2]^2\right]^0 + \dots \quad , \tag{6.46}$$

but also more complicated functions of scalars, such as

$$\sqrt{[\alpha_2 \times \alpha_2]^0} \sim \beta \quad . \tag{6.47}$$

The kinetic energy should be constructed in a similar way. The wave functions then also have to be built up as functions with good angular momentum $\psi_{IM}(\alpha_{\lambda\mu})$. As will be shown later in the section about the harmonic quadrupole vibrator, this is done most easily in second quantization.

In the principal-axes frame, which is often called the *intrinsic system*, on the other hand, rotational invariance is even easier to ensure: the energy must be independent of the Euler angles. The time derivatives of these angles of course can and should appear in the rotational kinetic energy. The deformation energy depends on β and γ exclusively. We thus expect an energy expression of the form

$$T(\beta, \gamma, \beta, \gamma, \theta_1, \theta_2, \theta_3) + V(\beta, \gamma) \quad . \tag{6.48}$$

For the wave functions, however, there are problems because of the nonuniqueness of the coordinate values as shown in the preceding section. There are in general six different points in the (β,γ) plane describing the same physical shape of the nucleus, plus an additional ambiguity in the Euler angles. It is thus necessary to examine the problem more closely.

Consider how many ways there are to choose an intrinsic coordinate system (again denoted by primed coordinates) that has its axes along the principal axes of the nucleus. Three steps are necessary to select such a system.

1. Choose the z' axis along any of the three principal axes, pointing in a positive or negative direction along that axis; there are six possibilities.
2. Now fix the y' axis by selecting one of the two remaining axes and again positive or negative orientation; four choices are possible.
3. The x' axis now is completely determined by the requirement of having a right-handed coordinate system.

The total number of different coordinate systems thus is 24.Mathematically they can be enumerated systematically by taking one of them as the "standard" system and constructing the others through transformations that perform a suitable permutation of the axes and the associated transformation of the collective coordinates. It is easy to convince oneself that all 24 of these transformations can be built up out of three basic ones $\hat{R}_k$, $k = 1,2,3$, which are usually chosen as follows.

$\hat{R}_1$: $\hat{R}_1(x',y',z') = (x',-y',-z')$. This corresponds to a rotation about the x' axis for an angle of π. It does not change β and γ, because the stretching of the nucleus along each axis is not changed by reflecting that axis. The Euler angles, however, are affected: $\hat{R}_1(\theta_1,\theta_2,\theta_3) = (\theta_1+\pi,\pi-\theta_2,-\theta_3)$. Note that $\hat{R}_1^2 = 1$.

$\hat{R}_2$: $\hat{R}_2(x',y',z') = (y',-x',z')$; a rotation of $\frac{1}{2}\pi$ about the z' axis. Therefore, $\hat{R}_2^4 = 1$. Examining the corresponding interchange of axes $x' \leftrightarrow y'$ in Fig. 6.3 shows that the effect on the deformation parameters is $\hat{R}_2(\beta,\gamma) = (\beta,-\gamma)$. The Euler angles obviously show only the rotation about the z' axis: $\hat{R}_2(\theta_1,\theta_2,\theta_3) = (\theta_1,\theta_2,\theta_3+\frac{\pi}{2})$.

$\hat{R}_3$: $\hat{R}_3(x',y',z') = (y',z',x')$, i.e., a cyclic permutation of the axes, which implies $\hat{R}_3^3 = 1$. In Fig. 6.3 this corresponds to $\hat{R}_3(\beta,\gamma) = (\beta,\gamma+\frac{2\pi}{3})$. Again, as for $\hat{R}_1$, it is not so trivial to derive the change in the Euler angles: $\hat{R}_3(\theta_1,\theta_2,\theta_3) = \left(\theta_1,\theta_2+\frac{\pi}{2},\theta_3+\frac{\pi}{2}\right)$.

The existence of these transformations, which describe ambiguities in the choice of the intrinsic system, now has an important consequence for collective wave functions: *a collective wave function must have the same value at all 24 points that are related by the 24 choices of the coordinate system.* Phrased in terms of the basic transformations, a collective wave function must fulfil:

$$
\begin{aligned}
(\hat{R}_1):\quad & \psi(\beta,\gamma,\theta_1,\theta_2,\theta_3) = \psi(\beta,\gamma,\theta_1+\pi,\pi-\theta_2,-\theta_3) \quad , \\
(\hat{R}_2):\quad & \psi(\beta,\gamma,\theta_1,\theta_2,\theta_3) = \psi\left(\beta,-\gamma,\theta_1,\theta_2,\theta_3+\tfrac{\pi}{2}\right) \quad , \\
(\hat{R}_3):\quad & \psi(\beta,\gamma,\theta_1,\theta_2,\theta_3) = \psi\left(\beta,\gamma+\tfrac{2\pi}{3},\theta_1,\theta_2+\tfrac{\pi}{2},\theta_3+\tfrac{\pi}{2}\right) \quad .
\end{aligned}
\tag{6.49}
$$

It is important to understand that this has nothing to do with a physical symmetry like, for example, parity. In that case, there are two physically distinct points in space, $\boldsymbol{r}$ and $-\boldsymbol{r}$, and if the Hamiltonian is parity invariant, we can choose wave functions fulfilling $\psi(\boldsymbol{r}) = \pi\psi(-\boldsymbol{r})$ with a parity quantum number π.

In contrast here, even for $\hat{R}_1$, which is very similar to parity in that it also fulfills $\hat{R}_1^2 = 1$, there is no parity-like quantum number. The difference is that the 24 different sets of coordinates do not describe physically related but distinct situations as $\boldsymbol{r}$ and $-\boldsymbol{r}$ do, but describe *one and the same physical configuration*. The different points in collective coordinate space correspond to only one physical situation and there is only one unique value of the wave function associated with all of them. So the term "symmetry" is used in a different sense here: there is no physical symmetry in the sense of that used in Chap. 2 but rather a degeneracy of the coordinates.

6.3 Surface Vibrations

6.3.1 Vibrations of a Classical Liquid Drop

It is instructive to examine the vibrations of a charged liquid drop in classical physics, because it gives particular insight into the origin of the important forces and their balance. The total energy of a charged drop with a sharp surface can be split up into kinetic energy, Coulomb energy, and surface energy:

$$E = T + E_C + E_S \quad , \tag{6.50}$$

where each term will depend on the deformation parameters $\alpha_{\lambda\mu}$ and the kinetic energy should also contain the collective velocities $\alpha_{\lambda\mu}$. Calculating these terms to reasonable complexity is possible in the limit of small $\alpha_{\lambda\mu}$. Expecting harmonic vibrations as the crudest approximation, we will keep terms up to second order. Let us now evaluate the contributions to the energy one by one.

The Coulomb energy of a proton charge distribution with density $\varrho(\boldsymbol{r})$ is in general given by

$$E_C = \frac{1}{2}\int d^3r \int d^3r' \, \frac{\varrho(\boldsymbol{r})\,\varrho(\boldsymbol{r}')}{|\boldsymbol{r}-\boldsymbol{r}'|} \quad . \tag{6.51}$$

For a nucleus with constant charge density equal to zero outside the nucleus and to ϱ_q inside, this can be rewritten as

$$E_C = \frac{1}{2}\varrho_q^2 \int_V d^3r \int_V d^3r' \frac{1}{|\boldsymbol{r}-\boldsymbol{r}'|} \quad , \tag{6.52}$$

or, in spherical coordinates,

$$E_C = \frac{1}{2}\varrho_q^2 \int d\Omega \int d\Omega' \int_0^{R(\Omega)} dr\, r^2 \int_0^{R(\Omega')} dr'\, r'^2 \frac{1}{|\boldsymbol{r}-\boldsymbol{r}'|} \quad . \tag{6.53}$$

The upper bounds for the radial integrations depend on the angle and are given by $R(\Omega)$ according to (6.53).

It is useful to split up the integrals into the energy of the unperturbed sphere, given by integration up to a constant radius of R_0, and the varying parts. Formally this can be achieved by writing

$$\int_0^{R(\Omega)} \mathrm{d}r \int_0^{R(\Omega')} \mathrm{d}r' = \left(\int_0^{R_0} \mathrm{d}r + \int_{R_0}^{R(\Omega)} \mathrm{d}r\right)\left(\int_0^{R_0} \mathrm{d}r' + \int_{R_0}^{R(\Omega')} \mathrm{d}r'\right) \quad , \tag{6.54}$$

and this can be recast into

$$\begin{aligned} &\int_0^{R_0} \mathrm{d}r \int_0^{R_0} \mathrm{d}r' + \int_0^{R_0} \mathrm{d}r \int_{R_0}^{R(\Omega')} \mathrm{d}r' + \int_{R_0}^{R(\Omega)} \mathrm{d}r \int_0^{R_0} \mathrm{d}r' + \int_{R_0}^{R(\Omega)} \mathrm{d}r \int_{R_0}^{R(\Omega')} \mathrm{d}r' \\ &\quad = \int_0^{R_0} \mathrm{d}r \int_0^{R_0} \mathrm{d}r' + 2 \int_{R_0}^{R(\Omega)} \mathrm{d}r \int_0^{R_0} \mathrm{d}r' + \int_{R_0}^{R(\Omega)} \mathrm{d}r \int_{R_0}^{R(\Omega')} \mathrm{d}r' \quad , \end{aligned} \tag{6.55}$$

where the last step utilized the interchangeability of the dummy variables r and r'. Inserting this into (6.53) yields

$$\begin{aligned} E_\mathrm{C} = E_\mathrm{C}^{(0)} &+ \varrho_q^2 \int \mathrm{d}\Omega \int \mathrm{d}\Omega' \int_{R_0}^{R(\Omega)} \mathrm{d}r\, r^2 \int_0^{R_0} \mathrm{d}r'\, r'^2 \frac{1}{|\boldsymbol{r} - \boldsymbol{r}'|} \\ &+ \frac{1}{2}\varrho_q^2 \int \mathrm{d}\Omega \int \mathrm{d}\Omega' \int_{R_0}^{R(\Omega)} \mathrm{d}r\, r^2 \int_{R_0}^{R(\Omega')} \mathrm{d}r'\, r'^2 \frac{1}{|\boldsymbol{r} - \boldsymbol{r}'|} \quad . \end{aligned} \tag{6.56}$$

The Coulomb energy of the undistorted sphere was abbreviated as

$$\begin{aligned} E_\mathrm{C}^{(0)} &= \frac{1}{2}\varrho_q^2 \int \mathrm{d}\Omega \int \mathrm{d}\Omega' \int_0^{R_0} \mathrm{d}r\, r^2 \int_0^{R_0} \mathrm{d}r'\, r'^2 \frac{1}{|\boldsymbol{r} - \boldsymbol{r}'|} \\ &= \frac{3}{5} \frac{Z^2 e^2}{R_0} \quad . \end{aligned} \tag{6.57}$$

Z is the charge number of the nucleus and is related to ϱ_q via

$$Ze = \varrho_q \frac{4\pi}{3} R_0^3 \quad . \tag{6.58}$$

To proceed further, the addition theorem for spherical harmonics is applied:

$$\frac{1}{|\boldsymbol{r} - \boldsymbol{r}'|} = 4\pi \sum_{\lambda=0}^{\infty} \sum_{\mu=-\lambda}^{\lambda} \frac{1}{2\lambda + 1} \frac{r_<^\lambda}{r_>^{\lambda+1}} Y_{\lambda\mu}^*(\theta, \phi)\, Y_{\lambda\mu}(\theta', \phi') \quad . \tag{6.59}$$

The expressions $r_<$ and $r_>$ stand for the smaller and larger of the two radii r and r', respectively. Now the first four-fold integral in (6.56) becomes

$$4\pi\varrho_q^2 \sum_{\lambda\mu} \frac{1}{2\lambda + 1} \int \mathrm{d}\Omega'\, Y_{\lambda\mu}(\Omega') \int \mathrm{d}\Omega\, Y_{\lambda\mu}^*(\Omega) \int_{R_0}^{R(\Omega)} \mathrm{d}r\, r^2 \int_0^{R_0} \mathrm{d}r'\, r'^2 \frac{1}{r_>} \quad . \tag{6.60}$$

The integration over Ω' can be done using

$$\int \mathrm{d}\Omega' \, Y_{\lambda\mu}(\Omega') = \sqrt{4\pi} \int \mathrm{d}\Omega' Y_{00}^* \, Y_{\lambda\mu}(\Omega') = \sqrt{4\pi}\, \delta_{\lambda,0}\, \delta_{\mu,0} \quad , \tag{6.61}$$

because of $Y_{00}(\Omega) = 1/\sqrt{4\pi}$ and the orthonormalization of the spherical harmonics. Thus the sum over λ and μ collapses to the term with $\lambda = 0$, $\mu = 0$ and the remaining function $Y_{00}(\Omega)$ yields only a factor $1/\sqrt{4\pi}$, reducing the integral to

$$4\pi \varrho_q^2 \int \mathrm{d}\Omega \int_{R_0}^{R(\Omega)} \mathrm{d}r \, r^2 \int_0^{R_0} \mathrm{d}r' \, r'^2 \frac{1}{r_>} \quad . \tag{6.62}$$

To perform the integration over r' we have to split up the integral into two parts because of the definition of $r_>$:

$$\begin{aligned} \int_0^{R_0} \mathrm{d}r' \, r'^2 \frac{1}{r_>} &= \int_0^{r} \mathrm{d}r' \, r'^2 \frac{1}{r} + \int_r^{R_0} \mathrm{d}r' \, r'^2 \frac{1}{r'} \\ &= \tfrac{1}{3} r^2 + \tfrac{1}{2} R_0^2 - \tfrac{1}{2} r^2 = \tfrac{1}{2}\left(R_0^2 - \tfrac{1}{3} r^2\right) \quad . \end{aligned} \tag{6.63}$$

The integration over r has also become simple now:

$$\int_{R_0}^{R(\Omega)} \mathrm{d}r \, r^2 \tfrac{1}{2}\left(R_0^2 - \tfrac{1}{3} r^2\right) = \frac{1}{6}\left(R_0^2 R^2(\Omega) - R_0^5 - \tfrac{1}{5} R^5(\Omega) - \tfrac{1}{5} R_0^5\right) \quad . \tag{6.64}$$

When we expand the difference between R_0 and $R(\Omega)$ to second order in $\alpha_{\lambda\mu}$, the full integral now becomes

$$\begin{aligned} 4\pi \varrho_q^2 &\int \mathrm{d}\Omega \int_{R_0}^{R(\Omega)} \mathrm{d}r \, r^2 \int_0^{R_0} \mathrm{d}r' \, r'^2 \frac{1}{r_>} \\ &\approx \frac{2}{3} \pi R_0^5 \varrho_q^2 \int \mathrm{d}\Omega \Big(2 \sum_{\lambda\mu} \alpha_{\lambda\mu}^* Y_{\lambda\mu}(\Omega) \\ &\quad + \sum_{\lambda\mu} \sum_{\lambda'\mu'} \alpha_{\lambda\mu}^* Y_{\lambda\mu}(\Omega)\, \alpha_{\lambda'\mu'} Y_{\lambda'\mu'}^*(\Omega) \Big) \\ &= \frac{2}{3} \pi R_0^5 \varrho_q^2 \Big(2\sqrt{4\pi}\, \alpha_{00} + \sum_{\lambda\mu} \alpha_{\lambda\mu}^* \alpha_{\lambda\mu} \Big) \\ &= -\frac{2}{3} \pi R_0^5 \varrho_q^2 \sum_{\lambda\mu} |\alpha_{\lambda\mu}|^2 \quad . \end{aligned} \tag{6.65}$$

Here the upper limit of the integration was inserted from (6.32), and then again the orthonormalization of the spherical harmonics was used. Finally the relation $\sqrt{4\pi}\alpha_{00} + \sum_{\lambda\mu} |\alpha_{\lambda\mu}|^2 = 0$ derived in Exercise 6.1 was inserted.

The second integral in (6.56) can be evaluated quite similarly. Keeping only terms up to second order in $\alpha_{\lambda\mu}$ leads to

$$\frac{1}{2}\varrho_q^2 \int \mathrm{d}\Omega \int \mathrm{d}\Omega' \int_{R_0}^{R(\Omega)} \mathrm{d}r\, r^2 \int_{R_0}^{R(\Omega')} \mathrm{d}r'\, r'^2 \frac{1}{|\boldsymbol{r}-\boldsymbol{r}'|}$$

$$= 2\pi R_0^5 \varrho_q^2 \sum_{\lambda\mu} \frac{\lambda}{2\lambda+1} |\alpha_{\lambda\mu}|^2 \quad . \tag{6.66}$$

Combining all of these terms we get the Coulomb energy to second order:

$$E_\mathrm{C} = E_\mathrm{C}^{(0)} \left(1 - \frac{5}{4\pi} \sum_{\lambda\mu} \frac{\lambda-1}{2\lambda+1} |\alpha_{\lambda\mu}|^2 \right) \quad . \tag{6.67}$$

Thus any small deformation from a sphere lowers the electrostatic energy of the nucleus (disregard the trivial case $\lambda = 1$, which in this order corresponds to translation and does not therefore affect the potential energy).

The surface energy of the nucleus E_S is given by

$$E_\mathrm{S} = \sigma \int_\mathrm{surface} \mathrm{d}S \quad , \tag{6.68}$$

with σ the surface tension and $\mathrm{d}S$ the surface element, in spherical coordinates equal to

$$\mathrm{d}S = \sqrt{1 + \frac{1}{R^2}\left(\frac{\partial R}{\partial \theta}\right)^2 + \frac{1}{R^2 \sin^2\theta}\left(\frac{\partial R}{\partial \phi}\right)^2}\, R^2 \sin\theta\, \mathrm{d}\theta\, \mathrm{d}\phi \quad . \tag{6.69}$$

If the multipole order is not too large, the derivatives should be small also and one may expand to first order to get

$$\mathrm{d}S \approx \left[R^2 + \frac{1}{2}\left(\frac{\partial R}{\partial \theta}\right)^2 + \frac{1}{2\sin^2\theta}\left(\frac{\partial R}{\partial \phi}\right)^2 \right] \sin\theta\, \mathrm{d}\theta\, \mathrm{d}\phi \quad . \tag{6.70}$$

Here again the definition of the nuclear surface according to (6.25) has to be inserted. To abbreviate the resulting formulas we use the notation

$$\eta = R_0 \sum_{\lambda\mu} \alpha_{\lambda\mu}^* Y_{\lambda\mu}(\Omega) \quad , \tag{6.71}$$

and with this the surface element becomes

$$\mathrm{d}S = \left\{ 1 + 2\eta + \eta^2 + \tfrac{1}{2}R_0^2 \left[\left(\frac{\partial \eta}{\partial \theta}\right)^2 + \frac{1}{\sin^2\theta}\left(\frac{\partial \eta}{\partial \phi}\right)^2 \right] \right\} \sin\theta\, \mathrm{d}\theta\, \mathrm{d}\phi \quad . \tag{6.72}$$

In analogy to the Coulomb energy we separate the surface energy of the unperturbed sphere

$$E_\mathrm{S}^{(0)} = 4\pi\sigma R_0^2 \quad , \tag{6.73}$$

to rewrite the surface energy as

$$E_S = E_S^{(0)} + \frac{1}{2}\sigma \int \sin\theta \, d\theta \, d\phi \left\{ -2\eta^2 + R_0^2 \left[\left(\frac{\partial \eta}{\partial \theta}\right)^2 + \frac{1}{\sin^2\theta}\left(\frac{\partial \eta}{\partial \phi}\right)^2 \right] \right\}$$

$$= E_S^{(0)} + \frac{1}{2}\sigma \int \sin\theta \, d\theta \, d\phi \left[-2\eta^2 + R_0^2 \sum_{\lambda\mu}\sum_{\lambda'\mu'} \alpha^*_{\lambda\mu}\alpha_{\lambda'\mu'} \right.$$

$$\left. \times \left(\frac{\partial Y_{\lambda\mu}}{\partial \theta}\frac{\partial Y^*_{\lambda'\mu'}}{\partial \theta} + \frac{1}{\sin^2\theta}\frac{\partial Y_{\lambda\mu}}{\partial \phi}\frac{\partial Y^*_{\lambda'\mu'}}{\partial \phi} \right) \right] \quad , \tag{6.74}$$

where the argument Ω of the spherical harmonics was suppressed for conciseness.

The aim now is to rewrite the derivatives of the spherical harmonics in such a way as to exploit some well-known simplifying relation. In fact this can be achieved by partial integration. For example, the θ integration in the first term may be rewritten

$$\int_0^\pi d\theta \left(\frac{\partial Y_{\lambda\mu}}{\partial \theta} \sin\theta \right) \frac{\partial Y^*_{\lambda'\mu'}}{\partial \theta}$$

$$= - \int_0^\pi \frac{\partial}{\partial \theta} d\theta \left(\frac{\partial Y_{\lambda\mu}}{\partial \theta}\sin\theta \right) Y^*_{\lambda'\mu'} + \left(\frac{\partial Y_{\lambda\mu}}{\partial \theta}\right)^2 \sin\theta \Big|_0^\pi$$

$$= - \int_0^\pi d\theta \left(\frac{\partial^2 Y_{\lambda\mu}}{\partial \theta^2} Y^*_{\lambda'\mu'} \sin\theta + \frac{\partial Y_{\lambda\mu}}{\partial \theta} Y^*_{\lambda'\mu'} \cos\theta \right) \quad . \tag{6.75}$$

Similar manipulations with the term including the ϕ derivatives finally lead to

$$E_S = E_S^{(0)} + \frac{1}{2}\sigma R_0^2 \sum_{\lambda\mu}\sum_{\lambda'\mu'} \left\{ \alpha^*_{\lambda\mu}\alpha_{\lambda'\mu'} \int d\Omega \right.$$

$$\left. \times \left[-2Y_{\lambda\mu}Y^*_{\lambda'\mu'} - Y^*_{\lambda'\mu'} \left(\frac{\partial^2}{\partial \theta^2} + \cos\theta \frac{\partial}{\partial \theta} + \frac{1}{\sin^2\theta}\frac{\partial^2}{\partial \phi^2} \right) Y_{\lambda\mu} \right] \right\} . \tag{6.76}$$

The differential operator appearing in this form is almost identical to the angular-momentum operator (a factor $-\hbar^2$ is missing), so we can utilize the eigenvalue equation

$$\left(\frac{\partial^2}{\partial \theta^2} + \cos\theta \frac{\partial}{\partial \theta} + \frac{1}{\sin^2\theta}\frac{\partial^2}{\partial \phi^2} \right) Y_{\lambda\mu}(\Omega) = -\lambda(\lambda+1) Y_{\lambda\mu}(\Omega) \tag{6.77}$$

to get the final result for the surface energy:

$$E_S = E_S^{(0)} + \frac{1}{2}\sigma R_0^2 \sum_{\lambda\mu}\sum_{\lambda'\mu'} \left\{ \alpha^*_{\lambda\mu}\alpha_{\lambda'\mu'}\delta_{\lambda\lambda'}\delta_{\mu\mu'}[\lambda(\lambda+1) - 2] \right\}$$

$$= E_S^{(0)} \left[1 + \frac{1}{8\pi} \sum_{\lambda\mu} (\lambda - 1)(\lambda+2)|\alpha_{\lambda\mu}|^2 \right] \quad . \tag{6.78}$$

Thus the surface energy shows a quadratic rise with deformation, which increases strongly with rising angular momentum, which is natural because more numerous smaller structures on the surface for higher angular momentum should lead to a stronger increase in the surface area. Again for $\lambda = 1$ the restriction to a pure translation leads to no change in the surface.

Finally the kinetic energy needs to be determined. This requires some additional assumption about what motion of the nuclear material is associated with a motion of the surface. Keeping to the spirit of the fluid model, we assume that there will be a local velocity field of the matter $\boldsymbol{v}(\boldsymbol{r}, t)$ and that the kinetic energy of the flow is given by

$$T = \int \mathrm{d}^3r \, \tfrac{1}{2} \varrho_m \, v^2(\boldsymbol{r}, t) \quad . \tag{6.79}$$

Here ϱ_m is the *mass density* inside the nucleus. The velocity field has to obey the boundary condition that at the nuclear surface it should agree with the velocity of the surface. Unfortunately this is not sufficient to fix the field; an additional assumption about the dynamics is necessary. In fluid mechanics the simplest natural case is that of an ideal, i.e., nonviscous, fluid for which the flow is irrotational,

$$\nabla \times \boldsymbol{v} = 0 \quad . \tag{6.80}$$

This *irrotational flow model* is often used in nuclear physics; although it is certainly not justified quantitatively (the viscosity of nuclear matter should be quite large at low excitation energies and the shell effects should also distort the dynamical behavior quite drastically), it provides a convenient limiting case to which more detailed pictures of nuclear dynamics can be compared.

For irrotational flow the velocity field can be expressed in terms of a potential,

$$\boldsymbol{v}(\boldsymbol{r}, t) = \nabla \Phi(\boldsymbol{r}, t) \quad , \tag{6.81}$$

and the assumed incompressibility of nuclear matter also demands

$$\nabla \cdot \boldsymbol{v}(\boldsymbol{r}, t) = 0 \quad . \tag{6.82}$$

Combining these two conditions we find a Laplace equation for the potential:

$$\Delta \Phi(\boldsymbol{r}, t) = 0 \quad . \tag{6.83}$$

In spherical coordinates the general solution, regular at the origin, is given by

$$\Phi(r, \theta, \phi, t) = \sum_{\lambda\mu} A_{\lambda\mu}(t) \, r^\lambda \, Y_{\lambda\mu}(\theta, \phi) \quad . \tag{6.84}$$

The coefficients $A_{\lambda\mu}(t)$ have to be determined from the boundary condition

$$\frac{\partial}{\partial t} R(\theta, \phi, t) = v_r \Big|_{r=R(\theta,\phi,t)} = \frac{\partial}{\partial r} \Phi(r, \theta, \phi) \Big|_{r=R(\theta,\phi,t)} \tag{6.85}$$

which expresses the equality of the radial component of the velocity on the surface to the velocity of the time-dependent surface itself (in principle the components should be evaluated along the surface normal; for small vibrations though, this deviates only slightly from the radial direction). Inserting the definition of the surface from (6.25) and the expansion of (6.84) leads to

$$R_0 \sum_{\lambda\mu} \alpha^*_{\lambda\mu} Y_{\lambda\mu}(\theta,\phi) = \sum_{\lambda\mu} A_{\lambda\mu} \lambda R^{\lambda-1}(\theta,\phi,t) Y_{\lambda\mu}(\theta,\phi) \quad . \tag{6.86}$$

For small oscillations we can replace $R(\theta,\phi,t)$ by R_0 on the right-hand side and then solve for the coefficients:

$$A_{\lambda\mu} = \frac{1}{\lambda} R_0^{2-\lambda} \alpha^*_{\lambda\mu} \quad . \tag{6.87}$$

These can now be inserted into the kinetic energy, which in spherical coordinates is given by

$$\begin{aligned} T &= \tfrac{1}{2}\varrho_m \int \mathrm{d}^3 r \; |\nabla \Phi|^2 \\ &= \tfrac{1}{2}\varrho_m \int \sin\theta \,\mathrm{d}\theta\,\mathrm{d}\phi\,\mathrm{d}r\, r^2 \left[\left(\frac{\partial \Phi}{\partial r}\right)^2 + \frac{1}{r^2}\left(\frac{\partial \Phi}{\partial \theta}\right)^2 + \frac{1}{r^2 \sin^2\theta}\left(\frac{\partial \Phi}{\partial \phi}\right)^2 \right] \end{aligned} \tag{6.88}$$

to yield

$$\begin{aligned} T = \tfrac{1}{2}\varrho_m \sum_{\lambda\mu} \sum_{\lambda'\mu'} A_{\lambda\mu} A^*_{\lambda'\mu'} \int_0^{R(\Omega)} \mathrm{d}r \, r^{\lambda+\lambda'} \int \mathrm{d}\Omega \\ \times \left(\lambda\lambda' Y_{\lambda\mu} Y_{\lambda'\mu'} + \frac{\partial Y^*_{\lambda\mu}}{\partial \theta} \frac{\partial Y_{\lambda'\mu'}}{\partial \theta} + \frac{1}{\sin^2\theta} \frac{\partial Y^*_{\lambda\mu}}{\partial \phi} \frac{\partial Y_{\lambda'\mu'}}{\partial \phi} \right) \quad . \end{aligned} \tag{6.89}$$

Again because of the smallness of the oscillations the upper boundary of the integral is set equal to R_0; then, like in the case of the surface energy, partial integration is used to exploit the eigenvalue relation for the angular-momentum operator. The final result is

$$\begin{aligned} T &= \tfrac{1}{2}\varrho_m \sum_{\lambda\mu} |A_{\lambda\mu}|^2 \lambda \, R_0^{2\lambda+1} \\ &= \tfrac{1}{2}\varrho_m R_0^5 \sum_{\lambda\mu} \frac{1}{\lambda} |\alpha_{\lambda\mu}|^2 \quad . \end{aligned}$$

This can be written in the familiar form

$$T = \tfrac{1}{2} \sum_{\lambda\mu} B_\lambda |\alpha_{\lambda\mu}|^2 \tag{6.90}$$

if the *collective mass parameters* B_λ are defined as

$$B_\lambda = \varrho_m R_0^5 / \lambda \quad . \tag{6.91}$$

If we define stiffness coefficientsfor the potential through

$$C_\lambda = (\lambda - 1)\left((\lambda+2) R_0^2 \sigma - \frac{3 e^2 Z^2}{2\pi (2\lambda+1) R_0} \right) \quad , \tag{6.92}$$

the kinetic and potential energy of the vibrating nucleus take the form

$$T = \sum_{\lambda\mu} \tfrac{1}{2} B_\lambda |\alpha_{\lambda\mu}|^2 = \sum_{\lambda} \frac{\sqrt{2\lambda+1}}{2} [\alpha_\lambda \times \alpha_\lambda]^0 \quad ,$$
$$V = \sum_{\lambda\mu} \tfrac{1}{2} C_\lambda |\alpha_{\lambda\mu}|^2 = \sum_{\lambda} \frac{\sqrt{2\lambda+1}}{2} [\alpha_\lambda \times \alpha_\lambda]^0 \quad . \tag{6.93}$$

This result shows that each single mode (characterized by λ and μ) behaves like a harmonic oscillator with both the mass parameters and the stiffness coefficients depending on the angular momentum. The second form given in each case shows their scalar character explicitly and can be seen immediately form $(\lambda\lambda 0|\mu - \mu 0) = (-1)^{\lambda-\mu}/2\lambda + 1$.

For the description of nuclear spectra, only quantization has to be added, which will be done in the next section.

The preceding discussion completes the first construction of a *nuclear model* and illustrates this crucial concept quite well. We have used physical ideas from another field of physics, which need not be exactly applicable but can give insight into the physics of the nucleus via its successes and shortcomings. In this case, the assumption of hydrodynamic behavior of the nuclear matter may or may not be adequate and this can be judged by looking at the predictions gained. All the parameters of the model are known from the liquid-drop mass formula. As happens often with such model ideas, we will find that the structures predicted, such as the excitation spectra constructed in the next section, are found in nature, but the detailed values of the parameters are not correct. In this case one may still employ the model, but determine the mass parameters and stiffness coefficients from experiments, hoping to still find the model useful for predicting additional pieces of data and for understanding the underlying physics. Eventually, however, each model has to be justified through a derivation within a more fundamental description of the nucleus.

6.3.2 The Harmonic Quadrupole Oscillator

The next step in the formulation of the surface vibration model is quantization. This can be done in strict analogy to the familiar harmonic oscillator. The method is outlined by: the definition of coordinates and canonically conjugate momenta, replacement of these by operators and postulation of canonical commutation relations, introduction of creation and annihilation operators. The last step is not strictly part of quantization, but rather an elegant solution method.

Restricting the discussion again to quadrupole deformations, we start off with a Lagrangian of the form

$$L = T - V = \tfrac{1}{2}\sqrt{5}\, B_2 [\alpha_2 \times \alpha_2]^0 - \tfrac{1}{2}\sqrt{5}\, C_2 [\alpha_2 \times \alpha_2]^0 \quad . \tag{6.94}$$

Following the outline given, we first derive the conjugate momentum to $\alpha_{2\mu}$:

$$\pi_{2\mu} = \frac{\partial L}{\partial \alpha_{2\mu}} = \frac{\partial}{\partial \alpha_{2\mu}} \frac{1}{2} B_2 \sum_{\mu'=-2}^{2} (-1)^{\mu'} \alpha_{2\mu'} \alpha_{2-\mu'} = B_2 (-1)^{\mu} \alpha_{2-\mu} \quad . \tag{6.95}$$

From this expression one may already deduce the principal properties of the conjugate momenta: $\pi_{2\mu}$ does not form a spherical tensor, because although it is defined in terms of the spherical tensor $\alpha_{2\mu}$, the components are exchanged; $\pi^*_{2\mu} = (-1)^\mu \pi_{2-\mu} = \alpha_{2\mu}$, however, is a spherical tensor. Since $\pi_{2\mu}$ must then transform under the complex conjugate representation,

$$\pi'_{2\mu} = \sum_{\mu'} \mathcal{D}^{(2)\,*}_{\mu\mu'}(\boldsymbol{\theta})\, \pi_{2\mu'} \quad , \tag{6.96}$$

the usual angular-momentum coupling, etc., can be transferred to the momenta, but if one wants to construct terms involving both momenta and coordinates, $\pi^*_{2\mu}$, then $\pi_{2\mu}$ should not be coupled with $\alpha_{2\mu}$. The kinetic energy can be rewritten in terms of the momenta as

$$T = \frac{\sqrt{5}}{2B_2}[\pi \times \pi]^0 = \frac{1}{2B_2} \sum_{\mu=-2}^{2} (-1)^\mu\, \pi_{2\mu}\, \pi_{2-\mu} \quad . \tag{6.97}$$

Quantization is now easy. $\alpha_{2\mu}$ and $\pi_{2\mu}$ are replaced by operators $\hat{\alpha}_{2\mu}$ and $\hat{\pi}_{2\mu}$. These operators must have the generalized property

$$\hat{\alpha}^*_{2\mu} = (-1)^\mu\, \hat{\alpha}_{2-\mu} \quad , \quad \hat{\pi}^*_{2\mu} = -(-1)^\mu\, \hat{\pi}_{2-\mu} \quad , \tag{6.98}$$

where $\alpha_{2\mu}$ and $\pi^*_{2\mu}$ are spherical tensors.

Note that there is an *additional minus sign* for $\pi^*_{2\mu}$ associated with the time reversal properties of the momentum operator: this is a purely quantum-mechanical property that has no analog in classical mechanics; compare the analogous difference between the classical momentum, which is a real quantity, and the quantum operator $-\mathrm{i}\hbar\nabla$. This additional sign does not affect the transformation properties under rotation.

Quantization can now be done by imposing the commutation relations

$$\begin{aligned} \left[\hat{\alpha}_{2\mu}, \hat{\alpha}_{2\mu'}\right] &= 0 \quad , \\ \left[\hat{\pi}_{2\mu}, \hat{\pi}_{2\mu'}\right] &= 0 \quad , \\ \left[\hat{\alpha}_{2\mu}, \hat{\pi}_{2\mu'}\right] &= \mathrm{i}\hbar\, \delta_{\mu\mu'} \quad . \end{aligned} \tag{6.99}$$

Analogously to the standard case of cartesian coordinates, the commutation relations can be fulfilled by setting

$$\hat{\pi}_{2\mu} = -\mathrm{i}\hbar \frac{\partial}{\partial \alpha_{2\mu}} \quad , \tag{6.100}$$

and creation and annihilation operators $\hat{\beta}^\dagger_{2\mu}$ and $\hat{\beta}_{2\mu}$ may be introduced:

$$\begin{aligned} \hat{\beta}^\dagger_{2\mu} &= \sqrt{\frac{B_2\omega_2}{2\hbar}}\, \hat{\alpha}_{2\mu} - \mathrm{i}\sqrt{\frac{1}{2B_2\hbar\omega_2}}(-1)^\mu\, \hat{\pi}_{2-\mu} \quad , \\ \hat{\beta}_{2\mu} &= \sqrt{\frac{B_2\omega_2}{2\hbar}}(-1)^\mu\, \hat{\alpha}_{2-\mu} + \mathrm{i}\sqrt{\frac{1}{2B_2\hbar\omega_2}}\, \hat{\pi}_{2\mu} \quad . \end{aligned} \tag{6.101}$$

The pseudoparticles that are created and annihilated by these operators are called *phonons* in analogy to the quanta of vibrations in solids.

The precise form of these definitions still requires some explanations. The factors are exactly identical to the cartesian case if, as usual, the frequency is defined as $\omega_2 = \sqrt{C_2/B_2}$. What is new is the particular combination of azimuthal quantum numbers and the factors of $(-1)^\mu$; these are necessary to make the operators into spherical tensor operators, so that, for example, $\hat{\alpha}_{2\mu}$ and $\hat{\pi}_{2-\mu}$ can only be combined in the way shown, and just this combination must appear in the creation operator for it to also become a spherical tensor and to have the desired property that $\hat{\beta}^\dagger_\mu$ creates a phonon with projection $+\mu$. $\hat{\beta}_{2\mu}$ on the other hand transforms like $\pi_{2\mu}$, and its form follows from that of $\hat{\beta}^\dagger_{2\mu}$ through Hermitian conjugation.

The commutation relations between the operators take the usual form for bosons:

$$[\hat{\beta}^\dagger_{2\mu}, \hat{\beta}^\dagger_{2\mu'}] = 0 \quad , \quad [\hat{\beta}_{2\mu}, \hat{\beta}_{2\mu'}] = 0 \quad , \quad [\hat{\beta}_{2\mu}, \hat{\beta}^\dagger_{2\mu'}] = \delta_{\mu\mu'} \quad , \tag{6.102}$$

and the Hamiltonian becomes

$$\hat{H} = \hbar\omega_2 \left(\sum_{\mu=-2}^{2} \hat{\beta}^\dagger_{2\mu}\hat{\beta}_{2\mu} + \frac{5}{2} \right) \quad . \tag{6.103}$$

Note that the operators in the sum are effectively coupled to zero angular momentum.

The results of Chap. 3 can now be applied directly. The operators $\hat{\beta}^\dagger_{2\mu}$ create *quadrupole phonons* with magnetic quantum number μ, and we can immediately enumerate the lowest states of the nucleus in this model. All the states will have positive parity because of the parity of the $\alpha_{2\mu}$.

If we introduce the phonon number operator

$$\hat{N} = \sum_{\mu=-2}^{2} \hat{\beta}^\dagger_{2\mu}\hat{\beta}_{2\mu} \tag{6.104}$$

with eigenvalue N, the energies of the states will be given by

$$E_N = \hbar\omega_2 \left(N + \tfrac{5}{2}\right) \quad , \tag{6.105}$$

as we are effectively dealing with five oscillators, corresponding to the different magnetic quantum numbers, which can be excited independently and have a zero-point energy of $\frac{1}{2}\hbar\omega_2$ each. N counts the total number of quanta present.

Other quantum numbers that will appear are the angular momentum λ and its projection μ, so that the states can be labelled provisionally by $|N\lambda\mu\rangle$. The lowest-lying states are as follows.

1. The nuclear ground state is the phonon vacuum $|N = 0, \lambda = 0, \mu = 0\rangle$. Its energy is the zero-point energy:

$$\hat{H}\,|0\rangle = \tfrac{5}{2}\hbar\omega_2|0\rangle \quad . \tag{6.106}$$

2. The first excited state is the multiplet (one-phonon state)

$$|N = 1, \lambda = 2, \mu\rangle = \hat{\beta}^\dagger_{2\mu}|0\rangle, \quad \mu = -2, \ldots, +2 \quad , \tag{6.107}$$

with angular momentum 2. The excitation energy above the ground state is $\hbar\omega_2$.

3. The second set of excited states is given by the two-phonon states with an excitation energy of $2\hbar\omega_2$. It is not sufficient, of course, to just construct them using $\hat{\beta}^\dagger_{2\mu}\hat{\beta}^\dagger_{2\mu'}$: they have to be coupled to good total angular momentum. Thus these states are

$$|N=2,\lambda\mu\rangle = \sum_{\mu'\mu''} (22\lambda|\mu'\mu''\mu)\ \hat{\beta}^\dagger_{2\mu'}\hat{\beta}^\dagger_{2\mu''}|0\rangle \quad . \tag{6.108}$$

Angular-momentum selection rules allow values of $\lambda = 0,1,2,3,4$. However, it turns out that not all of these values are possible. Exchanging μ' and μ'' in the Clebsch–Gordan coefficient and using a symmetry property of the Clebsch–Gordan coefficients

$$\left(j_1 j_2 J\,|m_1 m_2 M\right) = (-1)^{j_1+j_2-J}\left(j_2 j_1 J\,|m_2 m_1 M\right) \tag{6.109}$$

to symmetrize the expression we get

$$|N=2,\lambda\mu\rangle = \tfrac{1}{2}\sum_{\mu'\mu''}\left[1+(-1)^\lambda\right](22\lambda|\mu'\mu''\mu)\ \hat{\beta}^\dagger_{2\mu'}\hat{\beta}^\dagger_{2\mu''}|0\rangle \quad , \tag{6.110}$$

because the operators commute. Consequently the wave functions for odd values of λ vanish: such states do not exist. The two-phonon states are thus restricted to angular momenta 0, 2, and 4, forming the *two-phonon triplet*. This effect is an example of the interplay of angular-momentum coupling and symmetrization (or, for fermions, antisymmetrization), which for more complex applications can be formalized in the *coefficients of fractional parentage*. They will not be needed in this book.

In general, the higher states of the spherical vibrator are not of interest for analyzing experimental spectra; for real nuclei one rarely finds states above the triplet which might be interpreted in this model. However, they are useful as mathematical basis states for expanding the wave functions of more complex models. A full analytic construction of the eigenstates was given by Chacón and Moshinsky [Ch76a], who also introduced the required additional quantum numbers. The first systematic application of the model as discussed here was given by Scharff-Goldhaber and Weneser [Sc55].

The spectrum and the properties of vibrational nuclei will be examined in more detail in Sect. 6.3.5, after a derivation of the other important observables calculable in this model.

It is interesting to note the *symmetry group* of the harmonic oscillator model; this will not be exploited directly but is mainly useful for a comparison to the IBA model of Sect. 6.8. When we write the Hamiltonian as

$$\hat{H} = \frac{1}{2B_2}\sum_{\mu=-2}^{2}|\pi_{2\mu}|^2 + \frac{C_2}{2}\sum_{\mu=-2}^{2}|\alpha_{2\mu}|^2 \quad , \tag{6.111}$$

both terms appear as absolute squares of five-dimensional complex vectors, and such a quantity is invariant under unitary transformations in five dimensions, i.e., under the group U(5). This same group will appear in the vibrational limit of the IBA model.

6.3.3 The Collective Angular-Momentum Operator

The collective coordinates $\alpha_{\lambda\mu}$ have been defined in a highly abstract way. To construct proper eigenstates of total angular momentum we will also need a collective angular-momentum operator, which acts on these coordinates. It has to fulfil the following crucial properties, which are all apparent from the general discussion of angular momentum in Chap. 2.

1. It should be a spherical tensor operator $\hat{L}_\mu$, $\mu = -1, 0, 1$, of rank 1, constructed out of $\hat{\alpha}_{2\mu}$ and $\hat{\pi}^*_{2\mu}$.
2. It must fulfil the usual angular-momentum commutation rules.
3. More specifically, it must act on the collective coordinate operators in the correct way for an infinitesimal rotation; as was seen in Sect. 2.3.7 this requires

$$\left[\hat{L}_z, \hat{T}^k_q\right] = \hbar q \hat{T}^k_q \quad , \quad \left[\hat{L}_\pm, \hat{T}^k_q\right] = \hbar\sqrt{(k \pm q + 1)(k \mp q)}\, \hat{T}^k_{q\pm 1} \tag{6.112}$$

for any tensor operator $\hat{T}^k_q$ constructed out of the collective coordinate $\hat{\alpha}_{\lambda\mu}$ and the momentum operator $\hat{\pi}_{\lambda\mu}$. Note that if this property holds when substituting both $\hat{\alpha}_{\lambda\mu}$ and $\hat{\pi}^*_{\lambda\mu}$ for $\hat{T}^k_q$, it must be true also for all spherical tensors constructed from these by angular-momentum coupling, so that, e.g., the angular-momentum commutation rules for the $\hat{L}_\mu$ follow automatically.

We will discuss for the motivation the general form for such an operator and check the last property for the special case $\hat{T}_{\lambda\mu} = \hat{\alpha}_{\lambda\mu}$, which is sufficient to fix an overall normalizing factor of the operator. The calculation for $\hat{\pi}^*_{\lambda\mu}$ is quite analogous.

For the familiar case of cartesian coordinates the angular-momentum operator is given by $\hat{\boldsymbol{L}} = \boldsymbol{r} \times \hat{\boldsymbol{p}}$, which can also be written in terms of the angular-momentum coupling of the two tensor operators to a resulting angular momentum of one. Assuming the same form for the present case leads to a conjectured expression

$$\hat{L}_\mu = C \sum_{\mu'\mu''} (\lambda\lambda 1|\mu'\mu''\mu)\, \hat{\alpha}_{\lambda\mu'}\hat{\pi}^*_{\lambda\mu''} \tag{6.113}$$

with an as yet unspecified constant factor C. Note that we have used π^* because of the mentioned transformation properties of the conjugate momenta.

Inserting

$$\hat{\pi}^*_{\lambda\mu''} = -(-1)^{\mu''}\hat{\pi}_{\lambda-\mu''} = \mathrm{i}\hbar(-1)^{\mu''}\frac{\partial}{\partial\alpha_{\lambda-\mu''}} \tag{6.114}$$

into (6.133) we get for the z component ($\mu = 0$) of the angular momentum

$$\begin{aligned} \hbar\mu\alpha_{\lambda\mu} &= \left[C\sum_{\mu'}(\lambda\lambda 1|\mu' -\mu' 0)\mathrm{i}\hbar(-1)^{\mu'}\alpha_{\lambda\mu'}\frac{\partial}{\partial\alpha_{\lambda\mu'}}, \alpha_{\lambda\mu}\right] \\ &= \mathrm{i}\hbar(-1)^\mu C\,(\lambda\lambda 1|\mu -\mu 0)\,\alpha_{\lambda\mu} \quad , \end{aligned} \tag{6.115}$$

so that we should require

$$C = -\mathrm{i}(-1)^\mu \frac{\mu}{(\lambda\lambda 1|\mu{-}\mu 0)} \quad . \tag{6.116}$$

To make sense, of course, C must turn out to be independent of μ. The value of the special Clebsch–Gordan coefficient appearing here is

$$(\lambda\lambda 1|\mu{-}\mu 0) = \sqrt{3}\,(-1)^{\lambda-\mu} \frac{\mu}{\sqrt{(2\lambda+1)(\lambda+1)\lambda}} \quad , \tag{6.117}$$

so that

$$C = -\mathrm{i}\,\frac{(-1)^\lambda}{\sqrt{3}} \sqrt{(2\lambda+1)(\lambda+1)\lambda} \quad . \tag{6.118}$$

Specifically for quadrupole coordinates we find $C = -\mathrm{i}\sqrt{10}$ and the angular-momentum operator becomes

$$\hat{L}_\mu = -\mathrm{i}\sqrt{10}\left[\alpha \times \hat{\pi}^*\right]^1_\mu \quad . \tag{6.119}$$

It is straightforward but laborious to check that the operator fulfills the remaining requirements, needing essentially the extensive manipulation of Clebsch–Gordan coefficients and their special expressions.

It remains for us to write this operator in second-quantized form, which is done in the following exercise.

EXERCISE

6.3 Angular-Momentum Operator for Quadrupole Phonons in Second Quantization

Problem. Write the angular-momentum operator for quadrupole phonons in second-quantized form.

Solution. From the definition of the creation and annihilation operators given in the preceding section, we can in turn solve for the coordinates and momenta (let us stick to the special case $\lambda = 2$):

$$\begin{aligned} \hat{\alpha}_{2\mu} &= \sqrt{\frac{\hbar}{2B_2\omega_2}}\left[\hat{\beta}^\dagger_{2\mu} + (-1)^\mu \hat{\beta}_{2-\mu}\right] \quad , \\ \hat{\pi}_{2\mu} &= \mathrm{i}\sqrt{\tfrac{1}{2}\hbar B_2\omega_2}\left[(-1)^\mu \hat{\beta}^\dagger_{2-\mu} - \hat{\beta}_{2\mu}\right] \quad . \end{aligned} \tag{1}$$

The expression for the angular-momentum operator now becomes

$$\begin{aligned} \hat{L}_\mu &= -\sqrt{10}\,\frac{\hbar}{2}\sum_{\mu'\mu''}(221|\mu'\mu''\mu)\left[\hat{\beta}^\dagger_{2\mu'} + (-1)^{\mu'}\hat{\beta}_{2-\mu'}\right]\left[\hat{\beta}^\dagger_{2\mu''} - (-1)^{\mu''}\hat{\beta}_{2-\mu''}\right] \\ &= -\sqrt{10}\,\frac{\hbar}{2}\sum_{\mu'\mu''}(221|\mu'\mu''\mu)\left[\hat{\beta}^\dagger_{2\mu'}\hat{\beta}^\dagger_{2\mu''} - (-1)^{\mu'+\mu''}\hat{\beta}_{2-\mu'}\hat{\beta}_{-\mu''}\right. \\ &\quad \left. + (-1)^{\mu'}\hat{\beta}_{2-\mu'}\hat{\beta}^\dagger_{2\mu''} - (-1)^{\mu''}\hat{\beta}^\dagger_{2\mu'}\hat{\beta}_{2-\mu''}\right] \quad . \end{aligned} \tag{2}$$

Exercise 6.3

The first two terms in the parentheses each vanish when summed up over the Clebsch–Gordan coefficients because the symmetry relation (6.109) leads to a sign change when interchanging the summation indices (the operators commute). The last two terms, on the other hand, can be combined using the Boson commutation relations to yield

$$\begin{aligned}\hat{L}_\mu = &-\sqrt{10}\,\frac{\hbar}{2}\sum_{\mu'\mu''}(221|\mu'\mu''\mu)\left[(-1)^{\mu'}\hat{\beta}^\dagger_{2\mu''}\hat{\beta}_{2-\mu'} - (-1)^{\mu''}\hat{\beta}^\dagger_{2\mu'}\hat{\beta}_{2-\mu''}\right] \\ &+\sqrt{10}\,\frac{\hbar}{2}\sum_{\mu'}(-1)^{\mu'}\left(221|\mu'-\mu'0\right) \quad . \end{aligned} \tag{3}$$

The last term again vanishes because of (6.109), while in the other two the dummy summation indices μ' and μ'' can be interchanged using this same relation:

$$\hat{L}_\mu = \sqrt{10}\,\hbar\sum_{\mu'\mu''}(221|\mu'\mu''\mu)\,(-1)^{\mu'}\hat{\beta}^\dagger_{2\mu''}\hat{\beta}_{2-\mu'} \quad . \tag{4}$$

An interesting special case is for $\mu = 0$. The sum can then be evaluated using (6.117) in the form

$$\left(221|\mu'-\mu'0\right) = (-1)^{\mu'}\frac{\mu'}{\sqrt{10}} \tag{5}$$

to give the result

$$\hat{L}_0 = \hbar\sum_\mu \mu\hat{\beta}^\dagger_\mu\hat{\beta}_\mu \quad . \tag{6}$$

This formula simply counts the number of phonons in each of the oscillator states with projection μ and sums up their contributions to the total projection. Clearly the formalism is consistent in that $\hat{\beta}^\dagger_\mu$ creates a phonon with angular momentum projection μ.

6.3.4 The Collective Quadrupole Operator

Another important ingredient for the model is the electric quadrupole operator expressed in the coordinates $\alpha_{2\mu}$. On the one hand, it allows one to calculate predictions for the transition rates and static moments, but, on the other hand, it also makes the physical significance of the $\alpha_{2\mu}$ more precise. The Hamiltonian and its model are largely determined by symmetry arguments and would look similar if $\alpha_{2\mu}$ were any other nuclear excitation degree of freedom of quadrupole type that could be approximated by harmonic oscillations. In fact, the Hamiltonian could be constructed using only symmetry arguments without any reference to the exact nature of the deformations considered.

Setting up the quadrupole operator, however, requires a specification of the nuclear shape and charge distribution. We first derive the classical quadrupole tensor and then replace the $\alpha_{2\mu}(t)$ by operators. We will naturally use the simplest assumption about the charge distribution, namely that the charge density is uniform within the time-dependent nuclear shape given by the $\alpha_{2\mu}(t)$. The quadrupole tensor is then given by using the methods developped in Exercise 6.1:

$$\begin{aligned}
Q_{2\mu} &= \varrho_0 \int_{\text{nuclear volume}} \mathrm{d}^3 r\, r^2\, Y_{2\mu}(\Omega) \\
&= \varrho_0 \int \mathrm{d}\Omega\, Y_{2\mu}(\Omega) \int_0^{R(\Omega)} \mathrm{d}r\, r^4 \\
&= \frac{\varrho_0}{5} \int \mathrm{d}\Omega\, Y_{2\mu}(\Omega)\, R(\Omega)^5 \\
&= \frac{\varrho_0}{5} R_0^5 \int \mathrm{d}\Omega\, Y_{2\mu}(\Omega) \Big[1 + 5\Big(\alpha_{00} + \sum_{\mu'} \alpha_{2\mu'} Y_{2\mu'}^*(\Omega)\Big) \\
&\quad + 10 \sum_{\mu'\mu''} \alpha_{2\mu'}\alpha_{2\mu''} Y_{2\mu'}^*(\Omega) Y_{2\mu''}^*(\Omega)\Big] \quad . \qquad (6.120)
\end{aligned}$$

What was done here was to insert the definition of the deformed surface (including α_{00} to correct the volume of the nucleus according to Exercise 6.1), and then to expand the fifth power of the nuclear radius up to second order in the deformations. According to Exercise 6.1, α_{00} itself is of second order in the $\alpha_{2\mu}$, and this was taken into account in the omission of the higher-order terms. Now it remains to evaluate the integrals over one to three spherical harmonics. The integral over a single spherical harmonic was determined in (6.61). The integral over two spherical harmonics is directly given by the orthonormality relation, so that (6.120) can be rewritten as

$$Q_{2\mu} = \frac{\varrho_0}{5} R_0^5 \Big[5\alpha_{2\mu} + 10 \sum_{\mu'\mu''} \alpha_{2\mu'}\alpha_{2\mu''} \int \mathrm{d}\Omega\, Y_{2\mu}(\Omega) Y_{2\mu'}^*(\Omega) Y_{2\mu''}^*(\Omega)\Big] \, . \qquad (6.121)$$

To go further we need the formula for an integral over three spherical harmonics, which is easily obtained from the reduced matrix element given in the Appendix,

$$\begin{aligned}
&\int \mathrm{d}\Omega\, Y_{l_1 m_1}(\Omega)\, Y_{l_2 m_2}(\Omega)\, Y_{l_3 m_3}^*(\Omega) \\
&\quad = \langle l_3 || Y_{l_2} || l_1 \rangle\, (l_1 l_2 l_3 | m_1 m_2 m_3) \\
&\quad = \sqrt{\frac{(2l_1+1)(2l_2+1)}{4\pi(2l_3+1)}}\, (l_1 l_2 l_3 | m_1 m_2 m_3)\, (l_1 l_2 l_3 | 000) \quad . \qquad (6.122)
\end{aligned}$$

Since the right-hand side is a real number, we can also apply the complex conjugate of this equation

$$\int \mathrm{d}\Omega\, Y_{2\mu}(\Omega)\, Y_{2\mu'}^*(\Omega)\, Y_{2\mu''}^*(\Omega) = \sqrt{\frac{5}{4\pi}}\, (222|\mu'\mu''\mu)\, (222|000) \quad . \qquad (6.123)$$

The μ-dependent Clebsch–Gordan coefficient can be used to couple the $\alpha_{2\mu}$ to total angular momentum 2, while the constant one is given by

$$(222|000) = -\sqrt{\tfrac{2}{7}} \quad . \qquad (6.124)$$

In this way the quadrupole tensor becomes

$$Q_{2\mu} = \varrho_0 R_0^5 \left(\alpha_{2\mu} - \frac{10}{\sqrt{70\pi}} \left[\alpha \times \alpha\right]^2_\mu \right) \quad , \tag{6.125}$$

and it may be quantized simply by substituting the operator $\hat{\alpha}_{2\mu}$ for the classical deformation.

Express the charge density ϱ_0 in terms of the proton number and the nuclear volume as the quadrupole operator may finally be written as

$$\hat{Q}_{2\mu} = \frac{3Ze}{4\pi} R_0^2 \left(\hat{\alpha}_{2\mu} - \frac{10}{\sqrt{70\pi}} \left[\hat{\alpha} \times \hat{\alpha}\right]^2_\mu \right) \quad . \tag{6.126}$$

Note that if the second quantized version of this operator is inserted, it shows that the operator allows a change in the phonon number by 1 (to first order in $\alpha_{2\mu}$) or 2 (to second order), so that typically E2 transitions will strongly link states differing by one phonon, but also cause two-phonon transitions with less probability.

6.3.5 The Quadrupole Vibrational Spectrum

The angular momenta and energies of the first few excited states have already been given in Sect. 6.3.2 and are illustrated in Exercise 6.4. Here we will derive a few more properties of these states and then discuss their realization in nature. Remember that the states are denoted by $|N\,\lambda\mu\rangle$ with phonon number N and angular momentum quantum numbers λ and μ (this classification is unique only for the lowest few states).

As already mentioned, the energy spectrum by itself is not very specific, in the sense that any harmonic motion of quadrupole type would show the same structures. It is thus very important to investigate also the predictions for transition rates, but even before this the parameters of the model have to be fixed. These are two: the stiffness C_2 and the mass parameter (also called the collective inertia) B_2. Their combination $\hbar\omega_2 = \hbar\sqrt{C_2/B_2}$ can be read off immediately from the energy spectrum as the spacing between the ground state and the first excited state, which should be a 2^+ state and should have equal separation from the $0^+ - 2^+ - 4^+$ triplet above it. The remainder of the spectrum is also completely determined by this one parameter – one may find agreement with experiment or not, but one has no handle to adjust things further. To determine the second independent parameter, however, already requires some other measurable quantity. In the electromagnetic properties the nuclear radius appears as an additional quantity; this should not be fitted but assumed to be given by the standard experimental result $R_0 = r_0 A^{1/3}$.

Some easily derivable quantities are given here.

1. The mean deformation:

$$\langle \alpha_{2\mu} \rangle = \langle N\,\lambda'\mu' | \hat{\alpha}_{\lambda\mu} | N\,\lambda'\mu' \rangle = 0 \quad . \tag{6.127}$$

   That this quantity must vanish is clear mathematically from the fact that the operator $\hat{\alpha}_{\lambda\mu}$ changes the phonon number by ± 1 and thus cannot have a nonzero expectation value in a state of good phonon number. Intuitively this is a consequence of the fact that dynamically the nuclear surface spends as much time in positive deformations as in negative ones – similar to the vanishing of the average $\langle x \rangle$ for the one-dimensional oscillator.

2. It is more useful to regard the mean square deformation

$$\langle \beta^2 \rangle = \langle N \lambda \mu | \sum_\mu |\alpha_{2\mu}|^2 | N \lambda \mu \rangle \quad . \tag{6.128}$$

It can be evaluated using the standard methods of second quantization, but there is a much faster way using the virial theorem, which for the harmonic oscillator in both classical and quantum mechanics states that the average values of the kinetic and potential energies are equal to each other. In our case this should also apply, since each μ component acts like a harmonic oscillator, so that for an N-phonon state

$$\begin{aligned} E_N &= 2\langle V \rangle = \hbar\omega_2 \left(N + \tfrac{5}{2}\right) \\ &= C_2 \langle N \lambda \mu | \sum_\mu |\alpha_{2\mu}|^2 \, | N \lambda \mu \rangle = C_2 \langle \beta^2 \rangle \quad , \end{aligned} \tag{6.129}$$

which yields

$$\langle \beta^2 \rangle = \frac{\hbar}{B_2 \omega_2} \left(N + \tfrac{5}{2}\right) \quad . \tag{6.130}$$

The quantity $\langle \beta^2 \rangle$ indicates the softening of the nuclear surface caused by the vibrations. Although we used the assumption of a homogeneous charge distribution for setting up the quadrupole operator, these oscillations will effectively produce a diffuse surface, even though this effect should not be taken too seriously as an "explanation" of the surface diffuseness. The associated mean square deviation in the radius is $\langle \Delta R^2 \rangle = R_0^2 \langle \beta^2 \rangle$. We will look at numbers in an exercise. The mean square deformation of the ground state is a useful parameter, so we assign to it the notation

$$\beta_0^2 = \frac{5\hbar}{2 B_2 \omega_2} \quad . \tag{6.131}$$

3. A related and directly measurable quantity is the mean square charge radius, defined as the average of r^2 over the charge distribution, i.e., in this model with its homogeneous charge distribution but time-dependent surface,

$$\begin{aligned} \langle r^2 \rangle &= \frac{\varrho_0 \int d^3r \, r^2}{\varrho_0 \int d^3r} = \frac{3}{4\pi R_0^3} \int dr \, d\Omega \, r^4 \\ &= \frac{3R_0^2}{20\pi} \int d\Omega \Big(1 + \alpha_{00} Y_{00}(\Omega) + \sum_\mu \alpha_{2\mu} Y^*_{2\mu}(\Omega)\Big)^5 \quad . \end{aligned} \tag{6.132}$$

Using the same methods as for the quadrupole tensor, the integral may be reduced to

$$\langle r^2 \rangle = \frac{3}{5} R_0^2 + \frac{3}{4\pi} R_0^2 \langle \beta^2 \rangle \quad , \tag{6.133}$$

and one may now, for example, investigate its dependence on the phonon number N. An excitation of the nucleus leads to a change in the Coulomb field felt by the electrons, so that these show a slight but measurable energy shift.

4. The B(E2) values are relatively easily obtained if the quadratic term in the quadrupole operator is neglected. In this case

$$\hat{Q}_{2\mu} = \varrho_0 R_0^5 \sqrt{\frac{\hbar}{2B_2\omega_2}} \left(\hat{\beta}_\mu^\dagger + (-1)^\mu \hat{\beta}_{-\mu}\right) \quad , \tag{6.134}$$

so that only one-phonon transitions are allowed. The matrix element of the operator is trivial in the case of transitions between the ground state and the first excited state $|2_1^+ m\rangle = |N = 1, 2m\rangle = \hat{\beta}_m^\dagger |0\rangle$:

$$\langle 2_1^+ m|\hat{Q}_{2\mu}|0\rangle = \varrho_0 R_0^5 \sqrt{\frac{\hbar}{2B_2\omega_2}} \langle 0|\hat{\beta}_m \hat{\beta}_\mu^\dagger|0\rangle = \varrho_0 R_0^5 \sqrt{\frac{\hbar}{2B_2\omega_2}}\, \delta_{m\mu} \quad . \tag{6.135}$$

The Wigner–Eckart theorem states that

$$\langle 2_1^+ m|\hat{Q}_{2\mu}|0\rangle = \langle 2_1^+||\hat{Q}_2||0\rangle\,(022|0\mu m) = \langle 2_1^+||\hat{Q}_{2\mu}||0\rangle\,\delta_{m\mu} \quad , \tag{6.136}$$

so that in this simple case the reduced matrix element just differs from the standard one by the Kronecker symbol. Using $J_i = 0$ and $J_f = 2$, the B(E2) value reduces to

$$\begin{aligned} B\left(\mathrm{E2}, 0_1^+ \to 2_1^+\right) &= \frac{2J_f + 1}{2J_i + 1} \left|\langle 2_1^+||\hat{Q}_{2\mu}||0\rangle\right|^2 \\ &= \left(\varrho_0 R_0^5\right)^2 \frac{5\hbar}{2B_2\omega_2} \\ &= \left(\varrho_0 R_0^5\right)^2 \beta_0^2 \quad . \end{aligned} \tag{6.137}$$

This $B(E2)$-value therefore directly determines the mean square deformation, and together with the excitation energy suffices to fix all the parameters of the model.

5. The quadrupole moment of an excited state is given by

$$Q = \sqrt{\frac{16\pi}{5}} \langle N\lambda\mu = \lambda|\hat{Q}_{20}|N\lambda\mu = \lambda\rangle \tag{6.138}$$

(see Sect. 5.4.6), and in this case the linear term does not contribute because the phonon number is the same on both sides of the matrix element. Of the quadratic term only the combinations with one creation and one annihilation operator need to be considered, so that we can substitute

$$\begin{aligned} \hat{Q}_{20} \to &-\frac{10}{\sqrt{70\pi}} \varrho_0 R_0^5 \frac{\hbar}{2B_2\omega_2} \\ &\times \sum_\mu (-1)^\mu\,(222|\mu{-}\mu 0)\,\left(\hat{\beta}_\mu^\dagger \hat{\beta}_\mu + \hat{\beta}_\mu \hat{\beta}_\mu^\dagger\right) \quad . \end{aligned} \tag{6.139}$$

The operators may be commuted according to

$$\hat{\beta}_\mu^\dagger \hat{\beta}_\mu + \hat{\beta}_\mu \hat{\beta}_\mu^\dagger = 2\hat{\beta}_\mu^\dagger \hat{\beta}_\mu + 1 \quad , \tag{6.140}$$

and the constant term herein does not contribute because of

$$\sum_\mu (-1)^\mu \,(222|\mu{-}\mu 0) = 0 \tag{6.141}$$

(to show this once again use the symmetry relation (6.109)). So we are left with

$$Q = -\frac{8}{\sqrt{14}}\varrho_0 R_0^5 \frac{\hbar}{B_2\omega_2}\sum_\mu (-1)^\mu \,(222|\mu{-}\mu 0) \times \langle N\lambda\mu = \lambda|\hat{\beta}_\mu^\dagger \hat{\beta}_\mu|N\lambda\mu = \lambda\rangle \quad . \tag{6.142}$$

The sum can be evaluated only for specific cases. In the ground state the matrix element is zero, but for the first excited state we get

$$\langle 2_1^+ m = 2|\hat{\beta}_\mu^\dagger \hat{\beta}_\mu|2_1^+ m = 2\rangle = \delta_{\mu 2} \quad , \tag{6.143}$$

and because $(222|2{-}20) = \sqrt{\frac{2}{7}}$ the final result is

$$Q = -\frac{12}{35\pi}\beta_0^2 Z e R_0^2 \quad . \tag{6.144}$$

In the exercise we suggest applying this model to a specific nucleus. The quality of the description obtained is typical also for other nuclei and we can summarize the experimental situation in the following way.

Spectra like that predicted by the spherical vibrator model are approximated for a small number of nuclei near closed shells, which are generally believed to be spherical in their ground states. Usually one finds the predicted triplet of 0^+, 2^+, and 4^+ states at about twice the energy of the first excited 2^+ state, but the triplet is not degenerate. The quadrupole moments and B(E2) values are not described well by the model, and this is easily understood. The nondegeneracy of the triplet already indicates that there are deviations from harmonic motion, and this implies that the true states of the nucleus do not possess good phonon number. But then for the quadrupole moment, for example, the linear term contributes even in the diagonal matrix element, and even if the admixture of different phonon numbers is small, it may still yield contributions comparable to those of the second order term and completely change the result expected in the model. The harmonic oscillator with its high degree of mathematical symmetries is simply too idealized to make its properties robust with respect to perturbations; its main merit is a starting point for more sophisticated approximations.

EXERCISE

6.4 ^{114}Cd as a Spherical Vibrator

Problem. Apply the model of the spherical vibrator to the nucleus ^{114}Cd.

Solution. The experimental data are summarized on the right-hand side of Fig. 6.5 First we use the energy of the first excited 2^+ state and the B(E2) value of the transition to the ground state to compute C_2 and B_2 using the relations

$$E\left(2_1^+\right) = \hbar\sqrt{C_2/B_2} = 0.558 \text{ MeV} \quad ,$$
$$B\left(\text{E2}; 2_1^+ \to 0_1^+\right) = \left(\frac{3Ze}{4\pi}R_0^2\right)^2 \frac{\hbar}{2\sqrt{C_2 B_2}} = 1018\, e^2\,\text{fm}^4 \quad , \tag{1}$$

together with

$$Z = 48 \quad , \quad A = 114 \quad , \quad R_0 = 1.2 A^{1/3} \quad , \tag{2}$$

to get

$$C_2 = 41.3 \text{ MeV} \quad , \quad B_2 = 132\,\hbar^2/\,\text{MeV} \quad . \tag{3}$$

The spectrum shows that the two-phonon triplet (0_2^+, 2_2^+, 4_1^+) is almost degenerate near the theoretical energy, but the presence of an additional 0^+ and a 2^+ state close by already indicates strong anharmonic effects, if these are also to be interpreted as collective surface modes.

We now compare these with the predictions of the liquid-drop model [(6.91) and (6.92)]. The liquid-drop mass formula contains a surface term $a_S\, A^{2/3}$ with $a_S \approx 13\,\text{MeV}$, yielding a surface tension

$$\sigma = \frac{a_S A^{2/3}}{4\pi r_0^2 A^{2/3}} \approx 0.72 \text{ MeV fm}^{-2} \quad . \tag{4}$$

Together with $e^2 \approx 1.44$ MeV fm this yields

$$C_2^{\text{liquid drop}} \approx 42.4 \text{ MeV} \quad , \quad B_2^{\text{liquid drop}} \approx 11.09\,\hbar^2/\,\text{MeV} \quad . \tag{5}$$

Note that the density ϱ_m in (6.91) is the mass density. Thus the stiffness coefficient is described very well, while the mass parameter is off by an order of magnitude. This is not surprising, because for the mass parameter the probably unrealistic assumption of irrotational hydrodynamic flow was made.

The B(E2) values for the transitions from the triplet to the one-phonon state and from that to the ground state are

$$B\left(\text{E2}; 0_2^+, 2_2^+, 4_1^+ \to 2_1^+\right) = 2B\left(\text{E2}; 2_1^+ \to 0_1^+\right) = 2036\, e^2\,\text{fm}^4 \quad , \tag{6}$$

whereas the experimental values are smaller by factors of 2–10. The transition from the two-phonon state 2_2^+ to the ground state should be quite small, because it is a second-order effect. In fact, we find

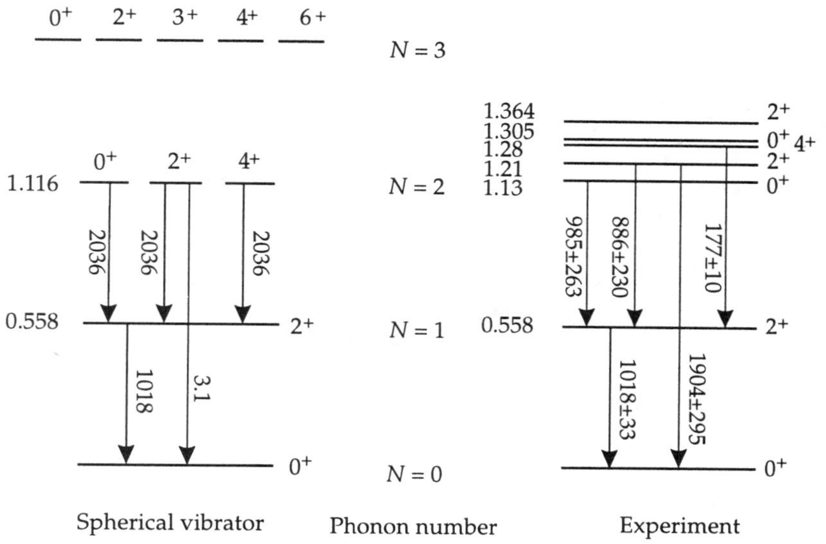

Fig. 6.5 Comparison of the spherical vibrator model with experimental data for ^{114}Cd. The energy levels are in MeV, while the B(E2) values, indicated next to the transition arrows, are given in e^2 fm^4. The data have been taken from Nucl. Data Sheets Vol. **35**, No. 3 (1982).

$$B\left(\mathrm{E2}; 2_2^+ \rightarrow 0_1^+\right) = \left(\frac{3Ze}{4\pi} R_0^2 \frac{10}{\sqrt{70\pi}}\right)^2 \frac{\hbar^2}{4C_2B_2} = 3.1\, e^2\,\mathrm{fm}^4 \quad , \tag{7}$$

which is very small compared to the experimental value of $1904\, e^2\,\mathrm{fm}^4$. Again this indicates anharmonic effects: if the wave functions contain admixtures of other phonon numbers, the first and much larger term in the quadrupole operator may contribute to this transition.

The mean square deformation of the state 2_1^+ is

$$\langle\beta^2\rangle\left(2_1^+\right) = \frac{\hbar}{\sqrt{C_2B_2}}\left(1 + \tfrac{5}{2}\right) = 0.047 \quad , \tag{8}$$

which also compares unfavorably with the experimental value of 0.193. The quadrupole moment of the state 2_1^+ is given by

$$Q\left(2_1^+\right) = -\frac{15}{7}\frac{ZeR_0^2\hbar^2}{B_2C_2} = -0.6\, e^2\,\mathrm{fm}^4 \quad . \tag{9}$$

Note that here, too, only the second-order term contributes, so that it is no surprise that the experimental value, $-3600\, e^2\,\mathrm{fm}^2$, is bigger by a factor of 10^4.

Finally, the surface diffuseness can be obtained from the mean square deformation of the ground state

$$\beta_0^2 = \frac{\hbar}{B_2\omega_2}\left(N + \tfrac{5}{2}\right) = 0.034 \quad \text{as} \quad \sqrt{\langle\Delta R^2\rangle} = 1.03\,\mathrm{fm} \quad , \tag{10}$$

which is surprisingly acceptable.

6.4 Rotating Nuclei

6.4.1 The Rigid Rotor

Another simple concept of the nucleus that describes some features of excited states quite well is the *rigid-rotor* model. As is well known from classical mechanics, the degrees of freedom of a rigid rotor are the three Euler angles, which describe the orientation of the body-fixed axes in space (the translational degrees of freedom can be ignored, since they do not lead to internal excitations of the nucleus).

Naturally the easiest description will be achieved if the principal axes of the nucleus are selected for determining the body-fixed system, since then the inertia tensor will be diagonal (refer also to the discussion in Sect. 6.2.4; the present case corresponds to constant values of β and γ). The classical kinetic energy of such a rotating rigid body is

$$E = \sum_{i=1}^{3} \frac{J_i'^2}{2\Theta_i} \quad , \tag{6.145}$$

where Θ_i is the moment of inertia about the ith principal axis of the nucleus. In general all three moments of inertia may be different. While this looks like a very simple formula, which could be quantized by replacing J' by $\hat{J}'$, there are actually two complicating factors.

One of them is indicated by the primes on the angular-momentum operators in (6.145). These operators cannot be identical with the standard angular-momentum operators rotating the system about space-fixed axes, because even if these axes were chosen to coincide with the body-fixed principal axes initially; they would no longer do this after the rotation. Instead, the rotation has to be done through operators that always rotate the nucleus about the instantaneous directions of the corresponding principal axes, and these *body-fixed angular-momentum operators* are denoted as $\hat{J}_i'$.

Their commutation relations look similar to those of the space-fixed operators, but with a crucial change in sign:

$$[\hat{J}_x', \hat{J}_y'] = -\mathrm{i}\hbar\hat{J}_z' \quad , \quad [\hat{J}_y', \hat{J}_z'] = -\mathrm{i}\hbar\hat{J}_x' \quad , \quad [\hat{J}_z', \hat{J}_x'] = -\mathrm{i}\hbar\hat{J}_y' \quad . \tag{6.146}$$

This is caused by the different definition of how the second rotation is applied. For the space-fixed operators, for example, $\hat{J}_x\hat{J}_y$ describes an infinitesimal rotation about the y axis followed by one about the x axis. In the body-fixed case, however, the second rotation will be about the original x axis *already rotated by the first operator into the x' axis*. The difference is illustrated in Figs. 6.6 and 6.7, while the consequences for the matrix elements are explored in Exercise 6.5.

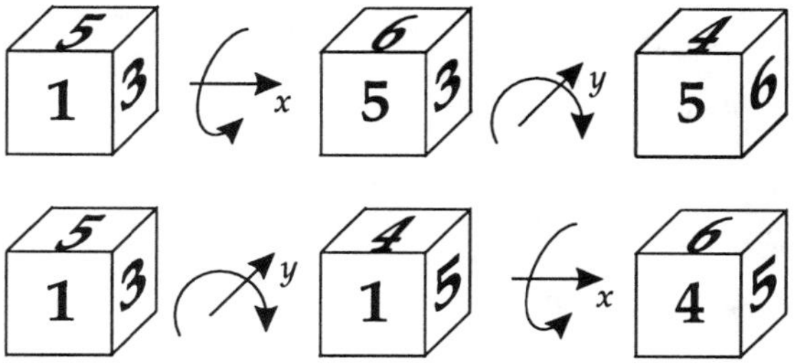

Fig. 6.6 A die is used to illustrate the effect of a rotation about the space-fixed axes; in the upper part a rotation about the x axis is followed by one about the y axis and in the lower part the order is reversed. Note that these are finite rotations about 90°, so that the difference is not simply a rotation about the z axis, as would be the case for infinitesimal rotations.

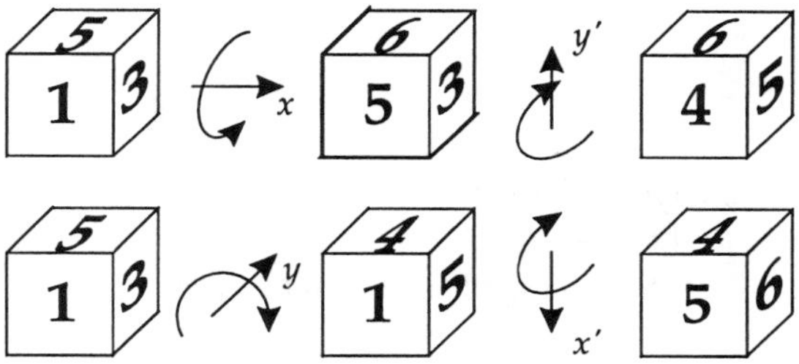

Fig. 6.7 The same case as for Fig. 6.6, but the second rotation is about the new axis in each case. Note that the *difference* between the final configurations is just reversed from Fig. 6.6, making the difference in sign in the commutation relations plausble.

EXERCISE

6.5 Angular-Momentum Operators in the Intrinsic System

Problem. Derive the representations of the body-fixed angular-momentum operators in the intrinsic system.

Solution. Since only the sign in the commutation relations differs, we can closely follow the developments in Sect. 2.3.2, noting only the differences as we go along. The basis is chosen to be diagonal in both $\hat{J}'^2$ and $\hat{J}'_z$:

$$\hat{J}'^2|jm\rangle = \hbar^2 \Lambda_j |jm\rangle \quad , \quad \hat{J}'_z|jm\rangle = \hbar m|jm\rangle \quad . \tag{1}$$

The shift operators are also defined as

$$\hat{J}'_+ = \hat{J}'_x + i\hat{J}'_y \quad , \quad \hat{J}'_- = \hat{J}'_x - i\hat{J}'_y \quad , \tag{2}$$

but now the commutation relations are

$$[\hat{J}'_z, \hat{J}'_\pm] = \mp\hbar\hat{J}'_\pm \quad , \tag{3}$$

Exercise 6.5

and we obtain the eigenvalue corresponding to $\hat{J}'_{\pm}|jm\rangle$ as

$$\hat{J}'_z\left(\hat{J}'_{\pm}|jm\rangle\right) = \hbar(m \mp 1)\hat{J}'_{\pm}|jm\rangle \quad , \tag{4}$$

showing that the directions of the shifts are interchanged. Using μ again for the largest eigenvalue, we must have

$$\hat{J}'_{-}|j\mu\rangle = 0 \quad , \tag{5}$$

and

$$\hat{J}'^2 = \tfrac{1}{2}\left(\hat{J}'_{+}\hat{J}'_{-} + \hat{J}'_{-}\hat{J}'_{+}\right) + \hat{J}'^2_z \tag{6}$$

may now be rewritten using the commutation relation

$$\left[\hat{J}'_{+}, \hat{J}'_{-}\right] = -2\hbar\hat{J}'_z \tag{7}$$

as

$$\hat{J}'^2 = \hat{J}'_{+}\hat{J}'_{+} - \hat{J}'^2_z + \hbar\hat{J}'_z \quad , \tag{8}$$

which agrees with the old formula except for the substitution $\hat{J}'_{-} \leftrightarrow \hat{J}'_{+}$, so that all the further calculations for the magnitude of μ are identical.

The next place where differences come in is the norm of the state $\hat{J}'_{\pm}|jm\rangle$ given by

$$\begin{aligned}\langle jm|\hat{J}'_{\mp}\hat{J}'_{\pm}|jm\rangle &= \langle jm|\hat{J}'^2 - \hat{J}'^2_z \pm \hat{J}'_z|jm\rangle \\ &= \hbar^2\left[j(j+1) - m^2 \pm m\right] \\ &= \hbar^2(j \mp m + 1)(j \pm m) \quad ,\end{aligned} \tag{9}$$

and the nonvanishing matrix elements of the shift operators are

$$\langle j, m \mp 1|\hat{J}'_{\pm}|jm\rangle = \hbar\sqrt{(j \mp m + 1)(j \pm m)} \quad , \tag{10}$$

with those of the cartesian components now given by

$$\begin{aligned}\langle j, m'|\hat{J}'_x|jm\rangle &= \frac{\hbar}{2}\Big[\sqrt{(j+m+1)(j-m)}\,\delta_{m',m+1} \\ &\qquad + \sqrt{(j-m+1)(j+m)}\,\delta_{m',m-1}\Big] \quad , \\ \langle j, m'|\hat{J}'_y|jm\rangle &= -\frac{\mathrm{i}\hbar}{2}\Big[\sqrt{(j-m+1)(j+m)}\,\delta_{m',m-1} \\ &\qquad - \sqrt{(j+m+1)(j-m)}\delta_{m',m+1}\Big] \quad , \\ \langle jm'|\hat{J}'_z|jm\rangle &= \hbar m\,\delta_{m'm} \quad .\end{aligned} \tag{11}$$

Note that compared with the space-fixed case the matrix element for $\hat{J}'_x$ is unchanged while that of $\hat{J}'_y$ differs only by an overall minus sign.

Finally it is useful to remark that

$$\hat{J}^2 = \hat{J}'^2 \quad , \tag{6.147}$$

which is plausible as this quantity is itself a scalar and thus does not depend on the particular axes defining the components of $\hat{\boldsymbol{J}}'$.

The second problem concerns which rotations are actually allowed. A classical rotor can rotate about any of its axes. In quantum mechanics, however, there is a special case if the nucleus has rotational symmetries *and no internal structure*. For example, a spherical nucleus cannot rotate, because any rotation leaves the surface invariant and thus by definition does not change the quantum-mechanical state. This may be hard to accept at first, but keep in mind that there is nothing "inside the nucleus" to change its position during the rotation in this limit and there is nothing marked on the surface to define the orientation! We thus can conclude that

- a spherical nucleus has no rotational excitations at all, and
- a nucleus with axial symmetry cannot rotate around the axis of symmetry.

The final decision about the validity of these statements has to come from experiment, of course; it will depend on whether other degrees of freedom are involved. We shall see that rotations about a symmetry axis are made possible by simultaneous dynamic deviations from axial symmetry.

The Hamiltonian for a rigid rotor thus comes in two variations:

$$\hat{H} = \sum_{i=1}^{3} \frac{\hat{J}_i'^2}{2\Theta_i} \tag{6.148}$$

for a triaxial nucleus with three different moments of inertia, and

$$\hat{H} = \frac{1}{2\Theta} \left(\hat{J}_x'^2 + \hat{J}_y'^2\right) \tag{6.149}$$

for nuclei with axial symmetry about the z axis.

Quantum numbers for the rotor will be generated by the space-fixed operators $\hat{J}^2$ and $\hat{J}_z$, because the energy of the nucleus does not depend on its orientation in space. The corresponding quantum numbers are conventionally called I and M, respectively, so that the eigenvalues are $\hbar^2 I(I+1)$ and $\hbar M$. In addition, if the nucleus is axially symmetric about the body-fixed z axis, $\hat{J}_z'$ will also generate a good quantum number, which is denoted by K (actually, K will turn out to be restricted to zero for the rigid rotor).

Before entering a discussion of the spectra resulting from these Hamiltonians, we derive the wave functions. In principle one may write down the explicit expressions for the body-fixed angular-omentum operators in terms of the Euler angles and then solve the resulting very complicated differential equations, but there is a much simpler way, which is based on the fact that the wave functions depend on the very coordinates which also describe the symmetry of the system. To explain this derivation it may be helpful to first examine a simpler case, namely that of the free particle.

For the free particle in one dimension the only degree of freedom is the x coordinate, and the Hamiltonian

$$\hat{H} = \frac{\hat{p}_x^2}{2m} \tag{6.150}$$

is translationally invariant, so that it commutes with the momentum operator, and the eigenstates of $\hat{H}$ are also eigenstates of $\hat{p}_x$, with an eigenvalue of, say, k. Call the wave function $\psi_k(x)$. Applying the translation operator transformation of Sect. 2.2 to go from x_0 to $x_0 + x$, we can write

$$\psi_k'(x_0) = \psi(x_0 - x) = \exp\left(-\tfrac{\mathrm{i}}{\hbar} x\, \hat{p}_x\right) \psi(x_0) = \mathrm{e}^{-ikx}\, \psi(x_0) \quad . \tag{6.151}$$

But this implies that if $\psi(x_0)$ is known, then the wave function is determined in all of space. Using $x_0 = 0$, substituting $x \to -x$, and writing $\psi(0) = \psi_0$, we have

$$\psi(x) = \psi_0\, \mathrm{e}^{ikx} \quad , \tag{6.152}$$

which is the standard plane-wave solution, but obtained purely from symmetry considerations and the general expression for finite transformations under that symmetry.

Now apply these ideas to rotation. Again we know how to rotate a wave function, and this will determine the values at arbitrary Euler angles in terms of those at a specific set. We deal with the triaxial case first: the wave function will have quantum numbers I and M. Denote the wave function by $\phi_{IM}(\boldsymbol{\theta})$. It can be rotated through another set of Euler angles $\boldsymbol{\theta}_1$ according to

$$\phi_{IM}'(\boldsymbol{\theta}) = \sum_{M'} \mathcal{D}^{(I)}_{M'M}(\boldsymbol{\theta}_1)\, \phi_{IM'}(\boldsymbol{\theta}) \quad . \tag{6.153}$$

But by definition

$$\phi_{IM}'(\boldsymbol{\theta}) = \phi_{IM}(\boldsymbol{\theta}') \quad , \tag{6.154}$$

where $\boldsymbol{\theta}'$ is that orientation which is carried over into $\boldsymbol{\theta}$ by the rotation $\boldsymbol{\theta}_1$ (Compared with the translation example, the correspondence is $x_0 \sim \boldsymbol{\theta}$, $x \sim \boldsymbol{\theta}_1$, and $x_0 - x \sim \boldsymbol{\theta}'$). Combining these two equations yields

$$\phi_{IM}(\boldsymbol{\theta}') = \sum_{M'} \mathcal{D}^{(I)}_{M'M}(\boldsymbol{\theta}_1)\, \phi_{IM'}(\boldsymbol{\theta}) \quad , \tag{6.155}$$

and as for the case of translation we need the wave function only at one specific set of arguments, for example at the point $\boldsymbol{\theta}' = \mathbf{0}$ (i.e., for all Euler angles equal to zero). Then one must assume $\boldsymbol{\theta} = \boldsymbol{\theta}_1$, and the relation (6.155) can be inverted using the unitarity of the rotation matrices to yield

$$\phi_{IM}(\boldsymbol{\theta}) = \sum_{M'} \mathcal{D}^{(I)*}_{MM'}(\boldsymbol{\theta})\, \phi_{IM'}(\mathbf{0}) \quad . \tag{6.156}$$

The wave function consequently is determined completely by the values $\phi_{IM'}(\mathbf{0})$. They can, of course, be normalized.

6.4.2 The Symmetric Rotor

For an axially symmetric nucleus, the wave function may be simplified further. The nucleus then does not change its orientation under a rotation given by θ_3, and when we look at the decomposition of the rotation matrices given in Sect. 2.3.3, this implies that M' is also a good quantum number, so that the sum in (6.156) must drop out. In fact, because of the definition of θ_3 as a rotation angle about the body-fixed z axis, $\hbar M'$ must be the eigenvalue of J_z', and should be denoted by K. The eigenfunction for the symmetric rotor is thus essentially

$$\phi_{IMK}(\boldsymbol{\theta}) = \mathcal{D}_{MK}^{(I)*}(\boldsymbol{\theta}) \quad . \tag{6.157}$$

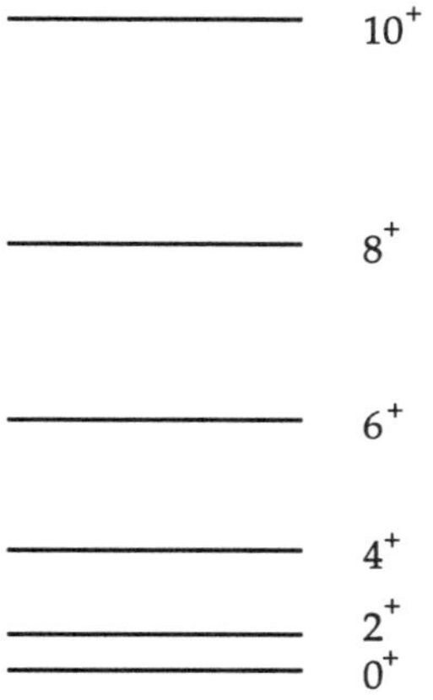

Fig. 6.8 The spectrum of a rigid symmetric rotor, showing the typical $I(I+1)$ dependence of the energies. Only even angular momenta appear.

This wave function still has to be normalized, and we will show shortly that the symmetries discussed in Sect. 6.2.4 will also cause a modification.

The energy of the axially symmetric rotating nucleus (6.149) can now be written in terms of the quantum numbers:

$$E = \frac{\hbar^2\left[I(I+1) - K^2\right]}{2\Theta} \quad , \tag{6.158}$$

where $\hat{J}_x'^2 + \hat{J}_y'^2 = \hat{J}^2 - \hat{J}_z'^2$ was used.

The application of the required symmetries as in Sect. 6.2.4 is straightforward, but must be modified slightly. In Sect. 6.2.4 the intrinsic axes could be chosen arbitrarily along the three principal axes defined by the nuclear shape. Here, in contrast, the z' axis is fixed by the condition that it be the symmetry axis of the nucleus, but it can still be chosen in two opposite directions. The x' and y' axes can be rotated arbitrarily around the z' axis because of the axial symmetry, which makes any direction perpendicular to the z' axis a principal axis. This second ambiguity is similar to $\hat{R}_2$, but with an arbitrary rotation angle η:

$$\hat{R}_2'(\theta_1, \theta_2, \theta_3) = (\theta_1, \theta_2, \theta_3 + \eta) \quad . \tag{6.159}$$

The semiexplicit form of the rotation matrices given at the end of Sect. 2.3.3 is sufficient to derive that

$$\phi_{IMK}(\theta_1, \theta_2, \theta_3 + \eta) = \mathrm{e}^{\mathrm{i}\eta K}\, \phi(\theta_1, \theta_2, \theta_3) \quad , \tag{6.160}$$

so that for invariance we need $K = 0$. This is a more formal derivation of the aforementioned impossibility of rotations about a symmetry axis.

$\hat{R}_3$ drops out because it changes the z' axis. The other basic transformation that should be applied is the inversion of the z' axis corresponding to $\hat{R}_1$ of Sect. 6.2.4. It acts on the Euler angles as

$$\hat{R}_1(\theta_1, \theta_2, \theta_3) = (\theta_1 + \pi, \pi - \theta_2, -\theta_3) \quad . \tag{6.161}$$

As $K = 0$, the inversion of the sign of θ_3 does not change the wave function, whereas the increment in θ_1 produces a phase factor of $\exp(\mathrm{i}M\pi)$. The remaining change due to θ_2 can be derived from the symmetry relation

$$d_{mm'}^{(j)}(\theta_2) = (-1)^{j-m}\, d_{m-m'}^{(j)}(\pi - \theta_2) \quad . \tag{6.162}$$

Together they yield

$$\hat{R}_1 \phi_{IMK}(\boldsymbol{\theta}) = (-1)^I \phi_{IMK}(\boldsymbol{\theta}) \quad , \tag{6.163}$$

and the implication is that the total angular momentum has to be even.

We can thus summarize the eigenstates of the rigid rotor. The states are

$$|IM\rangle, \quad I = 0, 2, 4, \ldots \quad , \quad M = -I, \ldots, +I \quad , \tag{6.164}$$

degenerate only trivially through the laboratory projection M. The eigenenergies are

$$E_I = \frac{\hbar^2 I(I+1)}{2\Theta} \quad . \tag{6.165}$$

The corresponding spectrum is that of a typical *rotational band* shown in Fig. 6.8. Although this sort of band appears in exceedingly many nuclei, the present case of only one such band it not very realistic, because any additional degree of freedom will lead to other bands based on excitations of those degrees of freedom. One such example, the coupling to vibrations in the β and γ coordinates, will be treated soon.

It remains for us to compute the normalization of the eigenstates. The normalization integral for the rotation matrices is given by

$$\int \mathrm{d}^3\theta \, \mathcal{D}^{(I_1)*}_{M_1 K_1}(\boldsymbol{\theta}_1) \, \mathcal{D}^{(I_2)}_{M_2 K_2}(\boldsymbol{\theta}_2) = \frac{8\pi^2}{2I_1 + 1} \delta_{I_1 I_2} \, \delta_{M_1 M_2} \, \delta_{K_1 K_2} \quad , \tag{6.166}$$

so that the normalized wave function is given by

$$\phi_{IMK}(\boldsymbol{\theta}) = \sqrt{\frac{2I+1}{8\pi^2}} \, \mathcal{D}^{(I)}_{MK}(\boldsymbol{\theta}) \quad . \tag{6.167}$$

The calculation of the transition probabilities is deferred to the discussion of the more general rotation–vibration model. Here we only cite the result

$$B\left(\mathrm{E2}; I_i \rightarrow I_f\right) = \left(\frac{3ZR_0^2}{4\pi}\right)^2 \beta_0^2 \left(1 + \frac{2}{7}\sqrt{\frac{5}{\pi}}\beta_0\right)^2 \tag{6.168}$$

for the only transitions present in this model. This result contains the deformation β_0 of the nucleus and not its moment of inertia, because the former appears in the quadrupole operator; we will later find a connection between the moment of inertia and the deformation (see Sect. 6.5.1), but this contains the (unknown) mass parameter, so one may as well accept a new parameter in the form of β_0. Clearly β_0 is directly determined by one transition probabilities, while one level spacing fixes the moment of inertia; together then these two values determine the model completely. In practice, of course, the simple angular-momentum dependence of the model will be approximated for only the lowest states, and the deviations for higher angular momentum indicate the change of deformation and/or moment of inertia with rotation.

6.4.3 The Asymmetric Rotor

A more interesting spectrum is exhibited by the asymmetric rotor, which was first investigated by Davydov and Filippov [Da58]. This model assumes a rotational Hamiltonian as in (6.148) with all three moments of inertia different. The eigenfunctions constructed in the preceding two chapters applied to only the case of axial symmetry, leading to K as a good quantum number. Without axial symmetry a more general solution as in (6.156) must be used, but it is better to use suitably symmetrized basis functions instead. Again we have to consider carefully which of the symmetries apply. Because all three intrinsic axes are now assigned to specific and distinct moments of inertia, they are physically distinct and only those symmetries that invert the individual axes should be considered (a change in axis selection cannot be compensated by adjusting the deformation in this case). Looking at their definition, it is clear that one of these is $\hat{R}_2^2$, which corresponds to $(x', y', z') \to (-x', -y', z')$ and transforms the Euler angles to $(\theta_1, \theta_2, \theta_3 + \pi)$. Applied to a rotation matrix it yields a factor of $\exp(\mathrm{i}K\pi)$, so K must be even. The other transformation is still $\hat{R}_1$, which transforms $\mathcal{D}_{MK}^{(I)*}(\boldsymbol{\theta})$ into $(-1)^I \mathcal{D}_{M-K}^{(I)*}(\boldsymbol{\theta})$. Since in this case the result is a wave function that does not differ by only a factor, this does not lead simply to a restriction of the quantum numbers, as in previous cases; instead, the symmetry must be satisfied by symmetrizing the wave function.

The symmetrized combinations to be used as basic functions are therefore defined by

$$\langle \boldsymbol{\theta} | IMK \rangle = \sqrt{\frac{(2I+1)}{(1+\delta_{K0})16\pi^2}} \left(\mathcal{D}_{MK}^{(I)*}(\boldsymbol{\theta}) + (-1)^I \mathcal{D}_{M-K}^{(I)*}(\boldsymbol{\theta}) \right) \quad , \tag{6.169}$$

so positive and negative values of K cannot be used independently. The factor $1 + \delta_{K0}$ takes into account the difference in normalization between $K = 0$ and the other states. The eigenfunction of the asymmetric rotor may now be expanded as

$$|IMi\rangle = \sum_K a_K^{Ii} |IMK\rangle \quad . \tag{6.170}$$

The index i enumerates the different eigenstates for the same I with expansion coefficients a_K^{Ii}. The index K runs over all nonnegative even numbers $0 \le K \le I$.

Note that for odd I the function with $K = 0$ vanishes, so that $K = 0$ is forbidden in this case. Furthermore, for $I = 1$ both projections $K = 0$ and $K = 1$ are forbidden (the latter because it is odd), so the model does not allow any state of $I = 1$. For $I = 0$ there is only one state, which must be the ground state. The third special case is $I = 3$, for which only $K = 2$ is permitted, so there is also only one state with $I = 3$.

In the general case the wave function is expanded in the set of basic functions and the Hamiltonian can be split up into a diagonal part and a remainder by taking the average of the x' and y' rotational energies:

$$\begin{aligned}\hat{H} &= \frac{1}{4}\left(\frac{1}{\Theta_1}+\frac{1}{\Theta_2}\right)\left(\hat{J}_1'^2+\hat{J}_2'^2\right)+\frac{\hat{J}_3'^2}{2\Theta_3}+\frac{1}{4}\left(\frac{1}{\Theta_1}-\frac{1}{\Theta_2}\right)\left(\hat{J}_1'^2-\hat{J}_2'^2\right)\\ &= \frac{1}{4}\left(\frac{1}{\Theta_1}+\frac{1}{\Theta_2}\right)\left(\hat{J}'^2-\hat{J}_3'^2\right)+\frac{\hat{J}_3'^2}{2\Theta_3}\\ &\quad+\frac{1}{2}\left(\frac{1}{\Theta_1}-\frac{1}{\Theta_2}\right)\left(\hat{J}_+'^2+\hat{J}_-'^2\right) \quad . \end{aligned} \tag{6.171}$$

The remainder cannot be explicitly diagonalized, and no generally useful results can be written down. There are, however, interesting special cases for $I = 2$ and $I = 3$. In the latter case, there is as mentioned only one wave function $|3M\rangle = |3MK = 2\rangle$, and the eigenvalue can be evaluated straightforwardly. The result is

$$E(3^+) = 2\hbar^2\left(\frac{1}{\Theta_1}+\frac{1}{\Theta_2}+\frac{1}{\Theta_3}\right) \quad . \tag{6.172}$$

For the two states with $I = 2$ the calculation is a bit more involved and is left to Exercise 6.6; the result is

$$\begin{aligned}E\left(2^+_{1,2}\right) &= \hbar^2\left(\frac{1}{\Theta_1}+\frac{1}{\Theta_2}+\frac{1}{\Theta_3}\right)\\ &\mp\left[\left(\frac{1}{\Theta_1}+\frac{1}{\Theta_2}+\frac{1}{\Theta_3}\right)^2-\frac{3(\Theta_1-\Theta_2)^2}{8\Theta_1^2\Theta_2^2}\right.\\ &\left.-3\left(\frac{1}{\Theta_1\Theta_2}+\frac{1}{\Theta_1\Theta_2}+\frac{1}{\Theta_2\Theta_3}\right)\right]^{1/2} \quad . \end{aligned} \tag{6.173}$$

This result appears at first to be not very enlightening, but if we compare it with the energy of the 3^+ state the interesting property

$$E\left(2^+_1\right)+E\left(2^+_2\right) = E\left(3^+\right) \tag{6.174}$$

is immediately apparent. This is a clear prediction of the model, which can be easily checked through experiment.

EXERCISE

6.6 States with Angular Momentum 2 in the Asymmetric Rotor Model

Problem. Calculate the states with angular momentum 2 in the asymmetric rotor model.

Solution. For angular momentum 2 there are two basis states, namely the two states with $K = 0$ and $K = 2$, which we denote by $|20\rangle$ and $|22\rangle$, respectively. The diagonal part of the Hamiltonian (6.173) yields the matrix elements

$$\begin{aligned}\langle 20|\hat{H}|20\rangle &= \frac{3\hbar^2}{2}\left(\frac{1}{\Theta_1}+\frac{1}{\Theta_2}\right) \quad ,\\ \langle 22|\hat{H}|22\rangle &= \frac{\hbar^2}{2}\left(\frac{1}{\Theta_1}+\frac{1}{\Theta_2}\right)+\frac{2\hbar^2}{\Theta_3} \quad . \end{aligned} \tag{1}$$

Exercise 6.6

For the nondiagonal matrix elements we note that from the matrix elements of the shift operators given in Exercise 6.5 we have

$$\hat{J}_-'^2|20\rangle = 2\hbar^2\sqrt{6}|22\rangle \quad , \tag{2}$$

so that

$$\langle 22|\hat{H}|20\rangle = \langle 20|\hat{H}|22\rangle = \hbar^2\sqrt{6}\left(\frac{1}{\Theta_1} - \frac{1}{\Theta_2}\right) \quad . \tag{3}$$

This determines the 2×2 matrix completely, and its eigenvalues λ can be obtained from the solution of the secular equation:

$$\det\begin{pmatrix} \frac{3\hbar^2}{2}\left(\frac{1}{\Theta_1} + \frac{1}{\Theta_2}\right) - \lambda & \hbar^2\sqrt{6}\left(\frac{1}{\Theta_1} - \frac{1}{\Theta_2}\right) \\ \hbar^2\sqrt{6}\left(\frac{1}{\Theta_1} - \frac{1}{\Theta_2}\right) & \frac{\hbar^2}{2}\left(\frac{1}{\Theta_1} + \frac{1}{\Theta_2}\right) + \frac{2\hbar^2}{\Theta_3} - \lambda \end{pmatrix} = 0 \quad . \tag{4}$$

A lengthy but trivial calculation now yields the result given in (6.173).

6.5 The Rotation–Vibration Model

6.5.1 Classical Energy

The most important special case of collective surface motion is that of well-deformed nuclei, whose energies have a deep axially deformed minimum like in the simple rotor model treated in Sect. 6.4, but with the additional feature of small oscillations around that minimum in both the β and γ degrees of freedom. As the harmonic vibrations are easier to deal with in cartesian coordinates, we set up the potential in (a_0, a_2) instead of in (β, γ). Assuming the potential minima to be at $a_0 = \beta_0$ and $a_2 = 0$, we introduce the deviations from the equilibrium deformation as

$$\xi = a_0 - \beta_0 \quad , \quad \eta = a_2 = a_{-2} \tag{6.175}$$

and expand the potential as

$$V(\xi, \eta) = \tfrac{1}{2}C_0\xi^2 + C_2\eta^2 \quad . \tag{6.176}$$

Note the additional factor of 2 in the η-dependent part, which takes into account that η really stands for the two coordinates a_2 and a_{-2}. Also the potential at the minimum was assumed to vanish; this is allowed since only energy differences are meaningful in this model.

The construction of the kinetic energy is much more involved. The harmonic kinetic energy of the vibrational model,

$$T = \tfrac{1}{2}B_2\sum_\mu |\dot{\alpha}_{2\mu}|^2 \quad , \tag{6.177}$$

has to be transformed into the new set of coordinates (ξ,η) and orientation angles and then must be quantized. As in the case of the rigid rotor in Sect. 6.4.1 the kinetic energy should be expressed in terms of the angular momenta about the *body-fixed* axes, because only in this way will the moments of inertia be constant during the rotation. This implies that for writing down the classical kinetic energy the angular velocities ω_i', $i=1,2,3$ about the intrinsic axes should be used.

As the kinetic energy does not depend on the orientation of the nucleus, only on its rate of change, we are free to evaluate it for a specially chosen orientation, for which it is natural to use the idea that the body-fixed axes are equal to the space-fixed ones. In this case we have simply

$$\alpha_0=\beta_0+\xi \quad , \quad \alpha_1=\alpha_{-1}=0 \quad , \quad \alpha_2=\alpha_{-2}=\eta \quad . \tag{6.178}$$

The desired transformation is now given by the chain rule of differentiation:

$$\dot{\alpha}_{2\mu}=\sum_k \frac{\partial\alpha_{2\mu}}{\partial\theta_k'}\omega_k'+\frac{\partial\alpha_{2\mu}}{\partial\xi}\dot{\xi}+\frac{\partial\alpha_{2\mu}}{\partial\eta}\dot{\eta} \quad , \tag{6.179}$$

and the only nontrivial derivative is that with respect to the angles of rotation θ_k' about the body-fixed axes. The results from Exercise 6.7, rewritten using ξ and η, are

$$\begin{aligned}
\dot{\alpha}_{20}&=\dot{\xi} \quad ,\\
\dot{\alpha}_{2\pm1}&=-\tfrac{\mathrm{i}}{2}\left[\sqrt{6}(\beta_0+\xi)+2\eta\right]\omega_1'\pm\tfrac{1}{2}\left[\sqrt{6}(\beta_0+\xi)-2\eta\right]\omega_2' \quad ,\\
\dot{\alpha}_{2\pm2}&=\dot{\eta}\mp 2\mathrm{i}\eta\omega_3' \quad .
\end{aligned} \tag{6.180}$$

Inserting these into (6.177), we get a decoupled expression in the sense that there are no mixed terms in the velocities:

$$\begin{aligned}
T&=\tfrac{1}{2}B\left(\dot{\xi}^2+2\dot{\eta}^2\right)+4B\eta^2\omega_3'^2\\
&\quad+\frac{B}{4}\left[\sqrt{6}(\beta_0+\xi)+2\eta\right]^2\omega_1'^2\\
&\quad+\frac{B}{4}\left[\sqrt{6}(\beta_0+\xi)-2\eta\right]^2\omega_2'^2 \quad .
\end{aligned} \tag{6.181}$$

This may be written more succinctly as

$$T=\tfrac{1}{2}B\left(\dot{\xi}^2+2\dot{\eta}^2\right)+\tfrac{1}{2}\sum_{k=1}^{3}\mathcal{J}_k\omega_k'^2 \tag{6.182}$$

with the *moments of inertia* as

$$\begin{aligned}
\mathcal{J}_1&=\frac{B}{2}\left[\sqrt{6}(\beta_0+\xi)+2\eta\right]^2=4B\beta^2\sin^2\left(\gamma-\tfrac{2}{3}\pi\right) \quad ,\\
\mathcal{J}_2&=\frac{B}{2}\left[\sqrt{6}(\beta_0+\xi)-2\eta\right]^2=4B\beta^2\sin^2\left(\gamma-\tfrac{4}{3}\pi\right) \quad ,\\
\mathcal{J}_3&=8B\eta^2=4B\beta^2\sin^2\gamma \quad .
\end{aligned} \tag{6.183}$$

The expressions in the (β,γ) notation are also given and can be summarized as

$$\mathcal{J}_k = 4B\beta^2 \sin^2\left(\gamma - \tfrac{2}{3}\pi k\right) \quad . \tag{6.184}$$

Note that they contain $\beta = \beta_0 + \xi$.

The kinetic energy thus neatly splits up into a vibrational part,

$$T_{\text{vib}} = \tfrac{1}{2}B\left(\xi^2 + 2\eta^2\right) \quad , \tag{6.185}$$

and a rotational part,

$$T_{\text{rot}} = \frac{B}{4}\left[\sqrt{6}(\beta_0 + \xi) + 2\eta\right]^2 \omega_1'^2 + \frac{B}{4}\left[\sqrt{6}(\beta_0 + \xi) - 2\eta\right]^2 \omega_2'^2 + 4B\eta^2\omega_3'^2 \quad . \tag{6.186}$$

Both are coupled by the dependence of the moments of inertia on the deformation parameters, while the vibrational kinetic energy contains simple constant mass parameters B for both degrees of freedom. Note the factor of 2 in the η-dependent contribution, which arises as usual from the fact that η represents both a_2 and a_{-2}.

At this point it simplifies things if only the lowest order in ξ and η is considered. The only moment of inertia where the dependence on ξ and η is not a small correction is $\mathcal{J}_3$, which is zero for $\eta = 0$. Physically this describes the fact that for $\eta = 0$ the nucleus is axially symmetric around the z' axis and, as mentioned in Sect. 4.3, cannot rotate about this axis. The dependence on η makes such rotations possible *dynamically*, and they are intimately coupled to the dynamic deviations from axial symmetry described by η.

In the other moments of inertia the dependence on ξ and η indicates the dynamic changes in the moments of inertia with deformation and gives rise to an additional coupling between rotations and vibrations. We will not examine this coupling in detail, so for further developments the moments of inertia are given by the lowest-order expressions

$$\mathcal{J} \equiv \mathcal{J}_1 = \mathcal{J}_2 = 3B\beta_0^2 \equiv \mathcal{J} \quad , \quad \mathcal{J}_3 = 8B\eta^2 \quad , \tag{6.187}$$

and the kinetic energy in its simplest form may be written as

$$T = \tfrac{1}{2}B\left(\xi^2 + 2\eta^2\right) + \tfrac{1}{2}\mathcal{J}\left(\omega_1'^2 + \omega_2'^2\right) + 4B\eta^2\omega_3'^2 \quad . \tag{6.188}$$

The classical rigid-body kinetic energy has a quite different energy dependence. Without even calculating it one can see that, for example, $\mathcal{J}_3$ depends on a_0 to lower order than on a_2, because increasing a_0 decreases the mean distance of nuclear matter from the z' axis strongly by stretching the nucleus, while a change in a_2 only redistributes the matter between the x' and y' axes and should leave $\mathcal{J}_3$ unchanged to first order. With the present result, however, increasing a_0 for $a_2 = 0$ does not affect the moment of inertia at all, as the nucleus remains axially symmetric and $\mathcal{J}_3 = 0$.

One should keep in mind that this result is based on the simple harmonic kinetic energy in the laboratory frame and therefore may be quite limited in application. Although this does lead to the correct suppression of rotation about symmetry axes, its main problem is the treatment of rotations and vibrations on the same footing, which makes both the vibrational masses and the moments of inertia depend on the single parameter B. Thus this model seems to imply that measuring the moments of inerta also determines the vibrational masses. Microscopically, as will become apparent in Sect. 9.3, this need not be true. In practical applications, however, the possibly wrong description of the vibrational masses can be hidden by the parameters in the potential. In any case the expression of (6.188) has been highly successful.

EXERCISE

6.7 The Time Derivatives of the $\alpha_{2\mu}$

Problem. Derive the time derivatives of the $\alpha_{2\mu}$ using (6.179).

Solution. In this exercise we will not restrict the discussions to small deviations from the equilibrium deformation as given by (6.175), but work with arbitrary a_0 and a_2. From the exponential representation of the rotations we get for small angles

$$\begin{aligned}\frac{\partial\alpha_{2\mu}}{\partial\theta'_k} &= \frac{\partial}{\partial\theta'_k}\exp\left(-\tfrac{\mathrm{i}\theta'_k}{\hbar}\hat{J}'_k\right)\alpha_{2\mu} \\ &= -\tfrac{\mathrm{i}}{\hbar}\hat{J}'_k\alpha_{2\mu} \\ &= -\tfrac{\mathrm{i}}{\hbar}\sum_\nu\langle 2\mu|\hat{J}'_k|2\nu\rangle\alpha_{2\nu} \quad ,\end{aligned} \tag{1}$$

in which the matrix elements of $\hat{J}'_k$ have to be inserted. These were derived in Exercise 6.5 and are given by

$$\begin{aligned}\langle j\,,m'|\hat{J}'_x|j\;m\rangle &= \tfrac{\hbar}{2}\big[\sqrt{(j\;+m+1)(j\;-m)}\;\delta_{m',m+1} \\ &\qquad + \sqrt{(j\;-m+1)(j\;+m)}\;\delta_{m',m-1}\big] \quad , \\ \langle j\,,m'|\hat{J}'_y|j\;m\rangle &= -\tfrac{\mathrm{i}\hbar}{2}\big[\sqrt{(j\;-m+1)(j\;+m)}\;\delta_{m',m-1} \\ &\qquad - \sqrt{(j\;+m+1)(j\;-m)}\;\delta_{m',m+1}\big] \quad , \\ \langle j\;m'|\hat{J}'_z|j\;m\rangle &= \hbar m\;\delta_{m'm} \quad .\end{aligned} \tag{2}$$

For the z' component $\langle 2\mu|\hat{J}'_3|2\nu\rangle = \hbar\mu\delta_{\mu\nu}$ and with the special expressions for the $\alpha_{2\mu}$ given above the only nonzero terms turn out to be

$$\frac{\partial\alpha_{2\pm2}}{\partial\theta'_z} = \mp 2\mathrm{i}a_2 \quad . \tag{3}$$

For the x' and y' components we immediately note that the contributions to the derivatives of α_{20} and $\alpha_{2\pm2}$ in (1) vanish, because the operators couple these components only to $\alpha_{2\pm1}$, which are zero themselves. For the remaining components the calculation is now trivial:

$$\begin{aligned}\frac{\partial\alpha_{2\pm1}}{\partial\theta'_x} &= -\tfrac{\mathrm{i}}{2}\left(\sqrt{6}\,a_0 + 2a_2\right) \quad , \\ \frac{\partial\alpha_{2\pm1}}{\partial\theta'_y} &= \pm\tfrac{1}{2}\left(\sqrt{6}\,a_0 - 2a_2\right) \quad ,\end{aligned} \tag{4}$$

and all the ingredients have been assembled, so that the final result is

Exercise 6.7

$$\begin{aligned}\alpha_{20} &= a_0 \quad ,\\ \alpha_{2\pm1} &= -\tfrac{\mathrm{i}}{2}\left(\sqrt{6}\,a_0 + 2a_2\right)\omega_1' \pm \tfrac{1}{2}\left(\sqrt{6}\,a_0 - 2a_2\right)\omega_2' \quad ,\\ \alpha_{2\pm2} &= a_2 \mp 2\mathrm{i}a_2\omega_3' \quad .\end{aligned} \tag{5}$$

6.5.2 Quantal Hamiltonian

The next step is the quantization. One is tempted simply to repeat the standard procedure followed in Sect. 6.3.2: write down the Lagrangian $L = T - V$, determine the conjugate momenta and the Hamiltonian, and then demand the canonical commutation relations between coordinates and momenta. In this case, however, this is not so simple, because the coordinates ξ, η, and $\boldsymbol{\theta}'$ define a *curvilinear coordinate system.* To see what this means and implies, examine the simple example of polar coordinates in the plane.

In classical mechanics the Lagrangian in cartesian coordinates,

$$L = \tfrac{m}{2}\left(x^2 + y^2\right) - V(x,y) \quad , \tag{6.189}$$

is transformed into polar coordinates (r,ϕ) to become

$$L = \tfrac{m}{2}\left(r^2 + r^2\phi^2\right) - V(r,\phi) \quad . \tag{6.190}$$

The conjugate momenta are

$$p_r = \frac{\partial L}{\partial r} = mr \quad , \quad p_\phi = \frac{\partial L}{\partial \phi} = mr^2\phi \quad , \tag{6.191}$$

and the classical Hamiltonian becomes

$$H = \frac{1}{2m}\left(p_r^2 + \frac{p_\phi^2}{r^2}\right) + V(r,\phi) \quad . \tag{6.192}$$

Going over to momentum operators by requiring

$$\left[r,\hat{p}_r\right] = \mathrm{i}\hbar \quad , \quad \left[\phi,\hat{p}_\phi\right] = \mathrm{i}\hbar \tag{6.193}$$

leads to

$$\hat{p}_r = -\mathrm{i}\hbar\frac{\partial}{\partial r} \quad , \quad \hat{p}_\phi = -\mathrm{i}\hbar\frac{\partial}{\partial \phi} \quad , \tag{6.194}$$

and inserting this into (6.192) yields the operator

$$\hat{H}_{\text{wrong}} = -\frac{\hbar^2}{2m}\left(\frac{\partial^2}{\partial r^2} + \frac{\partial^2}{\partial \phi^2}\right) + V(r,\phi) \quad , \tag{6.195}$$

which does not agree with the well-known result obtained by performing a coordinate transformation in the quantized cartesian expression:

$$\hat{H} = -\frac{\hbar^2}{2m}\left(\frac{\partial^2}{\partial x^2} + \frac{\partial^2}{\partial y^2}\right) + V(x,y)$$
$$= -\frac{\hbar^2}{2m}\left(\frac{1}{r}\frac{\partial}{\partial r}r\frac{\partial}{\partial r} + \frac{1}{r^2}\frac{\partial}{\partial \phi^2}\right) + V(r,\phi) \quad . \tag{6.196}$$

What is the reason for this discrepancy? In the classical expression the first part of the kinetic energy might be written either as p_r^2 or as $\frac{1}{r}p_r r p_r$, which are identical classically, but give rise to two different kinetic energy operators when p_r is made an operator which does not commute with r. Such ambiguities may arise whenever the kinetic energy contains coordinate-dependent factors, which is typically the case for curvilinear coordinates. We see from this simple example that the correct version can be selected only by basing the quantization on a cartesian-like coordinate system. "Cartesian" in this context means a system in which the kinetic energy is purely quadratic in the velocities with no coordinate-dependent coefficients.

In the general case, where the associated cartesian coordinate system may not even be known, the procedure to be followed has been developed by *Podolsky* [Po28] and consists in applying the formalism of curvilinear coordinates. In our case, the Hamiltonian written in the $\alpha_{2\mu}$ is cartesian and has been quantized in Sect. 6.3.2, so that we can also transform the kinetic energy operator given by

$$\hat{T} = \frac{1}{2B}\sum_{\mu=-2} 2(-1)^{\mu}\,\hat{\pi}_{2\mu}\,\hat{\pi}_{2-\mu} = -\frac{\hbar^2}{2B}\sum_{\mu=-2}^{2}\frac{\partial}{\partial\alpha_{2\mu}}\frac{\partial}{\partial\alpha_{2-\mu}} \tag{6.197}$$

to the new coordinates $(\xi, \eta, \boldsymbol{\theta}')$.

In this book only the simple case of the lowest-order energy according to (6.188) will be treated, so that $\xi, \eta \ll \beta_0$ is assumed. The transformation can be calculated most easily by using results from differential geometry. In curvilinear coordinates, x_i, the line element is generally given by the expression

$$\mathrm{d}s^2 = \sum_{ij} g_{ij}\,\mathrm{d}x_i\,\mathrm{d}x_j \tag{6.198}$$

with g_{ij} the *metric tensor*. The Laplacian operator in these coordinates then becomes

$$\Delta = \frac{1}{\sqrt{g}}\sum_{ij}\frac{\partial}{\partial x_i}\left(\sqrt{g}\,g_{ij}^{-1}\frac{\partial}{\partial x_j}\right) \quad , \tag{6.199}$$

where g the determinant of the metric tensor and g^{-1} is its inverse. The kinetic energy operator then becomes

$$\hat{T} = -\frac{\hbar^2}{2B}\Delta \quad . \tag{6.200}$$

The "line element" must be interpreted as the expression appearing in the classical kinetic energy:

$$T = \frac{1}{2}B\frac{\mathrm{d}^2 s}{\mathrm{d}t^2} = \frac{1}{2}B\sum_{\mu=-2}^{2}\frac{|\mathrm{d}\alpha_{2\mu}|^2}{\mathrm{d}t^2} \quad , \tag{6.201}$$

from which we read off

$$ds^2 = \sum_{\mu=-2}^{2} |d\alpha_{2\mu}|^2 d\xi^2 + 2d\eta^2 + \sum_k \frac{\mathcal{J}_k d\theta_k'^2}{B} \quad . \tag{6.202}$$

The metric tensor thus is diagonal with

$$g_{\xi\xi} = 1 \quad , \quad g_{\eta\eta} = 2 \quad , \quad g_{\theta_k'\theta_k'} = \mathcal{J}_k/B \quad , \quad k = 1,2,3 \quad , \tag{6.203}$$

and its inverse can be computed trivially as

$$g_{\xi\xi}^{-1} = 1 \quad , \quad g_{\eta\eta}^{-1} = \tfrac{1}{2} \quad , \quad g_{\theta_k'\theta_k'}^{-1} = B/\mathcal{J}_k \quad . \tag{6.204}$$

The determinant is

$$g = 2B^{-3}\mathcal{J}^2\mathcal{J}_3 \quad . \tag{6.205}$$

Inserting these expressions into the Laplacian is quite simple because of the diagonal structure and because they depend only on η. Thus in all terms except the η-dependent one, where its derivative must be considered, the $\sqrt{g}$ term cancels; this leads to a first derivative in η appearing:

$$\begin{aligned}\hat{T} &= \frac{-\hbar^2}{2B}\left(\frac{\partial^2}{\partial\xi^2} + \frac{1}{2}\frac{\partial^2}{\partial\eta^2} + \frac{1}{2\sqrt{g}}\frac{\partial\sqrt{g}}{\partial\eta}\frac{\partial}{\partial\eta} + \sum_k \frac{B}{\mathcal{J}_k}\frac{\partial^2}{\partial\theta_k'^2}\right) \\ &= \frac{-\hbar^2}{2B}\left(\frac{\partial^2}{\partial\xi^2} + \frac{1}{2}\frac{\partial^2}{\partial\eta^2} + \frac{1}{2\eta}\frac{\partial}{\partial\eta}\right) + \frac{\hat{J}_1'^2 + \hat{J}_2'^2}{2\mathcal{J}} + \frac{\hat{J}_3'^2}{16B\eta^2} \quad . \end{aligned} \tag{6.206}$$

An important additional point is the determination of the volume element. According to differential geometry the volume element is the product of the differentials of the variables multiplied by $\sqrt{|g|}$. In our case this yields

$$dV = 4|\eta|\, B^{-1}\mathcal{J}\, d\xi\, d|\eta|\, d\theta_1'\, d\theta_2'\, d\theta_3' \quad . \tag{6.207}$$

Note that here the absolute value of η has to appear, because the volume element must be positive. A constant factor in dV is unimportant, since it will be hidden in the normalization of the wave function. In addition, in practical calculations the rotation angles about the intrinsic axes θ_k' will be replaced by the Euler angles, since we have seen that these are more appropriate for global considerations and from the treatment of the rigid rotor in Sect. 6.4 it may be expected that the angular dependence of the wave function contains rotation matrices with the Euler angles as arguments. So we write the angular part in a abbreviated form as $d^3\theta$ and take the volume element as

$$dV = d\xi\, |\eta|\, d\eta\, d^3\theta \quad . \tag{6.208}$$

This implies that the solutions of the collective Schrödinger equation, which has now taken the form

$$\hat{H}_1(\xi,\eta,\boldsymbol{\theta})\,\psi(\xi,\eta,\boldsymbol{\theta}) = E\,\psi(\xi,\eta,\boldsymbol{\theta}) \quad , \tag{6.209}$$

with

$$\hat{H}_1(\xi,\eta,\boldsymbol{\theta}) = \frac{-\hbar^2}{2B}\left(\frac{\partial^2}{\partial\xi^2} + \frac{1}{2}\frac{\partial^2}{\partial\eta^2} + \frac{1}{2\eta}\frac{\partial}{\partial\eta}\right) + \frac{\hat{J}_1'^2 + \hat{J}_2'^2}{2\mathcal{J}} + \frac{\hat{J}_3'^2}{16B\eta^2} + \frac{1}{2}C_0\xi^2 + C_2\eta^2 \quad , \tag{6.210}$$

must be normalized according to

$$\int \mathrm{d}\xi\,|\eta|\,\mathrm{d}\eta\,\mathrm{d}^3\theta\;\psi^*(\xi,\eta,\boldsymbol{\theta})\,\psi(\xi,\eta,\boldsymbol{\theta}) = 1 \quad . \tag{6.211}$$

The same type of integration must be used in all matrix elements.

The index 1 in the Hamiltonian indicates that it is only a provisional form, because there is a trick to simplify both the volume element and the Schrödinger equation itself. Define a new wave function ϕ via

$$\phi(\xi,\eta,\boldsymbol{\theta}) = \sqrt{|\eta|}\;\psi(\xi,\eta,\boldsymbol{\theta}) \quad . \tag{6.212}$$

The matrix elements are then simplified to

$$\int \mathrm{d}\xi\,\mathrm{d}\eta\,\mathrm{d}^3\theta\;\phi^*(\xi,\eta,\boldsymbol{\theta})\,\phi(\xi,\eta,\boldsymbol{\theta}) = 1 \quad . \tag{6.213}$$

In the Schrödinger equation the additional factor leads to modifications only in the η derivatives. We have

$$\left(\frac{\partial^2}{\partial\eta^2} + \frac{1}{\eta}\frac{\partial}{\partial\eta}\right)\psi = \left(\frac{\partial^2}{\partial\eta^2} + \frac{1}{\eta}\frac{\partial}{\partial\eta}\right)\frac{\psi}{\sqrt{|\eta|}} = \frac{1}{\sqrt{|\eta|}}\left(\frac{\partial^2}{\partial\eta^2} + \frac{1}{4\eta^2}\right)\phi \quad , \tag{6.214}$$

where the signs due to the absolute value in the square root cancel. In consequence ϕ fulfills a new Schrödinger equation with the simple second-order derivative in η, but with an additional potential, which because of its dependence on η^{-2} can be combined with the $\hat{J}_3^2$ term:

$$\hat{H}(\xi,\eta,\boldsymbol{\theta})\,\phi(\xi,\eta,\boldsymbol{\theta}) = E\,\phi(\xi,\eta,\boldsymbol{\theta}) \quad , \tag{6.215}$$

with

$$\hat{H}(\xi,\eta,\boldsymbol{\theta}) = \frac{-\hbar^2}{2B}\left(\frac{\partial^2}{\partial\xi^2} + \frac{1}{2}\frac{\partial^2}{\partial\eta^2}\right) + \frac{\hat{J}_1'^2 + \hat{J}_2'^2}{2\mathcal{J}} + \frac{\hat{J}_3'^2 - \hbar^2}{16B\eta^2} + \frac{1}{2}C_0\xi^2 + C_2\eta^2 \quad . \tag{6.216}$$

This version of the Hamiltonian of the rotation–vibration model will be the starting point for studying the solutions in the next section.

From the preceding discussion the reader should bear in mind the ambiguities associated with quantization in curvilinear coordinates and especially that a non-constant volume element can in general be removed in favor of a different kinetic energy operator and a change in the potential energy. In any case, one has to be careful to use Hamiltonians, wave functions, and volume elements in a consistent way.

Before solving the Schrödinger equation given by (6.215) and (6.216), let us briefly note the more general expressions valid if the deformation dependence in the moments of inertia is not neglected. The derivation is quite analogous to what has been presented here, but of course significantly more complicated. The additional term in the Hamiltonian is the *rotation–vibration interaction*, which is given by

$$\begin{aligned}\hat{H}_{\text{vib-rot}} &= \frac{\hat{J}'^2 - \hat{J}_3'^2}{2\mathcal{J}}\left[\frac{2\eta^2}{\beta_0^2} - \frac{2\xi}{\beta_0} + \frac{3\xi^2}{\beta_0^2}\right] \\ &\quad + \frac{\hat{J}_+'^2 + \hat{J}_-'^2}{4\mathcal{J}}\left[2\sqrt{6}\,\frac{\xi\eta}{\beta_0^2} - \frac{2\sqrt{6}}{3}\frac{\eta}{\beta_0}\right] \quad . \end{aligned} \tag{6.217}$$

All the terms arise from an expansion of the deformation-dependent moments of inertia. This version of the Hamiltonian was first given by Faessler and Greiner [Fa62a–b], who found that by using only the elimination of the volume element as given above a simple analytic solution becomes possible.

Finally we also show the form of the Hamiltonian when expressed in β and γ as was done originally by Bohr [Bo52]:

$$\begin{aligned}\hat{H} &= -\frac{\hbar^2}{2B}\left[\frac{1}{\beta^4}\frac{\partial}{\partial\beta}\beta^4\frac{\partial}{\partial\beta} + \frac{1}{\beta^2}\frac{1}{\sin(3\gamma)}\frac{\partial}{\partial\gamma}\sin(3\gamma)\frac{\partial}{\partial\gamma}\right] \\ &\quad + \sum_{k=1}^{3}\frac{\hat{J}_k'^2}{2\mathcal{J}_k} + V(\beta,\gamma) \quad . \end{aligned} \tag{6.218}$$

Here the moments of inertia are written in the general form. This variant of the Hamiltonian can be obtained simply by performing the quantization in this set of variables. The volume element in this case becomes

$$dV = \beta^4 |\sin(3\gamma)|\, d\beta\, d\gamma\, d^3\theta \quad . \tag{6.219}$$

6.5.3 Spectrum and Eigenfunctions

We can now construct the eigenfunctions of the rotation–vibration model. They can be given analytically if the rotation–vibration interaction is neglected, so that the Hamiltonian of (6.216) is used. The solution proceeds through the familiar method of separation of variables; the term in η^{-2} is similar to the centrifugal potential and can be treated in the same way as in familiar problems. The rotational energy in the Hamiltonian will lead to the eigenfunctions of the rigid rotor, while the ξ-dependent and η-dependent parts can obviously be separated, so that a trial wave function takes the form

$$\psi(\xi,\eta,\boldsymbol{\theta}) = \mathcal{D}_{MK}^{(I)*}(\boldsymbol{\theta})\, g(\xi)\, \chi(\eta) \tag{6.220}$$

and additional quantum numbers will be added as needed. Insertion into the Schrödinger equation leads to

$$\left\{\frac{\hbar^2}{2\mathcal{J}}\left[I(I+1)-K^2\right]+\frac{\hbar^2}{16B}\frac{K^2-1}{\eta^2}-\frac{\hbar^2}{2B}\left(\frac{\partial^2}{\partial\xi^2}+\frac{1}{2}\frac{\partial}{\partial\eta^2}\right)\right.$$
$$\left.+\tfrac{1}{2}C_0\xi^2+C_2\eta^2\right\}g(\xi)\chi(\eta)=E\,g(\xi)\,\chi(\eta)\quad . \tag{6.221}$$

Note that K remains a good quantum number, reflecting the fact that the nucleus is axially symmetric around the body-fixed z axis. The next step is to complete the separation of the ξ and η dependence. Using E_0 as the separation constant, this results in

$$\begin{aligned}
E_0 g(\xi) &= \left(-\frac{\hbar^2}{2B}\frac{\mathrm{d}^2}{\mathrm{d}\xi^2}+\tfrac{1}{2}C_0\xi^2\right)g(\xi)\quad ,\\
(E-E_0)\chi(\eta) &= \left\{\frac{\hbar^2}{2\mathcal{J}}\left[I(I+1)-K^2\right]\right.\\
&\quad\left.+\frac{\hbar^2}{16B}\frac{(K^2-1)}{\eta^2}-\frac{\hbar^2}{4B}\frac{\mathrm{d}^2}{\mathrm{d}\eta^2}+C_2\eta^2\right\}\chi(\eta)\quad .
\end{aligned}\tag{6.222}$$

Solving these two differential equations is actually quite simple, as they should be familiar from elementary quantum mechanics. Apparently we have a pure harmonic oscillator in ξ, whose eigenenergies are given by

$$E_{0,n_\beta}=\hbar\omega_\beta\left(n_\beta+\tfrac{1}{2}\right)\quad ,\quad n_\beta=0,1,\ldots\quad , \tag{6.223}$$

with $\omega_\beta=\sqrt{C_0/B}$. The equation for η is known from the three-dimensional harmonic oscillator in spherical coordinates. With $\psi(r,\theta,\phi)=\frac{1}{r}u(r)Y_{lm}(\theta,\phi)$, the radial equation for the harmonic oscillator takes the form

$$\left[-\frac{\hbar^2}{2m}\frac{\mathrm{d}^2u}{\mathrm{d}r^2}+\frac{\hbar^2 l_K(l_K+1)}{2mr^2}+\tfrac{1}{2}m\omega_\gamma^2r^2\right]u(r)=E(r)\,u(r)\quad . \tag{6.224}$$

The reason for the notation l_K and ω_γ will become clear presently. This equation has eigenvalues characterized by a radial quantum number n_γ and given by

$$E_{n_\gamma}=\hbar\omega_\gamma\left(2n_\gamma+l_K+\tfrac{3}{2}\right)\quad ,\quad n_\gamma=0,1,\ldots\quad . \tag{6.225}$$

Comparing (6.222) and (6.224) we make the following identifications:

$$m\longrightarrow 2B\quad ,\quad l_K(l_K+1)\longrightarrow\tfrac{1}{4}\left(K^2-1\right)\quad ,\quad \omega_\gamma\longrightarrow\sqrt{C_2/B}\quad , \tag{6.226}$$

$$E_{n_\gamma}\longrightarrow E-E_0-\frac{\hbar^2}{2\mathcal{J}}\left(I(I+1)-K^2\right)\quad . \tag{6.227}$$

Solving the quadratic equation for l_K yields

$$l_K=\tfrac{1}{2}(-1\pm K)\quad . \tag{6.228}$$

Choosing either sign does not affect the value of $l_K(l_K+1)$; we may thus make the choice $l_K=\frac{1}{2}\left(|K|-1\right)$ to ensure a positive value for l_K in accord with its analog of angular momentum.

Before determining the associated wave functions, we note that the total energy is now given by

$$E_{n_\beta n_\gamma IK} = \hbar\omega_\beta\left(n_\beta + \tfrac{1}{2}\right) + \hbar\omega_\gamma\left(2n_\gamma + \tfrac{1}{2}|K| + 1\right) + \frac{\hbar^2}{2\mathcal{J}}\left[I(I+1) - K^2\right] . \quad (6.229)$$

For the eigenfunctions in the β direction the one-dimensional harmonic oscillator eigenfunctions can be used directly; since usually the evaluation of matrix elements is easier in the second-quantization formalism, we need no details and write them simply as $\langle\xi|n_\beta\rangle$. In the γ direction the result for the three-dimensional harmonic oscillator reads

$$u_{l_K n_\gamma}(\eta) = N_{l_K n_\gamma}\eta^{l_K} \mathrm{e}^{-\lambda\eta^2/2}\,{}_1F_1\left(-n_\gamma, l_K + \tfrac{3}{2}, \lambda\eta^2\right) \quad (6.230)$$

with $\lambda = 2B\omega_\gamma/\hbar$ corresponding to the factor $m\omega/\hbar$ in the oscillator. Using the analogies of (6.229), this translates into

$$\chi_{Kn_\gamma}(\eta) = N_{Kn_\gamma}\sqrt{|\eta|}\,\eta^{K/2}\,\mathrm{e}^{-\lambda\eta^2/2}\,{}_1F_1\left(-n_\gamma, l_K + \tfrac{3}{2}, \lambda\eta^2\right) \quad . \quad (6.231)$$

For completeness we also give the normalization factor:

$$N_{Kn_\gamma} = \frac{\sqrt{\lambda^{l_K+3/2}\Gamma\left(l_K + \frac{3}{2} + n_\gamma\right)}}{\sqrt{n_\gamma!}\;\Gamma\left(l_K + \frac{3}{2}\right)} \quad . \quad (6.232)$$

The only surprising part of the wave function is the absolute value in the factor $\sqrt{|\eta|}$. In the standard harmonic oscillator there is no problem, because the radial coordinate is always positive; here, however, η can also be negative, and one has to worry about this when symmetrizing the wave function. The argument for choosing the absolute value in this place is that such a factor must appear in the wave function because of the corresponding factor in the volume element of (6.208).

When the rotational wave function is added, the total wave function has the structure

$$\psi(\xi,\eta,\boldsymbol{\theta}) = \mathcal{D}^{(I)*}_{MK}(\boldsymbol{\theta})\;\chi_{Kn_\gamma}(\eta)\;\langle\beta|n_\beta\rangle \quad , \quad (6.233)$$

but we still have to apply to this expression the symmetrizations due to the ambiguities in the choice of intrinsic axes. Of the three fundamental symmetry operations studied in Sect. 6.2.4, $\hat{R}_3$ does not apply, because we have fixed the z' axis as the symmetry axis in the ground state, so that a permutation of the axes is not allowed (the same was true for the rigid rotor). Now $\hat{R}_1$ does not affect the (β,γ) coordinates, and its action on the rotation matrices is the same as for the rotor:

$$\hat{R}_1\mathcal{D}^{(I)*}_{MK}(\boldsymbol{\theta}) = (-1)^I\,\mathcal{D}^{(I)*}_{M-K}(\boldsymbol{\theta}) \quad . \quad (6.234)$$

Note the $-K$ in the argument, which comes from the $\mathrm{e}^{\mathrm{i}K\theta_3}$ dependence of the function (in the case of the rotor we had $K = 0$ always). Finally $\hat{R}_2$ inverts η and adds $\frac{1}{2}\pi$ to θ_3, so we pick up a factor $\mathrm{e}^{\mathrm{i}\pi K/2}$ from the rotational eigenfunction. The inversion of η yields an additional factor of $(-1)^{K/2}$ according to (6.231). The total result is

$$\hat{R}_2\psi(\xi,\eta,\boldsymbol{\theta}) = (-1)^K\;\psi(\xi,\eta,\boldsymbol{\theta}) \quad , \quad (6.235)$$

which immediately leads to the requirement that K should be even:

$$K = 0, \pm 2, \pm 4, \dots \quad . \tag{6.236}$$

As for the rigid rotor, $\hat{R}_1$ does not immediately restrict a quantum number. The correct symmetry can be achieved only by explicit symmetrization; the rotational part of the wave function should be given by

$$\mathcal{D}^{(I)*}_{MK} + (-1)^I \mathcal{D}^{(I)*}_{M-K} \tag{6.237}$$

with the other parts of the wave function not affected at all, as the η-dependent wave function contains only $|K|$. It is sufficient then to consider only positive values of K. Note that for $K = 0$ the symmetrized wave function vanishes in the case of odd I, so only even I are allowed in this case.

We are now in a position to summarize the results. Adding a normalization factor to the wave function (which is due to the rotational wave function, as the other ingredients are assumed to be normalized), gives

$$\begin{aligned} &\langle \xi\eta\boldsymbol{\theta} | IMKn_\beta n_\gamma \rangle \\ &\quad = \psi_{IMKn_\beta n_\gamma}(\xi\eta\boldsymbol{\theta}) \\ &\quad = \sqrt{\frac{(2I+1)}{16\pi^2(1+\delta_{K0})}} \left(\mathcal{D}^{(I)*}_{MK} + (-1)^I \mathcal{D}^{(I)*}_{M-K} \right) \chi_{Kn_\gamma}(\eta) \, \langle \xi | n_\beta \rangle \quad . \end{aligned} \tag{6.238}$$

The allowed values of the quantum numbers have been determined (note that for fixed K the angular momentum I must fulfil $I \geq K$, since K is its projection) to be

$$\begin{aligned} K &= 0,\ 2,\ 4, \dots \quad , \\ I &= \begin{cases} K,\ K+1,\ K+2, \dots & \text{for} \quad K \neq 0 \quad , \\ 0,\ 2,\ 4, \dots & \text{for} \quad K = 0 \quad , \end{cases} \\ M &= -I,\ -I+1, \dots, +I \quad , \\ n_\gamma &= 0,\ 1,\ 2, \dots \quad , \\ n_\beta &= 0,\ 1,\ 2, \dots \quad , \end{aligned} \tag{6.239}$$

and for completeness the energy formula is repeated:

$$E_{n_\beta n_\gamma IK} = \hbar\omega_\beta \left(n_\beta + \tfrac{1}{2}\right) + \hbar\omega_\gamma \left(2n_\gamma + \tfrac{1}{2}|K| + 1\right) + \frac{\hbar^2}{2\mathcal{J}} \left[I(I+1) - K^2 \right] . \tag{6.240}$$

The structure of the spectrum is shown in Fig. 6.9. The *bands* are characterized by a given set of (K, n_β, n_γ) and follow the $I(I+1)$ rule of the rigid rotor. The principal bands are:

1. the *ground-state* band (g.s.), made up of the states $|IM\,000\rangle$ with I even. The energies are given by $\hbar^2 I(I+1)/2\mathcal{J}$.
2. the *β band*, containing the states $|IM\,010\rangle$ with one quantum of vibration added in the β direction. It starts at $\hbar\omega_\beta$ above the ground state and also contains only even angular momenta.
3. the *γ band* is not, as the name seems to suggest, the band containing a quantum in the γ direction. Instead, it is characterized by $K = 2$, so that the states are

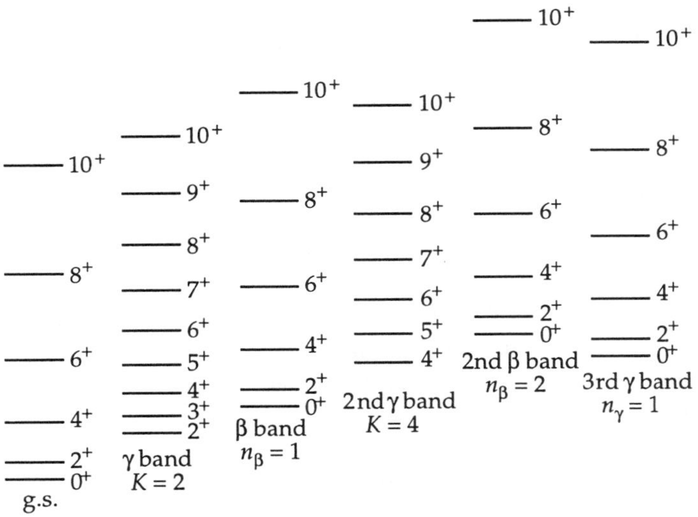

Fig. 6.9 Structure of the spectrum of the rotation–vibration model. The names of the bands and the quantum numbers are indicated below the bands.

given by $|IM\,200\rangle$. In the spectra it is easy to distinguish it from the β band, as it starts with a 2^+ and contains the odd angular momenta as well. The excitation energy of the band head $|2M\,200\rangle$ is given by

$$E_\gamma = \frac{\hbar^2}{\mathcal{J}} + \hbar\omega_\gamma \tag{6.241}$$

and clearly contains both a rotational and a γ-vibrational contribution, the latter leading to its name.

4. the next higher bands. These should be the additional γ bands with $K = 4$ and the one with $n_\gamma = 1$.

Why does the γ band with $K = 2$ but $n_\gamma = 0$ acquire an energy contribution from the γ vibration? This is because the moment of inertia around the z' axis of the nucleus behaves as $8B\eta^2$, leading to the appearance of a term reminiscent of a centrifugal potential in the Hamiltonian of (6.216). This causes a strong coupling between rotations and γ vibrations and physically expresses the fact that rotations with nonvanishing K become possible only in the presence of dynamical triaxial deformation.

6.5.4 Moments and Transition Probabilities

The collective quadrupole operator was calculated in Sect. 6.3.4 and found to be given by

$$\hat{Q}_{2\mu} = \frac{3Ze}{4\pi} R_0^2 \left(\hat{\alpha}_{2\mu} - \frac{10}{\sqrt{70\pi}} \left[\hat{\alpha} \times \hat{\alpha}\right]^2_\mu \right) \quad . \tag{6.242}$$

This expression has to be transformed to the intrinsic system, which is carried out in Exercise 6.8. The result is

$$\hat{Q}_{2\mu} = \frac{3Ze}{4\pi}R_0^2\left(\mathcal{D}^{(2)*}_{\mu 0}(\boldsymbol{\theta})\left\{\beta_0+\xi+\frac{2}{7}\sqrt{\frac{5}{\pi}}\left[(\beta_0+\xi)^2-2\eta^2\right]\right\}\right.$$
$$\left.+\left(\mathcal{D}^{(2)*}_{\mu 2}(\boldsymbol{\theta})+\mathcal{D}^{(2)*}_{\mu -2}(\boldsymbol{\theta})\right)\eta\left[1-\frac{4}{7}\sqrt{\frac{5}{\pi}}(\beta_0+\xi)\right]\right) \quad . \tag{6.243}$$

Recalling the definition of the quadrupole moment from Sect. 6.3.5, we have for a state in this model

$$Q_{IKn_\beta n_\gamma} = \sqrt{\frac{16\pi}{5}}\,\langle IM=I\,Kn_\beta n_\gamma|\hat{Q}_{20}|IM=I\,Kn_\beta n_\gamma\rangle \quad . \tag{6.244}$$

The result is calculated in Exercise 6.9 to lowest order, i.e., ignoring the ξ-dependent and η-dependent parts of the quadrupole operator,; it is independent of n_β and n_γ and is given by

$$Q_{IK} = Q_0\frac{3K^2-I(I+1)}{(I+1)(2I+3)}(1+\alpha) \quad . \tag{6.245}$$

Here

$$\alpha = \tfrac{2}{7}\sqrt{\tfrac{5}{\pi}}\,\beta_0 \tag{6.246}$$

was introduced as an abbreviation together with the *intrinsic quadrupole moment*

$$Q_0 = \frac{3ZeR_0^2}{\sqrt{5\pi}}\beta_0 \quad . \tag{6.247}$$

The physical import of this equation is quite interesting. In the limit considered, the nucleus is a rotating statically deformed shape. The quadrupole moment of this shape *in its principal axis frame* is just given by Q_0 (note that the correction factor $(1+\alpha)$ is not included). What does the angular-momentum-dependent factor describe? For the ground state with $I=K=0$ we get $Q=0$, and this merely expresses the fact that a state with equal probability of orientation of the nucleus in all spatial directions is spherically symmetric and shows no apparent deformation. For the first excited state with $I=2$ and $K=0$ we get $Q=-2Q_0/7$. If the intrinsic deformation is prolate, so that $Q_0>0$, the observed quadrupole momentum in the laboratory frame becomes negative, reflecting the fact that a twirled cigar produces an oblate apparent shape when time averaged.

We will not carry out the calculations for the matrix elements of the quadrupole operator in detail. Rather we will indicate some general properties apparent already from its functional form. Remembering that ξ and η were assumed to be small deviations from the equilibrium deformation, we find that the probability of the transitions will rapidly decrease with the powers of ξ and η responsible.

The dominant term is the one containing only β_0 and no ξ or η. Because of the latter it cannot connect states of different n_β or n_γ. This must be done by the first- and second-order terms. Unfortunately evaluating these in the model as presented here is relatively useless, because the terms neglected in the Hamiltonian are of the same order (they are mostly caused by the deformation dependence of the moments of inertia). Nevertheless we indicate the results in the following. They can be evaluated straightforwardly by integrating the reduced matrix elements of the quadrupole operator. In general we have

$$B\left(E2; I_i \rightarrow I_f\right) = \frac{2I_f+1}{2I_i+1}\left|\langle I_f||\hat{Q}||I_i\rangle\right|^2 \quad , \tag{6.248}$$

and a calculation left to Exercise 6.9 yields the final results:

$$\begin{aligned}
B(\mathrm{E2}; I_i \text{ g.s.band} \rightarrow I_f \text{ g.s.band}) &= \frac{5Q_0^2}{16\pi}(I_i 2 I_f|000)^2(1+\alpha)^2 \quad , \\
B(\mathrm{E2}; I_i\ \gamma\text{-band} \rightarrow I_f\ \gamma\text{-band}) &= \frac{5Q_0^2}{16\pi}(I_i 2 I_f|202)^2(1+\alpha)^2 \quad , \\
B(\mathrm{E2}; I_i\ \beta\text{-band} \rightarrow I_f\ \beta\text{-band}) &= \frac{5Q_0^2}{16\pi}(I_i 2 I_f|000)^2(1+\alpha)^2 \quad , \\
B(\mathrm{E2}; I_i\ \gamma\text{-band} \rightarrow I_f \text{ g.s.band}) &= \frac{5Q_0^2}{16\pi}(I_i 2 I_f|220)^2(1-2\alpha)^2\frac{3\hbar}{\omega_\gamma \mathcal{J}} \quad , \\
B(\mathrm{E2}; I_i\ \beta\text{-band} \rightarrow I_f \text{ g.s.band}) &= \frac{5Q_0^2}{16\pi}(I_i 2 I_f|000)^2(1+2\alpha)^2\frac{3\hbar}{2\omega_\beta \mathcal{J}} \quad .
\end{aligned} \tag{6.249}$$

The transition probabilities allow the determination of the intrinsic quadrupole moment and thus the deformation itself, whereas the energy spacing in the spectra determines the moment of inertia.

One simple consequence of the transition-rate formulas given here is worth mentioning: calculating the ratio of the $B(\mathrm{E2})$ values for the transitions from the γ-band head to the 2^+ and the 0^+ in the ground-state band, respectively, one finds immediately that it is just given by the ratio of the Clebsch–Gordan coefficients:

$$\frac{B(\mathrm{E2}; 2^+_\gamma \rightarrow 2^+_{\mathrm{g.s.}})}{2^+_\gamma \rightarrow 0^+_{\mathrm{g.s.}}} = \frac{(222|022)^2}{(022|022)^2} = \frac{10}{7} \quad . \tag{6.250}$$

This is one of the *Alaga rules*. Experimentally, it does not work very well, unless the rotation–vibration interaction is taken into account. As for the harmonic vibrator model, it appears that the rotation–vibration model in its simplest version contains too high a degree of symmetry, so that many matrix elements vanish and slight symmetry violations have drastic effects for the transition probabilities. One interesting case is also that of the transitions between the β and γ bands, which in this model must come from the second-order term in the quadrupole tensor (needing a product of a ξ and an η to change the phonon number in both directions), or by a corresponding term in the Hamiltonian which leads to a mixing of β and γ vibrations in the wave functions. In any case the transitions should be small, and this is a noticeable difference from the IBA model. This will be discussed briefly in Sect. 6.8.6.

EXERCISE

6.8 Transformation of the Quadrupole Operator

Problem. Transform the quadrupole operator to the coordinates $\boldsymbol{\theta}$, ξ, and η.

Solution. By definition, the Euler angles carry the space-fixed variables $\alpha_{2\mu}$ into the body-fixed ones $\alpha'_{2\mu}$, which have $\alpha'_{20} = \beta_0 + \xi$ and $\alpha'_{2\pm 2} = \eta$ as the only nonvanishing components. To go back to the space-fixed system requires the transformation

$$\alpha_{2\mu} = \sum_{\nu} \mathcal{D}^{(2)*}_{\mu\nu}(\boldsymbol{\theta})\, \alpha'_{2\nu} \quad . \tag{1}$$

This immediately yields the first-order term in (6.243). For the second-order term we note that since the two $\alpha_{2\mu}$ are coupled to a spherical tensor, it takes only one rotation matrix to transform them into the body-fixed frame:

$$[\alpha \times \alpha]^2_\mu = \sum_{\nu} \mathcal{D}^{(2)*}_{\mu\nu}(\boldsymbol{\theta})\, \left[\alpha' \times \alpha'\right]^2_\nu \quad , \tag{2}$$

and it remains for us to evaluate the angular-momentum coupling in the body-fixed frame. Considering the nonvanishing components we first calculate

$$\begin{aligned}
\left[\alpha' \times \alpha'\right]^2_0 &= \big((222|-220) + (222|2-20)\big)\eta^2 + (222|000)(\beta_0+\xi)^2 \quad , \\
\left[\alpha' \times \alpha'\right]^2_{\pm 1} &= 0 \quad , \\
\left[\alpha' \times \alpha'\right]^2_{\pm 2} &= \big((222|0\pm 2 \pm 2) + (222|\pm 20 \pm 2)\big)(\beta_0+\xi)\eta \quad .
\end{aligned} \tag{3}$$

The Clebsch–Gordan coefficients are given by

$$\begin{aligned}
(222|-220) &= (222|0 \pm 2 \pm 2) = (222|\pm 20 \pm 2) \\
&= -(222|000) = \sqrt{\frac{2}{7}} \quad .
\end{aligned} \tag{4}$$

It is now a simple matter of adding up the terms to arrive at the result of (2.49).

EXERCISE

6.9 Quadrupole Moments and Transition Probabilities

Problem. Derive the formula for the quadrupole moment, (6.245), and the matrix elements for the first three formulas of (6.249).

Exercise 6.9

Solution. To do this we need two general formulas from the Appendix. The first gives all the Clebsch–Gordan coefficients needed:

$$(I2I|M0M) = \frac{3M^2 - I(I+1)}{\sqrt{I(I+1)(2I+3)(2I-1)}} \quad . \tag{1}$$

The second one shows the result of an integral over three rotation matrices:

$$\int d^3\theta\, \mathcal{D}^{(I_3)}_{M_3K_3}(\boldsymbol{\theta}) \mathcal{D}^{(I_2)*}_{M_2K_2}(\boldsymbol{\theta})\, \mathcal{D}^{(I_1)*}_{M_1K_1}$$
$$= \frac{8\pi^2}{2I_3+1} \left(I_1I_2I_3|M_1M_2M_3\right) \left(I_1I_2I_3|K_1K_2K_3\right) \quad . \tag{2}$$

For the quadrupole moment we need to evaluate

$$Q_{IKn_\beta n_\gamma} = \sqrt{\frac{16\pi}{5}} \langle IM = I\; Kn_\beta n_\gamma | \hat{Q}_{20} | IM = I\; Kn_\beta n_\gamma \rangle \quad . \tag{3}$$

In the case where the vibrational quantum numbers of the initial and final states are the same, the ξ-dependent and η-dependent parts do not contribute, so that the quadrupole operator is reduced to

$$\hat{Q}_{20} \rightarrow \frac{3Ze}{4\pi} R_0^2\, \mathcal{D}^{(2)*}_{00}\; \beta_0(1+\alpha) \quad . \tag{4}$$

Introducing the intrinsic quadrupole moment we have

$$Q = Q_0(1+\alpha)\langle IM = IKn_\beta n_\gamma | \mathcal{D}^{(2)*}_{00} \rangle IM = IKn_\beta n_\gamma \quad . \tag{5}$$

The vibrational parts drop out because of normalization, so the matrix element becomes, after insertion of the rotational wave function,

$$Q = Q_0(1+\alpha) \frac{2I+1}{16\pi^2(1+\delta_{K0})}$$
$$\times \int d^3\theta \left(\mathcal{D}^{(I)}_{MK} + (-1)^I\, \mathcal{D}^{(I)}_{M\bar{K}} \right) \mathcal{D}^{(2)*}_{00} \left(\mathcal{D}^{(I)*}_{MK} + (-1)^I\, \mathcal{D}^{(I)*}_{M\bar{K}} \right) \quad . \tag{6}$$

For $K = 0$ the integral becomes

$$4 \int d^3\theta\, \mathcal{D}^{(I)}_{M0}\, \mathcal{D}^{(2)*}_{00}\, \mathcal{D}^{(I)*}_{M0} \quad , \tag{7}$$

whereas for $K \neq 0$ there are four terms containing the different combinations of K and $-K$. In the integration the mixed terms vanish because of (2), and the other two terms are equal. If we take into account the factor with δ_{K0} the net result for both cases is equal to

$$Q = Q_0\,(1+\alpha)\,(I2I|I0I)\,(I2I|K0K) \quad . \tag{8}$$

Finally, the Clebsch–Gordan coefficients can be inserted,

$$(I2I|I0I)(I2I|K0K) = \frac{\left[3I^2 - I(I+1)\right]\left[3K^2 - I(I+1)\right]}{I\,(I+1)\,(2I+3)\,(2I-1)} \quad , \tag{9}$$

and simplifying this fraction leads to the desired result.

Exercise 6.9

Now for the calculation of the transition probabilities. The procedure is quite similar to that for the quadrupole moment. First examine the transitions within a band, i.e., again ignore vibrational excitation. The calculation is then practically identical to the one for the quadrupole moment, except that the initial and final states have different angular momenta, and that we need a reduced matrix element:

$$
\begin{aligned}
&\langle I_f || \mathcal{D}^{(2)*}_{0} || I_i \rangle (I_i 2 I_f | M_i \mu M_f) \\
&= \frac{\sqrt{(2I_i+1)(I_f+1)}}{16\pi^2}(1+\delta_{K0}) \\
&\quad \times \int \mathrm{d}^3\theta \left(\mathcal{D}^{(I_f)}_{M_f K} + (-1)^{I_f} \mathcal{D}^{(I_f)}_{M\,-K} \right) \mathcal{D}^{(2)*}_{\mu 0} \left(\mathcal{D}^{(I_i)*}_{M_i K} + (-1)^{I_i} \mathcal{D}^{(I_i)*}_{M\,-K} \right) \\
&= \frac{\sqrt{(2I_i+1)(I_f+1)}}{8\pi^2} \frac{8\pi^2}{2I_f+1} (I_i 2 I_f | K 0 K)\,(I_i 2 I_f | M_i \mu M_f) \quad . \qquad (10)
\end{aligned}
$$

Here the last Clebsch–Gordan coefficient drops out in the reduced matrix element, and inserting this into the general formula leads to the given result.

For a transition accompanied by a change by one phonon in the β vibrations, the contributing parts of the quadrupole operator are given by

$$
\begin{aligned}
\hat{Q} &\to \sqrt{\frac{5}{16\pi}} Q_0 \mathcal{D}^{(2)*}_{\mu 0} \beta_0^{-1} \left(\xi + \frac{4}{7}\sqrt{\frac{5}{\pi}}\, \xi \beta_0 \right) \\
&= \sqrt{\frac{5}{16\pi}}\, Q_0\, \beta_0^{-1}\, \mathcal{D}^{(2)*}_{\mu 0} (1+2\alpha)\xi \quad , \qquad (11)
\end{aligned}
$$

and the integral over the Euler angles is as before, only with different angular momenta appearing in the result. For the ξ-vibrational part one may use second quantization:

$$
\xi = \sqrt{\frac{\hbar}{2B\omega_\beta}} \,(\hat{\beta}^\dagger + \hat{\beta}) \quad . \qquad (12)
$$

For the transition between the ground state band and the β band the creation operator contributes a matrix element of 1, so that above operator gets a contribution of

$$
\sqrt{\frac{\hbar}{2B\omega_\beta}} = \beta_0 \sqrt{\frac{3\hbar}{2\mathcal{J}\omega_\beta}} \quad , \qquad (13)
$$

leading to the final result.

For the transition from the γ to the ground-state band the calculation is similar. Because the wave function in this case, (6.231), cannot be simply replaced by second-quantization expressions, the integral over the wave functions has to be evaluated explicitly. This is mainly a matter of looking into function tables like [Ab64] and integral tables such as [Gr65a]. The η-dependent zero-point wave function is

$$
\chi_{00}(\eta) = \sqrt{\lambda |\eta|}\; \mathrm{e}^{-\lambda\eta^2/2}\; {}_1F_1\left(0, 1, \lambda\eta^2\right) \quad , \qquad (14)
$$

Exercise 6.9

and that for the γ-band head is

$$\chi_{20}(\eta) = \lambda\sqrt{|\eta|}\,\eta\,\mathrm{e}^{-\lambda\eta^2/2}\;{}_1F_1\left(0,2,\lambda\eta^2\right) \quad . \tag{15}$$

For the purpose of evaluating integrals, for example, the confluent hypergeometric function is too general; using more specific functions usually makes a much larger variety of integral formulas and recursion relations available. In this case we find that the appropriate special case is

$${}_1F_1(-n,\alpha+1,x) = (-1)^n\, n!\, L_n^{(\alpha)}(x) \tag{16}$$

with $L_n^{(\alpha)}(x)$ a *Laguerre polynomial.* The integral to be evaluated now becomes

$$\int\limits_{-\infty}^{\infty} \mathrm{d}\eta\; \chi_{00}(\eta)\,\eta\,\chi_{20}(\eta) = \lambda^{-1/2}\int\limits_{0}^{\infty}\mathrm{d}x\; x\,\mathrm{e}^{-x}\,L_0^{(0)}(x)\,L_0^{(1)}(x) \quad , \tag{17}$$

where $x = \lambda\eta^2$ was substituted and the symmetry of the integrand used. In [Gr65a] we find the integral

$$\int\limits_{0}^{\infty}\mathrm{d}x\;\mathrm{e}^{-x}\,x^{\alpha+\beta}\,L_m^{(\alpha)}(x)\,L_n^{(\beta)}(x) = (-1)^{m+n}\,(\alpha+\beta)!\binom{\alpha+m}{n}\binom{\beta+n}{m} \quad , \tag{18}$$

which in our case yields unity and a final value of

$$\frac{1}{\lambda} = \sqrt{\frac{\hbar}{2B\omega_\gamma}} \tag{19}$$

for the total integral. Taking into account that now the part of the quadrupole operator proportional to η is

$$\hat{Q} \rightarrow \sqrt{\frac{5}{16\pi}}\,Q_0\,\beta_0^{-1}\left(\mathcal{D}_{\mu 2}^{(2)*} + \mathcal{D}_{\mu -2}^{(2)*}\right)(1-2\alpha)\,\eta \quad , \tag{20}$$

leads to the desired result upon insertion of the integral and addition of a factor of two for the two contributions $\mu = \pm 2$, which turn out to be identical.

EXERCISE

6.10 ^{238}U in the Rotation–Vibration Model

Problem. Apply the rotation–vibration model to the nucleus ^{238}U using the subset of experimental data given in Fig. 6.10.

Solution. First use the band heads to determine the energy parameters. From the energy of the first 2^+ state we get

$$\frac{3\hbar^2}{\mathcal{J}} = 0.045\ \text{MeV} \quad \rightarrow \quad \mathcal{J} = 865\,\frac{\text{GeV fm}^2}{c^2} \quad . \tag{1}$$

A more understandable number is obtained by dividing this result by the total mass of the nucleus:

$$\frac{\mathcal{J}}{Am} = 3.88\,\text{fm}^2 \quad . \tag{2}$$

Note that the $I(I+1)$ rule in this case holds quite well up to the 10^+ state, the highest one given here, which should be at 0.825 MeV instead of the 0.776 MeV found experimentally. To find an even better overall description of the band, one might use a least-squares fit of the total band instead.

The other band heads give the parameters $\hbar\omega_\beta = 0.993\,\text{MeV}$ (from the second 0^+ state) and

$$\hbar\omega_\gamma = E_\gamma - \frac{\hbar^2}{\mathcal{J}} = 1.018\ \text{MeV} \quad , \tag{3}$$

so β and γ vibrational frequencies are comparable.

The ground-state deformation can be calculated from the transition probability

$$\mathcal{B} = B\left(\text{E2}; 0^+_{\text{g.s.}} \rightarrow 2^+_{\text{g.s.}}\right) \approx \frac{5Q_0^2}{16\pi}\,(022|000)^2 \quad . \tag{4}$$

Remember that you have to be careful which direction of the transition is given: the $B(\text{E2})$ values $I_i \rightarrow I_f$ are those for $I_f \rightarrow I_i$ multiplied by $(I_f+1)^2/(I_i+1)^2$.

Leaving out the term in α, as it provides only a minor correction, and using $(022|000) = 1$ (this is a trivial Clebsch–Gordan coefficient; it says that a tensor of angular momentum 2 multiplied by a scalar is still a tensor of angular momentum 2), we get

$$Q_0 = \sqrt{\frac{16\pi}{5}}\sqrt{\mathcal{B}} = 10.84\,e\,\text{b} \quad , \tag{5}$$

and taking into account the definition of Q_0

$$\beta_0 = \frac{4\pi}{3ZeR_0^2}\sqrt{\mathcal{B}} \quad . \tag{6}$$

Inserting $\mathcal{B} = 11.7\,e^2\,\text{b}^2 = 11.7 \times 10^4\,e^2\,\text{fm}^4$ and $R_0 = 1.2 \times 238^{1/3}\,\text{fm}$ leads to

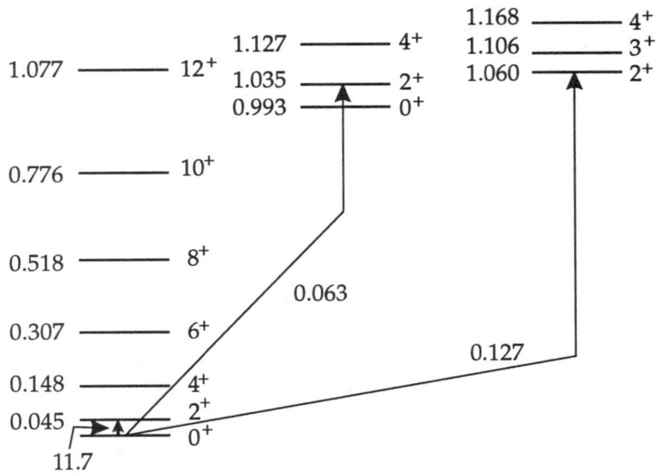

Fig. 6.10 Lowest experimental bands for the nucleus ^{238}U with selected transition probabilities. The energies written next to the levels are in MeV and the B(E2) values (next to the transition arrows) in $e^2\,\mathrm{b}^2$. Note that the arrows indicate the transition direction for the B(E2) values.

$$\beta_0 = 0.282 \quad . \tag{7}$$

This corresponds to quite a large ground-state deformation, supporting the view that this nucleus is a strongly deformed rotor.

Now that the parameters of the model have all been determined one may check what the predictions for the other observables are. For the transition to the β band there is a very simple factor if the correction in α is again neglected:

$$B\left(\mathrm{E2}; 0^+_{\mathrm{g.s.}} \to 2^+_\beta\right) = B\left(\mathrm{E2}; 0^+_{\mathrm{g.s.}} \to 2^+_{\mathrm{g.s.}}\right) \frac{3\hbar}{2\omega_\beta \mathcal{J}} \approx 0.79\, e^2\,\mathrm{b}^2 \quad . \tag{8}$$

The transition to the γ band also involves only trivial Clebsch–Gordan coefficients:

$$B\left(\mathrm{E2}; 0^+_{\mathrm{g.s.}} \to 2^+_\gamma\right) = B\left(\mathrm{E2}; 0^+_{\mathrm{g.s.}} \to 2^+_{\mathrm{g.s.}}\right) \frac{3\hbar}{\omega_\gamma \mathcal{J}} \frac{(022|022)^2}{(022|000)^2} \approx 1.58\, e^2\,\mathrm{b}^2 \quad . \tag{9}$$

Both of these are too large by about an order of magnitude, showing that the wave functions of this simple model should be mixed by, for example, the rotation–vibration interaction.

It is interesting to compare the moment of inertia determined from the rotational band with the classical rigid-body value. For a rotation about the y axis in the body-fixed frame the moment of inertia is given by

$$\int_{\mathrm{Volume}} \mathrm{d}^3 r\; \rho_0 \left(x^2 + z^2\right) \quad . \tag{10}$$

This can easily be evaluated to first order in β_0 using the methods for such integrals developed before (for example, Exercise 6.1); note only that

$$x = -\sqrt{\frac{2\pi}{3}}\, r\, \left(Y_{11}(\Omega) - Y_{1-1}(\Omega)\right) \quad , \quad z = \sqrt{\frac{4\pi}{3}}\, r\, Y_{20}(\Omega) \quad , \tag{11}$$

although in this case it is probably easier to insert the explicit expressions for the spherical harmonics. The result is

Exercise 6.10

$$\mathcal{J}_{\text{rigid}} = m\,A\,R_0^2\left(\frac{2}{5}+\frac{1}{10}\sqrt{\frac{5}{\pi}}\beta_0\right) \quad . \tag{12}$$

The first term in parentheses corresponds to the moment of inertia of a sphere, while the second one indicates the increase with deformation. Inserting numbers for the special case of ^{238}U, we get

$$\mathcal{J}_{\text{rigid}} \approx m\,A\,(15.3+1.4) \quad . \tag{13}$$

Comparing the factors in parentheses to the value of 3.88 from experiment, we see that the contribution for the spherical nucleus is much too large, whereas the deformation-dependent part is also off by a factor of two. That the classical result for a sphere cannot be correct is already to be expected from the fact that a sphere cannot rotate in this model; however, this result also suggests that the rotation of the nucleus is not like that of a rigid body. Only a fraction of the nucleons must actually participate in the motion, probably associated with the part of the shape deviating from the sphere.

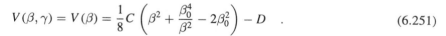

6.6 γ-Unstable Nuclei

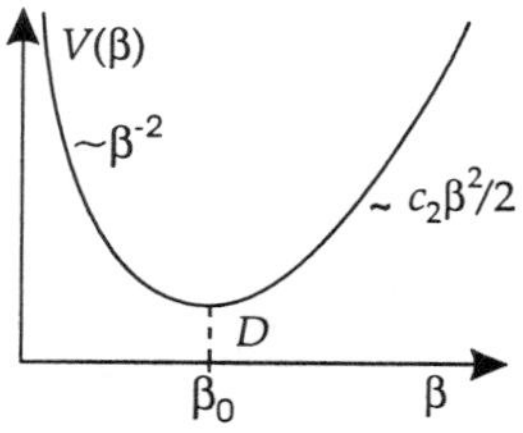

Fig. 6.11 Cut of the Wilets–Jean potential along the β direction. The potential is totally independent of γ and thus possesses axial symmetry. The potential parameters are illustrated as discussed in the text.

There is another special case in which an analytical solution of the collective model is possible. This is the so-called γ-unstable case, which was discovered by Wilets and Jean [Wi56]. In this model the potential-energy surface has a minimum at a finite β, which is totally flat along γ, effectively producing a ring-like structure in the (β,γ) plane. Nuclei that can be interpreted in this way are typically found in the region $Z>52$, $N<80$. As this model has not been widely used, we only give a brief sketch.

The specific form of the potential is

$$V(\beta,\gamma) = V(\beta) = \frac{1}{8}C\left(\beta^2+\frac{\beta_0^4}{\beta^2}-2\beta_0^2\right) - D \quad . \tag{6.251}$$

This potential is illustrated in Fig. 6.11

The three parameters determining the potential are the stiffness, C, in the β direction at the potential minimum, the location β_0 of the minimum, and the potential depth D. (The fourth parameter of the model is the mass parameter B with the usual meaning).

For this potential the Hamiltonian can be written in the intrinsic system as (see (6.218)):

$$\hat{H}_0 = -\frac{1}{2B\beta^2}\left(\frac{1}{\beta^2}\frac{\partial}{\partial\beta}\beta^4\frac{\partial}{\partial\beta}+\frac{1}{\sin(3\gamma)}\frac{\partial}{\partial\gamma}\sin(3\gamma)\frac{\partial}{\partial\gamma}\right.$$
$$\left. -\frac{1}{4}\sum_{k=1}^{3}\frac{I_k'^2}{\sin^2\left(\gamma-\frac{2\pi k}{3}\right)}\right)+V(\beta) \quad . \tag{6.252}$$

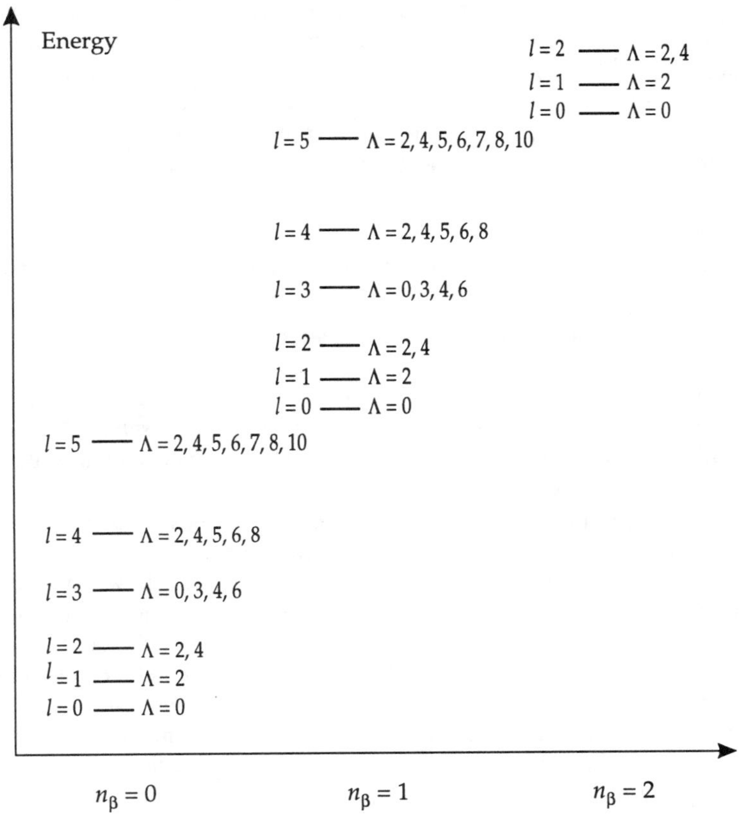

Fig. 6.12 The schematic energy spectrum of the Wilets–Jean potential.

As there is no symmetry axis in this problem, its eigenfunctions must be set up as expansions over K values, analogously to the treatment of the asymmetric rotor,

$$\Psi_{n_\beta \lambda IM} = \sqrt{\frac{2I+1}{8\pi^2}} \frac{\phi_{n_\beta \lambda}(\beta)}{\beta^2} \sum_{K=-I}^{I} t_{\lambda IK}(\gamma) \mathcal{D}^{I\ *}_{MK}(\theta_i) \quad , \tag{6.253}$$

and this leads to a separation of the β-dependent and γ-dependent parts.

The actual calculation of the wave functions is quite lengthy, so we give only the results for the spectrum. The eigenvalues are given by

$$\epsilon_{n_\beta \lambda} = \sqrt{\frac{B}{C}} \left(n_\beta + \frac{1}{2} + \frac{1}{2}\alpha_\lambda \right) - \tfrac{1}{4} C \beta_0^2 - D \tag{6.254}$$

with

$$\alpha_\lambda = \frac{1}{2} \sqrt{\sqrt{BC}\, \beta_0^4 + (2\lambda + 3)^2} \quad . \tag{6.255}$$

The quantum numbers are the numbers of phonons in the β direction and a *seniority* quantum number λ, both with values in $0, \ldots, \infty$. The allowed angular momenta for a given combination (n_β, λ) are limited by 2λ and some are omitted because of symmetry requirements, as in the case of the harmonic oscillator (mathematically, in fact, the present Hamiltonian can be related to the harmonic oscillator quite intimately, the only difference being the presence of the centrifugal-type term in β^{-2}). In Fig. 6.12 we show the resulting energy spectrum schematically: on every band head given by a fixed number of β phonons there is a sequence of states built with different seniority λ and a set of allowed angular momenta.

6.7 More General Collective Models for Surface Vibrations

6.7.1 The Generalized Collective Model

The types of surface motions considered up to now have been confined to three limiting cases: the pure harmonic oscillations around a spherical equilibrium shape, the rotations and vibrations based on a well-deformed ground state shape, and the γ-unstable case.

Although the rotation–vibration model describes a large class of nuclei quite well and the spherical vibrator is at least approximated by some nearly magic nuclei, the large majority of nuclei do not fit particularly well into one or the other of these quite abstract cases. Most nuclei with a spherical ground state do not show the full degeneracy in their spectra expected for a pure harmonic oscillator. Thus *anharmonicities* in either the potential or the kinetic energy have to be taken into account. For the deformed nuclei, again the minimum will usually not be so deep that the potential can be approximated by a harmonic oscillator in the ξ and η variables. Add to this the more exotic possibility of nuclei possessing more than one minimum in the potential-energy surface, and it becomes clear that there is a need to extend the collective model approach to much more complicated Hamiltonians.

Such an extension must allow for a larger class of functional forms for the terms in the Hamiltonian. Starting with an unspecified form such as

$$\hat{H} = \hat{H}\left(\alpha_{2\mu}, \pi_{2\mu}; C_1, C_2, \ldots, C_N\right) \quad , \tag{6.256}$$

with N parameters C_i, it is clear that the specific form must be carefully selected to minimize the number of unavoidable parameters. In nuclear physics there are not usually enough data to determine a larger number of parameters, and in any case fitting a model with an excessive number of parameters to an experiment makes the results of dubious value.

The belief that "anything can be fitted with a sufficient number of parameters", however, also oversimplifies the situation, because the data are usually strongly correlated. In a collective model, for example, there will be bands of the types we have discussed, and it should require a very strange potential energy surface to produce a sequence like $0^+-4^+-2^+$ built on the ground state! The danger is rather that if too many parameters are used, some of them will not be determined by the available data, and the theorist then has to be careful not to "predict" features caused by accidental values of these parameters. Thus it could happen that a secondary minimum appears in the (β, γ) plane, which does not affect the observed properties of states which are really seen experimentally and are all located in the ground state minimum: such a secondary minimum should not be taken seriously.

An even more important consideration historically was the need to be able to actually solve the model with the computers available at the time. Any more-general Hamiltonian certainly requires numerical diagonalization and the existence of a suitable set of basic functions plus a method for evaluating the matrix elements. These requirements already restrict the available models considerably.

An additional, more theoretical consideration is that of whether to construct potentials in the principal axes or the laboratory frame. In the former case, rotational invariance simply implies that the potential (and mass parameters) should not depend on the Euler angles, so that an arbitrary function $V(\beta, \gamma)$ could be used. With the symmetries in the (β, γ) plane (see Sect. 6.2.4), however, it is clear that $V(\beta, \gamma)$ must be invariant under the transformations

$$\gamma \to -\gamma, \quad \gamma \to \gamma + \frac{2\pi}{3} \quad , \tag{6.257}$$

and this can be fulfilled simply by making the potential depend on $\cos(3\gamma)$ instead of on γ itself. A useful form for the potential might thus be

$$V(\beta, \gamma) = \sum_{i,j=0}^{\infty} V_{ij} \beta^i \cos^j(3\gamma) \quad . \tag{6.258}$$

The term for $i = 0$ has to be excluded for $j \neq 0$, because it would have an indefinite limit for $\beta \to 0$.

In the laboratory frame, on the other hand, a scalar potential can be constructed by explicit couplings to zero angular momentum. Gneuss and Greiner [Gn71] used an expansion up to sixth order:

$$\begin{aligned} V(\alpha) = {} & C_2 [\alpha \times \alpha]^0 + C_3 \left[\alpha \times [\alpha \times \alpha]^2\right]^0 + C_4 \left([\alpha \times \alpha]^0\right)^2 \\ & + C_5 \left[\alpha \times [\alpha \times \alpha]^2\right]^0 [\alpha \times \alpha]^0 \\ & + C_6 \left(\left[\alpha \times [\alpha \times \alpha]^2\right]^0\right)^2 + D_6 \left([\alpha \times \alpha]^0\right)^2 \quad . \end{aligned} \tag{6.259}$$

It appears that many more terms with different intermediate couplings are possible for constructing a tensor of zero total angular momentum and prescribed order in α, but it can be shown that all of these may be reduced to powers of the two basic tensors $[\alpha \times \alpha]^0$ and $\left[\alpha \times [\alpha \times \alpha]^2\right]^0$ [No68].

In both cases a polynomial expansion is used. The reasons for this are not based on any fundamental requirements, an arbitrary function like $f(\beta)$ or $g([\alpha \times \alpha]^0)$ being rotationally invariant; however, the matrix elements of polynomial terms are easiest to evaluate. Hess and collaborators [He80, He81] used the basis functions of the harmonic oscillator given by Chacoń and Moshinsky [Ch76a, Ch77a] in the (β, γ) representation, whereas Gneuss and Greiner used a basis built up by continuing the construction in the laboratory system given in Sect. 6.3.2 to higher excitations.

In both cases one has to consider how many terms to take for the potential. We give this line of reasoning for the Gneuss–Greiner potential. Writing their potential in β and γ,

$$\begin{aligned} V(\beta, \gamma) = {} & C_2' \beta^2 + C_3' \beta^3 \cos(3\gamma) + C_4' \beta^4 \\ & + C_5' \beta^5 \cos(3\gamma) + C_6' \beta^6 \cos^2(3\gamma) + D_6' \beta^6 \quad , \end{aligned} \tag{6.260}$$

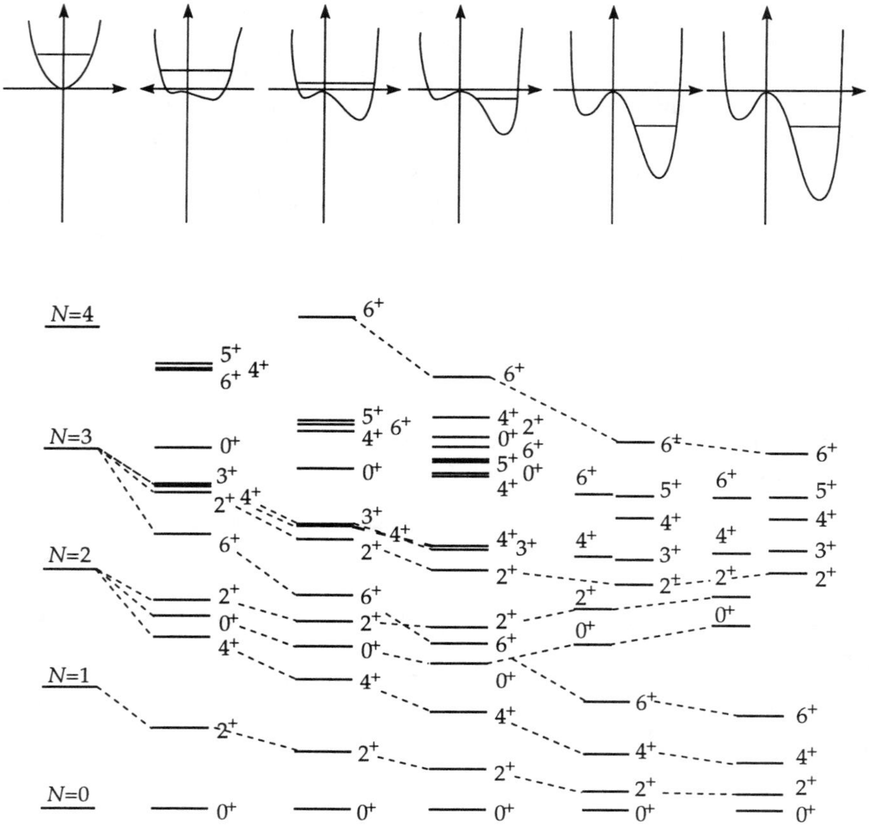

Fig. 6.13 A gradual transition of the potential energy surface from a spherical vibrator (left) to a well deformed nucleus (right), shown through the plots of the potential along the a_0 axis above, leads to a smooth restructuring of the spectra as shown in the lower part. The dashed lines indicate how some of the levels are related to each other.

where the constants were renamed to primed ones to absorb all the complicated numerical factors from the angular-momentum coupling, it is possible to discuss the flexibility inherent in this expression. To prescribe the properties of a minimum in the potential (location in β and γ, depth, and curvatures in both directions) requires five parameters, so it is clear that the six parameters in the above expression are sufficient to fix details of one ground-state minimum, but for the description of two minima there is much less freedom for fixing details. One should thus not expect too much detail from such a description. Nevertheless, this model proved to be quite useful in describing nuclei intermediate between spherical vibrator, deformed rotor, and triaxial nuclei, and in fact the gradual transition between these limiting cases can be shown in a convincing way. Fig. 6.13 shows an example of such a chain, in which the gradual development of the rotational bands out of the vibrational equidistant spectrum can be followed.

Up to now we have only discussed the potential energy and assumed that the kinetic energy is something trivial like the harmonic kinetic energy of the quadrupole oscillator. Microscopic theories indicate that this may be far from the truth, i.e., that the kinetic energy may show as complicated structures as the potential energy. Adding terms to the kinetic energy poses special problems, though.

1. A more complicated kinetic energy adds more parameters, which may be difficult to extract from experimental data.
2. There is a certain ambiguity as to what structures should be put into the kinetic or potential energy. For example, the transformation $\alpha \leftrightarrow \pi$ leaves the spectrum of the Hamiltonian invariant, but transforms a Hamiltonian with a simple kinetic energy and a complicated potential into one with the opposite characteristics. More complicated canonic transformations may be devised that mix coordinates and momenta and leave some part of the observable data unchanged. The crucial quantity not reasonably transformable in this way is the quadrupole operator, which must always contain the true deformation coordinates, but $B(\mathrm{E2})$-values and quadrupole moments are often not specific enough to allow the elimination all the ambiguities.
3. From a practical point of view, the potential provides an easily understandable description of nuclear structure: minima are assigned to ground or isomeric states, stiffness determines vibrational excitation energies, and barriers hinder transitions, etc. The kinetic energy plays a much less transparent role.

For all of these reasons, most of the studies using the more complicated Hamiltonians were based on a simple kinetic energy, usually of a purely harmonic type, and all the more complicated effects were assigned to the potential.

How is such a model actually applied in practice? A full computer code for fitting the model to experimental data is published in [Tr91b]. Because the general procedure followed is characteristic for many numerical solutions in nuclear structure physics, it is useful to take a closer look. The computer code solves the problem in the following steps.

1. Generate a set of basis functions. For the basis in this case the eigenfunctions of the harmonic quadrupole oscillator are used. The infinite set is truncated by allowing a maximum phonon number, which is equivalent to an energy cutoff. A general rule is that the less similarity there is between the harmonic oscillator potential and the potential to be calculated, the more basis functions are needed. The result of this step is a set of functions $|i\rangle$, $i = 1, \ldots, M$, where i just counts the wave functions as an index in the computer and stands for the full set of quantum numbers $N(i)$, etc.
2. Calculate the matrix elements of the Hamiltonian, i.e., $\langle i|\hat{H}|j\rangle, i,j = 1, \ldots, M$. This can be done by general formulas in the case of polynomial terms in the potential (for example, by expanding the momenta and coordinates into their second-quantization representation), but requires many complicated manipulations because of the symmetry and the angular-momentum coupling. It is important for the practical numerics that the Hamiltonian is scalar, so that the matrix elements vanish between states of different angular momentum or projection. In this way the total matrix splits up into submatrices for fixed angular momentum, and for each of these only a given projection need be considered – in practice one works with the reduced matrix elements instead. The net result is a numerical matrix of dimension up to typically several hundred, for each angular momentum.
3. Now a numerical algorithm for matrix diagonalization is used to determine the eigenvalues and eigenvectors. To each energy eigenvalue E_k is associated a

vector of coefficients a_{ki} giving the expansion of the eigenvector in the basis set $|i\rangle$.

4. The spectrum being given by step 4, it remains for us to calculate the other observables such as, for example, the $B(\mathrm{E2})$-values. This is now simply a matter of calculating the matrix elements of the quadrupole operator in the basis states, $\langle i||\hat{Q}||j\rangle$ and then summing over the expansions of the eigenstates of the full Hamiltonian:

$$\langle k||\hat{Q}||k'\rangle = \sum_{ij} a_{ki} a_{k'j} \langle k||\hat{Q}||k'\rangle \quad . \tag{6.261}$$

5. The results of the model for a fixed set of potential parameters are known at this point, and one has to compare them with experiment and then try to adjust the parameters to make the agreement better. In the computer code of [Tr91b] this is done through an algorithm for minimizing the χ^2 deviation between theory and experiment; the weighting of the different pieces of data for the χ^2 value, however, as well as the selection of an appropriate starting set of parameters, still needs human judgment and experience.

As an example of the type of analysis possible with this model we show the fit for the energy levels of some even-even Mercury isotopes in Figs. 6.14–18. These isotopes are especially interesting because there is evidence for *shape isomerism*, i.e., the existence of a secondary minimum in the potential-energy surface at a different deformation, which supports additional bands that do not fit into the band structures related to the principal ground-state minimum. The results illustrate also the limits of such a phenomenological model: one has to be very careful in the interpretation of the fitted potential-energy surfaces.

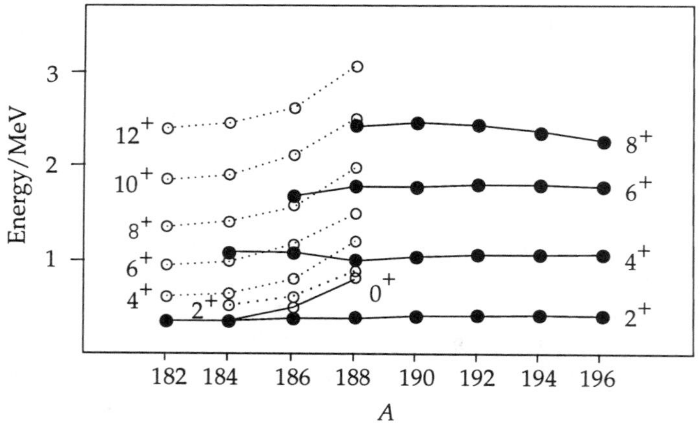

Fig. 6.14 Systematics of the low-lying states in the chain of Mercury isotopes. The states interpreted as belonging to a rotational band are indicated by open circles, and those belonging to a vibrational band by filled circles. Note that the secondary rotational band interpreted as being built on an isomeric secondary 0^+ state goes up in energy rapidly with increasing neutron number [Ha89].

The position of the isomeric minimum can be determined with confidence only if the experimental data that underly the fit contain such information. If transitions within the band built on the second minimum have been observed, one can extract the absolute value of the deformation, but not its sign; if only the spectrum is known the position of the minimum becomes highly uncertain and only the identification

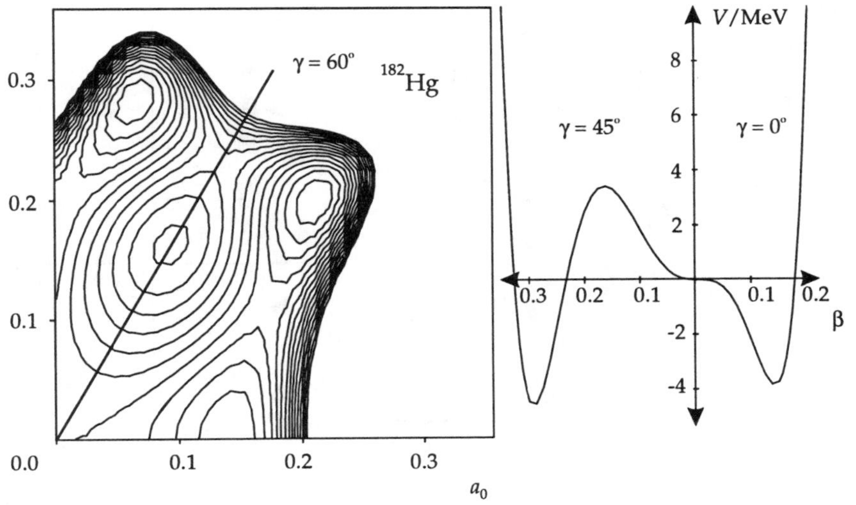

Fig. 6.15 Collective potential-energy surface of ^{182}Hg. For a description see Fig. 6.18.

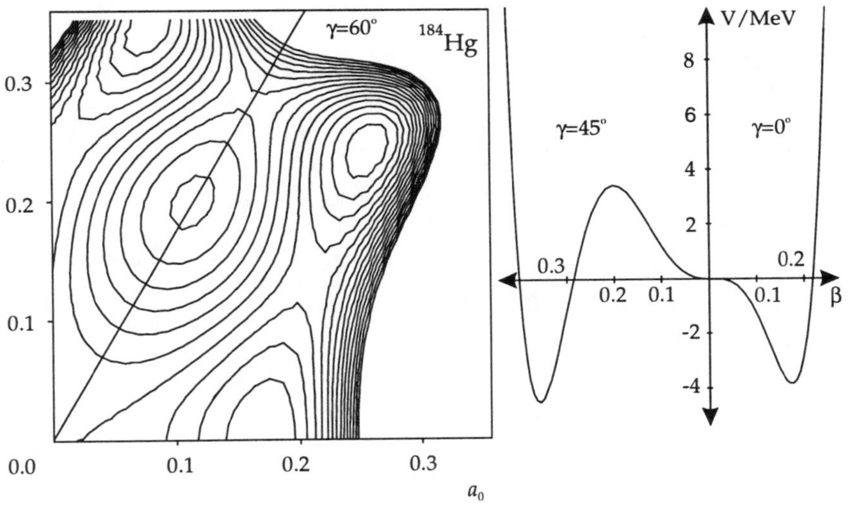

Fig. 6.16 Collective potential-energy surface of ^{184}Hg. For a description see Fig. 6.18.

of a rotation–vibrational, γ-unstable, or spherical vibrational spectra may offer clues as to its location. The fitting procedure will in these cases use the position of the minima in order to adjust finer – and probably unimportant – details in the observed data. Note that in the isotopes fitted, a secondary minimum appears only where the data demand it – there is no extrapolation from one isotope to the next, although one could in principle extend the model to encompass a smooth variation of collective behavior through such isotope chains.

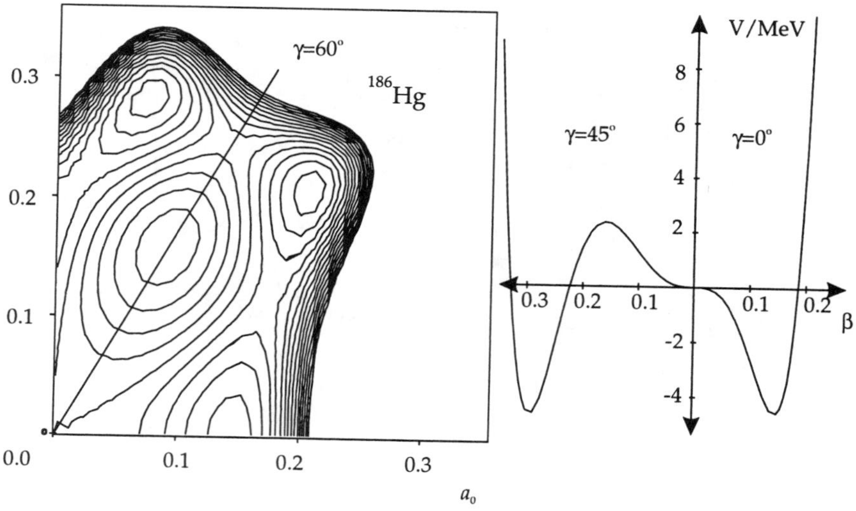

Fig. 6.17 Collective potential-energy surface of ^{186}Hg. For a description see Fig. 6.18.

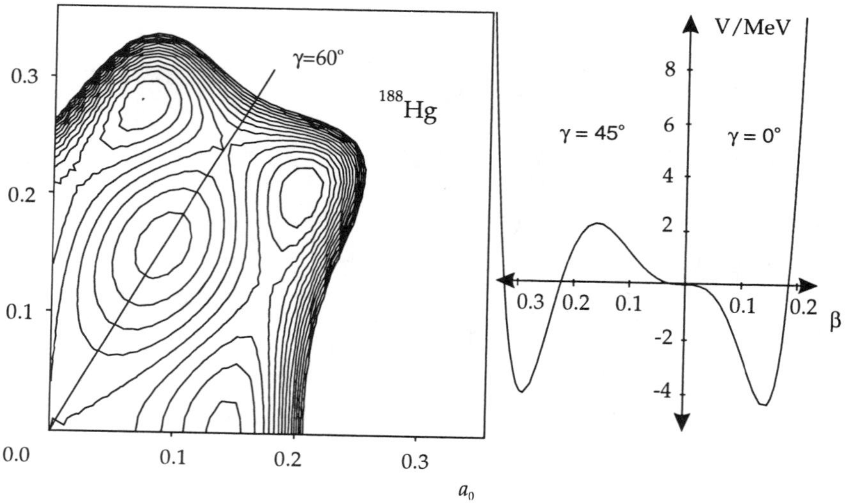

Fig. 6.18 Collective potential-energy surfaces describing the spectra and transition probabilities in the lighter of the Mercury isotopes [Tr91a]. On the left, isolines show the general layout of the minima, while cuts through the surfaces for $\gamma = 0°$ and $\gamma = 45°$ (the approximate angular position of the second minimum) show quantitative detail on the right. Note also the required symmetry around $\gamma = 60°$. In all the isotopes there is a second minimum in the potential-energy surface, which is located in the triaxial plane and rises slowly with mass number. The transition probabilities from this minimum to the ground-state minimum have not been observed yet, so the precise location of the minimum remains uncertain. As the heavier isotopes show no deformed band, the fit does not lead to a secondary minimum there and the surfaces are less interesting, so they are not shown here.

6.7.2 Proton–Neutron Vibrations

Collective states corresponding to relative motion between the protons and neutrons have long been known in the form of the *giant dipole resonance* to be discussed in Sect. 6.9. There are, however, also lower-lying collective states which may be interpreted in terms of the so-called *scissor mode* illustrated in Fig. 6.19.

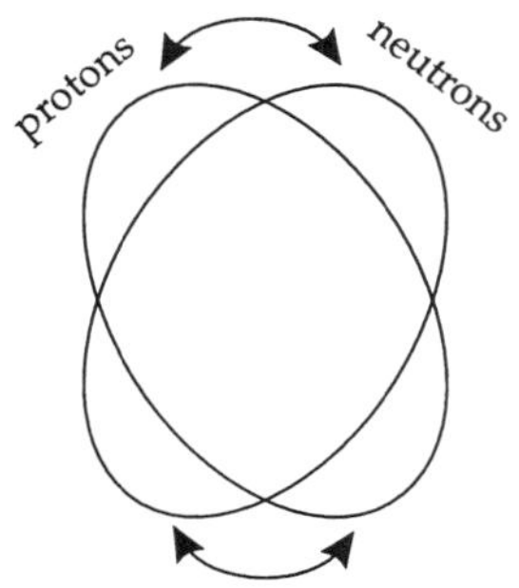

Fig. 6.19 The scissor mode corresponds to a relative oscillation of protons vs. neutrons, which is shown here with an exaggerated magnitude.

The natural way of generalizing the collective model to deal with such states is to introduce separate deformation variables for protons and neutrons, $\alpha^{\mathrm{p}}_{2\mu}$ and $\alpha^{\mathrm{n}}_{2\mu}$. These must, be coupled strongly, because the symmetry energy leads to a strong increase in energy when the two types of particle are separated spatially (for a quantitative discussion see Sect. 6.9). It is therefore more useful to introduce center-of-mass and relative coordinates according to

$$\alpha_{2\mu} = \frac{1}{A}(Z\alpha^{\mathrm{p}}_{2\mu} + N\alpha^{\mathrm{n}}_{2\mu}) \quad , \quad \xi_{2\mu} = \alpha^{\mathrm{p}}_{2\mu} - \alpha^{\mathrm{n}}_{2\mu} \quad . \tag{6.262}$$

One may then assume that vibrations in $\xi_{2\mu}$ will be of small amplitude and can be described by a stiff harmonic oscillator. The total Hamiltonian is constructed as

$$\hat{H} = \hat{H}_{\mathrm{c.m.}}(\alpha) + \hat{H}_{\mathrm{h.o.}}(\xi) + \hat{H}_{\mathrm{coupling}}(\alpha, \xi) \quad . \tag{6.263}$$

The spectrum of this total Hamiltonian is built up out of the spectrum of $\hat{H}_{\mathrm{c.m.}}$, which just contains the usual collective states, coupled to the 0^+ ground state of ξ. At an energy of $\hbar\omega_\xi$ above that, the spectrum is repeated, but with a richer structure caused by the coupling to the 2^+ first excited state in ξ, and so on. A particularly interesting point in this model is that it not only predicts the existence of scissor-vibrational states, but also allows magnetic transitions in the low-energy spectrum, if the coupling part $\hat{H}_{\mathrm{coupling}}$ is nonvanishing. Normally the collective magnetic moment of the nucleus is given by $\hat{\boldsymbol{M}} = (Z/A)g_N\hat{\boldsymbol{L}}$, because it is implied in the model that protons and neutrons move together, so that the magnetic moment is given by the fraction of protons in the nucleus times the magnetic moment of the proton, the magneton, One can show, however, that the interaction between α and ξ leads to a more general *tensorial g-factor* with the magnetic moment given by [Ma75]

$$\hat{M}_{1\mu} = \frac{Z}{A}g_N\hat{\boldsymbol{L}}_{1\mu} + \kappa\left[\alpha \times \hat{\boldsymbol{L}}\right]^1_\mu \quad , \tag{6.264}$$

and this is no longer diagonal, so it can cause transitions.

6.7.3 Higher Multipoles

In a similar way the model can be extended by allowing an additional octupole deformation $\alpha_{3\mu}$ or even a hexadecupole $\alpha_{4\mu}$. Again the Hamiltonian can be set up as a sum of the Hamiltonians of each multipole by itself, plus additional coupling terms. Mathematically this can be carried out relatively simply only for a spherical ground state, although even there the construction of the vibrational states in these higher multipoles is much more complicated. One may also consider a strongly deformed nucleus with the possibility of octupole vibrations added: then the octupole has to be described in the principal axes frame and leads to the existence of a seven-dimensional oscillator added to the usual rotation–vibrational model.

A problem with all of these generalizations is that the model necessarily contains a larger number of unknown parameters, so that the presently available experimental data can be easily fitted, but do not allow a unique determination of all the parameters or a strict test of the models.

6.8 The Interacting Boson Model

6.8.1 Introduction

We already discussed the geometric collective model which interprets the collective low-energy excitations of even–even nuclei as surface vibrations. This model is based on the geometrical picture of the nucleus as a liquid drop and neglects the internal nuclear structure as described by, for example, the shell model.

Furthermore we saw in Sect. 6.3.2 that certain types of collective excitation spectra can be formally described in terms of boson creation and annihilation operators for quadrupole phonons. From the point of view of the shell model one may think of these quadrupole phonons as the projection of the fermion space onto a collective bosonic *quasiparticle subspace*. For example, pairs of fermions may preferentially couple to total angular momentum of zero or two, and these pairs may to some approximation act like bosons. Assuming the appropriateness of this boson picture one can start phenomenologically by defining some bosons and setting up a general form for the Hamiltonian, with the model parameters to be fixed by adjusting them to get a satisfactory agreement between calculated and experimental data. This later stage of the model is quite similar to the geometric model, especially in its generalized form as discussed in Sect. 6.7.1, although with some interesting differences in detail.

Such a boson theory is the *interacting boson model* (IBM) [also known as the *interacting boson approximation* (IBA)], first proposed by Arima and Iachello [Ar75, Ar76]. An equivalent model, formulated slightly differently, has been proposed earlier by Janssen, Jolos, and Dönau [Ja74]. A very readable review of many aspects of the model can be found in the book [Ia87] and the review [Ca88], which the reader is advised to consult on any matter going beyond what is covered here. It will also help on the understanding of this model, especially its microscopic interpretation, if the reader has some acquaintance in with the material presented in Chap. 7.

The IBA includes two types of bosons with angular momentum $L = 0\hbar$ (s bosons) and $L = 2\hbar$ (d-bosons). The energy of each free boson is ϵ_s and ϵ_d, respectively. Motivated by the shell-model, we interpret these bosons as valence nucleon pairs counted from the next closed shell, i.e., the bosons are built up out of pairs of particles for less than half-filled shells and – due to the particle–hole equivalence (see the end of Sect. 7.3.2) – out of pairs of holes for more than half-filled shells. More involved versions of the IBA distinguish, for example, between proton and neutron pairs ("IBA2", [Ar77, Oh78]) or include f bosons with angular momentum $L = 3\hbar$ describing octupole degrees of freedom ("sdf-IBA", [Sc78]). Another extension consists in the introduction of angular-momentum-four bosons (g bosons) has done by many authors; we refer the reader to [Ca88] for details. For this book we concentrate on the simple IBA (now often called "IBA1"), deferring the more involved versions of the model to a brief discussion later.

In order to clarify the definition of the bosons we give two examples. First consider ${}^{110}_{48}\mathrm{Cd}_{62}$: the next closed shell for neutrons as well as for protons is 50, resulting in 12 neutrons (or 6 neutron bosons) and 2 proton holes (or 1 proton boson), for a total of 7 bosons. Another example is ${}^{156}_{64}\mathrm{Gd}_{92}$: the next closed shell for protons (neutrons) is 50 (80) resulting in a total of $14 + 12$ valence nucleons. Hence this isotope possesses a total of 13 bosons.

The interpretation of the bosons as pairs of nucleons in the valence shell is also the reason for one fundamental difference from the boson of the geometric model: in the IBA the number of bosons is fixed and given by the number calculated from the microscopic structure as in the examples. Thus for the description of collective states in one nucleus *only eigenstates of the Hamiltonian fulfilling* $n_s + n_d = N$ should be considered. This opens a very interesting possibility for this model: one may try to fit neighboring nuclei with the same Hamiltonian, varying only the number of bosons N. This has been successful in some cases. In comparison, in the geometric model there is no direct relation to nucleon number, but often a smooth variation of the potential-energy surface from one isotope to the neighboring one has been found.

Formally the states of the IBA are described in an occupation-number representation using creation operators $\hat{d}^\dagger_\mu$, $\hat{s}^\dagger$ and annihilation operators $\hat{d}_\mu$, $\hat{s}$ with the subscript μ denoting the angular-momentum projection ($\mu = -2, \ldots, 2$). These operators must fulfil the usual commutation relations

$$\left[\hat{s}, \hat{s}^\dagger\right] = 1 \quad , \quad \left[\hat{d}_\mu, \hat{d}^\dagger_\nu\right] = \delta_{\mu\nu} \quad , \tag{6.265}$$

with all other commutators vanishing. The total boson-number operator is given by

$$\hat{N} = \sum_\mu \hat{d}^\dagger_\mu \hat{d}_\mu + \hat{s}^\dagger \hat{s} = \hat{n}_d + \hat{n}_s \quad . \tag{6.266}$$

For the Hamiltonian a general form is selected with due consideration of invariances and limited to a reasonable number of low-order terms. The invariance requirements are:

- boson-number conservation – the eigenstates must be eigenstates of $\hat{N}$ also:

  $$\left[\hat{H}_{\mathrm{IBA}}, \hat{N}\right] = 0 \quad . \tag{6.267}$$

  Practically this can be achieved by having as many creation as annihilation operators in each term.

- angular-momentum conservation

  $$\left[\hat{H}_{\mathrm{IBA}}, \hat{L}_{1\mu}\right] = 0 \quad . \tag{6.268}$$

  As for the geometric model, this simply requires that all terms be explicitly coupled to a total angular momentum of zero.

The definition of the angular-momentum operator is identical to that for the geometric collective model (see Exercise 6.3), it is only by convention written slightly differently:

$$\hat{L}_\mu = \sqrt{10}\left[\hat{d}^\dagger \times \hat{\tilde{d}}1\right]^1_\mu \quad , \tag{6.269}$$

with the definition $\hat{\tilde{d}}_\mu \equiv (-1)^\mu \hat{d}_{-\mu}$. This is necessary because $\hat{d}_\mu$ is not a spherical tensor (cf. the same discussion concerning $\alpha_{2\mu}$ and $\hat{\pi}_{2\mu}$ in Sect. 6.3.3). The s boson of course does not contribute to the total angular momentum.

Note: In the IBA literature it is common to use a notation for the *scalar product* of two spherical tensors of the same angular momentum. It is defined as the standard coupling to a resulting angular momentum of zero, but without the denominator $\sqrt{2L+1}$ present in the Clebsch–Gordan coefficients, so that

$$\hat{R}^L \cdot \hat{S}^L = \sqrt{2L+1}\,\left[\hat{R} \times \hat{S}\right]^0 = \sum_{m=-L}^{L} (-1)^L \hat{R}^L_m \hat{S}^L_{-m} \quad . \tag{6.270}$$

The *square of a tensor operator* is analogously defined as

$$(\hat{T}^L)^2 = \hat{T}^L \cdot \hat{T}^L = \sqrt{2L+1}\,\left[\hat{T} \times \hat{T}\right]^0 = \sum_{m=-L}^{L} \hat{T}^{L\dagger}_m \hat{T}^L_m \quad . \tag{6.271}$$

In the last step $\hat{T}^{L\dagger}_m = (-1)^m\, \hat{T}^L_{-m}$ was used.

6.8.2 The Hamiltonian

One of the confusing things about the IBA for the newcomer is the coexistence of quite different formulations of the Hamiltonian. The two most widely used are based on multipole operators and upon the simplest second-quantization expansion, respectively. Naturally, the terms of both formulations are not simply identical, but are rather related by linear combination: this means that the coefficients are also related by a linear transformation.

The simplest form of the Hamiltonian is that based upon the phonon-number conserving terms, but with angular-momentum coupling also taken into account. For reasons of simplicity only one- and two-body terms are taken into account. With these restrictions the most general Hamiltonian can be written down immediately:

$$\begin{aligned}
\hat{H}_{\mathrm{IBA}} = {} & \epsilon_s \hat{n}_s + \epsilon_d \hat{n}_d \\
& + \sum_{L=0,2,4} \frac{1}{2}\sqrt{2L+1}\, C_L \left[\left[\hat{d}^\dagger \times \hat{d}^\dagger\right]^L \times \left[\hat{\tilde{d}} \times \hat{\tilde{d}}\right]^L\right]^0 \\
& + \frac{1}{\sqrt{2}} \tilde{V}_2 \left[\left[\hat{d}^\dagger \times \hat{d}^\dagger\right]^2 \times \left[\hat{\tilde{d}} \times \hat{\tilde{s}}\right]^2 + \left[\hat{d}^\dagger \times \hat{s}^\dagger\right]^2 \times \left[\hat{\tilde{d}} \times \hat{\tilde{d}}\right]^2\right]^0 \\
& + \frac{1}{2} \tilde{V}_0 \left[\left[\hat{d}^\dagger \times \hat{d}^\dagger\right]^0 \times \left[\hat{\tilde{s}} \times \hat{\tilde{s}}\right]^0 + \left[\hat{s}^\dagger \times \hat{s}^\dagger\right]^0 \times \left[\hat{\tilde{d}} \times \hat{\tilde{d}}\right]^0\right]^0 \\
& + U_2 \left[\left[\hat{d}^\dagger \times \hat{s}^\dagger\right]^2 \times \left[\hat{\tilde{d}} \times \hat{\tilde{s}}\right]^2\right]^0 \\
& + \frac{1}{2} U_0 \left(\hat{s}^\dagger \hat{s}^\dagger \hat{\tilde{s}} \hat{\tilde{s}}\right) \quad .
\end{aligned} \tag{6.272}$$

Remember that $\hat{\tilde{d}}_\mu$, not d_μ, is a spherical tensor. The analogous definition for the s boson was only introduced to have a symmetric notation for both types of bosons; this is not really necessary since we have simply $\hat{\tilde{s}} = \hat{s}$.

Only even angular momenta appear in the sum on the second line, since the other couplings vanish. This can be seen by exactly the same symmetry argument that led to the nonexistence of odd angular momenta in the two-phonon states of the harmonic oscillator (see Sect. 6.3.2).

The number of parameters of this Hamiltonian appears to be nine, but can be reduced using the condition of a fixed total number of bosons N, which implies that terms depending on the number of s bosons, n_s, can be rewritten in terms of n_d. Making use of this constraint and restructuring yields

$$\begin{aligned}\hat{H}_{\text{IBA}} &= \epsilon_s N + \frac{1}{2}U_0 N(N-1) + \epsilon N_d \\ &+ \sum_{L=0,2,4} \frac{1}{2}\sqrt{2L+1}C'_L \left[\left[\hat{d}^\dagger \times \hat{d}^\dagger\right]^L \times \left[\hat{\tilde{d}} \times \hat{\tilde{d}}\right]^L \right]^0 \\ &+ \frac{1}{\sqrt{2}}\tilde{V}_2 \left[\left[\hat{d}^\dagger \times \hat{d}^\dagger\right]^2 \times \left[\hat{\tilde{d}} \times \hat{\tilde{s}}\right]^2 + \left[\hat{d}^\dagger \times \hat{s}^\dagger\right]^2 \times \left[\hat{\tilde{d}} \times \hat{\tilde{d}}\right]^2 \right]^0 \\ &+ \frac{1}{2}\tilde{V}_0 \left[\left[\hat{d}^\dagger \times \hat{d}^\dagger\right]^0 \times \left[\hat{\tilde{s}} \times \hat{\tilde{s}}\right]^0 + \left[\hat{s}^\dagger \times \hat{s}^\dagger\right]^0 \times \left[\hat{\tilde{d}} \times \hat{\tilde{d}}\right]^0 \right]^0 \quad , \end{aligned} \tag{6.273}$$

where we used the definitions

$$\epsilon = (\epsilon_d - \epsilon_s) + \frac{1}{\sqrt{5}}U_2(N-1) - \frac{1}{2}U_0(2N-1) \tag{6.274}$$

and

$$C'_L = C_L + U_0\delta_{L0} - 2U_2\delta_{L2} \quad . \tag{6.275}$$

This result can be derived by trying to express all combinations of $\hat{s}^\dagger\hat{\tilde{s}} = \hat{n}_s$ through $N - \hat{n}_d$; for example, the last term of (6.272) can be rewritten as

$$\frac{U_0}{2}\hat{s}^\dagger\hat{s}^\dagger\hat{\tilde{s}}\hat{\tilde{s}} = \frac{U_0}{2}\hat{n}_s(\hat{n}_s - 1) \tag{6.276}$$

which can be shown either from the commutation relations or by simply noting that the particle-number operator in front of an annihilation operator "sees" a boson number reduced by 1. For the rest, the calculations are lengthy, because one has to explicitly expand the angular-momentum couplings to incorporate terms with $\hat{n}_d^2$ in those proportional to C_0 and C_2.

Since we are only interested in excitation energies, we can drop the first two terms, which contribute only a constant energy for a given nucleus. Thus six parameters are left for the IBA1 Hamiltonian in this form.

The last formulation to be discussed, and probably the most useful one for physical purposes, is the multipole form. Define a set of multipole operators via

$$\hat{T}_l = \left[\hat{d}^\dagger \times \hat{\tilde{d}}\right]^l \quad , \quad l = 0, \ldots, 4 \quad , \tag{6.277}$$

which are directly related to the number of d bosons

$$\hat{n}_d = \sqrt{5}\,\hat{T}_0 \tag{6.278}$$

and the angular-momentum operator

$$\hat{L}_\mu = \sqrt{10}\,\hat{T}_{1\mu} \quad . \tag{6.279}$$

In addition the set includes a quadrupole operator

$$\hat{Q}_{2\mu} = \left(\hat{d}^\dagger_\mu \hat{s} + \hat{s}^\dagger \hat{\tilde{d}}_\mu\right) - \frac{\sqrt{7}}{2}\hat{T}_{2\mu} \tag{6.280}$$

and the so-called "pairing operator"

$$\hat{P} = \tfrac{1}{2}\left(\left[\hat{\tilde{d}} \times \hat{\tilde{d}}\right]^0 - \hat{s}\hat{s}\right) \quad . \tag{6.281}$$

In terms of these the multipole form of the Hamiltonian is given by

$$\hat{H} = \epsilon''\hat{n}_d + a_0\hat{P}^\dagger\hat{P} + a_1\hat{L}^2 + a_2\hat{Q}^2 + a_3\hat{T}^2 + a_4\hat{T}_4^2 \quad . \tag{6.282}$$

The new set of coefficients is again related to those of the other formulations by linear transformation. In the literature, different sets of parameters distinguished by name and numerical factors can be found and the reader should refer to the specialized literature, notably [Ia87] and [Ca88], for details.

6.8.3 Group Chains

The next step is the determination of the eigenfunctions and eigenvalues, for which group-theoretical methods turn out to be extremely useful. Starting from the maximum symmetry group of the IBA Hamiltonian, which will prove to be U(6), we construct a chain of subgroups, as introduced in Sect. 4.2. The eigenvalues of the Casimir operator(s) of each subgroup serve as quantum numbers and classify the eigenstates completely.

In the following, we will not give all the details of the calculations, because space does not allow the presentation of all the required group-theoretical methods in this book. We thus give only an overview of the developments; although it requires a deeper understanding of group theory to derive the Casimir operators and quantum numbers in each case, it is relatively trivial but laborious to check the formulas given. Examples of this may be found in Exercises 6.11 and 6.12.

What is the symmetry group of the IBA? We note that the 36 operators $\hat{d}^\dagger_\mu \hat{d}_\nu$, $\hat{d}^\dagger_\mu \hat{s}$, $\hat{s}^\dagger \hat{d}_\nu$, and $\hat{s}^\dagger\hat{s}$ are the basic transformations for shifting a particle in a six-dimensional space spanned by the five states of the d boson with different projections and the one state of the s boson. Recalling the discussion in Sect. 4.3, it is clear that they will constitute the 36 generators of the group of transformations in a six-dimensional wave-function space, i.e., U(6). It is, however, much more useful to consider the angular-momentum-coupled products instead, which are given by

$$\begin{array}{cccc} \left[\hat{s}^\dagger \times \hat{\tilde{s}}\right]^0 & \left[\hat{d}^\dagger \times \hat{\tilde{s}}\right]^2_\mu & \left[\hat{s}^\dagger \times \hat{\tilde{d}}\right]^2_\mu & \left[\hat{d}^\dagger \times \hat{\tilde{d}}\right]^0 \\ \left[\hat{d}^\dagger \times \hat{\tilde{d}}\right]^1_\mu & \left[\hat{d}^\dagger \times \hat{\tilde{d}}\right]^2_\mu & \left[\hat{d}^\dagger \times \hat{\tilde{d}}\right]^3_\mu & \left[\hat{d}^\dagger \times \hat{\tilde{d}}\right]^4_\mu \end{array} \quad . \tag{6.283}$$

Counting the number of projections in each case shows that the total number of independent terms is still 36. This defines the Lie algebra of U(6).

Note that in the following derivations we will, as in Sect. 4.3, use angular-momentum operators not containing the physical scale factor $\hbar$.

The maximum symmetry group of the IBA Hamiltonian thus is U(6), and we now look for all the allowed chains of subgroups, which with their Casimir operators allow a complete classification of the eigenstates of the U(6). The easiest way is to construct the *subalgebras* of the Lie algebra by selecting smaller sets of operators from the Lie algebra given by (6.283), i.e., find smaller sets of linear combinations of these operators closed under commutation.

In Sect. 4.3 it was mentioned that a physically reasonable group chain should involve the groups $O(3) \supset O(2)$ to ensure the existence of angular-momentum quantum numbers (this already hints that $[\hat{d}^\dagger \times \hat{\tilde{d}}]^1$ should be included in the subalgebra). It can be shown that only three chains of the form $U(6) \supset \cdots \supset O(3) \supset O(2)$ are mathematically possible. The associated algebras can be obtained as follows.

A: Removing the 11 operators connected with the s boson we obtain 25 operators of the form $[\hat{d}^\dagger \times \hat{\tilde{d}}]^L_M$ which form the algebra of U(5). This is evident, because these operators just generate transformations in the five-dimensional subspace of the d bosons. The resulting space is quite analogous to that of the bosons in the vibrator model and it is thus not surprising that this case of the IBA describes vibrational nuclei. The next step is to discard the symmetric operator products, which are the ones with even angular momentum. Removing thus the 15 operators of the form $[\hat{d}^\dagger \times \hat{\tilde{d}}]^L_M$ with $L = 0, 2, 4$, we are left with 10 antisymmetric generators, which according to Exercise 4.3 generate the group O(5). Finally we drop the 7 operators of the form $[\hat{d}^\dagger \times \hat{\tilde{d}}]^3_M$, leaving the angular momentum operators $[\hat{d}^\dagger \times \hat{\tilde{d}}]^1_M$ which generate $O(3) \supset O(2)$. This completes the first group chain:

$$U(6) \supset U(5) \supset O(5) \supset O(3) \supset O(2) \quad . \tag{6.284}$$

B: For the second chain we consider the $1 + 3 + 5 = 9$ operators of monopole, dipole, and quadrupole character,

$$\left[\hat{s}^\dagger \times \hat{s}\right]^0 + \sqrt{5}\left[\hat{d}^\dagger \times \hat{\tilde{d}}\right]^0 \quad , \quad \hat{L}_\mu \quad , \quad \text{and} \quad \hat{Q}_\mu \quad , \tag{6.285}$$

which generate the algebra of U(3). The first operator equals the total number of bosons (the factor $\sqrt{5}$ serves to cancel the inverse factor present in the Clebsch–Gordan coefficients) and can be dropped, while the others are the angular-momentum and the quadrupole operators from (6.280) which together generate SU(3). Continuing with the angular momentum-groups by retaining the angular-momentum operator yields only the chain

$$U(6) \supset SU(3) \supset O(3) \supset O(2) \quad . \tag{6.286}$$

C: Finally we consider the $3 + 7 + 5 = 15$ operators

$$\hat{L}_\mu \quad , \quad \left[\hat{d}^\dagger \times \hat{\tilde{d}}\right]^3_\mu \quad , \quad \left[\hat{d}^\dagger \times \hat{s}\right]^2_\mu - \left[\hat{s}^\dagger \times \hat{d}\right]^2_\mu \quad , \tag{6.287}$$

which are generators of the algebra of O(6). Omitting the 5 operators of the last line leaves the generators of O(5), and continuing with the angular-momentum groups results in the group chain

$$\mathrm{U}(6) \supset \mathrm{O}(6) \supset \mathrm{O}(5) \supset \mathrm{O}(3) \supset \mathrm{O}(2) \quad . \tag{6.288}$$

6.8.4 The Casimir Operators

The next task we have is to find the Casimir operators for the group chains given above and their eigenvalues. Since the complete presentation of the necessary group-theoretical formalism is beyond the scope of this book we will in lieu of the calculational details concentrate on the physical meaning and essential properties of the derived results.

The common starting point for all group chains is U(6), and its Casimir operator is the number operator $\hat{N}$ (see Exercise 6.11). Its eigenvalue is the total number of bosons N and is constant for a fixed nucleus as mentioned.

Consider first the **group chain A**: U(6) $\supset$ U(5) $\supset$ O(5) $\supset$ O(3) $\supset$ O(2). In this case the Casimir operator of the algebra of U(5), which is spanned by the 25 generators $[\hat{d}^\dagger \times \hat{\tilde{d}}]^\lambda_\mu$, is given by the d boson number operator as

$$\hat{n}_d = \sum_\mu \hat{d}^\dagger_{2\mu} \hat{d}_{2\mu} \quad . \tag{6.289}$$

This is quite analogous to the case of U(6). Its eigenvalue is denoted by n_d with a range of $0 \leq n_d \leq N$. The Casimir operator of O(5) is

$$\hat{\Lambda}^2 = \frac{1}{2} \sum_{\mu,\nu} \hat{\Lambda}_{\mu\nu} \hat{\Lambda}^+_{\mu\nu} \quad \text{with} \quad \hat{\Lambda}_{\mu\nu} \equiv \hat{d}^\dagger_\mu \hat{\tilde{d}}_\nu - \hat{d}^\dagger_\nu \hat{\tilde{d}}_\mu \tag{6.290}$$

and has eigenvalues given by $\lambda(\lambda + 3)$. The physical interpretation of λ is that it gives the number of bosons which are *not* coupled pairwise to zero angular-momentum. λ is called the *seniority* and its range is given by

$$\lambda = n_d, n_d - 2, n_d - 4, \ldots, 1 \text{ or } 0 \quad . \tag{6.291}$$

We will reencounter the concept of seniority in the chapter on pairing (see Sect. 7.3.2).

The Casimir operators of O(3) and O(2) are, of course, the familiar $\hat{L}^2$ and $\hat{L}_z$ with eigenvalues $L(L+1)$ and M with $(-L \leq M \leq L)$.

Unfortunately these five quantum numbers are not enough to classify all states completely since there are several irreducible representations (irreps) of O(3) for a given irrep of O(5), or, in other words, more than one state of the same L for fixed quantum numbers N, n_d, and λ. In this case one says that the step from O(5) to O(3) is not fully decomposable; similar problems will appear also in the other group chains. In order to enumerate these surplus irreps the new label n_Δ

Table 6.1 Casimir operators, their eigenvalues, and the associated quantum numbers for the group chain A.

Group	U(6)	U(5)	O(5)	O(3)	O(2)
Casimir Operator	$\hat{N}$	$\hat{n}_d$	$\hat{\Lambda}^2$	$\hat{L}^2$	$\hat{L}_z$
Eigenvalue	N	n_d	$\lambda(\lambda+3)$	$L(L+1)$	M
Quantum Number	N	n_d	λ	L	M

is introduced. Physically it describes the number of d-boson triplets which are coupled to zero angular momentum . This yields the scheme given in Table 6.1.

For **group chain B** we need the Casimir operator of SU(3). It is found to be

$$\hat{C}^2 = 2\hat{Q}^2 + \frac{3}{4}\hat{L}^2 \quad . \tag{6.292}$$

The eigenvalue of $\hat{C}^2$ is given by $l^2 + lm + m + 3(l+m)$ with the quantum numbers l and m, whose range for a fixed N is given by

$$\begin{aligned}(l,m) = {} & (2N,0), (2N-4,2), \ldots, (2N-6,0), \\ & (2N-10,2), \ldots, (2N-12,0), (2N-16,2), \ldots \quad .\end{aligned} \tag{6.293}$$

For more detail concerning the group SU(3) and a derivation of these results, the reader is referred to the volume "Symmetries" of this series.

The Casimir operators of O(3) and O(2) have already been discussed above and in this case the step from SU(3) to O(3) is also not fully decomposable. That is why we have to introduce an additional label, which is commonly denoted by K and known as *Eliott's quantum number*. Its range is given by

$$K = \min(\lambda, \mu),\ \min(\lambda, \mu) - 2, \ldots, 1 \text{ or } 0 \quad . \tag{6.294}$$

For $K = 0$ the range of the angular momentum L is given by

$$L = \max(\lambda, \mu),\ \max(\lambda, \mu) - 2, \ldots, 1 \text{ or } 0 \quad , \tag{6.295}$$

and for $K \neq 0$ it is given by

$$L = K, K+1, K+2, \ldots, K + \max(\lambda, \mu) \quad . \tag{6.296}$$

Summarizing these results we get the scheme of Table 6.2. with the eigenstates denoted by $|N(l,m)KLM\rangle$.

Table 6.2 Casimir operators, their eigenvalues, and the associated quantum numbers for the group chain B.

Group	U(6)	SU(3)	O(3)	O(2)
Casimir Operator	$\hat{N}$	$\hat{C}^2$	$\hat{L}^2$	$\hat{L}_z$
Eigenvalue	N	$l^2 + lm + m^2 + 3(l+m)$	$L(L+1)$	M
Quantum Number	N	(l,m)	L	M

Devoting ourselves finally to the **group chain C** we first arrive at the subgroup O(6), whose 15 generators $\hat{\bar{\Lambda}}_{\mu\nu}$ are given by the set combining the generators of O(5) (discussed in connection with group chain A) and those transforming a d boson into an s boson or vice versa:

$$\hat{\Lambda}_{\mu\mu'} \quad , \quad \hat{d}^{\dagger}_{\mu}\hat{s} - \hat{s}^{\dagger}\hat{\tilde{d}}_{\mu'} \quad . \tag{6.297}$$

The Casimir operator is given by

$$\hat{\bar{\Lambda}}^2 = \frac{1}{2}\sum_{\mu\nu} \bar{\Lambda}_{\mu\nu}\bar{\Lambda}^{\dagger}_{\mu\nu} \tag{6.298}$$

and its eigenvalue by $\bar{\lambda}(\bar{\lambda}+4)$. As a rule the eigenvalue of one Casimir operator of the general group O(n) equals $l(l+n-2)$ [it is apparent that this reproduces the values for the groups O(3) and O(5)].

Obviously the principle of construction for $\hat{\bar{\Lambda}}^2$ is identical to the one used for building $\hat{\Lambda}^2$. The only extension is that because of the step from SO(5) to SO(6) the s boson is now included. This analogy is reflected in the physical significance of $\bar{\lambda}$, the *generalized seniority* with values in the range

$$\bar{\lambda} = N, N-2, N-4, \ldots, 1 \text{ or } 0 \quad . \tag{6.299}$$

The range of the quantum number for the next group in the chain, O(5), must then be given by

$$\lambda = \bar{\lambda}, \bar{\lambda}-1, \ldots, 0 \quad , \tag{6.300}$$

since it does not count the s bosons.

As mentioned above the transition from O(5) to O(3) is not fully decomposable and we need an additional quantum number which in this case is commonly called τ. Including the angular-momentum quantum numbers, we denote the diagonal eigenstates as $|N\bar{\lambda}\lambda\tau LM\rangle$ and the group-theoretical scheme is as in Table 6.3.

Table 6.3 Casimir operators, their eigenvalues, and the associated quantum numbers for the group chain C.

Group	U(6)	O(6)	O(5)	O(3)	O(2)
Casimir Operator	$\hat{N}$	$\hat{\bar{\Lambda}}^2$	$\hat{\Lambda}^2$	$\hat{L}^2$	$\hat{L}_z$
Eigenvalue	N	$\bar{\lambda}(\bar{\lambda}+4)$	$\lambda(\lambda+3)$	$L(L+1)$	M
Quantum Number	N	$\bar{\lambda}$	λ	L	M

EXERCISE

6.11 Casimir Operators of U(N)

Problem. Show that the operator $\hat{C} = \sum_{i=1}^{N} \hat{\beta}_i^\dagger \hat{\beta}_i$ is a Casimir operator of the algebra U(N), which is generated by the N^2 operators $\hat{\beta}_i^\dagger \hat{\beta}_k$ with $[\hat{\beta}_i, \hat{\beta}_k^\dagger] = \delta_{ik}$.

Solution. To prove that $\hat{C}$ is a Casimir operator it is sufficient and necessary to show that it commutes with all generators:

$$\begin{aligned}
[\hat{C}, \hat{\beta}_\alpha^\dagger \hat{\beta}_\beta] &= \sum_{i=1}^{N} \left[\hat{b}_i^\dagger \hat{b}_i, \hat{b}_\alpha^\dagger \hat{b}_\beta\right] \\
&= \sum_{i=1}^{N} \left(\hat{b}_i^\dagger \hat{b}_i \hat{b}_\alpha^\dagger \hat{b}_\beta - \hat{b}_\alpha^\dagger \hat{b}_\beta \hat{b}_i^\dagger \hat{b}_i\right) \\
&= \sum_{i=1}^{N} \left(\delta_{i\alpha} \hat{b}_i^\dagger \hat{b}_\beta - \delta_{i\beta} \hat{b}_\alpha^\dagger \hat{b}_i\right) \\
&= 0 \quad .
\end{aligned} \tag{1}$$

This completes the proof.

6.8.5 The Dynamical Symmetries

In the last section we derived the quantum numbers characterizing the eigenstates for the three possible group chains of U(6). In order to find a general solution for the Schrödinger equation, we must expand the eigenfunctions of $\hat{H}_{\mathrm{IBA}}$ in one of the three bases and hence diagonalize the Hamiltonian matrix for a given set of parameters. In the general case, this requires numerical solution.

Alternatively it is possible to obtain analytical solutions in some limiting cases corresponding to special restrictions on the parameters, which reduce the general Hamiltonian to a form that is diagonal in the eigenfunctions of one of the group chains. For this purpose we regard, for a given group chain, only those terms in $\hat{H}_{\mathrm{IBA}}$ that can be expressed by the Casimir operators of this group chain and set all other terms equal to zero. In this case we have a trivially diagonal basis and the Casimir operators can be replaced by their eigenvalues, so that simple analytical expressions for the energy eigenvalues result.

It is not easy to see from the mathematical restrictions whether these limiting cases are physically reasonable. It will be shown, however, that the resulting spectra correspond to certain limiting cases realized in experiment. To be more precise: the energy formulas for the group chains A, B, and C generate spectra typical for vibrational, rotational, and γ-unstable nuclei, respectively.

If the Hamiltonian $\hat{H}_{\mathrm{IBA}}$ is simplified in the way described, it is said to possess a *dynamical symmetry*. This has to be distinguished from the fundamental symmetries such as rotational invariance, which have to hold for all possible values of the parameters.

Before examining these limiting cases more closely, we first show that $\hat{H}_{\rm IBA}$ can be completely expressed in terms of the Casimir operators $\hat{N}$, $\hat{n}_d$, $\hat{\Lambda}^2$, $\hat{\tilde{\Lambda}}^2$, $\hat{Q}^2$, and $\hat{L}^2$ of the three group chains. The resulting form is close but not identical to the multipole form given in (6.282). For this purpose we go step by step through the most general form of the Hamiltonian and express each term through the above Casimir operators (in some cases the required computations are too lengthy to be given explicitly). For convenience, let us write down this most general form again:

$$\begin{aligned}\hat{H}_{\rm IBA} &= \epsilon_s \hat{n}_s + \epsilon_d \hat{n}_d \\ &+ \sum_{L=0,2,4} \frac{1}{2}\sqrt{2L+1}\, C_L \left[[\hat{d}^\dagger \times \hat{d}^\dagger]^L \times \left[\hat{\tilde{d}} \times \hat{\tilde{d}}\right]^L \right]^0 \\ &+ \frac{1}{\sqrt{2}} \tilde{V}_2 \left[[\hat{d}^\dagger \times \hat{d}^\dagger]^2 \times \left[\hat{\tilde{d}} \times \hat{\tilde{s}}\right]^2 + [\hat{d}^\dagger \times \hat{s}^\dagger]^2 \times \left[\hat{\tilde{d}} \times \hat{\tilde{d}}\right]^2 \right]^0 \\ &+ \frac{1}{2}\tilde{V}_0 \left[[\hat{d}^\dagger \times \hat{d}^\dagger]^0 \times [\hat{\tilde{s}} \times \hat{\tilde{s}}]^0 + [\hat{s}^\dagger \times \hat{s}^\dagger]^0 \times \left[\hat{\tilde{d}} \times \hat{\tilde{d}}\right]^0 \right]^0 \\ &+ U_2 \left[[\hat{d}^\dagger \times \hat{s}^\dagger]^2 \times \left[\hat{\tilde{d}} \times \hat{\tilde{s}}\right]^2 \right]^0 \\ &+ \frac{1}{2} U_0 \left(\hat{s}^\dagger \hat{s}^\dagger \hat{\tilde{s}} \hat{\tilde{s}}\right) \quad . \end{aligned} \tag{6.301}$$

The first term contains $\hat{n}_s$ and can trivially be expressed as

$$\hat{n}_s = \hat{s}^\dagger \hat{s} = \hat{N} - \hat{n}_d \quad . \tag{6.302}$$

The next term, $\hat{n}_d$, is itself a Casimir operator, so that we proceed to the term in the sum with $L = 0$, obtaining

$$\left[[\hat{d}^\dagger \times \hat{d}^\dagger]^0 \times \left[\hat{\tilde{d}} \times \hat{\tilde{d}}\right]^0 \right]^0 = \frac{1}{5} \hat{P}^\dagger \hat{P} = \frac{1}{5} \left[\hat{n}_d(\hat{n}_d + 3) - \hat{\Lambda}^2 \right] \quad . \tag{6.303}$$

For the other angular momenta, unfortunately, the explicit form of the angular-momentum coupling has to be used. After lengthy calculations one obtains for the terms proportional to C_2 and C_4

$$\left[[\hat{d}^\dagger \times \hat{d}^\dagger]^2 \times \left[\hat{\tilde{d}} \times \hat{\tilde{d}}\right]^2 \right]^0 = \frac{1}{7\sqrt{5}} \left[2\hat{n}_d(\hat{n}_d - 2) + 2\hat{\Lambda}^2 - \hat{L}^2 \right] \tag{6.304}$$

and

$$\left[[\hat{d}^\dagger \times \hat{d}^\dagger]^4 \times \left[\hat{\tilde{d}} \times \hat{\tilde{d}}\right]^4 \right]^0 = \frac{1}{35} \left[6\hat{n}_d(\hat{n}_d - 2) - \hat{\Lambda}^2 + \frac{5}{3}\hat{L}^2 \right] \quad . \tag{6.305}$$

For the term proportional to $\tilde{V}_2$ one gets similarly

$$\begin{aligned} &\left[[\hat{d}^\dagger \times \hat{d}^\dagger]^2 \times \left[\hat{\tilde{d}} \times \hat{\tilde{s}}\right]^2 + [\hat{d}^\dagger \times \hat{s}^\dagger]^2 \times \left[\hat{\tilde{d}} \times \hat{\tilde{d}}\right]^2 \right]^0 \\ &= \frac{1}{\sqrt{35}} \left[\hat{Q}^2 - \tfrac{1}{2}\hat{\Lambda}^2 + \hat{\tilde{\Lambda}}^2 - \tfrac{1}{8}\hat{L}^2 + \tfrac{1}{2}\hat{n}_d(7\hat{n}_d + 11) - \hat{N}(4\hat{n}_d + 10) \right] , \end{aligned} \tag{6.306}$$

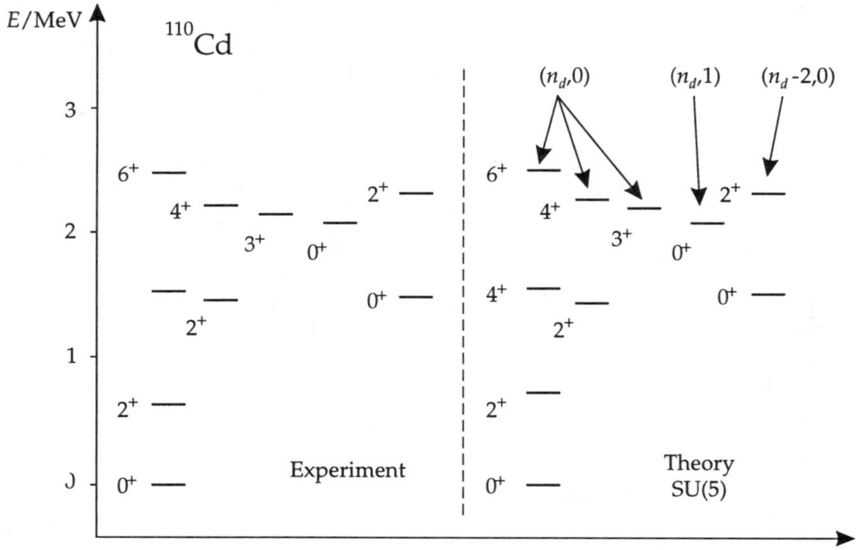

Fig. 6.20 Comparison of experimental spectra with theoretical results obtained in best-fit calculations within the vibrational limit for the nucleus $^{110}_{48}Cd_{62}$. For the theory, the quantum numbers λ and n_Δ are given in parentheses above the corresponding sets of levels.

while for that proportional to $\tilde{V}_0$ the calculation is

$$\begin{aligned}
&\left[\left[\hat{d}^\dagger \times \hat{d}^\dagger\right]^0 \times \left[\hat{s} \times \hat{s}\right]^0 + \left[\hat{s}^\dagger \times \hat{s}^\dagger\right]^0 \times \left[\tilde{\hat{d}} \times \tilde{\hat{d}}\right]^0 \right]^0 \\
&\quad = \left[\hat{d}^\dagger \times \hat{d}^\dagger\right]^0 \hat{s}\hat{s} + \hat{s}^\dagger \hat{s}^\dagger \left[\tilde{\hat{d}} \times \tilde{\hat{d}}\right]^0 \\
&\quad = \frac{1}{\sqrt{5}} \left(\hat{P}^\dagger \hat{s}\hat{s} + \hat{s}^\dagger \hat{s}^\dagger \hat{P}\right) \\
&\quad = \frac{1}{\sqrt{5}} \left[\Lambda^2 - \hat{\Lambda}^2 + \hat{n}_d\left(\hat{N} - \hat{n}_d + 1\right) + (\hat{n}_d + 4)(\hat{N} - \hat{n}_d) \right] \quad .
\end{aligned} \tag{6.307}$$

The term proportional to U_2 can be rewritten as

$$\left[\left[\hat{d}^\dagger \times \hat{s}^\dagger\right]^2 \times \left[\tilde{\hat{d}} \times \hat{s}\right]^2 \right]^0 = \left[\hat{d}^\dagger \times \tilde{\hat{d}}\right]^0 \hat{s}^\dagger \hat{s} = \frac{1}{\sqrt{5}} \hat{n}_d \left(\hat{N} - \hat{n}_d\right) \quad . \tag{6.308}$$

Finally we come to the last term, proportional to U_0, which is simple to rewrite:

$$\hat{s}^\dagger \hat{s}^\dagger \hat{s}\hat{s} = \left(\hat{s}^\dagger \hat{s}\right)\left(\hat{s}^\dagger \hat{s} - 1\right) = \left(\hat{N} - \hat{n}_d\right)\left(\hat{N} - \hat{n}_d - 1\right) \quad . \tag{6.309}$$

For the first step, $\hat{s}^\dagger \hat{s}$ is a number operator for s bosons, which sees the boson number reduced by 1.

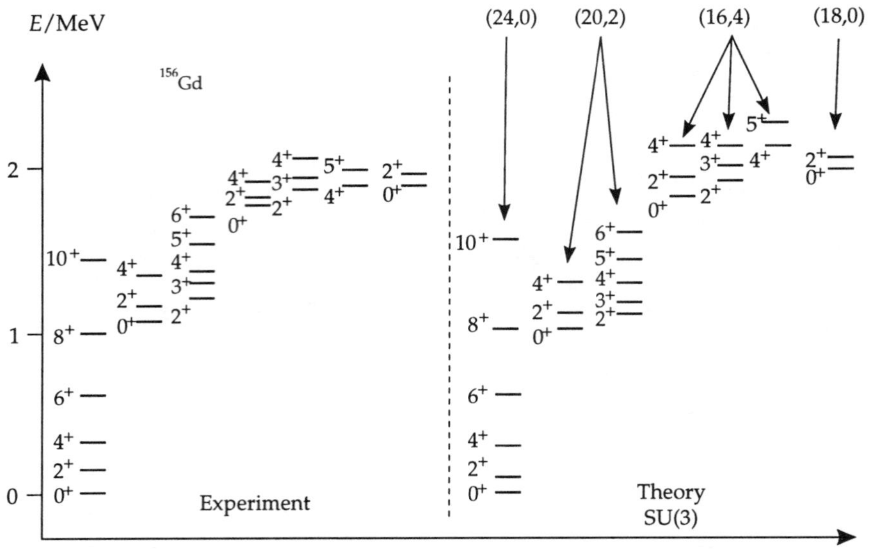

Fig. 6.21 Comparison of experimental spectra with theoretical results obtained in best-fit calculations within the rotational limit for the nucleus ${}^{156}_{64}\mathrm{Gd}_{92}$. The quantum numbers l and m are indicated in parentheses above the corresponding set of levels.

The net result of these laborious calculations is that $\hat{H}_{\mathrm{IBA}}$ can be expressed solely by the Casimir invariants of the three group chains in the form

$$\hat{H}_{\mathrm{IBA}} = \epsilon_N \hat{N} + \epsilon_{n_d} \hat{n}_d + C_{Nn_d} \hat{N} \hat{n}_d + C_{n_d} \hat{n}_d^2 + C_{\bar{\lambda}} \hat{\bar{\Lambda}}^2 + C_\lambda \hat{\Lambda}^2 + C_L \hat{L}^2 + C_Q \hat{Q}^2 \quad . \tag{6.310}$$

Note that the coefficients C_i and ϵ_k can be uniquely calculated from the parameters given in (6.272). They just express an additional way of parametrizing the Hamiltonian. Since for practical applications the parameters are fitted to experiment anyway, their precise relationships are not important here and will not be given in detail.

The Casimir operator of SU(3) was split according to

$$C^2 = 2\hat{Q}^2 + \frac{3}{4}\hat{L}^2 \quad , \tag{6.311}$$

since the term $\hat{Q}^2$ is interpreted as a quadrupole–quadrupole interaction and thus has immediate physical significance.

As already mentioned, one is usually interested only in the structure of one nucleus and in this case the value N is fixed. If additionally only excitation energies are of interest, all terms depending only on N may be dropped to yield

$$\hat{H}_{\mathrm{IBA}} = \epsilon \hat{n}_d + C_{n_d} \hat{n}_d^2 + C_{\bar{\lambda}} \hat{\bar{\Lambda}}^2 + C_\lambda \hat{\Lambda}^2 + C_L \hat{L}^2 + C_Q \hat{Q}^2 \tag{6.312}$$

with the definition

$$\epsilon \equiv \epsilon_{n_d} + N\, C_{n_d} \quad . \tag{6.313}$$

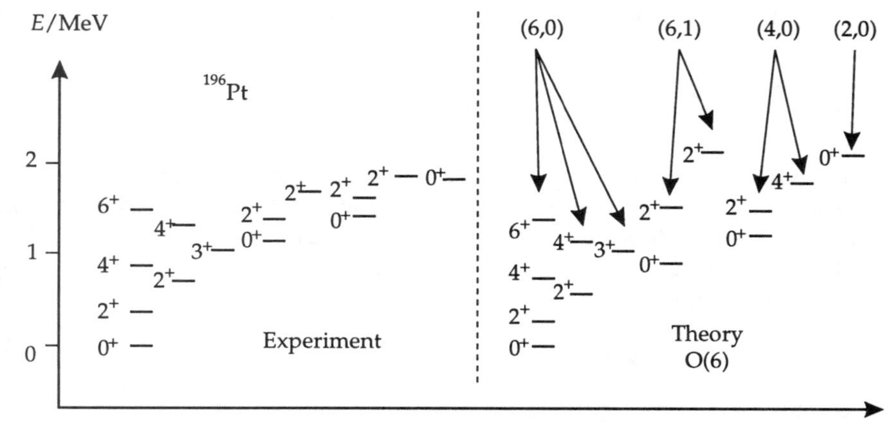

Fig. 6.22 Comparison of experimental energies and the results obtained in best-fit calculations within the γ-unstable limit for the nucleus $^{196}_{78}Pt_{118}$. The quantum numbers $\bar{\lambda}$ and τ are given in parentheses above the corresponding sets of levels.

The Casimir operator of O(2), $\hat{L}_z$, must not appear in the Hamiltonian, because rotational invariance does not allow a distinction between the different orientations of the angular-momentum vector.

Next we apply the various dynamical symmetry cases with the object of a comparison to experimental data.

Group chain A possesses the Casimir operators $\hat{N}$, $\hat{n}_d$, $\hat{\Lambda}^2$, $\hat{L}^2$, and $\hat{L}_z$. Only the terms with these operators can remain in the Hamiltonian in this case. The Hamiltonian is then reduced to

$$\hat{H}^{\mathrm{A}}_{\mathrm{IBA}} = \epsilon\,\hat{n}_d + C_{n_d}\,\hat{n}_d^2 + C_\lambda\,\hat{\Lambda}^2 + C_L\,\hat{L}^2 \quad . \tag{6.314}$$

Now we can immediately write down the energy eigenvalues:

$$E_{n_d\lambda L} = \epsilon\,n_d + C_{n_d}\,n_d^2 + C_\lambda\,\lambda(\lambda+3) + C_L\,L(L+1) \quad . \tag{6.315}$$

Comparing this result with the geometric model, we recognize that its leading term is identical to that of the harmonic oscillator. The main difference is that the number of (quadrupole) phonons is infinite in the geometric model, whereas it must be $\leq N$ in the IBA. Because of the similarity of the spectra, this limit of the IBA, the dynamical symmetry for group chain A, is referred to as the *vibrational limit*. Of course this limit of the IBA also contains corrections, which correspond to anharmonic effects in the geometric model.

Hence we conclude that the vibrational spectra of spherical nuclei should be well described by the above energy formula. As an example we show in Fig. 6.20 the calculated results obtained in a best fit of the parameters ϵ, C_{n_d}, C_λ, and C_L in comparison to the experimental data.

Next we discuss group chain B, whose eigenstates $|N(\lambda,\mu)KLM\rangle$ form a diagonal basis for $\hat{H}_{\mathrm{IBA}}$ if the relations

$$\epsilon = C_{n_d} = C_{\bar{\lambda}} = C_\lambda = 0 \tag{6.316}$$

hold. Hence the general Hamiltonian of (6.272) reduces to

$$\hat{H}^{\mathrm{B}}_{\mathrm{IBA}} = C_Q\,\hat{Q}^2 + C_L\,\hat{L}^2 = \tfrac{1}{2}C_Q\,\hat{C}^2 + \left(C_L - \tfrac{3}{8}C_Q\right)\hat{L}^2 \quad . \tag{6.317}$$

Inserting the eigenvalues of the Casimir operators leads to the energy eigenvalues E_{lmL}

$$E_{lmL} = \frac{C_Q}{2}\left[l^2 + m^2 + lm + 3(l+m)\right] + \left(C_L - \frac{3C_Q}{8}\right) L(L+1) \quad , \tag{6.318}$$

which contain the typical $L(L+1)$ structure of the rotor. Hence the dynamical symmetry of the group chain B is called the rotational limit. An example for the agreement with experimental data is given in Fig. 6.21.

Finally we come to the dynamical symmetry defined by the group chain C. In this case the Hamiltonian is simplified to

$$\hat{H}^{\mathrm{C}}_{\mathrm{IBA}} = C_{\bar{\lambda}}\, \hat{\bar{\Lambda}}^2 + C_\lambda\, \hat{\Lambda}^2 + C_L \hat{L}^2 \quad , \tag{6.319}$$

since only the Casimir operators of O(6), O(5), and O(3) are available. For the energy eigenvalues the result is

$$E_{\bar{\lambda}\lambda L} = C_{\bar{\lambda}}\, \bar{\lambda}(\bar{\lambda}+4) + C_\lambda\, \lambda(\lambda+3) + C_L\, L(L+1) \quad . \tag{6.320}$$

This energy formula is similar to that for γ-unstable nuclei in the geometric model (see Sect. 6.6). The characteristic for the spectra of these nuclei (see Fig. 6.22) is the sequence of the lowest states: above the ground state 0_1 there are, with equal spacing the 2_1 state and then the $(2_2, 4_1)$ doublet. Comparing this spectrum to that of the harmonic oscillator, it is significant that the 0_2 state of the two-phonon triplet $(0_2, 2_2, 4_1)$ is pushed to higher energies. Since the group chain C in earlier calculations has been applied with satisfying agreement to γ-unstable nuclei within a geometrical collective model, this dynamical symmetry is called the γ-unstable limit. Fig. 6.22 again gives an impression of the comparison to an experimental spectrum.

For the IBA there is also a published code [Sc91] which makes a numerical fit of the IBA Hamiltonian to experimental data and can be highly recommended for hands-on practical experimentation.

6.8.6 Transition Operators

Concluding the discussion of the IBA, we indicate how transition operators $\hat{T}^i_{LM}$ (i =electric, magnetic) are set up. The procedure is purely phenomenological as it is for the Hamiltonian and simply consists in writing down all appropriate angular momentum couplings of the s and d bosons, restricting the result to a reasonable number of low-order terms (usually only one-body terms are used). The coefficients in front of these coupling terms are treated as free parameters.

Following this scheme we get for the electric monopole operator

$$\hat{T}^{\mathrm{E}}_{00} = \alpha_0 \left[\hat{s}^\dagger \times \hat{s}\right]^0 + \gamma_0 \left[\hat{d}^\dagger \times \hat{\tilde{d}}\right]^0 \quad , \tag{6.321}$$

for the electric quadrupole operator

$$\hat{T}^{\mathrm{E}}_{2M} = \beta_2 \left(\left[\hat{d}^\dagger \times \hat{s}\right]^2_M + \left[\hat{s}^\dagger \times \hat{\tilde{d}}\right]^2_M\right) + \gamma_2 \left[\hat{d}^\dagger \times \hat{\tilde{d}}\right]^2_M \quad , \tag{6.322}$$

and for the electric hexadecupole operator

$$\hat{T}^{\mathrm{E}}_{4M} = \gamma_4 \left[\hat{d}^\dagger \times \hat{\tilde{d}}\right]^4_M \quad . \tag{6.323}$$

Note, however, that the monopole operator can be rewritten as

$$\hat{T}^{E}_{00} = \alpha_0 \hat{N} + \left(\frac{\gamma_0}{\sqrt{5}} - \alpha_0\right) \hat{n}_d \quad , \tag{6.324}$$

and is hence effectively proportional to the number of d bosons. The expressions for the magnetic operators are rather similar and we obtain for the magnetic dipole operator

$$\hat{T}^{\mathrm{M}}_{1M} = \gamma_1 \left[\hat{d}^\dagger \times \hat{\tilde{d}}\right]^1_M \tag{6.325}$$

and for the magnetic octupole operator

$$\hat{T}^{\mathrm{M}}_{3M} = \gamma_3 \left[\hat{d}^\dagger \times \hat{\tilde{d}}\right]^3_M \quad . \tag{6.326}$$

As $\hat{T}^{\mathrm{M}}_{1M}$ is proportional to the angular-momentum operator, it is purely diagonal and does not cause transitions. This is similar to the geometric model (see Sect. 6.7.2) and in disagreement with experiment, so that higher-order terms of the form

$$\left[\left[\hat{d}^\dagger \times \hat{\tilde{s}}\right]^2 \times \left[\hat{d}^\dagger \times \hat{\tilde{d}}\right]^1\right]^1_M \quad \text{or} \quad \left[\left[\hat{d}^\dagger \times \hat{\tilde{d}}\right]^2 \times \left[\hat{d}^\dagger \times \hat{\tilde{d}}\right]^1\right]^1_M \tag{6.327}$$

have to be added. Alternatively we can keep the first order but then have to introduce the isospin degree of freedom as discussed in the next section.

The calculation of transition amplitudes requires detailed group-theoretical calculations to determine the eigenstates first. It is found that although the basic features are similar to those of the geometric model, there are noticeable differences in detail such as strong transitions between the β and γ bands, which even dominate over the $\beta \rightarrow$ g.s. transitions and strong first-order deviations from the Alaga rules.

6.8.7 Extended Versions of the IBA

In this chapter we give the main ideas underlying the various extensions of the IBA with emphasis on IBA2 [Ar77, Oh78]. To make the hierarchy clearer, the version of the IBA treated up to now is often denoted as "IBA1".

The characteristic of the IBA2 is the distinction between proton and neutron bosons. Altogether twelve types of bosons are introduced, with

$$\hat{d}^\dagger_{\pi\mu} \quad , \quad \hat{d}^\dagger_{\nu\mu} \quad , \quad \hat{s}^\dagger_{\pi} \quad , \quad \hat{s}^\dagger_{\nu} \quad , \tag{6.328}$$

denoting the corresponding creation operators (π = proton, ν =neutron). For example the nucleus ${}^{110}_{48}\mathrm{Cd}_{62}$ has one π boson and six ν bosons, since for both types of nucleons the next closed shell is 50.

This distinction can be easily justified by microscopic considerations since especially for mid-heavy and heavy nuclei the protons and neutrons occupy different major shells. Hence the correlation between like nucleons is expected to be much stronger. Another motivation is the fact that magnetic collective nuclear properties, which from investigations within geometric collective models are known to arise essentially from the proton–neutron difference (see Sect. 6.7.2), are described properly (see below).

The maximum symmetry group for the IBA2 is the direct product $U_\pi(6) \times U_\nu(6)$, which guarantees that the proton and neutron boson numbers are separately conserved. For the Hamiltonian $\hat{H}_{\text{IBA2}}$ this implies that it must commute with the number operators for both boson types separately:

$$\left[\hat{H}_{\text{IBA2}}, \hat{N}_\pi\right] = \left[\hat{H}_{\text{IBA2}}, \hat{N}_\nu\right] = 0 \quad . \tag{6.329}$$

The total-number operator is

$$\hat{N} = \hat{N}_\pi + \hat{N}_\nu = \sum_\mu \hat{d}^\dagger_{\pi\mu}\hat{d}_{\pi\mu} + \hat{s}^\dagger_\pi \hat{s}_\pi + \sum_\mu \hat{d}^\dagger_{\nu\mu}\hat{d}_{\nu\mu} + \hat{s}^\dagger_\nu \hat{s}_\nu \tag{6.330}$$

and the angular-momentum operator is

$$\hat{L}_{1\mu} = \hat{L}_{\pi\mu} + \hat{L}_{\nu\mu} = \sqrt{10}\left(\left[\hat{d}^\dagger_\pi \times \tilde{\hat{d}}_\pi\right]^1_\mu + \left[\hat{d}^\dagger_\nu \times \tilde{\hat{d}}_\nu\right]^1_\mu\right) \quad . \tag{6.331}$$

An important advantage of the IBA2 compared to the IBA1 is the fact that the lowest order for the magnetic dipole operator

$$\hat{M}_{1\mu} = g_\pi \hat{L}_{\pi\mu} + g_\nu \hat{L}_{\nu\mu} \quad , \tag{6.332}$$

with $g_\pi \neq g_\nu$ adjustable parameters, yields nonvanishing transition matrix elements, because the angular momenta are not conserved separately. Thus an important origin of the collective magnetic properties is the missing colinearity of the magnetic dipole and the angular momentum operator, in analogy to the situation in the geometric model.

Next we discuss the construction of the IBA2 Hamiltonian as commonly used. With the same general restrictions based on invariance and simplicity as mentioned in the derivation of $\hat{H}_{\text{IBA}}$, the general form of $\hat{H}_{\text{IBA2}}$ reads

$$\hat{H}_{\text{IBA2}} = \hat{H}^\pi_{\text{IBA}} + \hat{H}^\nu_{\text{IBA}} + \hat{H}^{\nu\pi}_{\text{INT}} \quad , \tag{6.333}$$

with $\hat{H}^{\nu\pi}_{\text{INT}}$ a general interaction Hamiltonian. With an ansatz of this type the number of free parameters is obviously rather large. Practical experience and a comparison to microscopic theories have shown, however, that the strongly simplified Hamiltonian

$$\hat{H}_{\text{IBA2}} = \epsilon_\pi \hat{n}_{\pi d} + \epsilon_\nu \hat{n}_{\nu d} + \hat{V}_{\pi\pi} + \hat{V}_{\nu\nu} + \kappa\sqrt{5}\left[\hat{Q}_\pi \times \hat{Q}_\nu\right]^0 + \lambda \hat{M} \tag{6.334}$$

describes many essential collective nuclear structure properties satisfactorily. The meaning of the terms is discussed next.

The first and the second terms count the number of proton and neutron d bosons separately and hence describe the effect of the pairing interaction between like nucleons. The third and fourth terms describe the interaction between bosons of the same type:

$$\hat{V}_{ii} = \frac{1}{2} \sum_{L=0,2,4} C_{iL} \left[\left[\hat{d}_i^\dagger \times \hat{d}_i^\dagger \right]^L \times \left[\hat{\tilde{d}}_i \times \hat{\tilde{d}}_i \right]^L \right]^0 \quad , \tag{6.335}$$

with $i = \pi, \nu$, and are motivated by $\hat{H}_{\mathrm{IBA}}$. Practical applications of this Hamiltonian show, however, that reasonable agreement with experimental data can already be obtained with a small number of nonvanishing C_{iL}.

The fifth term in $\hat{H}_{\mathrm{IBA2}}$ is a proton–neutron quadrupole–quadrupole interaction. In analogy to the IBA1, $\hat{Q}_{i\mu}$ is given by

$$\hat{Q}_{i\mu} = \hat{d}_{i\mu}^\dagger \hat{s}_i + \hat{s}_i^\dagger \hat{\tilde{d}}_{i\mu} + \chi_i \left[\hat{d}_i^\dagger \times \hat{\tilde{d}}_i \right]_\mu^2 \quad , \tag{6.336}$$

with χ_i a free parameter.

Finally we come to the so-called *Majorana operator* which is defined by

$$\begin{aligned} \hat{M} &= \sqrt{5} \left[(\hat{s}_\nu^\dagger \hat{d}_\pi^\dagger - \hat{s}_\pi^\dagger \hat{d}_\nu^\dagger) \times (\hat{s}_\nu \hat{\tilde{d}}_\pi - \hat{s}_\pi \hat{\tilde{d}}_\nu) \right]^0 \\ &\quad + 2 \sum_{L=1,3} \xi_L \sqrt{2L+1} \left[[\hat{d}_\nu^\dagger \times \hat{d}_\pi^\dagger]^L \times \left[\hat{\tilde{d}}_\nu \times \hat{\tilde{d}}_\pi \right]^L \right]^0 \end{aligned} \tag{6.337}$$

with the adjustable parameters λ, ξ_1, and ξ_3. To understand the structure of this proton–neutron interaction term we have to compare the result of the IBA1 with those to be expected from the IBA2. We already saw that the IBA1 describes the experimental low-energy collective states rather well and we require that these states should be equally well reproduced by the IBA2 without adding additional states in this energy region. On the other hand, it is obvious that those states in the IBA2 which have exactly the same quantum numbers for protons and neutrons, i.e., are symmetric in the proton–neutron degree of freedom, correspond to the IBA1 states. That is the moment when the Majorana force (see Sect. 7.1.2) comes into play since it pushes the states which are nonsymmetric in the proton–neutron degree of freedom to higher energies while the proton–neutron symmetric states remain unaffected.

It is remarkable that there is experimental evidence for the existence of those nonsymmetric or mixed-symmetric states in the energy region of approximately 3MeV. In a geometrical picture (Sect. 6.7.2) these states are interpreted as the relative oscillations of proton and neutron surfaces.

Finally we want to briefly mention some other extensions of the IBA1.

- In the so-called *sdg*-IBA one uses hexadecupole bosons (*g* bosons) with an angular momentum of 4. Usually one neglects the microscopical motivation for the boson number in this case and N becomes a free parameter. When we do so the model is extended to higher spins, in contrast to the IBA1, where the maximum angular momentum is limited by $L_{\max} = 2N\hbar$ because of the limited number of phonons.
- In the *sdf*-IBA octupole bosons (*f* bosons) are introduced which possess an angular momentum of 3. Therefore one is able to describe bands with negative parity and asymmetric ground-state deformations.

- One can construct the boson analogy to the core-plus-particle model (see Sect. 8.1) in order to describe odd–even nuclei. These kinds of models are called interacting boson–fermion models (IBFM).

6.8.8 Comparison to the Geometric Model

The discussion of the solutions of the IBA has shown that it is generally applicable to similar phenomena, such as the geometric model. It is therefore interesting to highlight some of the differences and strong and weak points of each model here.

- Physical interpretation: the IBA offers the chance of a more direct foundation in terms of microscopic structure; this has, however, not been fully successful up to now. On the other hand, the cutoff in angular momentum due to the limited phonon number, which was originally thought to be an interesting prediction of the model, has never been confirmed.
- Practical applications: the geometric model in its more general formulation (Sect. 6.7.1) is mathematically very similar to the IBA so that it is not surprising that the fits to experimental data are of the same quality with similar numbers of parameters. The IBA has an advantage in being able to fit a group of neighboring nuclei with the same Hamiltonian by varying only the total boson number N, and this has been applied with reasonable success in a few cases; in the geometric model it should be possible in principle to fit a series of nuclei with smoothly mass-dependent parameters, but this has not been done yet in any systematic fashion. Another strong point of the IBA is that $\beta \longrightarrow \gamma$ transitions and deviations from the Alaga rule are predicted already to lowest order in the model while in the geometric model they result from higher-order corrections.
- One may ask whether the geometric model might not be regarded also as a mathematically motivated ansatz that uses quadrupole phonons to build up a general Hamiltonian similarly, but not identically, to the IBA. It is true that the Hamiltonian does not depend in any way on the geometric interpretation of these phonons; however, the transition operators were calculated from specific assumptions about the geometric form of the charge distribution and their parameters are completely determined, in contrast to the IBA. That these parameters can be used successfully in most cases is a clear success of the geometric interpretation.

There have been many papers devoted to finding a formal relation between the IBA and a geometric potential-energy surface $V(\beta, \gamma)$. For a list of references and a full discussion see [Ca88]. Here we only quote a general result, which gives some insight into this field. According to [Is81] and [Ia87] the Hamiltonian of (6.273) is associated with a potential-energy surface of the form

$$\begin{aligned} E(N, \beta, \gamma) = & \frac{N \epsilon_d \beta^2}{1 + \beta^2} \\ & + \frac{N(N-1)}{(1+\beta^2)^2} \left(\alpha_1 \beta^4 + \alpha_2 \beta^3 \cos(3\gamma) + \alpha_3 \beta^2 + \alpha_4 \right) \quad , \end{aligned} \tag{6.338}$$

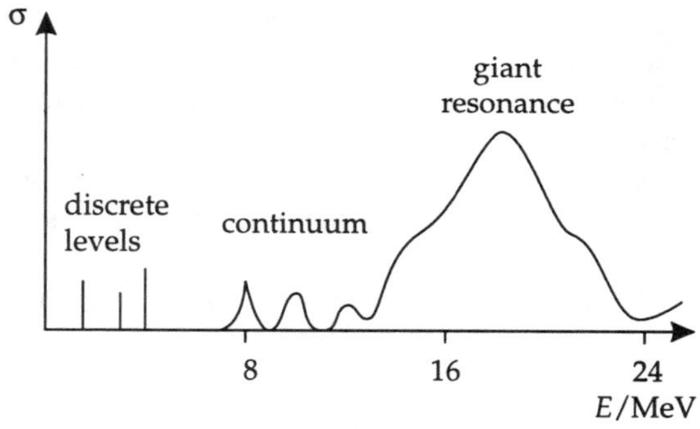

Fig. 6.23 Sketch of the photo-absorption cross section. Above the isolated low-energy peaks the continuum starts in at about 8 MeV, leading to the broad giant resonance.

with the αs simply related to the coefficients in the Hamiltonian. Note that the potential depends on the phonon number. One important feature this result shows is that γ only appears in a linear term in $\cos 3\gamma$, so that minima can occur only on the prolate or oblate axis while triaxial minima are not in the model, at least for the standard Hamiltonian.

Dieperink et al. [Di80] have given potentials for the three dynamic symmetry cases. The expressions are:

- U(5): $$E(N,\beta,\gamma) = \epsilon_d N \frac{\beta^2}{1+\beta^2} \quad ,$$
- SU(3): $$E(N,\beta,\gamma) = \kappa N(N-1) \frac{1+\frac{3}{4}\beta^4 - \sqrt{2}\beta^3\cos(3\gamma)}{(1+\beta^2)^2} \quad ,$$
- O(6): $$E(N,\beta,\gamma) = \kappa' N(N-1)\left(\frac{1-\beta^2}{1+\beta^2}\right)^2 \quad .$$

A simple examination of the minima of these potentials shows the physical significance of these limiting cases in a convincing way.

6.9 Giant Resonances

6.9.1 Introduction

Another kind of collective motion distinct from the surface excitations discussed up to now, is represented by the *giant resonances*, of which the oldest known and most studied is the *giant dipole resonance*. They appear as broad resonances in the nuclear photo effect (i.e., absorption of a photon by a nucleus) with a mean energy systematically decreasing from about 22 MeV for light nuclei to about 14 MeV for heavy ones. The width of the resonance is between 2 and 7 MeV, and it is often split up into several peaks.

To illustrate the circumstances better, Fig. 6.23 shows a typical photoabsorption cross section, which contains both the isolated peaks due to low-lying states, amongst them the collective bands studied up to now, and the continuum beginning in near 8 MeV, which is dominated by the peak of the giant dipole resonance.

The fraction of the total *dipole excitation strength* contained in the giant dipole resonance can be given more precisely through use of the *Thomas–Reiche–Kuhn sum rule*, which also provides a simple and illustrative example of sum rules in general.

Consider the transition probability for electric dipole transitions between states α and β. The corresponding absorption cross section is given by

$$\begin{aligned}\int_{\text{line}} \mathrm{d}E_\beta\, \sigma_{\text{abs}} &= \frac{4\pi^2(E_\beta - E_\alpha)}{\hbar c} \left| \int \mathrm{d}^3r\, \langle\beta|\hat{\rho}(\boldsymbol{r})|\alpha\rangle z \right|^2 \\ &= \frac{2\pi^2\hbar}{mc} f_{\beta\alpha} \quad ,\end{aligned} \tag{6.339}$$

which defines the *transition strength* $f_{\beta\alpha}$. We now use a single-particle picture and assume first that the density is that of one nucleon with charge q, so that

$$\begin{aligned}\int \mathrm{d}^3r\, \langle\beta|\hat{\rho}(\boldsymbol{r})|\alpha\rangle z &= \int \mathrm{d}^3r \int \mathrm{d}^3r'\, \psi_\beta^*(\boldsymbol{r}')\, q\, \delta(\boldsymbol{r}-\boldsymbol{r}')\, \psi_\alpha(\boldsymbol{r}')\, z \\ &= \int \mathrm{d}^3r'\, \psi_\beta^*(\boldsymbol{r}')\, q\, z'\, \psi_\alpha(\boldsymbol{r}') \\ &= q\, \langle\beta|z|\alpha\rangle \quad .\end{aligned} \tag{6.340}$$

The energy difference can be expressed via the nuclear Hamiltonian $\hat{H}$ generating the states α and β as eigenstates:

$$\begin{aligned}f_{\beta\alpha} &= \frac{2mq^2}{\hbar^2}(E_\beta - E_\alpha)\left|\langle\beta|z|\alpha\rangle\right|^2 = \frac{2mq^2}{\hbar^2}\langle\alpha|z|\beta\rangle(E_\beta - E_\alpha)\langle\beta|z|\alpha\rangle \\ &= \frac{mq^2}{\hbar^2}\left(\langle\alpha|z|\beta\rangle\langle\beta|[\hat{H},z]|\alpha\rangle - \langle\alpha|[\hat{H},z]|\beta\rangle\langle\beta|z|\alpha\rangle\right) \quad .\end{aligned} \tag{6.341}$$

The crucial point is that the summation over all final states β can now be done using the completeness of these states:

$$\sum_\beta f_{\beta\alpha} = \frac{mq^2}{\hbar^2}\langle\alpha|\left[z,\left[\hat{H},z\right]\right]|\alpha\rangle \quad . \tag{6.342}$$

Furthermore, the commutator remaining can be evaluated in a useful limiting case: if the Hamiltonian $\hat{H}$ consists of the sum of the kinetic energies of all nucleons plus a purely local potential, the only contribution to the commutator will come from the kinetic energy of the particle making the transition, so that we get

$$\left[z,\left[\hat{H},z\right]\right] = -\frac{\hbar^2}{2m}\left[z,\left[\frac{\partial^2}{\partial z^2},z\right]\right] = -\frac{\hbar^2}{2m}\left[z,2\frac{\partial}{\partial z}\right] = \frac{\hbar^2}{m} \quad , \tag{6.343}$$

making $\sum_\beta f_{\beta\alpha} = q^2$.

The total cross section summed over all final states is then given by

$$\sum_\beta \int \mathrm{d}E_\beta\, \sigma_{\text{abs}} = \frac{2\pi^2\hbar}{mc} q^2 \quad , \tag{6.344}$$

which has to be summed over the contribution of all nucleons. Using the effective charges Ne/A for protons and Ze/A for neutrons yields

$$\sum_{\text{nucleons}} \sum_{\beta} \int \mathrm{d}E_\beta \, \sigma_{\text{abs}} = \frac{2\pi^2\hbar}{mc} e^2 \left[Z(N/A)^2 + N(Z/A)^2 \right]$$
$$= \frac{2\pi^2\hbar e^2}{mc} \frac{NZ}{A}$$
$$\approx \frac{NZ}{A} \times 60\,\text{MeV mb} \quad . \tag{6.345}$$

Unfortunately the derivation is still not sufficient for practical purposes, because the assumption that z commutes with the potential in the Hamiltonian is violated noticeably in the nuclear case. The culprits are interaction terms containing a space exchange. We cannot derive quantitative detail here and relegate examining this problem in more detail to Exercise 6.12, mentioning only that the correction is about half the magnitude of the leading term in the Thomas–Reiche–Kuhn sum rule.

EXERCISE

6.12 Contributions to the Thomas–Reiche–Kuhn Sum Rule

Problem. Explain why a space exchange term in the Hamiltonian yields a contribution to the Thomas–Reiche–Kuhn sum rule.

Solution. As an example, take the case of a pure Majorana force given by

$$\hat{V} = \frac{1}{2} \sum_{ij} v\big(|\boldsymbol{r}_i - \boldsymbol{r}_j|\big) P^{\mathrm{M}} \quad . \tag{1}$$

The commutator with the z coordinate of a selected particle clearly can be nonvanishing only if either i or j is the index of this particle. Both possibilities lead to the same result because of the symmetry of the potential:

$$\left[z, [\hat{V}, z]\right] = 2 \sum_j \left[z, \left[v\big(|\boldsymbol{r} - \boldsymbol{r}_j|\big) P^{\mathrm{M}}, z\right]\right]$$
$$= 2z v\big(|\boldsymbol{r} - \boldsymbol{r}_j|\big) P^{\mathrm{M}} z - z^2 v\big(|\boldsymbol{r} - \boldsymbol{r}_j|\big) P^{\mathrm{M}} - v\big(|\boldsymbol{r} - \boldsymbol{r}_j|\big) P^{\mathrm{M}} z^2$$
$$= \left(2zz_j - z^2 - z_j^2\right) v\big(|\boldsymbol{r} - \boldsymbol{r}_j|\big) P^{\mathrm{M}} \quad . \tag{2}$$

The last step used the fact that P^{M} interchanges the position coordinates of the two interacting particles, so that $P^{\mathrm{M}} z = z_j P^{\mathrm{M}}$ in the term multiplying $v\big(|\boldsymbol{r} - \boldsymbol{r}_j|\big)$.

Experimentally it is found that the giant dipole resonance exhausts about 80–90% of the sum rule, so that it is clearly by far the most dominant excitation mechanism leading to dipole strength.

The basic theoretical idea underlying the interpretation of the giant dipole resonance in the collective model is that of a separation of the centers of mass of protons and neutrons, leading to a large dipole moment of the nucleus. Following a suggestion by Jensen, the first model formulation was due to Goldhaber and Teller, followed independently by the more elaborate one of Steinwedel and Jensen.

6.9.2 The Goldhaber–Teller Model

Goldhaber and Teller [Go48] gave the first theory of the giant resonance. They already recognized that a relative motion of the protons with respect to the neutrons must be responsible for this mode and proposed three basic possibilities.

1. A displacement between protons and neutrons with the restoring force proportional to the displacement and independent of the size of the nucleus and the specific proton affected. This case can be rejected because it leads to an A-independent energy of the resonance.
2. No separation between protons and neutrons on the surface of the nuclei but a difference in the densities in the inside, the restoring force being proportional to the gradient of the density difference. This is essentially the idea of the Steinwedel–Jensen model to be discussed in the next section, but was not further investigated because at that time the observed variation in the frequeny of the giant dipole resonance was best described by the following idea.
3. The protons and neutrons each form a spherical system interpenetrating but slightly replaced with respect to each other, as illustrated in Fig. 24. Clearly this is a pure dipole mode. It is the basis of what is now known as the Goldhaber–Teller model and is described now in more detail.

Call the displacement between the centers of the proton and neutron spheres ξ. How does this displacement change the energy of the system? Of the various terms in the Bethe–Weizsäcker formulation of the liquid drop energy, the only two that could change if the neutrons are displaced with respect to the protons are the symmetry and Coulomb energy. While the latter can easily be written as an integral involving the local charge density, the symmetry energy is known only in an integrated form

$$E_{\text{sym}} = a_{\text{sym}} \frac{(N-Z)^2}{A} \quad , \quad a_s \approx 20 \text{ MeV} \quad , \tag{6.346}$$

and has to be transformed into an integral in terms of local densities. A natural way to rewrite it is

$$E_{\text{sym}} = a_{\text{sym}} \int d^3r \frac{\left(\rho_{\text{n}}(\boldsymbol{r},t) - \rho_{\text{p}}(\boldsymbol{r},t)\right)^2}{\rho_0} \quad , \tag{6.347}$$

and this will actually be used in the Steinwedel–Jensen formulation.[3] Goldhaber and Teller in their original work also did not use this expression for the symmetry

[3] The reader should be aware of the fact that there are some ambiguities in this, which fortunately do not affect the present considerations. For example, the Fermi-gas model yields a $\rho_0^{-1/3}$ dependence instead of the ρ_0^{-1} seen here, and this indicates in general that

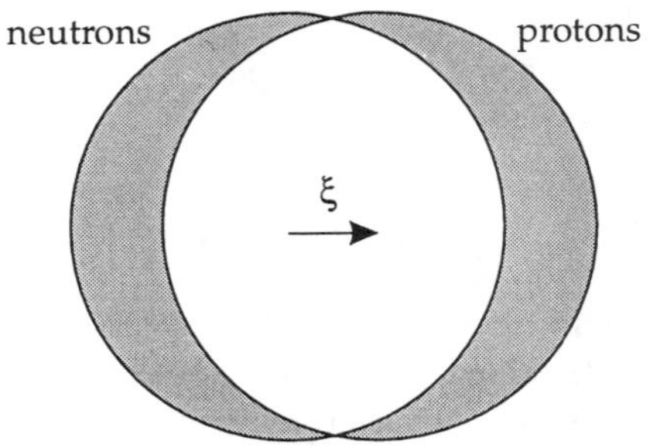

Fig. 6.24 Sketch of the geometrical assumptions in the Goldhaber–Teller model of the giant dipole resonance with an exaggerated separation distance ξ. The shaded regions show where the protons are separated from the neutrons.

energy. This appears reasonable even today, because they were dealing with regions of space containing *only* neutrons or *only* protons, so that the quadratic expansion about equal density of both species of particles may really not be applicable. Instead they assumed an energy ϕ needed to extract one proton (or neutron) from the regular environment of almost equal proton and neutron densities.

It is easy to calculate the change in "symmetry" energy: it is simply proportional to the shaded volume in Fig. 6.24, multiplied with the appropriate density factors. Straightforward integration shows that this volume is given by $\Delta V = \pi R^2 |\xi|$, and with the above assumption this yields

$$\Delta E = \phi(\rho_\mathrm{p} + \rho_\mathrm{n})\Delta V = \phi(Z + N)\frac{\Delta V}{V} = \frac{3}{4}\phi A \frac{|\xi|}{R} \quad . \tag{6.348}$$

This expression causes a slight problem: being linear instead of quadratic in $|\xi|$, it does not lead to harmonic oscillations. Clearly this behavior cannot be true for very small separations, where the diffuseness of the nuclear surface must play a role. Goldhaber and Teller simply assumed a quadratic dependence $E_\mathrm{sym} = \frac{1}{2}k\xi^2$ for small ξ fitted to join the linear dependence at a certain value of ξ, taken to be $\epsilon = 2$ fm. This yields

$$k\epsilon = \frac{3}{4}\frac{\phi A}{R} \quad , \tag{6.349}$$

determining k.

The Coulomb energy is not affected by the mechanism of the Goldhaber–Teller model, as the charge distribution is only translated, not altered in any other way.

The remaining ingredient in the model is the kinetic energy, which in this case is $\mu\xi^2/2$, with $\mu = ZNm/A$ the reduced mass of the proton and neutron two-body system. Together with the harmonic potential we now get the frequency of the giant resonance as

$$\hbar\omega = \hbar\sqrt{\frac{3a_\mathrm{sym}}{4\epsilon m}}\sqrt{\frac{A^2}{ZNR}} \approx \frac{45\ \mathrm{MeV}}{A^{1/6}} \quad . \tag{6.350}$$

---

it is not clear how to transcribe the denominator in the mass formula term. While in the present application the constancy of ρ_0 makes the denominator a constant overall factor, the difference between the various powers can be decisive in the case of a more realistic description of the nuclear surface with densities smoothly approaching zero.

The numerical result was obtained with, for simplicity, $Z = N$, and with Goldhaber's and Teller's value of 40 MeV for ϕ. We will compare it to the one in the Steinwedel–Jensen model.

6.9.3 The Steinwedel–Jensen Model

As mentioned, in the Steinwedel–Jensen model [St50] the giant dipole resonance is described as a dynamic polarization of the nucleus by a change in the local ratio of protons to neutrons while keeping the total density constant everywhere. The dynamic treatment is based on the assumption of *hydrodynamic irrotational flow* for the kinetic energy and on the liquid-drop model for the potential energy.

The principal physical quantities involved are the proton density $\rho_p(\boldsymbol{r}, t)$ and the neutron density $\rho_n(\boldsymbol{r}, t)$, both allowed to vary in time and space. The total density ρ_0 is assumed to be constant inside the nucleus:

$$\rho_0 = \rho_p(\boldsymbol{r}, t) + \rho_n(\boldsymbol{r}, t) \quad , \tag{6.351}$$

and for the simplest case of a spherical nucleus we will further regard a spherical shape only, so that the densities are defined for $|\boldsymbol{r}| \leq R_0$ and are tacitly taken to vanish outside.

For small oscillations we expand the varying densities around their equilibrium values:

$$\rho_p(\boldsymbol{r}, t) = \rho_p^0 + \eta(\boldsymbol{r}, t) \quad , \quad \rho_n(\boldsymbol{r}, t) = \rho_n^0 - \eta(\boldsymbol{r}, t) \quad , \tag{6.352}$$

with $\rho_p^0 = Z\rho_0/A$ and $\rho_n^0 = N\rho_0/A$. Conservation of the number of each kind of particle is assured by requiring

$$\int d^3r \; \eta(\boldsymbol{r}, t) = 0 \quad . \tag{6.353}$$

The restoring force is now mainly determined by the symmetry energy, which is used in the form given by (6.347). The Coulomb energy is also affected by a redistribution of the protons. This effect is, however, much smaller than the change in the symmetry energy, so it is neglected in the following.

Introducing the small deviation $\eta(\boldsymbol{r}, t)$ leads to

$$E_{\mathrm{sym}} = a_{\mathrm{sym}} \frac{(N-Z)^2}{A} + \frac{4a_s}{2\rho_0} \int d^3r \; \eta^2(\boldsymbol{r}, t) \quad . \tag{6.354}$$

The first term is the constant symmetry energy of the unperturbed nucleus and the terms linear in η vanish because of (6.353). The second term on the right is thus the only one to be used as the potential energy of the density vibrations. It remains for us to construct the kinetic energy.

For the kinetic energy the classical kinetic energy of a fluid should be used:

$$T = \frac{m}{2} \int d^3r \; \left(\rho_p \boldsymbol{v}_p^2 + \rho_n \boldsymbol{v}_n^2\right) \quad , \tag{6.355}$$

where $\boldsymbol{v}_p(\boldsymbol{r}, t)$ and $\boldsymbol{v}_n(\boldsymbol{r}, t)$ are the local flow velocities of the proton and neutron fluids, respectively. This expression also shows that *number densities*, not mass densities are used (this makes the factor m necessary). We transform to relative and center-of-mass velocities

$$\begin{aligned} \boldsymbol{v}(\boldsymbol{r},t) &= \boldsymbol{v}_{\mathrm{p}}(\boldsymbol{r},t) - \boldsymbol{v}_{\mathrm{n}}(\boldsymbol{r},t) \quad , \\ \boldsymbol{V}(\boldsymbol{r},t) &= \left(\rho_{\mathrm{p}}(\boldsymbol{r},t)\boldsymbol{v}_{\mathrm{p}}(\boldsymbol{r},t) + \rho_{\mathrm{n}}(\boldsymbol{r},t)\boldsymbol{v}_{\mathrm{n}}(\boldsymbol{r},t)\right)/\rho_0 \quad . \end{aligned} \tag{6.356}$$

The kinetic energy then becomes

$$T = \frac{m}{2}\int d^3r \,\left(\rho_0 V^2 + \rho_{\mathrm{red}} v^2\right) \quad , \tag{6.357}$$

where the reduced density

$$\rho_{\mathrm{red}}(\boldsymbol{r},t) = \frac{\rho_{\mathrm{p}}\rho_{\mathrm{n}}}{\rho_{\mathrm{p}}+\rho_{\mathrm{n}}} = \frac{\rho_{\mathrm{p}}^0\rho_{\mathrm{n}}^0}{\rho_0} + \frac{\rho_{\mathrm{n}}^0 - \rho_{\mathrm{p}}^0}{\rho_0}\eta - \frac{\eta^2}{\rho_0} \tag{6.358}$$

was introduced.

In the following we will do the derivation to lowest order only and also assume that the mean velocity $\boldsymbol{V}$ is zero, as is appropriate for a nucleus at rest in its ground state. In the kinetic energy only the first term of the reduced density has to be considered in this case, leading to

$$T = \frac{m}{2}\frac{ZN}{A^2}\rho_0 \int d^3r \, v^2 \quad , \tag{6.359}$$

while the potential energy is still given by the last term in (6.354). The Lagrangian of the system is thus given by

$$L = \frac{m}{2}\frac{ZN}{A^2}\rho_0 \int d^3r \, v^2 - \frac{4a_s}{\rho_0}\int d^3r \, \eta^2 \quad . \tag{6.360}$$

Note that for $\boldsymbol{V} = 0$ the velocities are determined by the relative velocity as

$$\boldsymbol{v}_{\mathrm{p}} = \frac{N}{A}\boldsymbol{v}, \qquad \boldsymbol{v}_{\mathrm{n}} = \frac{Z}{A}\boldsymbol{v} \quad . \tag{6.361}$$

Before we derive the equations of motion from the variational principle, some properties of the velocity fields have to be given. In addition these have to be severely constrained by some model assumptions, because a change in the densities could be effected by almost arbitrary velocity fields. A universal constraint is provided only by particle number conservation expressed in continuity equations:

$$\frac{\partial \rho_{\mathrm{p,n}}}{\partial t} + \nabla\cdot\left(\rho_{\mathrm{p,n}}\boldsymbol{v}_{\mathrm{p,n}}\right) = 0 \tag{6.362}$$

with the boundary condition that the flows through the surface vanish:

$$\boldsymbol{r}\cdot\boldsymbol{v}_{\mathrm{p,n}}\big|_{|\boldsymbol{r}|=R_0} = 0 \quad . \tag{6.363}$$

Rewriting the continuity equation for the protons to first order we get

$$\frac{\partial \rho_{\mathrm{p}}}{\partial t} = \frac{\partial \eta}{\partial t} = -\nabla\cdot\left[\left(\rho_{\mathrm{p}}^0 + \eta\right)\boldsymbol{v}_{\mathrm{p}}\right] \approx -\rho_{\mathrm{p}}^0\nabla\cdot\boldsymbol{v}_{\mathrm{p}} = -\frac{ZN}{A^2}\rho_0\nabla\cdot\boldsymbol{v} \quad . \tag{6.364}$$

In the last step, (6.361) was used.

The variational principle takes the form

$$\delta \int \mathrm{d}t \int \mathrm{d}^3 r \left(\frac{m}{2} \frac{ZN}{A^2} \rho_0 \boldsymbol{v}^2 - \frac{4a_s}{\rho_0} \eta^2 \right) = 0 \quad . \tag{6.365}$$

With respect to which variables should the variation be carried out? The underlying mechanical degrees of freedom are the time-dependent *displacements* of the particles, which determine both the flow velocities and the density variations. The motion of each element of the fluid can be characterized by its time-dependent position $\boldsymbol{s}(\boldsymbol{r}, t)$ if the initial position $\boldsymbol{s}(\boldsymbol{r}, t = 0)$ is given. In principle the displacements should be defined separately for protons and neutrons, but since these are trivially related, it is much more elegant to define it such that it follows the relative motion, so that

$$\boldsymbol{v} = \frac{\partial \boldsymbol{s}}{\partial t} \quad , \quad \delta \boldsymbol{v} = \frac{\partial \delta \boldsymbol{s}}{\partial t} \tag{6.366}$$

and the variation of the density distortion η is determined through (6.364) to be

$$\delta \eta = -\frac{ZN}{A^2} \rho_0 \nabla \cdot \delta \boldsymbol{s} \quad . \tag{6.367}$$

We can now carry out the variation, insert the two preceding equations for $\delta \boldsymbol{v}$ and $\delta \eta$, and perform a first partial integration (with respect to time in the first and to space in the second integrand):

$$\begin{aligned} 0 &= \delta \int \mathrm{d}t \, L \\ &= \int \mathrm{d}t \int \mathrm{d}^3 r \left(m \frac{ZN}{A^2} \rho_0 \boldsymbol{v} \cdot \delta \boldsymbol{v} - \frac{8a_s}{\rho_0} \eta \delta \eta \right) \\ &= \int \mathrm{d}t \int \mathrm{d}^3 r \left(m \frac{ZN}{A^2} \rho_0 \frac{\partial \boldsymbol{s}}{\partial t} \cdot \frac{\partial \delta \boldsymbol{s}}{\partial t} + 8a_s \frac{ZN}{A^2} \eta \nabla \cdot \delta \boldsymbol{s} \right) \\ &= \int \mathrm{d}t \int \mathrm{d}^3 r \left(-m \frac{ZN}{A^2} \rho_0 \frac{\partial^2 \boldsymbol{s}}{\partial t^2} - 8a_s \frac{ZN}{A^2} \nabla \eta \right) \cdot \delta \boldsymbol{s} \quad . \end{aligned} \tag{6.368}$$

Now the independence of the displacements at different points in space and time can be used, amounting to a dropping of the integrations:

$$m \frac{ZN}{A^2} \rho_0 \frac{\partial \boldsymbol{v}}{\partial t} = -8a_{\mathrm{sym}} \frac{ZN}{A^2} \nabla \eta \quad . \tag{6.369}$$

Taking the divergence of this equation and using the continuity equation via

$$\frac{\partial \nabla \cdot \boldsymbol{v}}{\partial t} = -\frac{A^2}{ZN \rho_0} \frac{\partial^2 \eta}{\partial t^2} \tag{6.370}$$

yields the equation of motion for η:

$$m \frac{\partial^2 \eta}{\partial t^2} = 8a_{\mathrm{sym}} \frac{ZN}{A^2} \nabla^2 \eta \quad . \tag{6.371}$$

This simple result shows that the density fluctuations fulfil a wave equation

$$\frac{1}{u^2}\frac{\partial^2\eta}{\partial t^2} = \nabla^2\eta \quad , \tag{6.372}$$

with the propagation speed of the density waves given by

$$u^2 = \frac{ZN}{A^2}\frac{8a_{\mathrm{sym}}}{m} \quad . \tag{6.373}$$

For $a_s \approx 23\,\mathrm{MeV}$ we can estimate $u \approx c/5$. Solving the wave equation is trivial with the methods learnt in Sect. 5.3.1: assuming a periodic time dependence $\eta(\boldsymbol{r},t) = \eta(\boldsymbol{r},0)\exp(\mathrm{i}\omega t)$ we get a Helmholtz equation,

$$\Delta\eta + k^2\eta = 0 \quad , \tag{6.374}$$

that can be solved in terms of the familiar solutions of good angular momentum

$$\eta_{klm}(\boldsymbol{r}) = B\, j_l(kr)\, Y_{lm}(\Omega) \quad , \quad k = \frac{\omega}{c} \quad . \tag{6.375}$$

The boundary condition is given by the stationarity of the nuclear surface: the velocity through the surface must vanish. Equation (6.369) shows that this amounts to

$$0 = \boldsymbol{e}_r \cdot \nabla\eta\Big|_{r=R} = \frac{\partial\eta}{\partial r}\Big|_{r=R} \quad \longrightarrow \quad j_l'(kR) = 0 \quad . \tag{6.376}$$

The eigenfrequencies are thus determined by the zeroes of the derivatives of the spherical Bessel functions. For $l = 1$ the lowest zero is ≈ 2.08, so that $k \approx 2.08/R$, and the energy of the state will be

$$\begin{aligned} \hbar\omega = \hbar uk &\approx \hbar\sqrt{\frac{ZN}{A^2}\frac{8a_{\mathrm{sym}}}{m}}\frac{2.08}{1.2\,\mathrm{fm}\,A^{1/3}} \\ &= \sqrt{\frac{4ZN}{A^2}}\frac{76.5\ \mathrm{MeV}}{A^{1/3}} \quad . \end{aligned} \tag{6.377}$$

The square-root factor was separated out, because it depends only slowly on the mass number and reduces to 1 for $Z = N$, while the characteristic $A^{-1/3}$ dependence is present in the second factor. The A dependence agrees quite well with the experimentally observed one.

6.9.4 Applications

Up to now the development of the theory has been purely classical, but since in this approximation we are dealing with a harmonic oscillator, it is easy to give a quantized version: simply introduce the creation and annihilation operators for *gions* with the correct angular momentum properties, $\hat{g}_{1\mu}^\dagger$ and $\hat{g}_{1\mu}$, allowing the expression of the Hamiltonian as

$$\hat{H} = \hbar\omega\left(\sum_\mu \hat{g}_\mu^\dagger\hat{g}_\mu + \tfrac{3}{2}\right) \quad . \tag{6.378}$$

Going this far in the theory makes sense only if states with more than one phonon are actually observed; the actual observation of such a two-phonon giant resonance

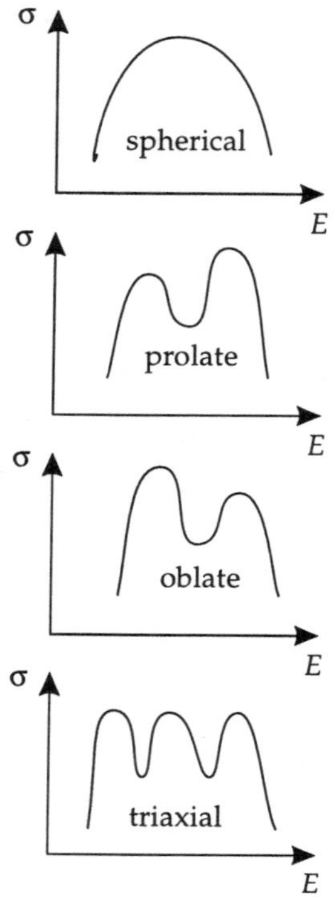

Fig. 6.25 Schematic illustration of the splitting of the giant dipole resonance depending on the deformation of the ground state.

has been achieved only quite recently (for a review see the article by Mordechai and Moore [Mor93]).

What happens if the giant dipole resonance is excited in a deformed nucleus? Formally this was developped in the framework of the *dynamic collective model* [Da64a, Da64b] for coupling the quanta of the giant dipole mode to the quadrupole surface phonons, but it is easy to understand the principles geometrically. Because the giant resonance has a much higher energy than the surface vibrations, it is much faster and one may use – at least as a first approximation – the adiabatic assumption of the giant resonance inside a fixed surface. The energy of the resonance was determined by the first zero of the derivative of the spherical Bessel function via $kR = 2.08$, so that it depends inversely on the nuclear radius. Now if the nucleus becomes deformed, the radius will be different along the three principal axes, and the resonance may split up into three peaks with the energy separation determined by the axis ratios (the correct formula for the energies of the peaks is not, however, quite so simple). If there is axial symmetry, two of these modes will coincide, and one of two peaks will be higher, corresponding actually to motion along two axes. The situation is illustrated in Fig. 6.25 for the four possible cases.

There is also a dynamic coupling between the giant resonance and the surface vibrations. Since the energy of the giant resonance is about ten times higher than the usual β and γ vibrations, this can be ignored in a first approximation by using the adiabatic approximation, i.e., by treating the giant resonance within the instantaneous shape of the nucleus. The correction can be formulated as an interaction between the gions and the surface-vibrational phonons, with the coupling strength parameters even derivable from the geometric picture. The net result is a further splitting of the giant resonance peak, leading to a more complicated structure of the total excitation function.

In experiment it turns out that in many cases the correlation between the nuclear deformation and the structure of the giant resonance is correct, but often also the resonance shows more structure. This may either indicate a more complicated coupling to the surface phonons, or noncollective effects. We will come back to the microscopic treatment of the giant resonances in Sect. 8.2.

The *quadrupole giant resonance* can be derived in the same model, by simply solving for the eigenenergies with $l = 2$. The corresponding zero of the derivative of the spherical Bessel function is $k = 3.342$, so the energy of the giant quadrupole resonance should be a factor of 1.6 above that of the dipole. Although this energy makes the theory more complicated because of the instability with respect to two-neutron emission, it is in broad qualitative agreement with experiment concerning the position of the resonance.

One principal problem of this collective treatment of the giant resonances is the lack of a theoretical prediction of the width. The giant resonances decay relatively rapidly because their energy exceeds the binding energy of the most loosely bound nucleon, leading to a large width. This can be taken into account more easily in a microscopic model, while the collective treatment has to be content with assuming some constant width for each of the eigenmodes associated with the resonance.

7. Microscopic Models

7.1 The Nucleon–Nucleon Interaction

7.1.1 General Properties

All microscopic models of the nucleus are based on some model of the basic interactions between nucleons. The word "model" must be used, since even at present there is still no exact and reliable theory such as exists for the electromagnetic interaction. Attempts to derive the nucleon–nucleon interaction from the quark model are not sufficiently mature to be used in nuclear models, so that there is still a plethora of model interactions that are used for various purposes. One class of interactions is quite successful in the description of nucleon–nucleon scattering processes, but does not fare well in the description of nuclei consisting of a larger number of nucleons. For the latter purpose *effective interactions* are employed, which are specially developed for use in Hartree–Fock and related calculations, but do not describe nucleon–nucleon scattering satisfactorily. They are thought to take into account some of the complicated correlations that characterize complex nuclei by introducing, e.g., a density dependence.

In this section we discuss the very general properties that an interaction should have in order to satisfy basic invariance requirements. We restrict the discussion to two-body interactions mostly; although three-body forces have been employed in nuclear theory it is not clear yet to what extent they are necessary. They do play a vital role in the case of the effective interactions and will be duly treated in that connection.

The nucleon–nucleon potential can depend on the positions, momenta, spins, and isospins of the two nucleons concerned:

$$v = v(\boldsymbol{r}, \boldsymbol{r}', \boldsymbol{p}, \boldsymbol{p}', \hat{\boldsymbol{\sigma}}, \hat{\boldsymbol{\sigma}}', \hat{\boldsymbol{\tau}}, \hat{\boldsymbol{\tau}}') \quad . \tag{7.1}$$

The functional form of v is restricted by the usual invariance requirements, which we will now examine one by one.

- **Translational invariance:** the dependence on the positions $\boldsymbol{r}$ and $\boldsymbol{r}'$ should only be through the relative distance $\boldsymbol{r} - \boldsymbol{r}'$. Because the individual vectors will not be needed further, we will henceforth use the notation $\boldsymbol{r}$ for this relative coordinate.
- **Galilei invariance:** the interaction potential should be independent of any transformation to another inertial frame of reference. This demands that the interaction should depend only on the relative momentum $\boldsymbol{p}_{\rm rel} = \boldsymbol{p} - \boldsymbol{p}'$. As with the relative spatial coordinate, we will denote this vector simply by $\boldsymbol{p}$.

- **Rotational invariance:** this condition cannot be used as simply to restrict the number of variables; instead it implies that all terms in the potential should be constructed to have a total angular momentum of zero.
- **Isospin invariance:** the isospin degrees of freedom enter only through the operators $\hat{\boldsymbol{\tau}}$ and $\hat{\boldsymbol{\tau}}'$, so that the only terms that are scalar under rotation in isospin space are those containing no isospin dependence, the scalar product $\hat{\boldsymbol{\tau}} \cdot \hat{\boldsymbol{\tau}}'$, or powers thereof. As $\hat{\boldsymbol{\tau}} \cdot \hat{\boldsymbol{\tau}} = 3$ for the nucleon with its isospin of $\frac{1}{2}$, the terms involving only one of the two operators are trivial. However, even the higher powers of $\hat{\boldsymbol{\tau}} \cdot \hat{\boldsymbol{\tau}}'$ are not independent, as can be seen from the properties of the Pauli matrices

$$\left[\hat{\tau}_i, \hat{\tau}_j\right] = 2\mathrm{i} \sum_k \varepsilon_{ijk} \hat{\tau}_k \quad , \quad \left\{\hat{\tau}_i, \hat{\tau}_j\right\} = 2\delta_{ij} \quad . \tag{7.2}$$

Adding these two equations we get

$$\hat{\tau}_i \hat{\tau}_j = \delta_{ij} + \mathrm{i} \sum_k \varepsilon_{ijk} \hat{\tau}_k \quad , \tag{7.3}$$

which leads to

$$\begin{aligned} \left(\hat{\boldsymbol{\tau}} \cdot \hat{\boldsymbol{\tau}}'\right)^2 &= \sum_{ij} \hat{\tau}_i \, \hat{\tau}_i' \, \hat{\tau}_j \, \hat{\tau}_j' \\ &= \sum_{ij} \left(\delta_{ij} + \mathrm{i} \sum_k \varepsilon_{ijk} \, \hat{\tau}_k\right) \left(\delta_{ij} + \mathrm{i} \sum_{k'} \varepsilon_{ijk'} \, \hat{\tau}_{k'}'\right) \quad . \end{aligned} \tag{7.4}$$

Expanding the right-hand side one finds that the terms with one ε vanish, because $\delta_{ij} \, \varepsilon_{ijk} = 0$, while the remainder produces

$$\begin{aligned} \left(\hat{\boldsymbol{\tau}} \cdot \hat{\boldsymbol{\tau}}'\right)^2 &= \sum_i \delta_{ii} - \sum_{ijkk'} \varepsilon_{ijk} \, \varepsilon_{ijk'} \, \hat{\tau}_k \, \hat{\tau}_{k'}' \\ &= 3 - 2 \sum_k \hat{\tau}_k \, \hat{\tau}_k' = 3 - 2 \, \hat{\boldsymbol{\tau}} \cdot \hat{\boldsymbol{\tau}}' \quad . \end{aligned} \tag{7.5}$$

Here the sum over the products with two εs could be reduced by noting that for a fixed pair of indices (i,j) the term is nonzero only if k and k' have the same value, which effectively runs through the range 1,2,3. However, each of these occurs twice: once for (i,j) and once for (j,i). The final result shows that all powers of $\hat{\boldsymbol{\tau}} \cdot \hat{\boldsymbol{\tau}}'$ can be reduced to the first-order product, so that only this needs to be taken into account.

Collecting the results obtained up to now, we may split the total potential into two terms:

$$v = v\left(\boldsymbol{r}, \boldsymbol{p}, \sigma, \sigma'\right) + \tilde{v}\left(\boldsymbol{r}, \boldsymbol{p}, \sigma, \sigma'\right) \, \hat{\boldsymbol{\tau}} \cdot \hat{\boldsymbol{\tau}}' \quad . \tag{7.6}$$

This separation into two terms according to the isospin dependence is only one of several alternative decompositions, which are all useful in different circumstances.

The first alternative is based on total isospin. Two nucleons with isospin $\frac{1}{2}$ may couple to a total isospin T of 0 (singlet) or 1 (triplet). The scalar product of the isospin operators can then be expressed as

$$\hat{\boldsymbol{\tau}} \cdot \hat{\boldsymbol{\tau}}' = 4\,\hat{\boldsymbol{t}} \cdot \hat{\boldsymbol{t}}' = 2\left[\left(\hat{\boldsymbol{t}} + \hat{\boldsymbol{t}}'\right)^2 - \hat{\boldsymbol{t}}_1^2 - \hat{\boldsymbol{t}}_2^2\right]$$
$$= 2\left[T(T+1) - \tfrac{3}{4} - \tfrac{3}{4}\right] = \begin{cases} -3 & \text{for the singlet} \\ +1 & \text{for the triplet} \end{cases} \quad . \tag{7.7}$$

This result allows the construction of projection operators onto the singlet or triplet, respectively, which are simply such linear combinations that they yield zero when applied to one of the two states and 1 when applied to the other:

$$\hat{P}_{T=0} = \tfrac{1}{4}\left(1 - \hat{\boldsymbol{\tau}} \cdot \hat{\boldsymbol{\tau}}'\right) \quad , \quad \hat{P}_{T=1} = \tfrac{1}{4}\left(3 + \hat{\boldsymbol{\tau}} \cdot \hat{\boldsymbol{\tau}}'\right) \quad . \tag{7.8}$$

The interaction between the nucleons can now be formulated also as a sum of a singlet and a triplet potential:

$$v = v_{T=0}\left(\boldsymbol{r}, \boldsymbol{p}, \sigma, \sigma'\right) \hat{P}_{T=0} + v_{T=1}\left(\boldsymbol{r}, \boldsymbol{p}, \sigma, \sigma'\right) \hat{P}_{T=1} \quad . \tag{7.9}$$

A third formulation is based on isospin exchange. The two-particle wave function with good isospin is

$$|TT_3\rangle = \sum_{t_3 t_3'} \left(\tfrac{1}{2}\tfrac{1}{2}T|t_3 t_3' T_3\right) |\tfrac{1}{2}t_3\rangle|\tfrac{1}{2}t_3'\rangle \quad , \tag{7.10}$$

and using the symmetry of the Clebsch–Gordan coefficients

$$(j_1 j_2 J\,|m_1 m_2 M) = (-1)^{j_1+j_2-J}(j_2 j_1 J\,|m_2 m_1 M) \quad , \tag{7.11}$$

we see that an exchange of the two isospin projections t_3 and t_3' corresponds to a change of sign for $T = 0$ and no change for $T = 1$. Because of (7.7), the isospin exchange operator $\hat{P}_\tau$ which produces the correct sign change can be expressed as

$$\hat{P}_\tau = \tfrac{1}{2}\left(1 + \hat{\boldsymbol{\tau}} \cdot \hat{\boldsymbol{\tau}}'\right) \quad , \tag{7.12}$$

and the isospin dependence of the nucleon–nucleon interaction may be formulated in a third way:

$$v = v_1\left(\boldsymbol{r}, \boldsymbol{p}, \sigma, \sigma'\right) + v_2\left(\boldsymbol{r}, \boldsymbol{p}, \sigma, \sigma'\right) \hat{P}_\tau \quad . \tag{7.13}$$

Note that while many of these manipulations apply to the nucleon spin as well, the spin dependence of the potential cannot be constrained to such an extent because the spin operators can be coupled to zero total angular momentum also with the position or momentum vectors. The general definition of the spin exchange operator can of course be done in analogy to (7.12) and is as useful as for isospin.

- **Parity invariance:** the requirement for the potential is

$$v\left(\boldsymbol{r}, \boldsymbol{p}, \hat{\boldsymbol{\sigma}}, \hat{\boldsymbol{\sigma}}', \boldsymbol{\tau}, \boldsymbol{\tau}'\right) = v\left(-\boldsymbol{r}, -\boldsymbol{p}, \hat{\boldsymbol{\sigma}}, \hat{\boldsymbol{\sigma}}', \boldsymbol{\tau}, \boldsymbol{\tau}'\right) \quad , \tag{7.14}$$

  which can be fulfilled by using only terms containing an even power of $\boldsymbol{r}$ and $\boldsymbol{p}$ together.

- **Time reversal invariance:** it requires

$$v\left(\boldsymbol{r}, \boldsymbol{p}, \hat{\boldsymbol{\sigma}}, \hat{\boldsymbol{\sigma}}', \boldsymbol{\tau}, \boldsymbol{\tau}'\right) = v\left(\boldsymbol{r}, -\boldsymbol{p}, -\hat{\boldsymbol{\sigma}}, -\hat{\boldsymbol{\sigma}}', \boldsymbol{\tau}, \boldsymbol{\tau}'\right) \quad , \tag{7.15}$$

  so that an even number of $\boldsymbol{p}$s and $\hat{\boldsymbol{\sigma}}$s combined are allowed in each term.

We can now collect the types of terms allowed by the preceding considerations. There is still a bewildering variety of such terms, so that it is natural to start with the lowest powers, using such terms as

$$\begin{aligned}&\hat{\boldsymbol{\sigma}} \cdot \hat{\boldsymbol{\sigma}}' \quad , \quad (\boldsymbol{r} \cdot \hat{\boldsymbol{\sigma}})\,(\boldsymbol{r} \cdot \hat{\boldsymbol{\sigma}}') \quad , \\ &\hat{\boldsymbol{L}} \cdot \hat{\boldsymbol{S}} = -\mathrm{i}\hbar(\boldsymbol{r} \times \hat{\boldsymbol{p}}) \cdot (\hat{\boldsymbol{\sigma}} + \hat{\boldsymbol{\sigma}}') \quad .\end{aligned} \tag{7.16}$$

All of these can be combined with arbitrary functions of $\boldsymbol{r}$ and $\boldsymbol{p}$. While it is known that the spatial dependence of the potential is quite pronounced, less is known about the momentum dependence. Interactions in practical use contain at most a $\hat{p}^2$ term.

7.1.2 Functional Form

Historically the first attempts to formulate a nucleon–nucleon interaction used a central momentum-independent potential, but spin and isospin dependence were soon recognized to be essential. The potential takes the general form

$$v = v_0(r) + v_\sigma(r)\,\hat{\boldsymbol{\sigma}} \cdot \hat{\boldsymbol{\sigma}}' + v_\tau(r)\,\hat{\boldsymbol{\tau}} \cdot \hat{\boldsymbol{\tau}}' + v_{\sigma\tau}\left(\hat{\boldsymbol{\sigma}} \cdot \hat{\boldsymbol{\sigma}}'\right)\left(\hat{\boldsymbol{\tau}} \cdot \hat{\boldsymbol{\tau}}'\right) \quad , \tag{7.17}$$

or in the traditional formulation using exchange operators:

$$v = v_{\mathrm{W}}(r) + v_{\mathrm{M}}\hat{P}_r + v_{\mathrm{B}}\hat{P}_\sigma + v_{\mathrm{H}}\hat{P}_r\,\hat{P}_\sigma \quad . \tag{7.18}$$

The indices stand for Wigner, Majorana, Bartlett, and Heisenberg. The isospin exchange operator $\hat{P}_\tau$ was used originally in the last term, but this is a trivial substitution because for fermions an exchange of all coordinates must simply change the sign of the wave function $\hat{P}_r\,\hat{P}_\sigma\,\hat{P}_\tau = -1$, so that $\hat{P}_r\,\hat{P}_\sigma = -\hat{P}_\tau$.

Of course it is possible to relate the expressions (7.17) and (7.18) to each other by writing the exchange operators in (7.18) in terms of spin and isospin operators and then comparing coefficients.

Another important ingredient of a nucleon–nucleon interaction that was found to be necessary to explain the properties of the deuteron is the *tensor force*. It contains the term $(\boldsymbol{r} \cdot \hat{\boldsymbol{\sigma}})(\boldsymbol{r} \cdot \hat{\boldsymbol{\sigma}}')$, but in such a combination that the average over the angles vanishes. The full expression is

$$S_{12} = \left(v_0(r) + v_1(r)\,\hat{\boldsymbol{\tau}} \cdot \hat{\boldsymbol{\tau}}'\right)\left[\frac{(\boldsymbol{r} \cdot \hat{\boldsymbol{\sigma}})(\boldsymbol{r} \cdot \hat{\boldsymbol{\sigma}}')}{r^2} - \frac{1}{3}\hat{\boldsymbol{\sigma}} \cdot \hat{\boldsymbol{\sigma}}'\right] \quad . \tag{7.19}$$

EXERCISE

7.1 The Angular Average of the Tensor Force

Problem. Show that the tensor force has a vanishing angular average.

Exercise 7.1

Solution. When insert we the vector $\boldsymbol{r}$ in spherical coordinates,

$$\boldsymbol{r} = r(\sin\theta\cos\phi, \sin\theta\sin\phi, \cos\theta) \quad , \tag{1}$$

the angular average of the numerator inside the parentheses becomes

$$\begin{aligned} I &= \frac{1}{4\pi}\int \mathrm{d}\Omega\, (\boldsymbol{r}\cdot\hat{\boldsymbol{\sigma}})(\boldsymbol{r}\cdot\hat{\boldsymbol{\sigma}}') \\ &= \frac{r^2}{4\pi}\int \sin\theta\,\mathrm{d}\theta\,\mathrm{d}\phi\, \left(\hat{\sigma}_x\sin\theta\cos\phi + \hat{\sigma}_y\sin\theta\sin\phi + \hat{\sigma}_z\cos\theta\right) \\ &\quad\times \left(\hat{\sigma}'_x\sin\theta\cos\phi + \hat{\sigma}'_y\sin\theta\sin\phi + \hat{\sigma}'_z\cos\theta\right) \quad . \end{aligned} \tag{2}$$

After expanding the product, the integration over ϕ yields π for the integrals over $\sin^2\phi$ or $\cos^2\phi$, and zero for the mixed products, so that we are left with

$$I = \frac{r^2}{4}\int \sin\theta\,\mathrm{d}\theta\, \left(\hat{\sigma}_x\hat{\sigma}'_x\sin^2\theta + \hat{\sigma}_y\hat{\sigma}'_y\sin^2\theta + 2\hat{\sigma}_z\hat{\sigma}'_z\cos^2\theta\right) \quad , \tag{3}$$

and the remaining integrations are carried out easily. The result is

$$I = \frac{r^2}{3}\hat{\boldsymbol{\sigma}}\cdot\hat{\boldsymbol{\sigma}}' \tag{4}$$

and this clearly leads to a cancellation with the second term in braces in the tensor potential of (7.19).

7.1.3 Interactions from Nucleon–Nucleon Scattering

The preceding section has given an overview of the possible functional forms that can be used to build up a nucleon–nucleon interaction. The natural way to find out which of these terms are actually necessary and to determine the parameters seems to be a study of nucleon–nucleon scattering. In this way a large number of interactions have been constructed; although nucleon–nucleon scattering can be described satisfactorily by these, they have not been as useful for nuclear structure calculations. It appears that the presence of many other nucleons inside nuclei modifies the scattering behavior to such an extent that it is more appropriate to use *phenomenological effective interactions*, which typically depend on the local density of nuclear matter. Such interactions will be studied in the next section. Here, however, we concentrate on what is known from scattering data.

Some basic features emerging from simple low-energy nucleon–nucleon scattering are:

- the interaction has a short range of about 1 fm,
- within this range, it is attractive with a depth of about 40 MeV for the larger distances, whereas
- there is strong repulsion at shorter distances ≤ 0.5 fm;
- it depends both on spin and isospin of the two nucleons.

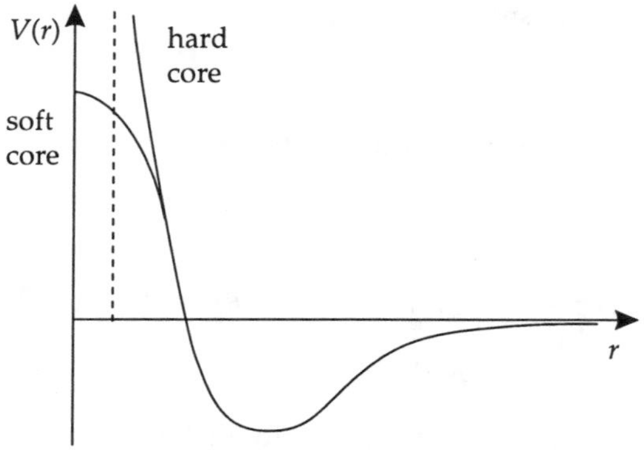

Fig. 7.1 Schematic plot of the radial dependence of the nucleon–nucleon potential.

The distance dependence is indicated schematically in Fig. 7.1.

The idea of Yukawa that the nucleon–nucleon interaction is mediated by pions just as the Coulomb interaction is caused by the exchange of (virtual) photons leads to the *one-pion exchange potential (OPEP)*. Taking the correct invariance properties of the pion field with respect to spin, isospin, and parity into account and regarding the nucleon as a static source of the pion field leads to a nucleon–nucleon interaction of the form

$$v_{\mathrm{OPEP}}(\boldsymbol{r}-\boldsymbol{r}',\hat{\boldsymbol{\sigma}},\hat{\boldsymbol{\sigma}}',\hat{\boldsymbol{\tau}},\hat{\boldsymbol{\tau}}') = -\frac{f^2}{4\pi\mu}(\hat{\boldsymbol{\tau}}\cdot\hat{\boldsymbol{\tau}}')\,(\hat{\boldsymbol{\sigma}}\cdot\nabla)(\hat{\boldsymbol{\sigma}}'\cdot\nabla')\,\frac{\mathrm{e}^{-\mu|\boldsymbol{r}-\boldsymbol{r}'|}}{|\boldsymbol{r}-\boldsymbol{r}'|}\,, \tag{7.20}$$

where f is the coupling strength and μ the mass of the pion. This expression can be obtained by assuming the standard Klein–Gordon Lagrangian for the pion and adding an interaction with the nucleon field ψ according to

$$\hat{H}_{\mathrm{int}} = \frac{f}{\mu}\int \mathrm{d}^3r\;\psi^\dagger(\boldsymbol{r})\,\hat{\boldsymbol{\tau}}\cdot\left(\hat{\boldsymbol{\sigma}}\cdot\nabla\boldsymbol{\phi}(\boldsymbol{r})\right)\psi(\boldsymbol{r}) \quad .$$

Note that the pion field $\boldsymbol{\phi}$ is a vector in isospin, which in this formula forms a scalar product with the isospin operator $\hat{\boldsymbol{\tau}}$ of the nucleon. Inserting this expression with a static point-particle distribution for the nucleon field into the wave equation of the pion yields the pion field of a nucleon located at $\boldsymbol{r}'$ as

$$\boldsymbol{\phi}(\boldsymbol{r}') = -\frac{f}{4\pi\mu}\hat{\boldsymbol{\tau}}(\hat{\boldsymbol{\sigma}}\cdot\nabla)\,\frac{\mathrm{e}^{-\mu|\boldsymbol{r}-\boldsymbol{r}'|}}{|\boldsymbol{r}-\boldsymbol{r}'|} \quad , \tag{7.21}$$

and the energy of a second nucleon in this field at position $\boldsymbol{r}$ is then obtained by a reinsertion of this result into $\hat{H}_{\mathrm{int}}$, yielding (7.20).

Evaluating the gradient operations in this equation leads to a lengthier but more transparent form of the OPEP:

$$\begin{aligned} v_{\mathrm{OPEP}}&(r,\hat{\boldsymbol{\sigma}},\hat{\boldsymbol{\sigma}}',\hat{\boldsymbol{\tau}},\hat{\boldsymbol{\tau}}') \\ &= \frac{1}{3}\mu c^2\frac{f^2}{4\pi\hbar c}(\hat{\boldsymbol{\tau}}\cdot\hat{\boldsymbol{\tau}}')\times\left[\hat{\boldsymbol{\sigma}}\cdot\hat{\boldsymbol{\sigma}}' + S_{12}\left(1+\frac{3}{\mu r}+\frac{3}{\mu^2r^2}\right)\right]\frac{\mathrm{e}^{-\mu r}}{\mu r} \\ &\quad - \frac{4\pi}{3\mu^2}\mu c^2\frac{f^2}{4\pi\hbar c}(\hat{\boldsymbol{\tau}}\cdot\hat{\boldsymbol{\tau}}')\,(\hat{\boldsymbol{\sigma}}\cdot\hat{\boldsymbol{\sigma}}')\,\delta(r) \quad , \end{aligned} \tag{7.22}$$

where the abbreviation $r = |\boldsymbol{r} - \boldsymbol{r}'|$ was used, and S_{12} represents the tensor force as in (7.19). The δ-function term results from the application of a ∇^2 on the r^{-1} potential.

The OPEP potential shows some, but not all, features of a realistic nucleon–nucleon interaction:

- it contains spin- and isospin-dependent parts as well as a tensor potential,
- the dominant radial dependence is of Yukawa type.

Other properties, however, show that it is not sufficient:

- there is no spin–orbit coupling and
- there is no short–range repulsion.

Conceptually also it is clear that this potential cannot be the whole story. The contributions of the exchange of more than one pion and that of other mesons must modify the interaction. The latter point is remedied in the so-called *one-boson exchange potentials* (OBEP), which correspond to sums over potentials of the complexity of (7.20). We only indicate some general features here. The type of interaction term generated by each boson depends on its parity and angular momentum and is summarized in Table 7.1.

Table 7.1. Mesons included in the OBEP interaction.

Type of meson	Physical meson	Interaction terms
scalar	"σ meson"	$1,\ \hat{\boldsymbol{L}} \cdot \hat{\boldsymbol{s}}$
pseudoscalar	$\pi,\ \eta,\ \eta'$	S_{12}
vector	$\rho,\ \omega,\ \phi$	$1,\ \hat{\boldsymbol{\sigma}} \cdot \hat{\boldsymbol{\sigma}}',\ S_{12},\ \hat{\boldsymbol{L}} \cdot \hat{\boldsymbol{s}}$

For the isovector mesons π and ρ, additional factors of $\hat{\boldsymbol{\tau}} \cdot \hat{\boldsymbol{\tau}}'$ appear in the interaction.

An important fact is that the σ meson, which is essential for a reasonable description of the scattering data, is not seen in experiment. It is usually argued that it may be too broad a resonance or that it really represents an approximation to the dominant contribution in two-meson exchange. Nevertheless, it is useful to include σ-meson exchange in such interactions and we will also see it appear in the relativistic mean-field model. If its mass together with the coupling constants for all the mesons are fitted to experimental scattering data, these data can be reproduced quite well in the OBEP potential.

Before going over to the effective potentials, let us just briefly mention two of the most popular sophisticated parametrized versions of nucleon–nucleon interactions. These are constructed out of the basic interaction terms with a larger number of free parameters and then fitted to experimental scattering. A typical example is the *Hamada–Johnston* potential [Ha62], which has the general form

$$V(r) = V_{\mathrm{C}}(r) + V_{\mathrm{T}}(r) S_{12} + V_{\mathrm{LS}}(r)\, \hat{\boldsymbol{L}} \cdot \hat{\boldsymbol{s}} + V_{\mathrm{LL}}(r) L_{12} \quad , \tag{7.23}$$

with

$$L_{12} = \hat{\boldsymbol{L}}^2(\hat{\boldsymbol{\sigma}} \cdot \hat{\boldsymbol{\sigma}}') - \tfrac{1}{2}\left[(\hat{\boldsymbol{\sigma}} \cdot \hat{\boldsymbol{L}})(\hat{\boldsymbol{\sigma}}' \cdot \hat{\boldsymbol{L}}) + (\hat{\boldsymbol{\sigma}}' \cdot \hat{\boldsymbol{L}})(\hat{\boldsymbol{\sigma}} \cdot \hat{\boldsymbol{L}})\right] \quad . \tag{7.24}$$

The functional form of the radial parts is inspired by the meson exchange potentials; for example,

$$V_C(r) = c_0 \mu c^2 (\hat{\boldsymbol{\tau}} \cdot \hat{\boldsymbol{\tau}}')(\hat{\boldsymbol{\sigma}} \cdot \hat{\boldsymbol{\sigma}}') \, \frac{\mathrm{e}^{-\mu r}}{\mu r} \left(1 + a_\mathrm{c} \frac{\mathrm{e}^{-\mu r}}{\mu r} + b_\mathrm{c} \frac{\mathrm{e}^{-2\mu r}}{\mu r^2}\right) \quad , \tag{7.25}$$

and similarly for the other terms. A different approach is used in the *Reid soft-core* and *Reid hard-core* potentials [Re68]. They are parametrized differently for the various spin and isospin combinations, for example,

$$V_{^1D}(r) = -\frac{10.6\,\mathrm{MeV}}{\mu r} \left(\mathrm{e}^{-\mu r} + 4.939\,\mathrm{e}^{-2\mu r} + 154.7\,\mathrm{e}^{-6\mu r}\right) \quad . \tag{7.26}$$

For those combinations for which they can contribute, spin–orbit and tensor contributions with similar radial behavior are also included. For the hard-core potentials, there is a hard-core radius r_c (different for each spin and isospin combination), below which the potential is set to infinity.

7.1.4 Effective Interactions

While the interactions discussed in the previous section describe nucleon–nucleon scattering exceedingly well, they are not often used for typical nuclear structure calculations. On the one hand, their complicated structure makes the necessary evaluations of matrix elements difficult, and on the other hand the results are not particularly encouraging. It seems that in nuclei the interaction is modified by complicated many-body effects to such an extent that it becomes more profitable to employ *effective interactions*, which do not describe nucleon–nucleon scattering well but are thought to include many of the effects of many-body correlations and should strictly be used only in Hartree–Fock and similar nuclear structure calculations, not for nucleon–nucleon scattering. In principle they may be valid only inside a certain space of shell-model states. Often the functional form of such an interaction is selected with a view to ease of computation, as will be seen below.

Probably the biggest step towards computability of single-particle models is made if the interactions are assumed to be *local*, i.e., their spatial dependence contains a factor $\delta(\boldsymbol{r} - \boldsymbol{r}')$. In this case the exchange term in the Hartree–Fock equations becomes similar in mathematical form to the direct term and the Hartree–Fock equations remain differential equations instead of becoming integral equations. Unfortunately a simple force of this type is not adequate except for simple qualitative calculations; the major step towards a successful interaction was made by introducing a momentum dependence. The matrix element of a potential in momentum space shows the reason:

$$\langle \boldsymbol{p}|v|\boldsymbol{p}'\rangle = \frac{1}{(2\pi\hbar)^3} \int \mathrm{d}^3r \; v(\boldsymbol{r}) \, \mathrm{e}^{-\frac{\mathrm{i}}{\hbar}(\boldsymbol{p}-\boldsymbol{p}')\cdot\boldsymbol{r}} \tag{7.27}$$

is a momentum-independent constant for a δ potential, but a finite range leads to a momentum dependence of a width inversely proportional to the range in coordinate space. The lowest-order momentum-dependence (second order because of time-reversal invariance) is

$$v(\boldsymbol{p},\boldsymbol{p}') = v_0 + v_1\left(\boldsymbol{p}^2 + \boldsymbol{p}'^2\right) + v_2\, \boldsymbol{p}\cdot\boldsymbol{p}' \quad , \tag{7.28}$$

and, disregarding normalization factors, this corresponds to the following expression in coordinate space:

$$v(\boldsymbol{r}) = v_0\,\delta(\boldsymbol{r}) + v_1\left(\hat{\boldsymbol{p}}^2\delta(\boldsymbol{r}) + \delta(\boldsymbol{r})\hat{\boldsymbol{p}}'^2\right) + v_2\,\hat{\boldsymbol{p}}\cdot\delta(\boldsymbol{r})\,\hat{\boldsymbol{p}}' \quad , \tag{7.29}$$

where the momentum operators are now derivatives. This motivates the momentum dependence typically included in effective interactions.

We can now give an overview over some well-known effective interactions. First there is a class of highly simplified but not necessarily local potentials useful for schematic calculations, for example,

- the Gauss potential $v(\boldsymbol{r}) = -V_0\,\mathrm{e}^{-r^2/r_0^2} \quad ,$
- the Hulthén potential $v(\boldsymbol{r}) = -V_0\,\dfrac{\exp(-r/r_0)}{1-\exp(-r/r_0)} \quad ,$
- the contact potential $v(\boldsymbol{r}) = -V_0\,\delta(\boldsymbol{r}/r_0) \quad .$

Usually $V_0 \approx 50\,\mathrm{MeV}$ and $r_0 \approx 1,\ldots,2\,\mathrm{fm}$ are used.

Probably the most widely used interaction in Hartree–Fock calculations are the forces of Skyrme type [Sk56, Sk59, Va70, Va72, En75, Be75, Gi80a, Gi80b]. Their distinguishing characteristic is the addition of a three-body interaction, i.e., the total interaction looks like

$$\hat{V} = \sum_{i<j} \hat{v}_{ij}^{(2)} + \sum_{i<j<k} \hat{v}_{ijk}^{(3)} \quad . \tag{7.30}$$

The two-body interaction contains momentum dependence as well as spin-exchange contributions and a spin-orbit force:

$$\begin{aligned}\hat{v}_{ij}^{(2)} &= t_0(1 + x_0\hat{P}_\sigma)\,\delta(\boldsymbol{r}_i - \boldsymbol{r}_j)\\ &\quad + \tfrac{1}{2}t_1\left(\delta(\boldsymbol{r}_i - \boldsymbol{r}_j)\hat{k}^2 + \hat{k}'^2\delta(\boldsymbol{r}_i - \boldsymbol{r}_j)\right) t_2\hat{\boldsymbol{k}}'\cdot\delta(\boldsymbol{r}_i - \boldsymbol{r}_j)\hat{\boldsymbol{k}}\\ &\quad + \mathrm{i}W_0(\hat{\boldsymbol{\sigma}}_i + \hat{\boldsymbol{\sigma}}_j)\cdot\hat{\boldsymbol{k}}'\times\delta(\boldsymbol{r}_i - \boldsymbol{r}_j)\,\hat{\boldsymbol{k}} \quad .\end{aligned} \tag{7.31}$$

Here, instead of the operator of relative momentum the related expressions

$$\hat{\boldsymbol{k}} = \frac{1}{2\mathrm{i}}(\nabla_i - \nabla_j) \quad , \quad \hat{\boldsymbol{k}}' = -\frac{1}{2\mathrm{i}}(\nabla_i - \nabla_j) \tag{7.32}$$

are used with the additional convention that $\hat{\boldsymbol{k}}'$ *acts on the wave function to its left.*

The three-body interaction is a purely local potential

$$\hat{v}_{ijk}^{(3)} = t_3\,\delta(\boldsymbol{r}_i - \boldsymbol{r}_j)\,\delta(\boldsymbol{r}_j - \boldsymbol{r}_k) \quad . \tag{7.33}$$

The Skyrme forces contain six parameters t_0, t_1, t_2, t_3, x_0, and W_0, which are fitted to reproduce properties of finite nuclei within a Hartree–Fock calculation. We will give more details in connection with the Hartree–Fock method in Sect. 7.2.8; let us only conclude the definition here by mentioning that the three-body force in spin-saturated nuclei can be shown to lead to a density-dependent two-body potential

$$\hat{v}_{ij}^{(3)} = \tfrac{1}{6}t_3(1 + P_\sigma)\,\delta(\boldsymbol{r}_i - \boldsymbol{r}_j)\,\rho\left(\tfrac{1}{2}(\boldsymbol{r}_i + \boldsymbol{r}_j)\right) \quad . \tag{7.34}$$

A further class of interactions, the *modified Skyrme forces*, is based on this formulation of the three-body force and replaces the linear dependence on the density by a power dependence

$$\hat{v}_{ij,\mathrm{modified}}^{(3)} = \tfrac{1}{6}t_3(1 + P_\sigma)\,\delta(\boldsymbol{r}_i - \boldsymbol{r}_j)\rho^\lambda\left(\tfrac{1}{2}(\boldsymbol{r}_i + \boldsymbol{r}_j)\right) \quad , \tag{7.35}$$

with $\lambda = \frac{1}{3}$ a common choice. This modification was introduced mainly to improve the compressibility properties of nuclear matter, which is too "stiff" against compression for the standard Skyrme forces [Kö76, Kr80].

The Skyrme forces have been widely used because they contain sufficient physics to allow a convincing quantitative description of heavy nuclei while being sufficiently simple mathematically to make computation feasible.

The *Gogny force* [Go73] has a similar structure as the Skyrme forces with the local two-body contributions replaced by Gaussians combined with spin and isospin exchange.

The next class of forces are not intended for full Hartree–Fock calculations, but only for calculating the interaction between particles and holes excited from the Hartree–Fock ground state. This means that the force need not describe the bulk properties of nuclei, such as saturation, and simpler expressions are quite adequate. An example are forces of *Migdal* type (familiar from Fermi-liquid theory), which contain terms like

$$V_0\,\delta(\boldsymbol{r}_i - \boldsymbol{r}_j)\big(f + f'\,\hat{\boldsymbol{\tau}}_i \cdot \hat{\boldsymbol{\tau}}_j + g\,\hat{\boldsymbol{\sigma}}_i \cdot \hat{\boldsymbol{\sigma}}_j + g'\,\hat{\boldsymbol{\sigma}}_i \cdot \hat{\boldsymbol{\sigma}}_j\,\hat{\boldsymbol{\tau}}_i \cdot \hat{\boldsymbol{\tau}}_j\big) \quad . \tag{7.36}$$

The *surface-delta interaction* (see, e.g., [Gr65b]) addresses similar applications with the idea that because of the Pauli principle the interaction between particles and holes is peaked near the surface of the nucleus, thus motivating the replacement

$$\delta(\boldsymbol{r}_i - \boldsymbol{r}_j) \rightarrow \delta(\boldsymbol{r}_i - \boldsymbol{r}_j)\,\delta(r_i - R_0) \tag{7.37}$$

with R_0 the nuclear radius.

Finally there is a class of interactions like the *Sussex potential* which are not written as potentials with a functional form at all; instead one uses the matrix elements of the interaction in a basis of oscillator wave functions as unknowns in a shell-model calculation and the (very large) set of matrix elements then describes the effective nucleon–nucleon potential numerically. Such interactions should of course then be used only for similar calculations in the same space of single-particle wave functions.

7.2 The Hartree–Fock Approximation

7.2.1 Introduction

Microscopic models are models that describe the structure of the nucleus in terms of the degrees of freedom of its microscopic constituents – the nucleons. To do this a microscopic Hamiltonian $\hat{H}$ is needed, which contains some suitable form of nucleon–nucleon interaction such as discussed in the previous sections. For most of these models the starting point is a nonrelativistic Hamiltonian containing only two-body interactions, and a general form is provided in second quantization by

$$\hat{H} = \sum_{ij} t_{ij}\, \hat{a}_i^\dagger \hat{a}_j + \tfrac{1}{2} \sum_{ijkl} v_{ijkl}\, \hat{a}_i^\dagger \hat{a}_j^\dagger \hat{a}_l \hat{a}_k \quad , \tag{7.38}$$

where the indices i, j, k, and l label the single-particle states in some complete orthonormal basis and the v_{ijkl} are the matrix elements of the nucleon–nucleon interaction. All indices run over all available states, i.e., normally from 1 to ∞. An eigenstate of this Hamiltonian can be expanded as a sum over states which all have the same total number of nucleons, but with the nucleons occupying the available single-particle states in all possible combinations. Formally we can write

$$|\Psi\rangle = \sum_{i_1, i_2, \ldots, i_A} c_{i_1, i_2 \ldots i_A}\, \hat{a}_{i_1}^\dagger \hat{a}_{i_2}^\dagger \cdots \hat{a}_{i_A}^\dagger |0\rangle \quad , \tag{7.39}$$

where the indices i_n, $n = 1, \ldots, A$, serve to select a subset of A single particle states occupied in that particular term in the sum from among the infinite number of available states.

The number of states in the sum of (7.39) is staggering. Even if we take into account the fact that the ordering of the indices is unimportant, the number of terms in the sum will be given by how many ways there are to choose A occupied single-particle states from N_c available ones, which is clearly

$$\binom{N_\mathrm{c}}{A} \quad , \tag{7.40}$$

if the number of available states is cut off to N_c. Obviously the problem is not tractable in general. There are two alternatives for proceeding: either restrict the number of particles and states, or replace the general wave function of (7.39) by a much simpler approximate one. The first approach is used in the *shell model*, which assumes that only a small number of nucleons outside a closed shell and confined to few states determine the nuclear properties. The most prominent case of the second approach is the *Hartree–Fock approximation*, which we will now develop. In essence the sum over all the different occupations is replaced by a single term; then the single-particle wave functions cannot be chosen arbitrarily but are prescribed by the method.

7.2.2 The Variational Principle

Restricting the functional form of the wave functions is a very useful approach to solving complicated problems such as (7.38). In this case, the restricted wave function can in general no longer be an exact eigenfunction. To develop this formally, examine the eigenvalue problem

$$\hat{H}\,\Psi_k = E_k\,\Psi_k \quad . \tag{7.41}$$

The unknown (but normalized) solutions Ψ_k are assumed to span the Hilbert space as usual. We seek the solution in the form of a wave function Φ which is somehow restricted in its functional form, so that all possible Φs span a subspace of the Hilbert space available with the Ψ_ks. Now whatever its form, any function Φ can certainly be expanded in the Ψ_k, i.e., we can write

$$\Phi = \sum_k c_k \Psi_k, \qquad \text{with } \sum_k |c_k^2| = 1 \quad . \tag{7.42}$$

Now assume that the index k runs from 0 upwards and the states are ordered in ascending energy, so that $k = 0$ denotes the ground state of the Hamiltonian. Then the expectation value of the Hamiltonian in Φ will be

$$\langle \Phi|\hat{H}|\Phi\rangle = \sum_k |c_k|^2 E_k = E_0 + \sum_{k>0} |c_k|^2 (E_k - E_0) \quad . \tag{7.43}$$

and this value will always be larger than or equal to E_0, with the minimum realized if $c_0 = 1$ and $c_k = 0$ for $k > 0$. This yields the exact solution, but if $|\Phi\rangle$ is restricted, it will not be possible to have all the coefficients for $k > 0$ vanish, and one can only minimize these terms. Any admixture of states $k > 0$ to Φ will increase its energy, so that we can conclude that *an optimal approximation to the ground state for the Hamiltonian $\hat{H}$ is attained for that wave function Φ whose energy expectation value is minimal.* Formulated as a variational principle it can be written as

$$\delta\langle \Phi|\hat{H}|\Phi\rangle = 0 \quad . \tag{7.44}$$

It is clear from (7.43) that what is being minimized more precisely is the mean deviation of the wave function from the true ground-state wave function with the contributions of the different states being weighted with their excitation energy.

In the above derivation of the variational principle it was assumed that the trial wave function Φ is normalized. Sometimes it is convenient to allow nonnormalized functions, but then normalization has to be imposed as a subsidiary condition in the variational principle:

$$\delta\big(\langle \Phi|\hat{H}|\Phi\rangle - E\,\langle \Phi|\Phi\rangle\big) = 0 \quad . \tag{7.45}$$

Why the variational parameter was denoted as E will be clear in a moment.

The variation in (7.45) can be carried out either with respect to $|\Phi\rangle$ or $\langle\Phi|$, since they correspond to two different degrees of freedom, like a number and its complex conjugate. Since the resulting equations are hermitian conjugates of each other, it suffices to examine one of these cases. Varying $\langle\Phi|$ yields

$$\langle\delta\Phi|\hat{H}|\Phi\rangle - E\langle\delta\Phi|\Phi\rangle = 0 \quad . \tag{7.46}$$

If Φ were an unrestricted wave function, then $|\delta\Phi\rangle$ would be an arbitrary vector and it could be concluded that

$$\hat{H}|\Phi\rangle - E|\Phi\rangle = 0 \quad , \tag{7.47}$$

since this vector is perpendicular to arbitrary variations in this case and thus must vanish. This shows that for unrestricted wave functions the Schrödinger equation is recovered and explains the notation E. In the case of interest here, of course, this conclusion is wrong and Φ is not an eigenfunction of $\hat{H}$. In this case (7.45) implies that $\hat{H}|\Phi\rangle - E|\Phi\rangle$ should be orthogonal to the subspace allowed for the solution.

7.2.3 The Slater-Determinant Approximation

The method of finding an optimum approximation within a restricted space of wave functions now is clear, and only the set of allowed wave functions still has to be selected. The choice that leads to the Hartree–Fock approximation is that of a *single Slater determinant*, i.e., a wave function of the form

$$|\Psi\rangle = \hat{a}_1^\dagger \hat{a}_2^\dagger \cdots \hat{a}_A^\dagger |0\rangle \quad , \tag{7.48}$$

or more concisely

$$|\Psi\rangle = \prod_{i=1}^{A} \hat{a}_i^\dagger |0\rangle \quad . \tag{7.49}$$

Here the index of the creation operators refers to a set of single particle states with the corresponding wave functions $\phi_i(\boldsymbol{r})$, $i = 1, \ldots, A$, to be determined from the variational principle.

Before starting with the derivation of the Hartree–Fock conditions, it is worthwhile to examine what the approximation means. To understand the concept, first ignore the antisymmetrization of the particles and take the simplest case of two particles. The approximation would then consist in using a wave function restricted to $\psi_1(\boldsymbol{r}_1)\psi_2(\boldsymbol{r}_2)$ instead of a more general $\psi(\boldsymbol{r}_1, \boldsymbol{r}_2)$. The probability density for finding particle number 1 at $\boldsymbol{r}_1$ *and* particle number 2 at $\boldsymbol{r}_2$ (the *joint probability*) is then given by $|\psi_1(\boldsymbol{r}_1)|^2|\psi_2(\boldsymbol{r}_2)|^2$ and is the product of the individual probabilities, i.e., that for finding particle number 1 at $\boldsymbol{r}_1$ irrespective of where particle 2 is located, times the same for particle 2. The two particles thus are *independent* of each other.

What is excluded in this approximation to the wave function is a *correlation* between the particles. For example, if there is a repulsive interaction between them, one would expect the joint probability to be reduced compared to the simple product whenever $|\boldsymbol{r}_1 - \boldsymbol{r}_2|$ becomes comparable to the interaction radius. The deviation between the joint probability and the product of the single probabilities is called the *correlation function*.

Taking into account the Fermi antisymmetrization changes these conclusions somewhat. The product wave function now has to be

$$\psi_{12}(\boldsymbol{r}_1, \boldsymbol{r}_2) = \frac{1}{\sqrt{2}} \left[\psi_1(\boldsymbol{r}_1)\psi_2(\boldsymbol{r}_2) - \psi_2(\boldsymbol{r}_1)\psi_1(\boldsymbol{r}_2) \right] \quad , \tag{7.50}$$

so that the joint probability becomes

$$\begin{aligned} |\psi_{12}(\boldsymbol{r}_1, \boldsymbol{r}_2)|^2 = \tfrac{1}{2} \big(& |\psi_1(\boldsymbol{r}_1)|^2 |\psi_2(\boldsymbol{r}_2)|^2 + |\psi_2(\boldsymbol{r}_1)|^2 |\psi_1(\boldsymbol{r}_2)|^2 \\ & - \psi_1^*(\boldsymbol{r}_1)\psi_2^*(\boldsymbol{r}_2)\psi_1(\boldsymbol{r}_2)\psi_2(\boldsymbol{r}_1) \\ & - \psi_1^*(\boldsymbol{r}_2)\psi_2^*(\boldsymbol{r}_1)\psi_1(\boldsymbol{r}_1)\psi_2(\boldsymbol{r}_2) \big) \quad . \end{aligned} \tag{7.51}$$

Clearly this is not the same as the product of the single probabilities. In fact if we set $\boldsymbol{r}_1 = \boldsymbol{r}_2 = \boldsymbol{r}$ the result is not $|\psi_1(\boldsymbol{r})|^2 |\psi_2(\boldsymbol{r})|^2$ but zero: the probability for finding two fermions at the same location vanishes, so that there is always a strong correlation (in the classical sense) between them. Since this effect is always present for fermions and does not need any interaction between the particles, however, it is better to still refer to the two particles as uncorrelated (or independent) in the quantum-mechanical sense, and to call correlations only the deviations from the joint probability expected with antisymmetrized product wave functions, i.e., of the Slater-determinant form.

7.2.4 The Hartree–Fock Equations

Now we are ready to investigate the variational principle applied to the Hamiltonian of (7.38) with a trial wave function as in (7.49). How can the wave function be varied? It has already been pointed out that the single-particle wave functions $\phi_j(\boldsymbol{r})$ are the objects to be determined. They can be varied via an infinitesimal admixture of other wave functions from the complete set:

$$\delta\phi_j(\boldsymbol{r}) = \sum_{k \neq j} \delta c_{jk} \, \phi_k(\boldsymbol{r}) \quad . \tag{7.52}$$

Equivalently the creation operators can be varied:

$$\delta \hat{a}_j^\dagger = \sum_{k \neq j} \delta c_{jk} \, \hat{a}_k^\dagger \quad . \tag{7.53}$$

Insert the varied operator into the wave function of (7.54). The result of varying only this one single-particle wave function is

$$|\Psi + \delta\Psi_j\rangle = |\Psi\rangle + \sum_{k \neq j} \delta c_{jk} \, \hat{a}_1^\dagger \hat{a}_2^\dagger \cdots \hat{a}_{j-1}^\dagger \hat{a}_k^\dagger \hat{a}_{j+1}^\dagger \cdots \hat{a}_A^\dagger |0\rangle \quad . \tag{7.54}$$

Now if the index k of the newly mixed-in wave function is among the indices of occupied states $1, \ldots, A$, the variation vanishes, because the corresponding creation operator appears twice in the product. This leads to two important consequences.

- The only nonvanishing variations of a single Slater determinant are admixtures of unoccupied wave functions to any of the occupied ones.

- Any replacement of the occupied wave functions by linear combinations among themselves leaves the total many-particle wave function invariant.

The latter property is also familiar from the fact that replacing a column in a determinant by a linear superposition with other columns does not change its value; here applied to the Slater determinant. This makes it clear that *we cannot expect the variational principle to determine the single-particle states themselves; only the total space of occupied states – disregarding any linear transformation of the states – has physical meaning.*

The variation of the wave function according to (7.54) may also be expressed more concisely by using an annihilation operator to depopulate the wave function j:

$$|\Psi + \delta\Psi_j\rangle = |\Psi\rangle + \sum_k \sigma c_{jk}\, \hat{a}_k^\dagger \hat{a}_j |\Psi\rangle \quad , \quad k > A \quad , \quad j \leq A \quad , \tag{7.55}$$

with $\sigma = \pm 1$ keeping track of the sign changes caused by permuting the operators to put them in front of all the other operators in $|\Psi\rangle$. For the variation, actually, it is not necessary to keep the sign or the sum over j and k, as all of these variations are independent; it suffices to demand that the expectation value of H be stationary with respect to variations of the form

$$|\delta\Psi\rangle = \varepsilon \hat{a}_k^\dagger \hat{a}_j |\Psi\rangle \tag{7.56}$$

as a function of the parameter ε for *arbitrary* values of the indices fulfilling $k > A$ and $j \leq A$.

Notation: Throughout this chapter it will be necessary to distinguish three types of indices: those referring to occupied or unoccupied states only, and others that are unrestricted. To facilitate the notation we use the following convention:

- The indices i, j and their subscripted forms i_1, j_2, etc., refer to occupied states only, i.e., they take values from 1 through A exclusively, where A is the number of occupied states: $i, j = 1, \ldots, A$.
- The indices m, n and their subscripted forms refer to unoccupied single-particle states only: $m, n = A + 1, \ldots, \infty$.
- The letters k and l are reserved for unrestricted indices: $k, l = 1, \ldots, \infty$.

We can now rephrase the variational problem in this notation. The Hamiltonian becomes

$$\hat{H} = \sum_{k_1 k_2} t_{k_1 k_2}\, \hat{a}_{k_1}^\dagger \hat{a}_{k_2} + \frac{1}{2} \sum_{k_1 k_2 k_3 k_4} v_{k_1 k_2 k_3 k_4}\, \hat{a}_{k_1}^\dagger \hat{a}_{k_2}^\dagger \hat{a}_{k_4} \hat{a}_{k_3} \quad , \tag{7.57}$$

and the fundamental variations take the form

$$|\delta\Psi\rangle = \varepsilon \hat{a}_m^\dagger \hat{a}_i |\Psi\rangle \quad . \tag{7.58}$$

As the variation does not change the normalization of the wave function to first order (clear from $\delta\langle\Phi|\Phi\rangle \approx \langle\delta\Phi|\Phi\rangle + \langle\Phi|\delta\Phi\rangle = 0$), one may use the variation without Lagrange multiplier for the norm. Conventionally the bra vector is varied, so we have to take the Hermitian conjugate of $|\delta\Phi\rangle$,

$$\langle \delta\Psi | = \langle \Psi | \varepsilon^* \hat{a}_i^\dagger \hat{a}_m \quad . \tag{7.59}$$

The variational equation now becomes

$$0 = \langle \delta\Psi | \hat{H} | \Psi \rangle = \varepsilon^* \langle \Psi | \hat{a}_i^\dagger \hat{a}_m \hat{H} | \Psi \rangle \quad , \tag{7.60}$$

where the Hamiltonian of (7.57) is to be inserted. The arbitrary factor ε^* can then be dropped. The calculation of the resulting matrix elements is given as an exercise below; the final result is

$$t_{mi} + \frac{1}{2} \sum_j \left(v_{mjij} - v_{mjji} - v_{jmij} + v_{jmji} \right) = 0 \quad . \tag{7.61}$$

From the definition of the matrix elements it is clear that they do not change their value if the first index is exchanged with the second and simultaneously the third with the fourth, so the equation can be simplified to

$$t_{mi} + \sum_j \left(v_{mjij} - v_{mjji} \right) = 0 \quad . \tag{7.62}$$

The notation is further simplified by abbreviating the antisymmetrized matrix element as

$$\bar{v}_{k_1 k_2 k_3 k_4} = v_{k_1 k_2 k_3 k_4} - v_{k_1 k_2 k_4 k_3} \quad , \tag{7.63}$$

so that we get

$$t_{mi} + \sum_{j=1}^{A} \bar{v}_{mjij} = 0 \quad . \tag{7.64}$$

To understand the physics contained in this equation, remember the range the individual indices vary over: i denotes an occupied state ($i \leq A$), m an unoccupied one ($m > A$), and j sums over all occupied states, which has been emphasized in (7.64). The equation thus demands that the single-particle states should be chosen such that the matrix elements

$$h_{kl} = t_{kl} + \sum_{j=1}^{A} \bar{v}_{kjlj} \tag{7.65}$$

vanish between occupied and unoccupied states. If we allow (k, l) to refer to arbitrary combinations of states, this defines a single-particle operator $\hat{h}$, which is usually called the *single-particle* or *Hartree–Fock Hamiltonian.* As expected, the condition (7.64) characterizes the set of occupied states and not those states individually.

To formulate these conditions more concisely and to prepare for further developments, the operator $\hat{h}$ is split up into four parts corresponding to different index ranges: $\hat{h}_{\text{pp}}$ (particle–particle) denotes those matrix elements with both indices referring to unoccupied single-particle states, $\hat{h}_{\text{hh}}$ (hole–hole) has both states occupied, and $\hat{h}_{\text{ph}}$ and $\hat{h}_{\text{hp}}$ denote the appropriate mixed cases. Formally then the matrix is decomposed into

$$\hat{h} = \begin{pmatrix} \hat{h}_{\rm hh} & \hat{h}_{\rm hp} \\ \hat{h}_{\rm ph} & \hat{h}_{\rm pp} \end{pmatrix} \quad , \tag{7.66}$$

or, in a more sloppy notation (the matrices have to be filled up to full size by adding zeroes),

$$\hat{h} = \hat{h}_{\rm pp} + \hat{h}_{\rm hh} + \hat{h}_{\rm ph} + \hat{h}_{\rm hp} \quad , \tag{7.67}$$

and the conditions become

$$\hat{h}_{\rm ph} = 0 \quad \text{and} \quad \hat{h}_{\rm hp} = 0 \quad . \tag{7.68}$$

The block matrix decomposition makes it apparent how these conditions may actually be fulfilled in practice: the states may be chosen such as to make $\hat{h}$ itself diagonal,

$$h_{kl} = t_{kl} + \sum_{j=1}^{A} \bar{v}_{kjlj} = \varepsilon_k \delta_{kl} \tag{7.69}$$

with *single-particle energies* ε_k. This defines one convenient choice of the single-particle states $\phi_k(\boldsymbol{r})$ themselves. Writing the *Hartree–Fock equations* (7.69) in configuration space makes their physical content even more apparent:

$$\varepsilon_k \phi_k(\boldsymbol{r}) = -\frac{\hbar^2}{2m} \nabla^2 \phi_k(\boldsymbol{r}) + \left(\int d^3r' \; v(\boldsymbol{r}' - \boldsymbol{r}) \sum_{j=1}^{A} |\phi_j(\boldsymbol{r}')|^2 \right) \phi_k(\boldsymbol{r})$$
$$- \sum_{j=1}^{A} \phi_j(\boldsymbol{r}) \int d^3r' \; v(\boldsymbol{r}' - \boldsymbol{r}) \phi_j^*(\boldsymbol{r}') \phi_k(\boldsymbol{r}') \quad . \tag{7.70}$$

The equations are quite similar in form to Schrödinger equations for each of the single-particle states. The second term on the right-hand side is the *average potential*

$$U(\boldsymbol{r}) = \int d^3r' \; v(\boldsymbol{r}' - \boldsymbol{r}) \sum_{j=1}^{A} |\phi_j(\boldsymbol{r}')|^2 \quad , \tag{7.71}$$

which has the simple interpretation of the potential generated by the density distribution of nucleons. The last term is the *exchange term*; together with the average potential it defines the *mean field*. The Hartree–Fock approximation is often called the mean-field approximation.

The exchange term of course makes the problem quite a bit more complicated than a simple one-particle Schrödinger equation. Since it changes the Schrödinger equation into an integral equation, it is much harder to deal with in practice, and various approximations are employed. One popular approach is that of using zero-range interactions like the Skyrme forces for which, as we will see shortly, the exchange term can be combined with the direct one. For forces with a finite range – unavoidable in the case of the Coulomb interaction – some approximation must be employed.

The Hartree–Fock equations show a mechanism by which the nucleons themselves can produce a strong central field in a nucleus in analogy to the central

Coulomb field in the atom, which is not produced by the electrons. What makes this possible? Clearly the assumption of the Hartree–Fock approximation was the neglection of correlations, arising from direct particle–particle scattering not mediated by the mean field. This is reasonable if the scattering is prohibited by the Pauli principle, so that an inordinately large amount of energy is needed for the process. The correlations are, however, important for the excited states and this will be the topic in Sect. 8.2.

The full two-body interaction to the extent that it is not included in the mean field is called the *residual interaction*. One important part of it is taken into account by the pairing force (Sect. 7.5) and other contributions lead to correlated excitations as mentioned.

The Hartree–Fock equations form a *self-consistent* problem in the sense that the wave functions determine the mean field, while the mean field in turn determines the wave functions. In practice this leads to iterative solutions in which one starts from an initial guess for the wave functions, such as harmonic-oscillator ones and determines the mean field from them. Solving the Schrödinger equations then yields a new set of wave functions, and this process is repeated until, hopefully, convergence is achieved.

EXERCISE

7.2 Matrix Elements in the Variational Equation

Problem. Evaluate the matrix elements appearing in the variational equations (7.60).

Solution. The techniques explained in the chapter on second quantization do not apply directly here, because we are not dealing with a vacuum expectation value. Whether the state corresponding to a certain index is occupied or not, is thus very important for deciding how an operator will act on $|\Psi\rangle$.

The kinetic energy contribution is

$$\sum_{k_1 k_2} t_{k_1 k_2} \langle \Psi | \hat{a}_i^\dagger \hat{a}_m \hat{a}_{k_1}^\dagger \hat{a}_{k_2} | \Psi \rangle \quad . \tag{1}$$

The basic procedure is the same as for vacuum expectation values. We note that both $\hat{a}_i^\dagger$ and $\hat{a}_m$ yield zero when applied to $|\Psi\rangle$, because i refers to an occupied state in $|\Psi\rangle$ and m to an unoccupied one. These two will thus be commuted through to the rear.

Commuting $\hat{a}_m$ to the back yields a nonvanishing contribution only from $\hat{a}_{k_1}^\dagger$, so that the intermediate result is

$$\sum_{k_1 k_2} t_{k_1 k_2} \delta_{m k_1} \langle \Psi | \hat{a}_i^\dagger \hat{a}_{k_2} | \Psi \rangle \quad , \tag{2}$$

and the remaining matrix element is simply $\delta_{i k_2}$, so that the result, after eliminating the sums, becomes t_{mi}.

Exercise 7.2

The evaluation of the potential matrix elements is somewhat more involved. Starting from

$$\tfrac{1}{2}\sum_{k_1k_2k_3k_4} v_{k_1k_2k_3k_4} \langle\Psi|\hat{a}_i^\dagger \hat{a}_m \hat{a}_{k_1}^\dagger \hat{a}_{k_2}^\dagger \hat{a}_{k_4} \hat{a}_{k_3}|\Psi\rangle \quad , \tag{3}$$

it is clear that permuting $\hat{a}_i^\dagger$ and $\hat{a}_m$ to the rear will again produce the desired result, but now each of them produces a nonvanishing commutator with either one of the two operators of opposite type following them. This leads to the following combinations of Kronecker symbols replacing the matrix element:

$$\delta_{mk_1}\delta_{ik_3} - \delta_{mk_1}\delta_{ik_4} - \delta_{mk_2}\delta_{ik_3} + \delta_{mk_2}\delta_{ik_4} \quad . \tag{4}$$

The signs were determined by counting how many permutations of the operators were needed to bring them directly in front of the operator with which they are commuted to leave a Kronecker symbol. Inserting this into (3), we find that in each term two of the sums are eliminated through the Kronecker symbols, and to combine the remaining sums it is advantageous to relabel the summation indices as $k_1, k_2 \to k$ and $k_3, k_4 \to k'$. This yields

$$\tfrac{1}{2}\sum_{kk'} \left(v_{mkik'} - v_{mkk'i} - v_{kmik'} + v_{kmk'i}\right) \langle\Psi|\hat{a}_k^\dagger \hat{a}_{k'}|\Psi\rangle \quad . \tag{5}$$

The surviving matrix element can be nonvanishing only for $k = k'$. As $\hat{a}_k^\dagger \hat{a}_k$ is the particle-number operator its matrix element is 1 if the state k is occupied and zero otherwise. The sum in (5) thus can be restricted to occupied states and we can do this in the notation used here by renaming $k = k'$ to j:

$$\tfrac{1}{2}\sum_{j} \left(v_{mjij} - v_{mjji} - v_{jmij} + v_{jmji}\right) \quad , \tag{6}$$

which is the final result.

7.2.5 Applications

Let us now investigate the properties of the states of the many-body system in the Hartree–Fock approximation. The ground state of course is given by (remember that the index m refers to occupied states only)

$$|\mathrm{HF}\rangle = \prod_m \hat{a}_m^\dagger |0\rangle \quad . \tag{7.72}$$

Its energy can be evaluated easily by using the methods developped above:

$$\begin{aligned} E_{\mathrm{HF}} &= \langle \mathrm{HF}|\hat{H}|\mathrm{HF}\rangle \\ &= \sum_m t_{mm} + \tfrac{1}{2}\sum_{mn} \bar{v}_{mnmn} \\ &= \sum_m \varepsilon_m - \tfrac{1}{2}\sum_{mn} \bar{v}_{mnmn} \quad . \end{aligned} \tag{7.73}$$

It is important to relize that the energy of the Hartree–Fock ground state is not simply the sum of the individual single-particle energies, but has an additional contribution from the potential interactions. The mathematical reason is that the Hamiltonian of the many-particle system is not the sum of the single-particle Hamiltonians, but contains the interactions weighted differently.

It appears a simple matter now to construct excited states based on the Hartree–Fock ground state: these should simply be given by the particle-hole excitations of various orders. In principle this is not quite true, because the mean field also depends on the states actually occupied, which in turn makes the single-particle Hamiltonian and the single-particle states themselves change depending on the particular occupation. In practice, however, this problem is usually ignored with the argument that for a heavier nucleus the change of state of one single particle will change the mean field only negligibly, and the associated change of the single-particle states is even smaller. We thus construct excited states as one-particle/one-hole (1p1h) excitations

$$|mi\rangle = \hat{a}_m^\dagger \hat{a}_i |\mathrm{HF}\rangle \quad , \tag{7.74}$$

two-particle/two-hole (2p2h) excitations

$$|mnij\rangle = \hat{a}_m^\dagger \hat{a}_n^\dagger \hat{a}_i \hat{a}_j |HF\rangle \quad , \tag{7.75}$$

and so on, with unchanged single-particle states. The expectation value of the energy of such states can easily be calculated. For the one-particle/one-hole excitations, for example, one obtains

$$\begin{aligned} E_{mi} &= \langle mi|\hat{H}|mi\rangle \\ &= E_{\mathrm{HF}} + t_{mm} - t_{ii} + \sum_{\substack{k=1 \\ k\neq i}}^{A} (\bar{v}_{mnmn} - \bar{v}_{inin}) \\ &= E_{\mathrm{HF}} + \varepsilon_m - \varepsilon_i - \bar{v}_{mimi} \end{aligned} \tag{7.76}$$

simply by writing down the sums with this difference in occupation (note that $\bar{v}_{mmmm} = 0$ for antisymmetry).

Thus in addition to the expected contribution from the single-particle energies of the two particles involved, there is also one arising from the change in the mean field. Unfortunately the result is really only an expectation value, as there are also off-diagonal matrix elements

$$\langle mi|\hat{H}|nj\rangle = -\bar{v}_{mjni} \quad , \tag{7.77}$$

so in principle a matrix should be diagonalized. In many cases, though, the potential contribution can be neglected and the particle-hole states can be treated as approximate eigenstates of the problem.

A similar analysis also sheds light on the physical meaning of the single-particle energies ε_k. Compare the energy of the nucleus with A nucleons to that of the nucleus with $A-1$ nucleons, for example, with one particle removed from the occupied state j. The latter is described by the wave function

$$|j\rangle = \hat{a}_j|\mathrm{HF}\rangle \quad , \tag{7.78}$$

where again the dependence of the wave functions themselves on the occupation of the states was neglected. Its energy is given by

$$E_j = \sum_{i \neq j} t_{ii} + \tfrac{1}{2} \sum_{i_1, i_2 \neq j} \bar{v}_{i_1 i_2 i_1 i_2} \tag{7.79}$$

and the difference from the ground-state energy becomes

$$\begin{aligned} E_j - E_{\mathrm{HF}} &= -t_{jj} - \tfrac{1}{2} \sum_i \bar{v}_{ijij} - \tfrac{1}{2} \sum_i \bar{v}_{jiji} \\ &= -t_{jj} - \sum_i \bar{v}_{ijij} \\ &= -\varepsilon_j \quad . \end{aligned} \tag{7.80}$$

Here the symmetry of the matrix elements $\bar{v}_{ijij} = \bar{v}_{jiji}$ was used. Thus *the single-particle energy indicates the energy required to remove a particle from the nucleus.* This is the contents of *Koopman's theorem*, which is discussed in more detail in [Kö71] and [Ba71].

Note that this argument cannot be used recursively: if it were reapplied to remove another nucleon, the potential interactions between the two nucleons would be treated incorrectly (otherwise the total energy of the ground state should be given by $\sum_i \varepsilon_i$, which was seen to be wrong).

The results of Hartree–Fock calculations can thus be used not only to predict the bulk properties of the nuclear ground state, such as the binding energy, mean square radius, surface thickness, and so on, but also for the description of excited states. Furthermore, the wave functions obtained in this way can be used as a basis for treating pairing (Sect. 7.5) and to describe collective states as coherent superpositions of particle-hole excitations (Sect. 8.2).

7.2.6 The Density Matrix Formulation

The Hartree–Fock equations take a particularly simple form when expressed in terms of the one-particle density matrix. This formalism is very elegant for formal manipulations, so that it is well worthwhile to get to know it.

Given a many-particle state $|\Phi\rangle$, the *one-particle density matrix* is defined as

$$\rho_{kl} = \langle \Phi | \hat{a}_l^\dagger \hat{a}_k | \Phi \rangle \quad , \tag{7.81}$$

where k and l run over the one-particle basic states. $|\Phi\rangle$ need not be a simple Slater determinant built out of these states but can be a general superposition of such Slater determinants. Note that the one-particle density matrix depends both on the state $|\Phi\rangle$ and on the single-particle basis defining the operators $\hat{a}_k^\dagger$ and $\hat{a}_l$. It is customary to use the shorter term "density matrix" for the one-particle density matrix if no confusion with other types of density matrix is possible.

The following elementary properties of the density matrix are easily derived:

- ρ_{kl} is hermitian:

$$\rho_{lk} = \langle \Phi | \hat{a}_k^\dagger \hat{a}_l | \Phi \rangle = \langle \Phi | (\hat{a}_l^\dagger \hat{a}_k)^\dagger | \Phi \rangle = \rho_{kl}^* \quad . \tag{7.82}$$

- Expectation values of single-particle operators such as

$$\hat{t} = \sum_{kl} t_{kl}\, \hat{a}_k^\dagger \hat{a}_l \tag{7.83}$$

can be calculated via

$$\langle \Phi | \hat{t} | \Phi \rangle = \sum_{kl} t_{kl} \langle \Phi | \hat{a}_k^\dagger \hat{a}_l | \Phi \rangle = \sum_{kl} t_{kl}\, \rho_{lk} \quad , \tag{7.84}$$

which can be rewritten using matrix trace notation:

$$\langle \Phi | \hat{t} | \Phi \rangle = \mathrm{Tr}\{t\rho\} \quad . \tag{7.85}$$

Here t in parentheses stands for the matrix t_{kl} representing the operator $\hat{t}$.[1]

- If the state $|\Phi\rangle$ is a simple Slater determinant, the form of the density matrix is quite restricted. We first regard the case that $|\Phi\rangle$ is built out of the same single-particle states as those contained in the single-particle basis defining the density matrix. Then we must have

$$\rho_{kl} = \begin{cases} \delta_{kl} & \text{for } k \text{ and } l \text{ occupied in } |\Phi\rangle \\ 0 & \text{otherwise} \end{cases} \quad . \tag{7.86}$$

Thus ρ is diagonal in this case with ones and zeroes on the diagonal depending on whether the corresponding single-particle state is occupied or empty. Furthermore it fulfills the fundamental relation

$$\rho^2 = \rho \quad , \tag{7.87}$$

which follows immediately from the special form of the matrix. But this relation continues to hold in the more general case. If the single-particle states that make up $|\Phi\rangle$ are not included in the basis defining ρ, they may in any case be expanded in those using some unitary matrix U,

$$\hat{\beta}_k = \sum_{k'} U_{kk'}\, \hat{a}_{k'} \quad , \tag{7.88}$$

where $\hat{\beta}_k$ now denotes the second quantization operator for these states occupied in $|\Phi\rangle$. The density matrix $\tilde{\rho}$ defined in the basis of the $\hat{\beta}_k$ now is given by

$$\tilde{\rho} = U\rho U^\dagger \quad , \tag{7.89}$$

so that conversely $\rho = U^\dagger \tilde{\rho} U$. Since $\tilde{\rho}$ fulfills (7.86), we get

$$\rho^2 = U^\dagger \tilde{\rho} U U^\dagger \tilde{\rho} U = U^\dagger \tilde{\rho}^2 U = U^\dagger \tilde{\rho} U = \rho \quad . \tag{7.90}$$

So (7.86) holds for a Slater determinant $|\Phi\rangle$ no matter what single-particle basis is used for defining ρ.[2]

---

[1] The following notation is used: $\hat{t}$ denotes an operator, t the corresponding matrix, and t_{kl} the elements of the matrix.

[2] Do not confuse this with the analogous equation for a general density matrix. A general density matrix fulfilling $\rho^2 = \rho$ describes a *pure state*. This has nothing to do with Hartree–Fock theory.

- In the case of a Slater determinant $|\Phi\rangle$, (7.86) implies that ρ is a projection operator. From the special form it takes if expressed in the single-particle states contained in $|\Phi\rangle$ it is clear that it projects onto the space of occupied single-particle states.

The last property of the density matrix allows a simple formulation of the decomposition of a matrix into particle–particle, particle–hole, etc., contributions (related to a given Slater determinant). Returning to the Hartree–Fock case, the hole–hole part of the single-particle Hamiltonian in matrix notation can immediately be written as

$$h_{\mathrm{hh}} = \rho h \rho \quad . \tag{7.91}$$

To find similar expressions for the other parts of h, note that the matrix $\sigma = 1 - \rho$ projects onto the space of empty single-particle states (check that $\sigma^2 = \sigma$ and find its meaning as for ρ). This yields immediately

$$h_{\mathrm{hp}} = \rho h \sigma \quad , \quad h_{\mathrm{ph}} = \sigma h \rho \quad , \quad h_{\mathrm{pp}} = \sigma h \sigma \quad . \tag{7.92}$$

The Hartree–Fock conditions were that $h_{\mathrm{hp}} = h_{\mathrm{ph}} = 0$ and can be rewritten now as

$$\sigma h \rho = 0 \quad , \quad \rho h \sigma = 0 \quad , \tag{7.93}$$

and upon inserting the definition of σ,

$$h\rho - \rho h \rho = 0 \quad , \quad \rho h \rho - \rho h = 0 \quad , \tag{7.94}$$

from which we conclude that $\rho h = h \rho$ or

$$[\rho, h] = 0 \quad . \tag{7.95}$$

It is clear why formulations using the density matrix are often more elegant: there is no need to separate occupied and empty single-particle states in this approach; this distinction is handled by the density matrix. Simple matrix manipulations can then replace cumbersome sum expressions that have the additional problem of having to distinguish these different index ranges.

7.2.7 Constrained Hartree–Fock

The Hartree–Fock equations as such yield an approximate wave function corresponding to a minimum in the expectation value of the energy. In practice the result is not unique. To understand this, consider for example the problem of fission. The energy of the nucleus is minimal for its ground state, but for heavy nuclei the energy of two separate fragments at a distance is even lower; in addition, it is known that many nuclei also possess a *shape-isomeric* state, i.e., a local minimum in the energy at relatively large deformation.

Which of these solutions is obtained by iterating the Hartree–Fock equations depends on the initial configuration, as illustrated in Fig. 7.2.

In contrast to the phenomenological single-particle models with parametrized shapes, the Hartree–Fock minimum may contain arbitrary deformations and also violate symmetries if these are allowed by the numerical procedure and if such a symmetry is not present in the initial configuration. For example, a deformed

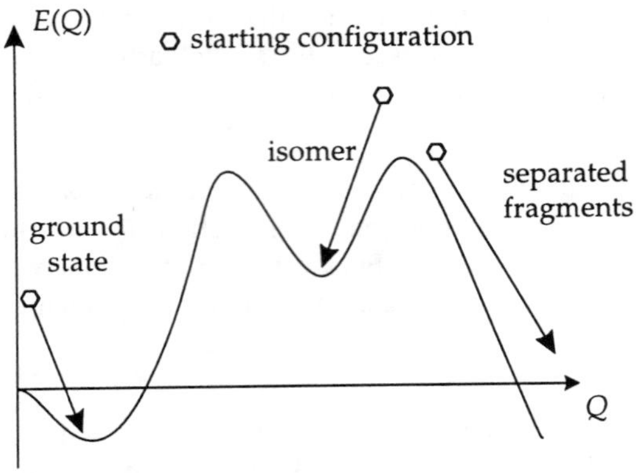

Fig. 7.2 Schematic sketch of the way the Hartree–Fock equations converge to different minima depending on the deformation contained in the initial wave functions. In this case, three different minima will be approached, with the separate fragments going to as large a separation as is allowed by the numerical procedure.

nucleus can acquire an additional asymmetric octupole-like deformation, provided the numerical method does not enforce reflection symmetry and the initial state is not ideally symmetric: in the latter case the iteration may not be able to "decide" in which direction to deviate from reflection symmetry and will be frozen in a symmetric shape.

The way to determine the whole curve in Fig. 7.2 is to add a constraint to the variational principle. For fission, for example, a natural constraint is given by the quadrupole moment of the nucleus, so that we consider the variational problem

$$\delta\langle \mathrm{HF}|\hat{H} - \lambda\hat{Q}|\mathrm{HF}\rangle = 0 \quad . \tag{7.96}$$

The operator $\hat{Q}$ is the sum of the quadrupole moments of the single particles.

In using this approach one may keep λ constant and then obtain a deformed state for each value of λ; unfortunately it is then difficult to find states near maxima in the energy, so that it is often better to demand a given expectation value of $\hat{Q}$ and vary λ during the iteration accordingly. Such calculations have been done extensively and yield quite reasonable fission potentials and deformed ground states. These should not be compared directly to the potentials used in a collective model, because the microscopic wave functions contain some uncertainty in the quadrupole moment – after all, only the *expectation value* is fixed – corresponding to zero-point oscillations in the collective ground state. The Hartree–Fock energy thus should always be higher than the collective potential-energy surface.

At present it is still too expensive to do Hartree–Fock calculations with more than one constraint, so that for the study of potential-energy surfaces depending on more than one deformation parameter the phenomenological models are preferable.

7.2.8 Alternative Formulations and Three-Body Forces

In many applications it is more convenient to use the variational principle itself instead of the Hartree–Fock equations derived from it. This may be due to reasons of simplicity or for the following practical consideration: normally the equations are solved numerically, for example by expressing the wave functions as discretized functions on a numerical grid, such as

$$\phi(x) \to \phi(x_j) \quad , \quad x_j = j\,\Delta x \quad , \quad j = 1, \ldots, N_x \quad . \tag{7.97}$$

In this case there is an additional approximation in the wave functions: we seek an optimal solution with a Slater determinant built out of such discretized single-particle states. Instead of discretizing the eigenvalue problem for the single-particle Hamiltonian of (7.69), it is better to discretize the expectation value of the Hamiltonian itself and vary with respect to the true unknowns, the $\phi(x_j)$. Usually there are only small differences between the two methods, but conceptually the second one should provide a better approximation.

A case in which this method is also applied is in the application of Hartree-Fock to Skyrme forces, where there is an additional three-body potential. If the treatment of two-body potentials is appropriately generalized, such a potential leads to a second-quantized operator

$$\frac{1}{6} \sum_{k_1 k_2 k_3 k_4 k_5 k_6} v_{k_1 k_2 k_3 k_4 k_5 k_6}\, \hat{a}^\dagger_{k_1} \hat{a}^\dagger_{k_2} \hat{a}^\dagger_{k_3} \hat{a}_{k_4} \hat{a}_{k_5} \hat{a}_{k_6} \quad , \tag{7.98}$$

and with exactly the same methods as before the expectation value of such a term may be evaluated to yield

$$\frac{1}{6} \sum_{ijk} \bar{v}_{ijkijk} \quad , \tag{7.99}$$

where $\bar{v}_{ijkijk}$ is the antisymmetrized combination

$$\bar{v}_{ijkijk} = v_{ijkijk} + v_{ijkjki} + v_{ijkkij} - v_{ijkjik} - v_{ijkikj} - v_{ijkkji} \quad . \tag{7.100}$$

In the next section this formalism will be applied, which has the interesting consequence that for a zero-range force the expectation value of the Skyrme Hamiltonian can be expressed as an integral over a local *energy density functional*.

7.2.9 Hartree–Fock with Skyrme Forces

The principal purpose of the discussion of three-body forces in the preceding section was the application to Skyrme forces. Recall the definition of the Skyrme interaction as a sum of two- and three-body parts,

$$\hat{H} = \sum_i \hat{t}_i + \sum_{i<j} v^{(2)}_{ij} + \sum_{i<j<k} v^{(3)}_{ijk} \quad ; \tag{7.101}$$

the two body part was given by

$$\begin{aligned} v^{(2)}_{12} &= t_0(1 + x_0 \hat{P}_\sigma)\, \delta(\boldsymbol{r}_1 - \boldsymbol{r}_2) \\ &\quad + \tfrac{1}{2} t_1 \left(\delta(\boldsymbol{r}_1 - \boldsymbol{r}_2) \hat{k}^2 + \hat{k}'^2 \delta(\boldsymbol{r}_1 - \boldsymbol{r}_2) \right) \\ &\quad + t_2\, \hat{\boldsymbol{k}}'^2 \cdot \delta(\boldsymbol{r}_1 - \boldsymbol{r}_2)\, \hat{\boldsymbol{k}} + \mathrm{i} W_0 (\hat{\boldsymbol{\sigma}}_1 + \hat{\boldsymbol{\sigma}}_2) \cdot \hat{\boldsymbol{k}}' \times \delta(\boldsymbol{r}_1 - \boldsymbol{r}_2)\, \hat{\boldsymbol{k}} \quad , \end{aligned} \tag{7.102}$$

and the three-body part by

$$v^{(3)}_{123} = t_3\, \delta(\boldsymbol{r}_1 - \boldsymbol{r}_2)\, \delta(\boldsymbol{r}_2 - \boldsymbol{r}_3) \quad . \tag{7.103}$$

Remember that the operators $\hat{\boldsymbol{k}}$ and $\hat{\boldsymbol{k}}'$ were defined as

$$\hat{\boldsymbol{k}} = \frac{1}{2\mathrm{i}}(\nabla_1 - \nabla_2) \quad , \quad \hat{\boldsymbol{k}}' = -\frac{1}{2\mathrm{i}}(\nabla_1 - \nabla_2) \tag{7.104}$$

and that $\hat{\boldsymbol{k}}'$ *acts to the left.*

To obtain the Hartree–Fock equations, we have to evaluate the expectation value of the Hamiltonian in a Slater determinant $|\mathrm{HF}\rangle$. It is given by

$$\begin{aligned} E &= \langle \mathrm{HF}|\hat{H}|\mathrm{HF}\rangle \\ &= \sum_i \langle i|\hat{t}|i\rangle + \tfrac{1}{2}\sum_{ij}\langle ij|\bar{v}^{(2)}|ij\rangle + \tfrac{1}{6}\sum_{ijk}\langle ijk|\bar{v}^{(3)}|ijk\rangle \quad . \end{aligned} \tag{7.105}$$

We will see that it can be rewritten as a spatial integral over a *Hamiltonian density,*

$$E = \int \mathrm{d}^3r \; \hat{\mathcal{H}}(\boldsymbol{r}) \quad . \tag{7.106}$$

The Hamiltonian density, in turn, is a functional of certain densities obtained by summing over the single-particle states. If we write them as

$$\phi_i(\boldsymbol{r}, \sigma, q) \quad , \tag{7.107}$$

where $\sigma = \pm\frac{1}{2}$ denotes the spin projection, $q = \pm\frac{1}{2}$ the isospin (the notation $q = \mathrm{p}$, n will also be used for fixed values of the index q), and i thus only enumerates the different orbital functions, it will be useful to define the following densities:

1. number densities for protons and neutrons (depending on the value of q),

$$\rho_q(\boldsymbol{r}) = \sum_{i,\sigma} |\phi_i(\boldsymbol{r}, \sigma, q)|^2 \quad , \tag{7.108}$$

2. kinetic energy densities for protons and neutrons

$$\tau_q(\boldsymbol{r}) = \sum_{i,\sigma} |\nabla\phi_i(\boldsymbol{r}, \sigma, q)|^2 \quad , \tag{7.109}$$

and

3. spin–orbit current

$$\boldsymbol{J}_q(\boldsymbol{r}) = -\mathrm{i}\sum_{i,\sigma,\sigma'} \phi_i^*(\boldsymbol{r}, \sigma, q)\left[\nabla\phi_i(\boldsymbol{r}, \sigma', q) \times \langle\sigma|\hat{\boldsymbol{\sigma}}|\sigma'\rangle\right] \quad . \tag{7.110}$$

Note that the kinetic-energy density does not include the factor $\hbar^2/2m$ and also does not involve the customary combination $\phi^*\nabla^2\phi$. The more symmetric combination used here is equivalent for the total (integrated) Hamiltonian, as partial integration shows. The spin-current densities appear to be quite complicated; inserting a single wave function, however, shows it to describe essentially a term of the form $\rho\boldsymbol{k}\times\boldsymbol{\sigma}$, i.e., the cross product of the local momentum with the spin, weighted by the local probability density.

The sums should be taken over all occupied single-particle states. We now make the assumption that the space of occupied states is invariant under time-reversal. This is clearly true for the most important application, namely static properties of even–even nuclei. Formally, the assumption means that if a state $\phi_i(\boldsymbol{r},\sigma,q)$ is occupied, the time-reversed state obtained with the time-reversal operator $\hat{\mathcal{T}}$ is also occupied. Using the result from Exercise 2.4 that $\hat{\mathcal{T}} = -\mathrm{i}\sigma_y$ for the spinor part combined with complex conjugation for the orbital wave functions, we get

$$\hat{\mathcal{T}}\phi_i(\boldsymbol{r},\sigma,q) = -\mathrm{i}\sum_{\sigma'}\langle\sigma|\sigma_y|\sigma'\rangle\phi_i^*(\boldsymbol{r},\sigma',q) = -2\sigma\phi_i^*(\boldsymbol{r},-\sigma,q) \quad . \tag{7.111}$$

The consequence is that expectation values containing a spin operator vanish:

$$\sum_{i\sigma\sigma'}\phi_i^*(\boldsymbol{r},\sigma,q)\langle\sigma|\hat{\boldsymbol{\sigma}}|\sigma'\rangle\phi_i(\boldsymbol{r},\sigma',q) = 0 \quad . \tag{7.112}$$

To see this use the operator $\frac{1}{2}\left(1+\hat{\mathcal{T}}\right)$ to time-reversal symmetrize the expression

$$\begin{aligned}
&\sum_i\phi_i^*(\boldsymbol{r},\sigma,q)\phi_i(\boldsymbol{r},\sigma',q)\\
&\quad = \tfrac{1}{2}\sum_i\left(\phi_i^*(\boldsymbol{r},\sigma,q)\phi_i(\boldsymbol{r},\sigma',q) + 4\sigma\sigma'\phi_i(\boldsymbol{r},-\sigma,q)\phi_i^*(\boldsymbol{r},-\sigma',q)\right)\\
&\quad = \tfrac{1}{2}\delta_{\sigma\sigma'}\sum_i\left(|\phi_i(\boldsymbol{r},\sigma,q)|^2 + |\phi_i(\boldsymbol{r},-\sigma,q)|^2\right)\\
&\quad = \tfrac{1}{2}\delta_{\sigma\sigma'}\rho_q(\boldsymbol{r}) \quad .
\end{aligned} \tag{7.113}$$

Now the desired expectation value can be evaluated:

$$\begin{aligned}
&\sum_{i\sigma\sigma'}\phi_i^*(\boldsymbol{r},\sigma,q)\,\langle\sigma|\hat{\boldsymbol{\sigma}}|\sigma'\rangle\,\phi_i(\boldsymbol{r},\sigma',q)\\
&\quad = \tfrac{1}{2}\sum_{\sigma\sigma'}\delta_{\sigma\sigma'}\rho_q(\boldsymbol{r})\langle\sigma|\hat{\boldsymbol{\sigma}}|\sigma'\rangle\\
&\quad = \tfrac{1}{2}\rho_q(\boldsymbol{r})\,\mathrm{Tr}\{\hat{\boldsymbol{\sigma}}\} = 0 \quad .
\end{aligned} \tag{7.114}$$

The last equation follows from the vanishing trace of the Pauli matrices. This calculation also makes it clear that if the condition of having both time-reversed states occupied is not fulfilled, the formalism will become much more complicated, because the single scalar density will have to be replaced by a density matrix with spin indices.

The derivation of the energy functional is given in Exercise 7.3 to sufficient detail to enable the reader to work out the whole problem if desired. Here we summarize the results. The final result for the energy density, including the kinetic energy, then becomes

$$\begin{aligned}
\hat{\mathcal{H}}(\boldsymbol{r}) = {}& \frac{\hbar^2}{2m}\tau + \tfrac{1}{2}t_0\left[\left(1+\tfrac{1}{2}x_0\right)\rho^2 - \left(x_0+\tfrac{1}{2}\right)\left(\rho_\mathrm{n}^2+\rho_\mathrm{p}^2\right)\right]\\
&+ \tfrac{1}{4}(t_1+t_2)\rho\tau + \tfrac{1}{8}(t_2-t_1)(\rho_\mathrm{n}\tau_\mathrm{n}+\rho_\mathrm{p}\tau_\mathrm{p})\\
&+ \tfrac{1}{16}(t_2-3t_1)\rho\nabla^2\rho + \tfrac{1}{32}(3t_1+t_2)\left(\rho_\mathrm{n}\nabla^2\rho_\mathrm{n}+\rho_\mathrm{p}\nabla^2\rho_\mathrm{p}\right)\\
&+ \tfrac{1}{16}(t_1-t_2)\left(\boldsymbol{J}_\mathrm{n}^2+\boldsymbol{J}_\mathrm{p}^2\right) + \tfrac{1}{4}t_3\rho_\mathrm{n}\rho_\mathrm{p}\rho\\
&- \tfrac{1}{2}W_0\left(\rho\nabla\cdot\boldsymbol{J} + \rho_\mathrm{n}\nabla\cdot\boldsymbol{J}_\mathrm{n} + \rho_\mathrm{p}\nabla\cdot\boldsymbol{J}_\mathrm{p}\right) \quad .
\end{aligned} \tag{7.115}$$

Coulomb effects are not included in this formula; the Coulomb energy with a suitable approximation for Coulomb exchange has to be added separately.

A much simpler form is valid for nuclei with $N = Z$, for which, in the absence of Coulomb effects, one may assume $\rho_\mathrm{p} = \rho_\mathrm{n} = \frac{1}{2}\rho$ and similarly for the other densities. Then we have

$$\hat{\mathcal{H}}(\boldsymbol{r}) = \frac{\hbar^2}{2m}\tau + \tfrac{3}{8}t_0\rho^2 + \tfrac{1}{16}t_3\rho^3 + \tfrac{1}{16}(3t_1 + 5t_2)\rho\tau \\ + \tfrac{1}{64}(9t_1 - 5t_2)(\nabla\rho)^2 - \tfrac{3}{4}W_0\rho\nabla\cdot\boldsymbol{J} + \tfrac{1}{32}(t_1 - t_2)J^2 \quad . \tag{7.116}$$

In both cases the energy expression is the starting point for deriving the Hartree–Fock equations by variation with respect to the single-particle wave functions. This yields equations of integro-differential form as in the standard Hartree–Fock case.

EXERCISE

7.3 The Skyrme Energy Functional

Problem. Calculate the energy associated with a Slater determinant for the case of the Skyrme force.

Solution. Let us demonstrate the methods used by evaluating some of the individual contributions to the Skyrme energy functional. The exchange contribution is taken into account by adding an exchange operator in the matrix element:

$$\begin{aligned}\bar{v}_{k_1k_2k_3k_4} &= v_{k_1k_2k_3k_4} - v_{k_1k_2k_4k_3}\,, \\ &= \langle k_1k_2|\hat{v}|k_3k_4\rangle - \langle k_1k_2|\hat{v}|k_4k_3\rangle \\ &= \langle k_1k_2|v(1 - P_{3\leftrightarrow 4})|k_3k_4\rangle \quad ,\end{aligned} \tag{1}$$

where $P_{3\leftrightarrow 4}$ exchanges the two particles in the ket. It can be expressed as the product of spatial exchange $\hat{P}_\mathrm{M}$ (Majorana), spin exchange $\hat{P}_\sigma$, and isospin exchange $\hat{P}_\tau$. Replacing the generic indices k by the triples (i,σ,q), the term proportional to t_0 yields

$$V_0 = \tfrac{1}{2}\sum_{ij\sigma\sigma'qq'}\langle ij\sigma\sigma'qq'|t_0\,\delta(\boldsymbol{r}_1 - \boldsymbol{r}_2)\big(1 + x_0\hat{P}_\sigma\big)\big(1 - \hat{P}_\mathrm{M}\hat{P}_\sigma\hat{P}_\tau\big)|ij\sigma\sigma'qq'\rangle \,. \tag{2}$$

The Majorana exchange operator $\hat{P}_\mathrm{M}$ reduces to a factor of 1, since the δ function forces the wave functions to be evaluated at the same point in space. For the isospin exchange operator, one has to assume that the wave functions retain pure isospin, so that there can be an overlap in isospin space with the other single-particle wave function only if the isospin is the same: $\hat{P}_\tau \to \delta_{qq'}$. Using the spin-exchange operator in the form $\hat{P}_\sigma = \frac{1}{2}(1 + \hat{\boldsymbol{\sigma}}_1\cdot\hat{\boldsymbol{\sigma}}_2)$, we get

$$\begin{aligned}V_0 &= \tfrac{1}{2}t_0\sum_{ij\sigma\sigma'qq'}\langle ij\sigma\sigma'qq'|\delta(\boldsymbol{r}_1 - \boldsymbol{r}_2)\Big[1 + \tfrac{1}{2}x_0(1 + \hat{\boldsymbol{\sigma}}_1\cdot\hat{\boldsymbol{\sigma}}_2)\Big] \\ &\quad\times\Big[1 - \tfrac{1}{2}\delta_{qq'}(1 + \hat{\boldsymbol{\sigma}}_1\cdot\hat{\boldsymbol{\sigma}}_2)\Big]|ij\sigma\sigma'qq'\rangle \\ &= \tfrac{1}{2}t_0\sum_{ij\sigma\sigma'qq'}\langle ij\sigma\sigma'qq'|\delta(\boldsymbol{r}_1 - \boldsymbol{r}_2) \\ &\quad\times\Big[1 + \tfrac{1}{2}x_0 - \tfrac{1}{2}\delta_{qq'} - \tfrac{x_0}{4}\delta_{qq'}(1 + \hat{\boldsymbol{\sigma}}_1\cdot\hat{\boldsymbol{\sigma}}_2)^2\Big]|ij\sigma\sigma'qq'\rangle \quad .\end{aligned} \tag{3}$$

Exercise 7.3

In the last step the fact was utilized that terms linear in $\hat{\boldsymbol{\sigma}}_1 \cdot \hat{\boldsymbol{\sigma}}_2$ vanish (see (7.114)), while those proportional to $(\hat{\boldsymbol{\sigma}}_1 \cdot \hat{\boldsymbol{\sigma}}_2)^2$ yield a factor of 3 as is easily derived from the relation

$$(\hat{\boldsymbol{\sigma}} \cdot \hat{\boldsymbol{\sigma}}')^2 = 3 - 2\hat{\boldsymbol{\sigma}} \cdot \hat{\boldsymbol{\sigma}}' \tag{4}$$

already derived in (7.5), and noting that only the constant 3 contributes. There is now no explicit exchange term anymore, and the wave functions from left and right can be combined into densities:

$$\begin{aligned} V_0 &= \tfrac{1}{2}t_0 \sum_{ij\sigma\sigma'qq'} \int \mathrm{d}^3r \, |\phi_i(\boldsymbol{r},\sigma,q)|^2 |\phi_j(\boldsymbol{r},\sigma',q')|^2 \left[1 + \tfrac{1}{2}x_0 - \delta_{qq'}\left(x_0 + \tfrac{1}{2}\right)\right] \\ &= \tfrac{1}{2}t_0 \sum_{ij\sigma qq'} \int \mathrm{d}^3r \, \rho_q(\boldsymbol{r})\rho_{q'}(\boldsymbol{r}) \left[1 + \tfrac{1}{2}x_0 - \delta_{qq'}\left(x_0 + \tfrac{1}{2}\right)\right] \quad . \end{aligned} \tag{5}$$

The contribution to the energy density becomes

$$\hat{\mathcal{H}}_0(\boldsymbol{r}) = \tfrac{1}{2}t_0 \left[\left(1 + \tfrac{1}{2}x_0\right)\rho^2(\boldsymbol{r}) - \left(x_0 + \tfrac{1}{2}\right)\left(\rho_\mathrm{p}^2(\boldsymbol{r}) + \rho_\mathrm{n}^2(\boldsymbol{r})\right)\right] \tag{6}$$

with the total density $\rho = \rho_\mathrm{p} + \rho_\mathrm{n}$.

For the term proportional to t_1 the same arguments can be applied to the exchange operators. Because the second contribution in this term is just the hermitian conjugate of the first, it suffices for us to consider only that one in detail. Inserting the definition of $\hat{\boldsymbol{k}}$ and using the above considerations for $(1 - \hat{P}_\mathrm{M}\hat{P}_\sigma\hat{P}_\tau)$, we get

$$\begin{aligned} V_1 = &-\tfrac{1}{16}t_1 \sum_{ij\sigma\sigma'qq'} \langle ij\sigma\sigma'qq'|\delta(\boldsymbol{r}_1 - \boldsymbol{r}_2)\left(\nabla_1^2 + \nabla_2^2 - 2\nabla_1\cdot\nabla_2\right) \\ &\times \left(1 - \tfrac{1}{2}\delta_{qq'} - \tfrac{1}{2}\hat{\boldsymbol{\sigma}}_1\cdot\hat{\boldsymbol{\sigma}}_2\,\delta_{qq'}\right)|ij\sigma\sigma'qq'\rangle + \text{h.c.} \end{aligned} \tag{7}$$

Using the same arguments that led to (7.114), we can see that some of the terms involving Pauli matrices do not contribute and the expression reduces to

$$\begin{aligned} V_1 = &-\tfrac{1}{16}t_1 \sum_{ij\sigma\sigma'qq'} \langle ij\sigma\sigma'qq'|\delta(\boldsymbol{r}_1 - \boldsymbol{r}_2)\left(\nabla_1^2 + \nabla_2^2 - 2\nabla_1\cdot\nabla_2\right)|ij\sigma\sigma'qq'\rangle \\ &\times \left(1 - \tfrac{1}{2}\delta_{qq'}\right) \\ &- \tfrac{1}{16}t_1 \sum_{ij} \langle ij\sigma\sigma'qq'|\delta(\boldsymbol{r}_1 - \boldsymbol{r}_2)(\nabla_1\cdot\nabla_2)\left(\hat{\boldsymbol{\sigma}}_1\cdot\hat{\boldsymbol{\sigma}}_2\right)|ij\sigma\sigma'qq'\rangle\delta_{qq'} + \text{h.c.} \\ \equiv &\; V_{11} + V_{12} + \text{h.c.} \quad . \end{aligned} \tag{8}$$

For the first term we may use time-reversal invariance to get

$$\nabla^2\rho(\boldsymbol{r}) = 2\sum_{i\sigma q}\phi_i^*\nabla^2\phi_i + 2\tau(\boldsymbol{r}) \quad , \tag{9}$$

and, similarly,

$$\begin{aligned} \sum_{i\sigma q}\phi_i^*\nabla\phi_i &= \tfrac{1}{2}\sum_{i\sigma q}\left(\phi_i^*(\boldsymbol{r},\sigma,q)\nabla\phi_i(\boldsymbol{r},\sigma,q) + \phi_i(\boldsymbol{r},-\sigma,q)\nabla\phi_i^*(\boldsymbol{r},-\sigma,q)\right) \\ &= \tfrac{1}{2}\nabla\rho \end{aligned} \tag{10}$$

Exercise 7.3

to obtain

$$V_{11} = -\tfrac{1}{16}t_1 \int d^3r \, \Big\{ \rho\nabla^2\rho - 2\rho\tau - \tfrac{1}{2}(\nabla\rho)^2 - \tfrac{1}{2}\sum_q \Big[\rho_q\nabla^2\rho_q - 2\rho_q\tau_q - \tfrac{1}{2}(\nabla\rho_q)^2 \Big] \Big\} . \quad (11)$$

Integrating by parts reduces it to

$$V_{11} = \tfrac{1}{16}t_1 \int d^3r \, \Big(2\rho\tau - \rho_n\tau_n - \rho_p\tau_p - \tfrac{3}{2}\rho\nabla^2\rho + \tfrac{3}{4}\rho_n\nabla^2\rho_n + \tfrac{3}{4}\rho_p\nabla^2\rho_p \Big) \quad .(7.117)$$

To evaluate V_{12}, one may use the vector identity

$$(\nabla_1 \cdot \nabla_2)(\hat{\boldsymbol{\sigma}}_1 \cdot \hat{\boldsymbol{\sigma}}_2) = \tfrac{1}{3}(\nabla_1 \cdot \hat{\boldsymbol{\sigma}}_1)(\nabla_2 \cdot \hat{\boldsymbol{\sigma}}_2) + \tfrac{1}{2}(\nabla_1 \times \hat{\boldsymbol{\sigma}}_1) \cdot (\nabla_2 \times \hat{\boldsymbol{\sigma}}_2) + \sqrt{5}\Big[[\nabla_1 \times \hat{\boldsymbol{\sigma}}_1]^2 \times [\nabla_2 \times \hat{\boldsymbol{\sigma}}_2]^2 \Big]^0 . \quad (13)$$

Note that in the last expression the square brackets stand for angular-momentum coupling. This expression can be reduced further only if axial symmetry is also assumed, which leads to a vanishing of all but the second term because the others will average to zero in the integration process. The final result is

$$V_{12} = -\tfrac{1}{32}t_1 \sum_{ij\sigma\sigma'qq'} \langle ij\sigma\sigma'qq' | \delta(\boldsymbol{r}_1 - \boldsymbol{r}_2)(\nabla_1 \times \hat{\boldsymbol{\sigma}}_1)(\nabla_2 \times \hat{\boldsymbol{\sigma}}_2) | ij\sigma\sigma'qq' \rangle \delta_{qq'} = \tfrac{1}{32}t_1 \int d^3r \, \big[\boldsymbol{J}_n^2(\boldsymbol{r}) + \boldsymbol{J}_p^2(\boldsymbol{r}) \big] \quad , \quad (14)$$

with $\boldsymbol{J}_{p,n}$ the proton and neutron contibutions to the total current density. The total t_1-dependent contribution to the energy density including the hermitian conjugate contributions now is

$$\mathcal{H}_1 = \tfrac{1}{16}\Big[4\rho\tau - 2\rho_n\tau_n - 2\rho_p\tau_p - 3\rho\nabla^2\rho + \tfrac{3}{2}\rho_n\nabla^2\rho_n + \tfrac{3}{2}\rho_p\nabla^2\rho_p + \boldsymbol{J}_n^2 + \boldsymbol{J}_p^2 \Big] \quad . \quad (15)$$

These detailed derivations should be sufficient to illustrate the methods used in the calculation and since no new ideas are used for deriving the others, we skip those calculations, completing this exercise only by a look at the particularly interesting case of the three-body term.

The matrix element

$$V_3 = \tfrac{1}{6} \sum_{ijk} \langle ijk | \tilde{v}_{123} | ijk \rangle \quad (16)$$

requires insertion of an operator antisymmetrizing over all permutations of the single-particle coordinates. The permutations of (123) appearing with a positive sign are (123), (231), and (312), while negative contributions come from (213), (132), and (321). Denoting by $\hat{P}(12)$ the operator exchanging particles 1 and 2, the correct operator is

$$\big[1 + \hat{P}(12)\hat{P}(23) + \hat{P}(13)\hat{P}(23) - \hat{P}(21) - \hat{P}(23) - \hat{P}(31) \big] \quad . \quad (17)$$

Exercise 7.3

Now each of the $\hat{P}$ operators is replaced by the combination of $\hat{P}_{\mathrm{M}}\hat{P}_{\sigma}\hat{P}_{\tau}$. Because of the δ functions, $\hat{P}_{\mathrm{M}}$ can be omitted, while, for example, $\hat{P}_{\sigma}(12)$, leads to $\frac{1}{2}(1 + \hat{\sigma}_1 \cdot \hat{\sigma}_2)$ in which the term with Pauli matrices can be dropped because of (7.114) as before. Thus we get

$$\begin{aligned}\tilde{v}_{123} &= t_3\,\delta(\boldsymbol{r}_1 - \boldsymbol{r}_2)\,\delta(\boldsymbol{r}_2 - \boldsymbol{r}_3) \\ &\quad \times \left[1 + \tfrac{1}{4}\hat{P}_\tau(12)\hat{P}_\tau(23) + \tfrac{1}{4}\hat{P}_\tau(13)\hat{P}_\tau(23)\right. \\ &\quad \left. - \tfrac{1}{2}\hat{P}_\tau(12) - \tfrac{1}{2}\hat{P}_\tau(23) - \tfrac{1}{2}\hat{P}_\tau(31)\right] \quad . \end{aligned} \tag{18}$$

Relabelling the indices shows that the second and third term yield the same contribution to the sum, and the same holds true for the last three terms. Also the isospin exchange operators result in Kronecker symbols as before, so we get

$$V_3 = \frac{1}{6} t_3 \sum_{ijk} \int d^3r\; |\phi_i|^2 |\phi_j|^2 |\phi_k|^2 \left(1 + \tfrac{1}{2}\delta_{qq'}\delta_{qq'} - \tfrac{3}{2}\delta_{qq'}\right) \quad . \tag{19}$$

Inserting the definition of the densities leads to the final result

$$\begin{aligned}\mathcal{H}_3 &= \tfrac{1}{6} t_3 \left[\rho^3 + \tfrac{1}{2}(\rho_{\mathrm{n}}^3 + \rho_{\mathrm{p}}^3) - \tfrac{3}{2}\rho(\rho_{\mathrm{n}}^2 + \rho_{\mathrm{p}}^2)\right] \\ &= \tfrac{1}{4} t_3 \rho_{\mathrm{n}} \rho_{\mathrm{p}} \rho \quad . \end{aligned} \tag{20}$$

7.3 Phenomenological Single-Particle Models

7.3.1 The Spherical-Shell Model

Historically the shell structure of the nucleus was not predicted by *ab initio* theoretical considerations based on the Hartree–Fock method. Instead, the experimental evidence for shell closure in analogy to the inert gases of atomic structure led to the phenomenological postulation of mean-field potentials [Ma48, Fe49, Ha49, Ma49, Ma50, Ma55]. which could later be explained in terms of self-consistent fields. The principal difference to atomic structure was recognized clearly: there is no dominating field generated by an external source corresponding to the Coulomb field of the nucleus, which suffices to explain many features of atoms without recourse to the much more complicated effects of the electron–electron interaction. In nuclei the mean field is exclusively produced by the nucleon–nucleon interaction.

This type of model of noninteracting particles in a mean potential is often called the *independent particle model*.

The most important piece of experimental information on shell structure is the existence of *magic numbers*. If the number of protons or neutrons is one of the magic values, the nucleus turns out to be especially stable; more specifically, it is characterized by

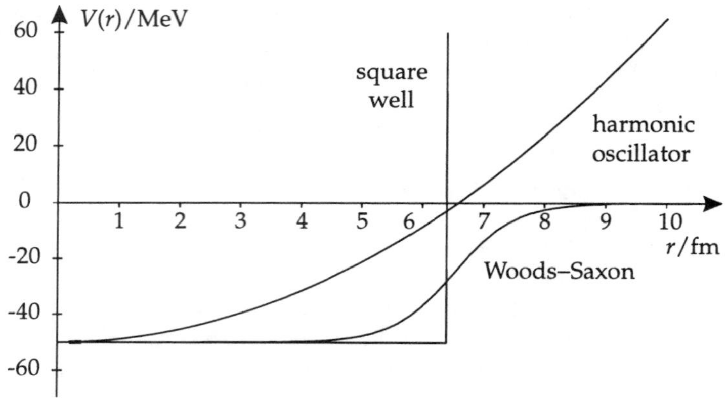

Fig. 7.3 Sketch of the functional form of three popular phenomenological shell-model potentials: Woods–Saxon, harmonic oscillator, and the square well. The parameters are applicable to ^{208}Pb. Note that for the harmonic oscillator the absolute value of the energy is unimportant, as it does have the natural limit of vanishing energy for large distances; for this figure it was adjusted to agree with the other potentials in the center of the nucleus.

- a larger total binding energy of the nucleus,
- a larger energy required to separate a single nucleon,
- a higher energy of the lowest excited states, and
- a large number of isotopes or isotones with the same magic number for protons (neutrons)

(all of these in comparison to neighboring nuclei in the table of nuclides). The lower magic numbers are the same for protons and neutrons, namely 2, 8, 20, 28, 50, and 82, whereas the next number, 126, is established only experimentally for neutrons. Theoretically one would expect additional magic numbers near 114 for protons and 184 for neutrons (the exact prediction depends on the theory) leading to *superheavy nuclei* [Gr69, Ni69, Fi72, Ra74] (for an extended treatment, see [Ir75, Ku89]), but these have not been confirmed in experiment, although there are hints of an increase of lifetimes in the heaviest elements observed up to now. We will come to this question in Sect. 9.2.

A phenomenological shell model thus is based on the Schrödinger equation for the single-particle levels:

$$\left(-\frac{\hbar^2}{2m}\nabla^2 + V(\boldsymbol{r})\right)\psi_i(\boldsymbol{r}) = \varepsilon_i\,\psi_i(\boldsymbol{r}) \quad , \tag{7.118}$$

with a prescribed potential $V(\boldsymbol{r})$.

What kind of function should be chosen for $V(\boldsymbol{r})$? It should be relatively constant inside heavier nuclei to explain the constant density suggested by the fact that the nuclear radius behaves as

$$R = r_0 A^{1/3} \quad , \tag{7.119}$$

but go to zero quite rapidly outside the nuclear surface. Popular and successful choices are given below (assuming spherical symmetry).

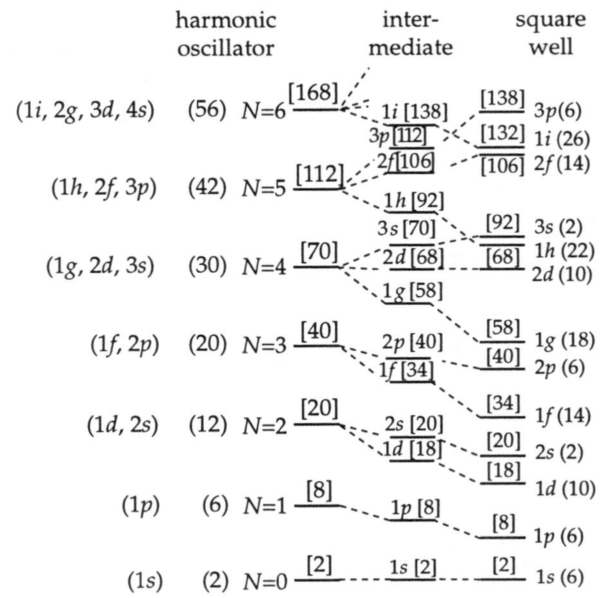

Fig. 7.4 Level structures in the harmonic-oscillator (*left*), Woods–Saxon (*center*), and square-well potentials (*right*). For each level the conventional quantum number designation is given consisting of the radial quantum number and the angular momentum. The number of particles in each state, including the two-fold spin degeneracy, is indicated in parantheses for the level, while the total occupation up to that level is given in square brackets.

- The *Woods–Saxon* potential [Wo54, Ro68]

$$V(r) = -\frac{V_0}{1 + \exp[(r - R)/a]} \quad . \tag{7.120}$$

Typical values for the parameters are: depth $V_0 \approx 50\text{MeV}$, radius $R \approx 1.1\ \text{fm}A^{1/3}$, and surface thickness $a \approx 0.5$ fm. Although this potential follows a similar form as the experimental nuclear density distributions and has the general behavior discussed above, it has the practical disadvantage of not leading to analytic forms for the wave functions. It is thus used usually in those cases where the asymptotic form of the wave functions is important and calculational expense not prohibitive.

- The *harmonic-oscillator* potential

$$V(r) = \tfrac{1}{2} m\omega^2 r^2 \tag{7.121}$$

with $\hbar\omega \approx 41\,\text{MeV} \times A^{-1/3}$ typically. This is very convenient for computation but as it goes to infinity instead of zero at large distances, it clearly does not produce the correct large-distance behavior of the wave functions, which fall off as $\exp(-k^2r^2)$ instead of $\exp(-kr)$ and there are no scattering states at all. The $A^{-1/3}$ dependence of the oscillator constant implies that on the nuclear surface, i.e., for $R = r_0A^{1/3}$, there is the same potential regardless of the value of A, thus simulating the constant depth of the potential well.

- Finally, for some applications, the *square-well* potential

$$V(r) = \begin{cases} -V_0 & \text{for} \quad r \le R \\ \infty & \text{for} \quad r > R \end{cases} \quad . \tag{7.122}$$

This is a compromise combining finite range with moderate computational difficulties. Note that the wave functions vanish for $r \geq R$ and are thus not realistic in this region. Although the wave functions can be given analytically, with the present computational equipment usually the Woods–Saxon is preferred in cases where the oscillator potential does not suffice.

All three potentials are sketched in Fig. 7.3. Since they are all spherically symmetric, the wave functions contain a factor $Y_{lm}(\Omega)$. For the harmonic oscillator in spherical coordinates the complete wave functions should be familiar from elementary quantum mechanics; they are given by

$$\psi_{nlm}(r,\Omega) = \sqrt{\frac{2^{n+l+2}}{n!(2n+2l+1)!!\sqrt{\pi}x_0^3}} \times \frac{r^l}{x_0^l} L_n^{l+1/2}(r^2/x_0^2)\, \mathrm{e}^{\mathrm{i}(-r^2/2x_0^2)}\, Y_{lm}(\Omega) \tag{7.123}$$

with $x_0 = \sqrt{\hbar/m\omega}$. The symbol $L_n^m(x)$ stands for the generalized Laguerre polynomial. The eigenenergies are determined by the principal quantum number $N = 2(n-1) + l$ as

$$E_N = \hbar\omega \left(N + \tfrac{3}{2}\right) \tag{7.124}$$

and are $\frac{1}{2}(N+1)(N+2)$-fold degenerate. The associated spectrum and occupation numbers (including spin) are indicated in Fig. 7.4. Clearly only the first three magic numbers are reproduced. A property of the spectrum that will be important later is that the shells of this model contain only states of the same parity, which is given by $(-1)^l$. The angular momentum is denoted by the letters $s, p, d, f, g, \ldots$ as in atomic physics.

For the square-well potential the wave functions have the form

$$\psi(r) \propto j_l(kr)\, Y_{lm}(\Omega) \quad , \quad r \leq R \quad , \tag{7.125}$$

and the energies follow from the matching condition $j_l(kR) = 0$, which can be solved only numerically. The resulting spectrum is also indicated in Fig. 7.4. It is less degenerate but still does not reproduce the experimental magic numbers.

Finally, for the Woods–Saxon potential, the radial wave functions can be obtained only numerically. The energies lie between the harmonic oscillator and the square-well results and also fail in comparison with experiment.

EXERCISE

7.4 The Eigenfunctions of the Harmonic Oscillator

Problem. Determine the eigenfunctions of the harmonic oscillator in Cartesian and cylindrical coordinates.

Solution. In Cartesian coordinates the Schrödinger equation is

Exercise 7.4

$$\left[-\frac{\hbar^2}{2m}\left(\frac{\partial^2}{\partial x^2}-\frac{\partial^2}{\partial y^2}-\frac{\partial^2}{\partial z^2}\right)+\frac{\omega^2}{2}\left(x^2+y^2+z^2\right)\right]\psi(x,y,z)$$
$$= E\,\psi(x,y,z) \quad . \tag{1}$$

As the Hamiltonian in this case is simply the sum of three one-dimensional harmonic-oscillator Hamiltonians, separation of variables is trivial and the solution is simply a product of one-dimensional harmonic-oscillator wave functions

$$\phi_n(\xi) = N_n\,\mathrm{e}^{-\frac{1}{2}\xi^2}\,\mathrm{H}_n(\xi) \quad , \quad N_n = \left(\sqrt{\pi}\,n!\,2^n\right)^{-1/2} \quad , \quad \xi = x\sqrt{\frac{m\omega}{\hbar}} \tag{2}$$

with oscillator quantum numbers n_x, n_y, and n_z,

$$\psi_{n_x n_y n_z}(x,y,z) = \phi_{n_x}(x)\,\phi_{n_y}(y)\,\phi_{n_z}(z) \quad . \tag{3}$$

This corresponds to the given number of excitations in the respective coordinate direction and an energy of

$$E_{n_x n_y n_z} = \hbar\omega\left(n_x+n_y+n_z+\tfrac{3}{2}\right) \quad . \tag{4}$$

Note that now all the states with a principal quantum number $N = n_x + n_y + n_z$ are degenerate.

In cylindrical coordinates the problem is a bit more complicated. The Schrödinger equation takes the form

$$\left[-\frac{\hbar^2}{2m}\left(\frac{\partial^2}{\partial z^2}+\frac{\partial^2}{\partial \rho^2}+\frac{1}{\rho}\frac{\partial}{\partial \rho}+\frac{\partial^2}{\partial \phi^2}\right)+\frac{\omega^2}{2}\left(z^2+\rho^2\right)\right]\psi(z,\rho,\phi)$$
$$= E\,\psi(z,\rho,\phi) \quad . \tag{5}$$

The usual separation of variables $\psi(z,\rho,\phi) = \xi(z)\,\chi(\rho)\,\eta(\phi)$ leads to

$$0 = \left(\frac{\mathrm{d}^2}{\mathrm{d}\phi^2}+\mu^2\right)\eta(\phi)$$
$$0 = \left(\frac{\mathrm{d}^2}{\mathrm{d}\rho^2}+\frac{1}{\rho}\frac{\mathrm{d}}{\mathrm{d}\rho}-\frac{m^2\omega^2}{\hbar^2}\rho^2-\frac{\mu^2}{\rho^2}+\frac{2mA}{\hbar^2}\right)\chi(\rho) \tag{6}$$
$$0 = \left(\frac{\mathrm{d}^2}{\mathrm{d}z^2}+-\frac{m\omega^2}{\hbar^2}z^2+\frac{2m(E-A)}{\hbar^2}\right)\zeta(z) \quad .$$

Here μ^2 and A are separation constants. It is of course not surprising that because of axial symmetry the azimuthal quantum number $\mu = 0, \pm 1, \pm 2, \ldots$ appears in the ϕ-dependent part

$$\eta_m(\phi) = \frac{1}{\sqrt{2\pi}}\mathrm{e}^{\mathrm{i}m\phi} \quad . \tag{7}$$

It was denoted as μ instead of m to distinguish it from the nucleon mass. The z-dependent equation is again that of a simple one-dimensional harmonic oscillator,

Exercise 7.4

so that we find the solution as given above for the Cartesian case with a quantum number n_z. The eigenenergy is given by

$$E = \hbar\omega\left(n_z + \tfrac{1}{2}\right) + A \quad . \tag{8}$$

Various methods are available for solving the ρ-dependent equation. The procedure usually followed in quantum-mechanics textbooks is to look at the asymptotic behavior of the wave functions near the origin and at infinity, extract suitable functional factors, and then (hopefully) find that the rest is given by polynomials that must satisfy a cutoff condition in order to obtain a normalizable wave function. In practice, whenever it is expected that an analytical solution does exist at all, one can abbreviate part of these steps by looking into mathematical handbooks.

For $\rho \to 0$ the equation becomes

$$0 = \left(\frac{\mathrm{d}^2}{\mathrm{d}\rho^2} + \frac{1}{\rho}\frac{\mathrm{d}}{\mathrm{d}\rho} - \frac{\mu^2}{\rho^2}\right)\chi(\rho) \quad , \tag{9}$$

and this may be solved with $\chi(\rho) = \rho^\alpha$; inserting yields $\alpha = |\mu|$ (the negative sign has to be discarded since the wave function must go to zero at the origin. For $\rho \to \infty$, on the other hand, the equation approaches

$$0 = \left(\frac{\mathrm{d}^2}{\mathrm{d}\rho^2} + \frac{1}{\rho}\frac{\mathrm{d}}{\mathrm{d}\rho} - \frac{m^2\omega^2}{\hbar^2}\rho^2\right)\chi(\rho) \quad , \tag{10}$$

and the experience with the one-dimensional oscillator suggests a trial with $\chi(\rho) = \exp(-\beta\rho^2)$. Substitution yields the condition

$$\left(4\beta^2 - \frac{m^2\omega^2}{\hbar^2}\right)\rho^2 - 4\beta = 0 \quad , \tag{11}$$

so that in the limit we get $\beta = m\omega/2\hbar$. Introducing the abbreviation $k = m\omega/\hbar$ we will thus use the decomposition

$$\chi(\rho) = \mathrm{e}^{-\frac{1}{2}k\rho^2}\,\rho^{|\mu|}\,L(\rho) \quad . \tag{12}$$

At this point we can abbreviate things by looking for a suitable orthogonal polynomial. Instead of trying to find one satisfying the differential equation remaining after the substitution (actually in this case it does not work, because it will turn out that the polynomial depends on $k\rho^2$, not on ρ alone), one may look in the tables for one fulfilling an appropriate orthogonality condition. The physical normalization is

$$\int_0^\infty \mathrm{d}\rho\,\rho\,\mathrm{e}^{-k\rho^2}\,\rho^{2\mu}\,L^2(\rho) = 1 \quad . \tag{13}$$

Among orthogonal polynomials the only ones defined on an infinite interval are the Laguerre and Hermite polynomials; the orthogonality relation that is found to be similar is that of the generalized Laguerre polynomials $L_n^\alpha(x)$ given by

$$\int_0^\infty \mathrm{d}x\,\mathrm{e}^{-x}\,x^\alpha\,L_n^\alpha(x)^2 = 1 \quad . \tag{14}$$

This suggests identifying $x \longrightarrow k\rho^2$ and $\alpha \longrightarrow |\mu|$, so that the final trial function is

$$\chi(\rho) = \mathrm{e}^{-\frac{1}{2}k\rho^2}\, \rho^{|\mu|}\, L_n^{|\mu|}(k\rho^2) \quad . \tag{15}$$

Inserting this into the differential equation and reducing everything to the new variable x produces

$$x\frac{\mathrm{d}^2}{\mathrm{d}x^2}L_n^{|\mu|}(x) + (|\mu| + 1 - x)\frac{\mathrm{d}}{\mathrm{d}x}L_n^{|\mu|}(x) - \left(\frac{|\mu|+1}{2} + \frac{mA}{2k\hbar^2}\right)L_n^{|\mu|}(x) = 0 \quad , \tag{16}$$

which can be compared with the differential equation for the generalized Laguerre polynomials

$$x\frac{\mathrm{d}^2}{\mathrm{d}x^2}L_n^{\alpha}(x) + (\alpha + 1 - x)\frac{\mathrm{d}}{\mathrm{d}x}L_n^{\alpha}(x) + n\,L_n^{\alpha}(x) = 0 \quad , \tag{17}$$

yielding not only the same identification for α, but also a condition for A:

$$A = \frac{\hbar^2 k}{m}(2n + |\mu| + 1) = \hbar\omega(2n + |\mu| + 1) \quad . \tag{18}$$

The new quantum number n can take values from $0, \ldots, \infty$. In order to make clear that it is associated with the ρ direction, we denote it by n_ρ from now on.

What this argument lacks, of course, is the proof that these are the only normalizable solutions; to see this one would have had to convert the differential equation into one of hypergeometric type and then show that the only normalizable solution is one with a special choice of the coefficients leading just to the Laguerre polynomials.

To summarize the results, the eigenfunctions of the harmonic oscillator in cylindrical coordinates are given by

$$\psi_{n_z n_\rho \mu}(z, \rho, \phi) = N \exp\left[-\tfrac{1}{2}k^2(z^2 + \rho^2)\right] \mathrm{H}_{n_z}(kz)\, \rho^{|\mu|}\, L_{n_\rho}^{|\mu|}(k\rho^2)\, \mathrm{e}^{\mathrm{i}\mu\phi} \tag{19}$$

with N an unspecified normalization constant that can be determined easily with the help of the integral formulas given, and $k = m\omega/\hbar$. The energy of the levels is given by

$$E = \hbar\omega\left(n_z + 2n_\rho + |\mu| + \tfrac{3}{2}\right) \quad . \tag{20}$$

Note that the number of "quanta" n_ρ in the ρ direction counts twice in the energy formula because it contains two oscillator directions, and that the angular-momentum projection contributes to the energy because of the centrifugal potential.

Of course the degeneracy of the levels is the same independent of the coordinate system used, and the principal quantum number N can be split up in three ways:

$$N = n_x + n_y + n_z = n_z + 2n_\rho + |\mu| = 2n + l \quad . \tag{21}$$

The various coordinate systems are useful in different situations; we will see immediately that the spherical basis makes the spin–orbit coupling diagonal, while deformed nuclei are often treated more simply in the cylindrical basis.

The basic insight that made the phenomenological single-particle model a viable tool in nuclear physics was the inclusion of a strong spin-orbit force by Goeppert-Mayer and Jensen [Ma55]. It couples the spin and orbital angular momentum of each individual nucleon and so corresponds to the jj-coupling limit of atomic theory. The additional term in the single-particle potential is thus

$$C\,\hat{\boldsymbol{l}}\cdot\hat{\boldsymbol{s}} \quad . \tag{7.126}$$

In the spherical case this term is diagonal and its value can be computed from

$$\hat{\boldsymbol{l}}\cdot\hat{\boldsymbol{s}} = \tfrac{1}{2}\left[\left(\hat{\boldsymbol{l}}+\hat{\boldsymbol{s}}\right)^2 - \hat{\boldsymbol{l}}^2 - \hat{\boldsymbol{s}}^2\right] = \tfrac{1}{2}\hbar^2[j(j+1) - l(l+1) - s(s+1)] \quad . \tag{7.127}$$

Of immediate interest is the splitting of the two levels with $j = l \pm \frac{1}{2}$. It is given by

$$E_{j=l+1/2} - E_{j=l-1/2} = C\,\hbar^2\left(l + \tfrac{1}{2}\right) \quad . \tag{7.128}$$

Experimentally one finds that the state with $j = l + \frac{1}{2}$ is lower in energy, so that the spin–orbit coupling term must have a negative sign.

There is still the possibility to have an r-dependent coefficient C. For the purely phenomenological approach the experimental data do not provide enough information to fix such an r dependence, so that usually a constant is assumed. Attempts to derive a spin–orbit coupling from microscopic considerations have been successful only in the relativistic meson-field formulation (see Sect. 7.4); the reativistic spin–orbit coupling inherent in the Dirac equation turns out to be too small.

Typical values for $|C|$ are in the range of 0.3 to 0.6 MeV$/\hbar^2$ (for a more detailed parametrization see the next section), and a spectrum for the harmonic oscillator with spin–orbit coupling is shown in Fig. 7.5. The levels are labelled by the radial quantum number n, orbital angular momentum l, and total angular momentum j as nlj. Each of these states is degenerate $2j+1$-fold with projections $\Omega = -j, \ldots, +j$. Clearly the magic numbers are now described correctly!

Let us now discuss the applications of this model and its confirmation by experimental data. The properties of the single-particle states themselves can be examined through *pickup and stripping reactions*, which allow the determination of the binding energy and angular momentum of the particles near the Fermi level. The experiments of this type tend to agree with the predictions of the model near closed shells.

The ground states of nuclei should be constructed by filling the single-particle levels up to the Fermi level. In principle the Coulomb potential should be added for the protons. Since for stable nuclei the Fermi levels for protons and neutrons should be equal (otherwise beta decay makes them equal), it lifts the single-particle states, so that the Fermi level for the protons should be at the same position as that for the neutrons, although these are more particles. Usually the influence of the Coulomb potential on the spectrum is ignored, so that the same spectrum is used and simply filled to a different Fermi level. Note, however, that for more sophisticated parametrizations (see the next section) the spin–orbit force is somewhat different for the two types of particles, so that the level schemes also differ in detail.

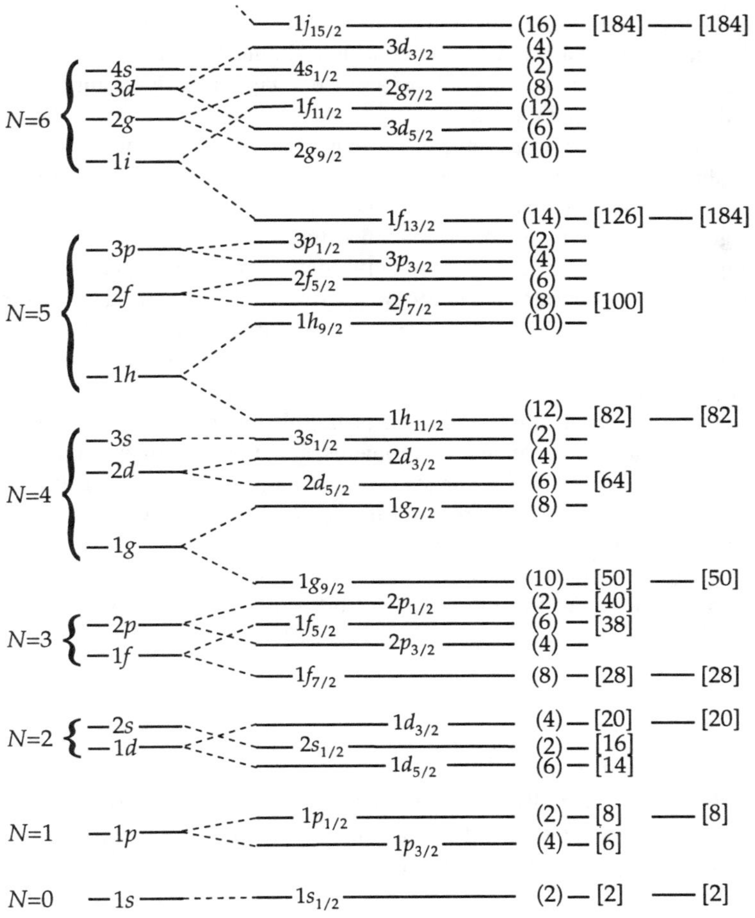

Fig. 7.5 Level scheme for a harmonic oscillator with spin–orbit coupling. The states of the pure harmonic oscillator plotted on the left split up to form the structure on the right. The number of particles in each level and the total number are indicated as in Fig. 7.3.2. On the extreme right the magic numbers corresponding to major shell closures are also given.

We can distinguish two types of shell closures. If all the projections Ω belonging to the same j state are filled, we call this a *closed j shell*. If there is a larger gap in the level scheme to the next unfilled j shell, i.e., if a magic number is reached, this corresponds to a major shell closure. In both cases the angular momentum of the nucleus should be zero. To see this simply note that a filled j shell must have vanishing angular momentum, because acting on it with any angular-momentum operator only transforms the substates with different projections amongst themselves, leaving the many-body state invariant.

We thus have as the first consequence that all nuclei corresponding to filled j shells in both protons and neutrons must have angular momentum $I = 0$. This agrees completely with experiment.

The next simplest case is that of adding a single nucleon to such a nucleus. In this case the nuclear angular momentum should result from the angular momentum of this single particle alone. This also turns out to be true *near the magic numbers.* The case of one hole in a j shell may be treated similarly.

If there is more than one nucleon in a partially filled j shell, the model predicts that all the angular-momentum couplings should be degenerate. For example, two particles in a j shell can couple to a nuclear angular momentum of $I = 0, 2, \ldots, 2j$ and all of these possibilities should lead to a degenerate ground state. Experimentally this is not true: the nuclear ground state is never degenerate, and this must be due to the residual interaction, which lifts the degeneracy of the pure single-particle model. One can give certain rules (such as the *Nordheim rules* as to which angular momentum is preferred. Since these are strictly outside the scope of the single-particle model, we do not give details and only note one example which will become important later: for two particles of the same isospin in a j shell the angular momenta always couple to zero, and in this case the residual interaction is of the *pairing* type (see Sect. 7.5). In addition we will show that $I = 0$ holds for all even–even nuclei.

A further interesting prediction of the phenomenological single-particle model concerns the nuclear magnetic moments. For the case of one nucleon outside a j shell it is expected that it carries the total magnetic moment of the nucleus. Since the details are not too interesting in this context we only mention that the results agree with experiment only near major shell closures.

Of more immediate interest is the question of the quadrupole moment. Again for one particle outside a j shell it should be determined only by the wave function of this particle. The result (see Exercise 7.5) is

$$Q_0 = -e\overline{r^2}\,\frac{2j-1}{2j+1} \quad . \tag{7.129}$$

This implies that Q_0 should vanish whenever $j = \frac{1}{2}$ and should be negative otherwise. Experimentally one finds that for most nuclei the values are positive and much (up to a hundred times) larger. Reasonable agreement with experiment is achieved only near major shell closures.

EXERCISE

7.5 The Quadrupole Moment of a Nucleus

Problem. Evaluate the quadrupole moment of a nucleus due to a particle in a state n, j, l.

Solution. Writing the wave function as

$$\psi_{nlj\Omega} = f_n(r) \sum_{ms} \left(l\tfrac{1}{2}j\,|ms\Omega\right) Y_{lm}(\theta,\phi)\,\chi_s \tag{1}$$

with some unspecified radial function $f_n(r)$, we can insert it into the definition of the quadrupole moment

Exercise 7.5

$$
\begin{aligned}
Q_0 &= e\sqrt{\frac{16\pi}{5}}\langle njl\Omega = j|r^2 Y_{20}(\theta,\phi)|njl\Omega = j\rangle \\
&= e\sqrt{\frac{16\pi}{5}}\left(\int \mathrm{d}r\, r^2 f_n^2(r)\, r^2\right) \\
&\quad\times \sum_{mm'ss'} \left(l\tfrac{1}{2}j|msj\right)\left(l\tfrac{1}{2}j|m's'j\right) \chi_s^\dagger \chi_{s'} \int \mathrm{d}(\cos\theta)\,\mathrm{d}\phi\; Y_{lm}^* Y_{20} Y_{lm'} \quad . \qquad (2)
\end{aligned}
$$

The arguments for the spherical harmonics are (θ, ϕ) in all cases and will also be suppressed below. The radial integral is just the mean square radius $\overline{r^2}$ of the wave function, and the angular integral can be evaluated using

$$
\int \mathrm{d}(\cos\theta)\,\mathrm{d}\phi\; Y_{lm}^* Y_{LM} Y_{l'm'} = (l'Ll|m'Mm)\sqrt{\frac{(2l'+1)(2L+1)}{4\pi(2l+1)}}\,(l'Ll|000)\,, \qquad (3)
$$

yielding for this special case

$$
\begin{aligned}
\int \mathrm{d}(\cos\theta)\,\mathrm{d}\phi\; Y_{lm}^* Y_{20} Y_{lm'} &= (l2l|m'0m)\sqrt{\frac{5}{4\pi}}(l2l|000) \\
&= \delta_{m'm}\,(l2l|m0m)\,(l2l|000)\,\sqrt{\frac{5}{4\pi}} \quad . \qquad (4)
\end{aligned}
$$

Together with the orthogonality of the spinors, $\chi_s^\dagger \chi_{s'} = \delta_{ss'}$, we get the result

$$
Q_0 = 2e\overline{r^2}\sum_{ms}\left(l\tfrac{1}{2}j|msj\right)^2 (l2l|m0m)\,(l2l|000) \qquad (5)
$$

and upon inserting the Clebsch–Gordan coefficients

$$
Q_0 = -e\overline{r^2}\,\frac{2j-1}{2j+1} \qquad (6)
$$

for both $j = l \pm \frac{1}{2}$.

Finally a calculation of the electromagnetic transition probabilities necessarily leads to results close to the Weisskopf estimates; again the experimental values off the major shell closures tend to be much larger, as we have already seen in connection with the collective model.

In summary one may say that the spherical phenomenological single-particle model succeeds in the explanation of the magic shell closures and of the properties of nuclei nearby, but fails for the nuclei between major shell closures. If we look back at the discussion of rotational states in Sect. 6.3–4 it appears that the model must be modified to account for nuclear deformations. This will be discussed in the next section.

7.3.2 The Deformed-Shell Model

The generalization of the phenomenological shell model to deformed nuclear shapes was first given by S. G. Nilsson [Ni55] and also in [Go56, Mo55], so this version is also often referred to as the *Nilsson model.* The principal idea is to make the oscillator constants different in the different spatial directions:

$$V(\boldsymbol{r}) = \frac{m}{2}\left(\omega_x^2 x^2 + \omega_y^2 y^2 + \omega_z^2 z^2\right) \quad . \tag{7.130}$$

Before finding a more convenient parametrization, consider for a moment what the associated nuclear shape will be. For the spherical-oscillator shell model we had seen that the nuclear surface has a constant potential value independent of A. In the deformed case the density distribution should follow the potential in the same way, so that we may define a geometric nuclear surface as consisting of all the points (x, y, z) with

$$\tfrac{1}{2} m \bar{\omega}_0^2 R^2 = \frac{m}{2}\left(\omega_x^2 x^2 + \omega_y^2 y^2 + \omega_z^2 z^2\right) \quad , \tag{7.131}$$

where $\hbar\bar{\omega}_0 = 41\,\mathrm{MeV} \times A^{-1/3}$ is the oscillator constant for the equivalent spherical nucleus. This describes an ellipsoid with axes X, Y, and Z given by

$$\bar{\omega}_0 R = \omega_x X = \omega_y Y = \omega_z Z \quad . \tag{7.132}$$

The condition of incompressibility of nuclear matter requires that the volume of the ellipsoid be the same as that of the sphere, implying $R^3 = XYZ$, and this imposes a condition on the oscillator frequencies:

$$\bar{\omega}_0^3 = \omega_x \omega_y \omega_z \quad . \tag{7.133}$$

Now assume axial symmetry around the z axis, i.e., $\omega_x = \omega_y$, and a small deviation from the spherical shape given by a small parameter δ. We define

$$\omega_x^2 = \omega_y^2 = \omega_0^2\left(1 + \tfrac{2}{3}\delta\right) \quad , \quad \omega_z^2 = \omega_0^2\left(1 - \tfrac{4}{3}\delta\right) \quad , \tag{7.134}$$

which clearly fulfills the volume conservation condition of (7.133) to first order with $\omega_0 = \bar{\omega}_0$. Volume conservation can be fulfilled to second order using

$$\bar{\omega}_0^6 = \left(1 - \tfrac{4}{3}\delta\right)\left(1 + \tfrac{2}{3}\delta\right)^2 \omega_0^6 \quad , \tag{7.135}$$

leading in second order to

$$\omega_0 \approx \left(1 + \tfrac{2}{9}\delta^2\right)\bar{\omega}_0 \quad . \tag{7.136}$$

Utilizing the explicit expression for the spherical harmonic Y_{20}, we may also write the potential as

$$V(\boldsymbol{r}) = \tfrac{1}{2} m \omega_0^2 r^2 - \beta_0 m \omega_0^2\, r^2\, Y_{20}(\theta, \phi) \quad , \tag{7.137}$$

where β_0 is related to δ via

$$\beta_0 = \frac{4}{3}\sqrt{\frac{4\pi}{5}}\,\delta \quad . \tag{7.138}$$

It is important to realize that the deformed shapes obtained in this way are different from those in the collective model, where the radius and not, as here, the potential, was expanded in spherical harmonics (see Sect. 6.1.1). The difference becomes pronounced for large deformations, where the collective-model shapes develop a neck and even separate: at $\theta = \frac{\pi}{2}$ we have the largest negative value $Y_{20}\left(\frac{\pi}{2}, \phi\right) = -\frac{1}{2}\sqrt{5/4\pi}$, so that

$$R\left(\tfrac{\pi}{2}, \phi\right) = R_0\left(1 + \alpha_{20}\left(\tfrac{\pi}{2}, \phi\right)\right) \tag{7.139}$$

becomes zero for $\alpha_{20} = 2\sqrt{4\pi/5}$ there, producing a fissioned shape. In contrast, the shapes in the deformed-shell model always remain ellipsoidal, albeit with arbitrarily long stretching.

We can now write down the Hamiltonian of the model:

$$\begin{aligned}\hat{H} = & -\frac{\hbar}{2m}\nabla^2 + \frac{m\omega_0^2}{2}\, r^2 - \beta_0 m\omega_0^2\, r^2\, Y_{20}(\theta, \phi) \\ & - \hbar\bar{\omega}_0\kappa\left(2\,\hat{\boldsymbol{l}}\cdot\hat{\boldsymbol{s}} + \mu\,\hat{\boldsymbol{l}}^2\right) \quad .\end{aligned} \tag{7.140}$$

The spin–orbit term is conventionally parametrized with a constant κ, and there is also a new $\hat{\boldsymbol{l}}^2$ term parametrized by μ, which is introduced phenomenologically to lower the energy of the single-particle states closer to the nuclear surface in order to correct for the steep rise in the harmonic-oscillator potential there. Both κ and μ may be different for protons and neutrons and also depend on the nucleon number, and there are various parametrizations in the literature assuming a variation with A or $A^{1/3}$, or with the major shell [Gu66, Ni69, Se67]. The details are very important when extrapolating the model to superheavy nuclei; we only mention here that the values are of the order of 0.05 for κ and 0.3 for μ.

The Hamiltonian may be diagonalized in the basis of the harmonic oscillator using either spherical or cylindrical coordinates depending on the application. In spherical coordinates the spin–orbit and $\hat{\boldsymbol{l}}^2$ terms are diagonal, but the Y_{20} term couples orbital angular momenta differing by ± 2 (this coupling was neglected in Nilsson's original work owing to computer limitations of the time). In cylindrical coordinates, on the other hand, the deformed oscillator potential is diagonal and the angular-momentum terms must be diagonalized numerically. In any case, with modern computing resources neither is a problem.

It is worthwhile to study the quantum numbers resulting from both approaches (for details see Exercise 7.4). Consider first the spherical oscillator without spin effects. The energy levels in the spherical basis are given by

$$\varepsilon = \hbar\omega_0\left(N + \tfrac{3}{2}\right) \tag{7.141}$$

with the principal quantum number $N = 2(n_r - 1) + l$, radial quantum number n_r, angular-momentum quantum number l, and projection m. In the cylindrical basis they are replaced by

$$\varepsilon = \hbar\omega_z\left(n_z + \tfrac{1}{2}\right) + \hbar\omega_\rho\left(2n_\rho + |m| + 1\right) \quad , \tag{7.142}$$

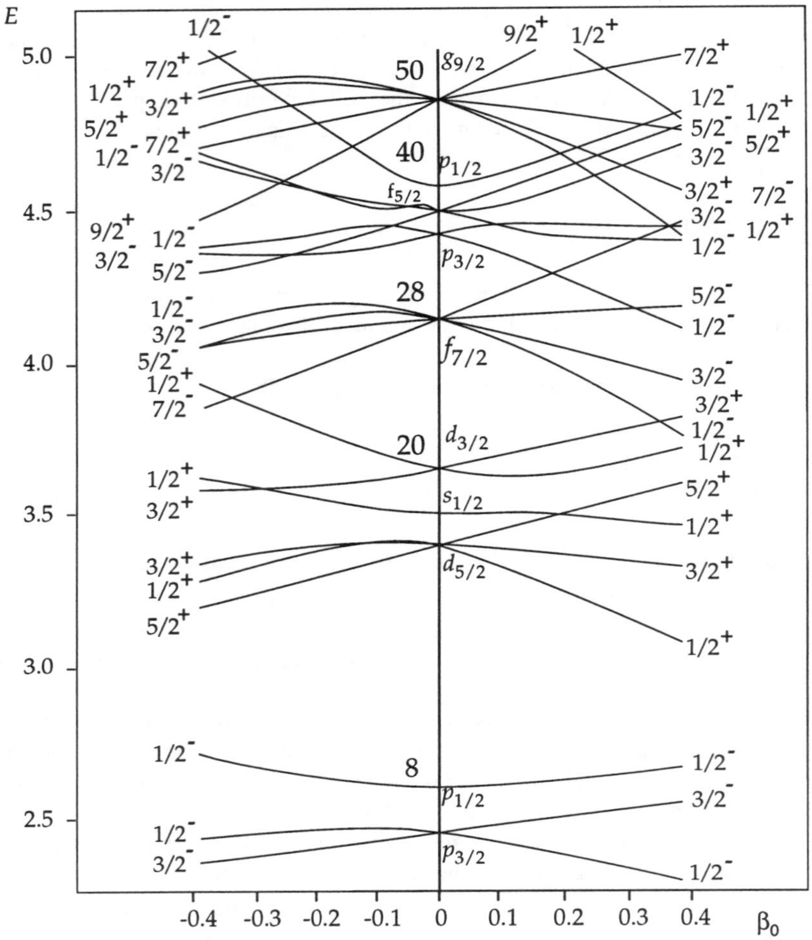

Fig. 7.6 Lowest part of the level diagram (Nilsson diagram) for the deformed shell model. The single-particle energies are plotted as functions of deformation and are given in units of $\hbar\bar{\omega}_0$. The quantum numbers Ω^π for the individual levels and l_j for the spherical ones are indicated as are the magic numbers for the spherical shape.

where n_z is the number of quanta in the z direction, n_ρ is that of radial excitations, and m is again the angular momentum projection on the z axis. For the spherical shape the levels will be grouped according to the principal quantum number N (with the splitting by the spin–orbit force then determined through the total angular momentum j), but the behavior with deformation depends on how much of the excitation is in the z direction. For prolate deformation, the potential becomes shallower in this direction, and the energy contributed by n_z excitations decreases. The cylindrical quantum numbers are thus helpful in understanding the splitting for small deformations.

For very large deformations, on the other hand, the influence of the spin–orbit and $\hat{l}^2$ terms becomes less important and one may classify the levels according to the cylindrical quantum numbers. It has thus become customary to label the single-particle levels with the set $\Omega^\pi[Nn_z m]$. Ω, the projection of total angular momentum, and the parity π are good quantum numbers while N, n_z, and m are only approximate and may be determined for a given level only by looking at

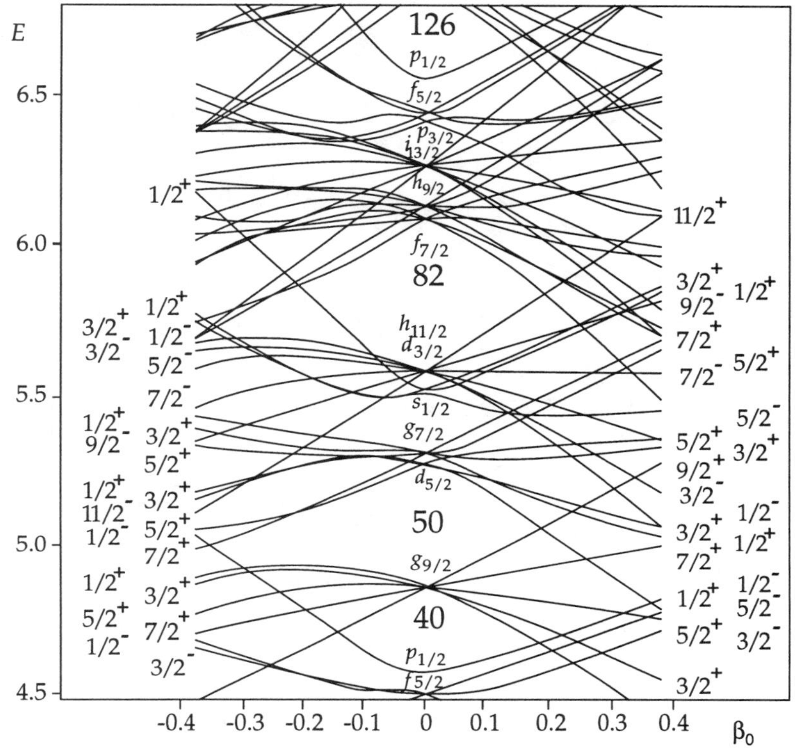

Fig. 7.7 Upper part of the Nilsson diagram. Inidividual quantum numbers are indicated only up to the spherical shell at magic number 82.

its behavior near the spherical state or, for m, at large deformation (in practice a calculation will show what the dominant basis function is in the expansion of the diagonalized state).

Exercise 7.6 explores how the calculation is actually carried out in the cylindrical basis.

Figures 7.6–7 show the resulting levels as functions of deformation. At first sight this *Nilsson diagram* appears to be a confusing mixture of intersecting levels, yet a number of interesting features can be observed. The highly degenerate spherical levels split up into the individual state pairs characterized by $\pm\Omega$ and the parity, which is determined by the orbital angular momentum in the case of a spherical shape. For the spherical levels the magic numbers and the conventional nomenclature for orbital and total angular momentum are also indicated.

The projection $|\Omega|$ and the parity are indicated for all the levels arising from spherical multiplets below the magic number 82. The way in which the levels diverge can be understood quite easily: states with a larger projection should have smaller quantum number n_z, so that for oblate deformation, where the frequency in the z direction increases, they are lowered with respect to the other states and the opposite happens for prolate deformation.

This systematic behavior of the levels is made more complicated by *avoided energy-level crossings*. As a general rule, levels with the same (exact) quantum numbers should never cross if they are plotted as functions of a single parameter.

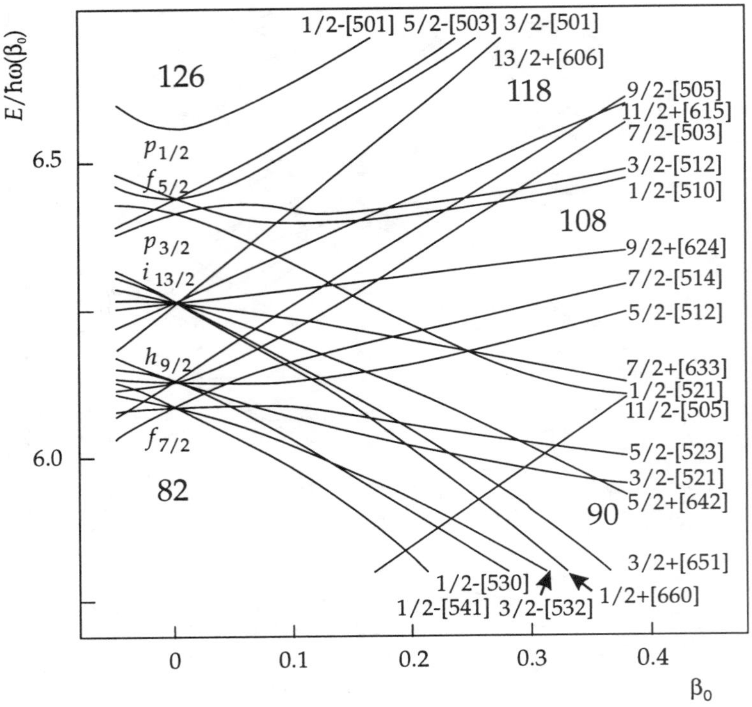

Fig. 7.8 An excerpt from the Nilsson diagram above the shell for Lead to show more detail. The single-particle states are labelled with the approximate quantum numbers $[Nn_zm]$ in addition to Ω^π, and the magic numbers are indicated both for sphericity and for deformed shell closures.

This was first noticed by von Neumann and Wigner [Ne29], who investigated how the number of degrees of freedom of a general Hermitian matrix is reduced if the matrix is constrained to have two equal eigenvalues; they found that it is in general not sufficient to vary a single parameter to reduce the matrix to this special case. Thus a degeneracy should always be caused by a symmetry which produces additional quantum numbers to distinguish the states.

If two levels with the same quantum numbers get close to each other, they are instead repelled as can be seen easily from the two-level problem demonstrated in Exercise 7.7. In the Nilsson diagram of Fig. 7.6 the effect is clearly observable, for example, for the $1/2^-$ level coming from the $f_{5/2}$ spherical multiplet below magic number 40. Going towards negative deformations it is first repelled by the $1/2^-$-level from the $p_{1/2}$ state above and then by the one from the $p_{3/2}$ below. It is not, however, forbidden from crossing the $9/2^+$ coming from above.

To show some more detail, Fig. 7.8 displays an excerpt from the Nilsson diagram above the shell for Lead. In this figure the levels are also labelled with the asymptotic quantum numbers in brackets according to $\Omega^\pi[Nn_zm]$, where N is the spherical principal oscillator quantum number, n_z is the number of phonons in the z direction, and m is the projection of the orbital angular momentum. The latter two become good quantum numbers in the limit of large deformation, where the spin–orbit coupling is negligible compared to the deformation in the oscillator potential. Note the slopes of the levels at large deformation, which clearly show the effect of decreasing frequency in the z direction, leading to an n_z-dependent decrease in the energy.

Another intersting point to note in this figure is the existence of gaps in the level scheme at larger deformation, which in some cases are comparable in magnitude to the minor spherical shell gaps. In these gaps the appropriate "magic numbers" are also indicated. More on the importance of these shell structures at larger deformations will be given in Chap. 9.

A problematic point in practice is the existence of different deformation parameters even within this one model. They reflect different ways of parametrization, depending on whether the potential is expanded in spherical harmonics (these parameters are often called ε, ε_2 corresponding to the quadrupole as discussed here and ε_4 to the hexadecupole), while others start with a ratio of axes or frequencies (δ, η, and so on). They are, of course, related quite trivially. Probably the best source of information on all the various nuclear shape parametrizations – collective and single-particle – is the one by Hasse and Myers [Ha88], which the reader is advised to consult for these problems.

One point that is of immediate importance is how the deformation parameter β_0 used above is related to that of the collective model with the same name but corresponding to α_{20}. As mentioned, the shapes allowed in the two models are quite different but agree sufficiently for small deformation to allow a comparison. One way would be to evaluate the quadrupole moments of the shapes and then identify deformation parameters leading to the same quadrupole moment. Instead we take the simpler way of comparing the axis ratios in the two models (to lowest order, of course, it leads to the same results).

In the collective model, the axis lengths R_z and $R_x = R_y$ are given by

$$\begin{aligned} R_z &= R_0\left(1+\alpha_{20}Y_{20}(\theta=0)\right) = R_0\left(1+\alpha_{20}\sqrt{\frac{5}{4\pi}}\right) \quad , \\ R_x &= R_0\left(1+\alpha_{20}Y_{20}(\theta=\tfrac{\pi}{2})\right) = R_0\left(1-\frac{1}{2}\alpha_{20}\sqrt{\frac{5}{4\pi}}\right) \quad , \end{aligned} \tag{7.143}$$

so that for the ratio we get to first order

$$\frac{R_z}{R_x} = \frac{1+\alpha_{20}\sqrt{\frac{5}{4\pi}}}{1-\frac{1}{2}\alpha_{20}\sqrt{\frac{5}{4\pi}}} \approx 1+\frac{3}{2}\alpha_{20}\sqrt{\frac{5}{4\pi}} \quad . \tag{7.144}$$

In the Nilsson model, on the other hand, the axis ratio is given by the relation

$$\tfrac{1}{2}m\omega_0^2\left(1-2\beta_0\,Y_{20}(\theta=0)\right)R_z^2 = \tfrac{1}{2}m\omega_0^2\left(1-2\beta_0\,Y_{20}(\theta=\tfrac{\pi}{2})\right)R_x^2 \quad , \tag{7.145}$$

from which, again to first order,

$$\frac{R_z^2}{R_x^2} = \frac{1+\frac{1}{2}\beta_0\sqrt{\frac{5}{\pi}}}{1-\beta_0\sqrt{\frac{5}{\pi}}} \approx 1+\frac{3}{2}\alpha_{20}\sqrt{\frac{5}{\pi}} \quad . \tag{7.146}$$

Taking the square root adds a factor of $\frac{1}{2}$ to the first-order term, so that the two expressinons agree to first order. This shows that the notation β_0 was appropriate to show the first-order equality to the β_0 of the collective model.

This type of level diagram is essential for understanding many of the properties of deformed nuclei. We will come back to the determination of the ground-state deformation in connection with the modern method of shell corrections in Sect. 9.2.2. Once the ground-state deformation is known, however, one can draw some conclusions about the angular momentum and parity of the nucleus. For even–even nuclei, pairing (see Sect. 7.5) lets all the nucleon pairs in the $\pm\Omega$ level pairs add up to a total angular momentum of zero, but in odd nuclei the angular momentum and parity of the state the single unpaired nucleon occupies determine the properties of the ground state. Thus at a fixed deformation one counts up the level scheme for protons or neutrons to find the level with the odd particle. Ideally the spin of the nucleus is then given by $I = \Omega$ and the parity is given by the parity of the single-particle state. In practice usually several levels are close together at that deformation: in this case one of them yields the ground-state properties, while the other quantum-number combinations can be found among the low-energy excitations.

In closing this chapter, let us mention that there are, of course, also generalizations of the Woods–Saxon and other realistic potentials for deformed shapes. This topic will be picked up again in Sect. 9.1 when we discuss large-scale collective motion.

EXERCISE

7.6 Single-Particle Energies in the Deformed Oscillator

Problem. Calculate the single-particle energies of the deformed-oscillator shell model, including spin–orbit coupling, but without the $\hat{l}^2$ term.

Solution. This problem cannot be solved completely analytically; but as it illustrates some of the most often used methods in nuclear structure calculations, we will present it as far as possible.

For the diagonalization of a Hamiltonian there are essentially two numeric approaches: the *basis expansion method*, which expands the unknown wave functions in a basis given by the eigenstates of a similar Hamiltonian, or the purely numeric approach, in which finite difference approximations or some other general numerical method such as splines are used. It is characteristic of the latter method that the expansion has little connection to the physical situation; it can usually be used for quite arbitrary problems. Although the increasing power of computers has made the latter methods more widespread, we will here discuss the basis expansion because of its greater physical content.

The choice of a basis is the crucial first step. For the spherical harmonic oscillator we have seen three different systems of basic functions based on Cartesian, cylindrical, and spherical coordinates, respectively. Each of these has certain advantages and drawbacks for the problem at hand.

- In Cartesian coordinates the oscillator potential is diagonal but the angular momentum does not yield any quantum number.

Exercise 7.6

- In cylindrical coordinates the oscillator potential is also diagonal, as long as axial symmetry around the z axis is maintained, and the angular-momentum projection μ is defined.
- In spherical coordinates the spin–orbit coupling is diagonal, but not the oscillator potential itself.

The choice of the basis set is thus a matter of whether the spin–orbit coupling or the deformation of the potential is more important. In practice this depends on deformation: near spherical shapes the spin–orbit coupling splits the levels much more than the deformation, while for large deformation the cylindrical basis is closer to the true states; this is apparent in the "asymptotic quantum numbers" discussed above. Here we will discuss the case of the cylindrical basis, because it is slightly more complicated and thus more interesting.

The cylindrical basis as discussed in Exercise 7.4 can be written in Dirac notation as $|n_z n_\rho \mu s\rangle$, where the spin projection s was added. Because the wave functions are products of functions of the different coordinates, we can also split them up as

$$|n_z n_\rho \mu s\rangle = |n_z\rangle |n_\rho \mu\rangle |\mu\rangle |s\rangle \quad , \tag{1}$$

where the kets on the right denote the z-, ρ-, ϕ-, and spin-dependent wave functions, respectively.

The Hamiltonian of the deformed shell model can be split into the part generating the basis functions and the spin–orbit part,

$$\hat{H} = \hat{H}_0 + \hat{H}_{\mathrm{so}} \tag{2}$$

with

$$\begin{aligned} \hat{H}_0 &= -\frac{\hbar^2}{2m}\left(\frac{\partial^2}{\partial z^2} + \frac{\partial}{\partial \rho^2} + \frac{1}{\rho}\frac{\partial}{\partial \rho} + \frac{1}{\rho^2}\frac{\partial}{\partial \phi^2}\right) + \tfrac{1}{2} m\omega_z^2 z^2 + \tfrac{1}{2} m\omega_\rho^2 \rho^2 \\ \hat{H}_{\mathrm{so}} &= -2\hbar\bar{\omega}_0 \kappa\, \hat{\boldsymbol{l}} \cdot \hat{\boldsymbol{s}} \\ &= -2\hbar\bar{\omega}_0 \kappa \left[\tfrac{1}{2}\left(\hat{L}_+ \hat{s}_- + \hat{L}_- \hat{s}_+\right) + \hat{L}_z \hat{s}_z\right] \quad . \end{aligned} \tag{3}$$

Here the difference of the axes in the z and ρ directions was expressed by different oscillator frequencies ω_z and ω_ρ; the spin–orbit potential was written in terms of angular-momentum shift operators because of their simpler matrix elements in a basis with good projection.

As promised, the operator $\hat{H}_0$ is diagonal in the basis:

$$\begin{aligned} &\langle n_z' n_\rho' \mu' s' | \hat{H}_0 | n_z n_\rho \mu s\rangle \\ &\quad = \left[\hbar\omega_z\left(n_z + \tfrac{1}{2}\right) + \hbar\omega_\rho(2n_\rho + |\mu| + 1)\right] \delta_{n_z' n_z} \delta_{n_\rho' n_\rho} \delta_{\mu' \mu} \delta_{s' s} \quad . \end{aligned} \tag{4}$$

So the nontrivial problem is the spin–orbit coupling. For this the first step is to transform the orbital angular-momentum operators into cylindrical coordinates. A lengthy calculation using the usual rules of partial differentiation yields

Exercise 7.6

$$\begin{aligned}\hat{L}_+ &= -\hbar e^{i\phi}\left(\rho\frac{\partial}{\partial z} - z\frac{\partial}{\partial\rho} - i\frac{z}{\rho}\frac{\partial}{\partial\phi}\right) \quad ,\\ \hat{L}_- &= \hbar e^{-i\phi}\left(\rho\frac{\partial}{\partial z} - z\frac{\partial}{\partial\rho} + i\frac{z}{\rho}\frac{\partial}{\partial\phi}\right) \quad ,\\ \hat{L}_z &= -i\hbar\frac{\partial}{\partial\phi} \quad .\end{aligned} \tag{5}$$

The spin operators have the simple matrix elements

$$\langle s'|\hat{s}_+|s\rangle = \hbar\,\delta_{s',s+1} \quad , \quad \langle s'|\hat{s}_-|s\rangle = \hbar\,\delta_{s',s-1} \quad , \quad \langle s'|\hat{s}_z|s\rangle = \hbar s\,\delta_{s's} \quad , \tag{6}$$

so that the first decomposition of the spin–orbit matrix element is

$$\begin{aligned}\langle n_z'n_\rho'\mu's'|\hat{H}_{\mathrm{so}}|n_zn_\rho\mu s\rangle = &-\hbar\bar{\omega}_0\kappa\langle n_z'n_\rho'\mu'|\hat{L}_+|n_zn_\rho\mu\rangle\delta_{s',s-1}\\ &-\hbar\bar{\omega}_0\kappa\langle n_z'n_\rho'\mu'|\hat{L}_-|n_zn_\rho\mu\rangle\delta_{s',s+1}\\ &-\hbar\bar{\omega}_0\kappa s\langle n_z'n_\rho'\mu'|\hat{L}_z|n_zn_\rho\mu\rangle\delta_{s's} \quad .\end{aligned} \tag{7}$$

The factor $\hbar$ from the spin operators was omitted (as will be the $\hbar$ contained in the orbital angular-momentum operators) because of the convention introduced by Nilsson [Ni55] to include these factors in the definition of κ.

The last matrix element is trivial because $\hat{L}_z$ is diagonal,

$$\langle n_z'n_\rho'\mu'|\hat{L}_z|n_zn_\rho\mu\rangle = \hbar\mu\delta_{n_z'n_z}\delta_{n_\rho'n_\rho}\delta_{s's} \quad , \tag{8}$$

while the others can be reduced to products of the matrix elements

$$\begin{aligned}&\langle n_z'|z|n_z\rangle \quad , \quad \langle n_z'|\frac{d}{dz}|n_z\rangle \quad , \quad \langle n_\rho'\mu\pm 1|\rho|n_\rho\mu\rangle \quad ,\\ &\langle n_\rho'\mu\pm 1|\frac{d}{d\rho}|n_\rho\mu\rangle \quad , \quad \langle n_\rho'\mu\pm 1|\frac{1}{\rho}|n_\rho\mu\rangle\end{aligned} \tag{9}$$

which we now have to calculate. Those for the z direction are, in fact, quite simple: since this is the straightforward one-dimensional harmonic oscillator, second quantization can be used. Expressing z and $\partial/\partial z$ in terms of creation and annihilation operators yields immediately

$$\begin{aligned}\langle n_z'|z|n_z\rangle &= \sqrt{\frac{\hbar}{2m\omega}}\left(\sqrt{n_z}\,\delta_{n_z',n_z-1} + \sqrt{n_z+1}\,\delta_{n_z',n_z+1}\right) \quad ,\\ \langle n_z'|\frac{d}{dz}|n_z\rangle &= \sqrt{\frac{m\omega}{2\hbar}}\left(\sqrt{n_z}\,\delta_{n_z',n_z-1} - \sqrt{n_z+1}\,\delta_{n_z',n_z+1}\right) \quad .\end{aligned} \tag{10}$$

For the radial coordinate the calculation is a bit more involved, and we have not even normalized the wave functions yet! Let us write the normalized eigenfunctions as

$$\chi_{n_\rho\mu}(\rho) = N_{n_\rho\mu}^{-1}\,e^{-\frac{1}{2}k\rho^2}\,\rho^{|\mu|}\,L_{n_\rho}^{|\mu|}(k\rho^2) \quad . \tag{11}$$

Looking into [Gr65a], one finds that few integrals with Laguerre polynomials are given, and even these are written as complicated expressions in Γ functions, hypergeometric functions, and so on. An alternative procedure to calculate the required

Exercise 7.6

integrals is as follows. Both the matrix elements and the normalization integral can be written as

$$K_{a,b} = \int_0^\infty \mathrm{d}\rho \, \chi_{n'_\rho\mu'}(\rho) \, \rho^a \, \frac{\mathrm{d}^b}{\mathrm{d}\rho^b} \chi_{n_\rho\mu}(\rho) \, \rho \quad . \tag{12}$$

Substituting $x = k\rho^2$ and inserting (11) leads to

$$K_{a,b} = \frac{2^{b-1} N^{-1}_{n'_\rho\mu'} N^{-1}_{n_\rho\mu}}{k^{(a+|\mu|+|\mu'|-b+2)/2}} \times \int_0^\infty \mathrm{d}x \, \mathrm{e}^{-x/2} \, x^{(|m'|+a+b)/2} \, L^m_{n'_\rho}(x) \, \frac{\mathrm{d}^b}{\mathrm{d}x^b} \left(\mathrm{e}^{-x/2} \, x^{|\mu|} \, L^m_{n_\rho}(x) \right) \quad . \tag{13}$$

We have available the orthogonality integral of the Laguerre polynomials,

$$I^m_{n'n} = \int_0^\infty \mathrm{d}x \, \mathrm{e}^{-x} \, x^m \, L^m_{n'}(x) \, L^m_n(x) = \frac{(n+m)!}{n!} \, \delta_{n'n} \tag{14}$$

(note that this applies to the same upper index of the two polynomials). The normalization integral can immediately be reduced to this integral for the special case $\mu = \mu'$, $n_\rho = n'_\rho$:

$$1 = K_{0,0} = \frac{N^{-2}_{n_\rho\mu}}{2k^{|m|+1}} I^\mu_{n_\rho n_\rho} = \frac{N^{-2}_{n_\rho\mu}}{2k^{|m|+1}} \frac{(n_\rho + |m|)!}{n_\rho!} \quad , \tag{15}$$

from which the normalization constant is

$$N_{n_\rho\mu} = \sqrt{\frac{(n_\rho + |\mu|)!}{2 n_\rho! k^{|\mu|+1}}} \quad . \tag{16}$$

The other integrals correspond to $a = 1$, $b = 0$, and $b = 1$, $a = 0$, respectively. Let us briefly indicate how to calculate them (take only the case with $\mu + 1$ on the left side of the matrix element, the other case can be converted into this by symmetry).

1. The integral containing one additional ρ can be rewritten using the recursion relation

$$L^m_n(x) = L^{m+1}_n(x) - L^{m+1}_{n-1}(x) \quad , \tag{17}$$

leading to

$$\frac{N^{-1}_{n'_\rho\mu+1} N^{-1}_{n_\rho\mu}}{2k^{m+2}} \int_0^\infty \mathrm{d}x \, \mathrm{e}^{-x} \, x^{|\mu|+1} \, L^{|\mu+1|}_{n'_\rho} \, L^{|\mu|}_{n_\rho}(x) = \delta_{n'_\rho n_\rho} \sqrt{k^{-1}(n_\rho + |\mu| + 1)} - \delta_{n'_\rho, n_\rho - 1} \sqrt{k^{-1} n_\rho} \quad . \tag{18}$$

2. The integral containing the derivative is a bit more complicated, because the derivative does not act solely on the Laguerre polynomial. Carrying out the differentiation produces a sum of three terms:

Exercise 7.6

$$
\begin{aligned}
&\langle n'_\rho \mu + 1 | \frac{\mathrm{d}}{\mathrm{d}\rho} | n_\rho \mu \rangle \\
&= -k \langle n'_\rho \mu + 1 | \rho | n_\rho \mu \rangle + |\mu| \langle n'_\rho \mu + 1 | \frac{1}{\rho} | n_\rho \mu \rangle \\
&+ \frac{N^{-1}_{n'_\rho \mu+1} N^{-1}_{n_\rho \mu}}{2k^{|\mu|+1}} \int\limits_0^\infty \mathrm{d}x \, \mathrm{e}^{-x} x^{|\mu|+1} \, L^{|\mu+1|}_{n'_\rho} \frac{\mathrm{d}}{\mathrm{d}x} L^{|\mu|}_{n_\rho}(x) \quad . \qquad (19)
\end{aligned}
$$

The first two matrix elements correspond to points 1 and 2, respectively, while using the formula

$$
\frac{\mathrm{d}}{\mathrm{d}x} L^m_n(x) = -L^{m+1}_{n-1}(x) \quad , \qquad (20)
$$

reduces the last integral to $-2\delta_{n'_\rho, n_\rho - 1}\sqrt{k n_\rho}$.

3. The integral containing $1/\rho$ need not actually be evaluated, because it drops out in the summation of the various parts of the total matrix element (the term appearing explicitly cancels with the one resulting from point 2).

It remains for us to multiply the matrix elements for the different coordinate directions and then to add all the terms. In this calculation it turns out to be advantageous to introduce a new quantum number $N_\rho = 2n_\rho + |\mu|$. The reason is that in the matrix elements both $|\mu|$ and μ appear and this new notation allows the combination of the different terms. The final result is

$$
\begin{aligned}
\langle n'_z N'_\rho \mu' s' | \hat{L} \cdot \hat{s} | n_z N_\rho \mu s \rangle = {} & \delta_{\mu', \mu+1} \delta_{s', s-1} \\
& \times \frac{1}{2} \left[\delta_{n'_z, n_z - 1} \delta_{N'_\rho, N_\rho + 1} \sqrt{\frac{n_z (N_\rho + \mu + 2)}{2}} \right. \\
& \left. + \delta_{n'_z, n_z + 1} \delta_{N'_\rho, N_\rho - 1} \sqrt{\frac{(n_z + 1)(N_\rho - \mu)}{2}} \right] \\
& + \delta_{\mu', \mu-1} \delta_{s', s+1} \\
& \times \frac{1}{2} \left[\delta_{n'_z, n_z - 1} \delta_{N'_\rho, N_\rho + 1} \sqrt{\frac{(n_z)(N_\rho - \mu + 2)}{2}} \right. \\
& \left. + \delta_{n'_z, n_z + 1} \delta_{N'_\rho, N_\rho - 1} \sqrt{\frac{(n_z + 1)(N_\rho + \mu)}{2}} \right] \\
& + s \mu \delta_{n'_z n_z} \delta_{N'_\rho N_\rho} \delta_{\mu' \mu} \delta_{s' s} \quad . \qquad (21)
\end{aligned}
$$

Although this is a lengthy expression, note that the selection rules are quite restrictive: each term contributes in a different combination of quantum numbers on the left and right, and the quantum numbers cannot be different by more than 1 in each case. So the actual matrix is quite sparse.

At this point numerical calculation has to take over, and we can only indicate how this would have to be done. The first step is to set up the set of basis functions. The only good quantum number is $\Omega = \mu + s$, so states of different Ω are not coupled by the Hamiltonian (check that the selection rules in the spin–orbit term preserve this!). In addition, the states with $\pm\Omega$ are degenerate (Kramers degeneracy,

Table 7.2. Quantum numbers of the harmonic oscillator states with $N < 5$ in the cylindrical coordinate basis.

No.	$n_z + N_\rho$	n_z	N_ρ	n_ρ	μ	s
1	0	0	0	0	0	+1/2
2	1	1	0	0	0	+1/2
3	2	2	0	0	0	+1/2
4	2	0	2	1	0	+1/2
5	3	3	0	0	0	+1/2
6	3	1	2	1	0	+1/2
7	4	4	0	0	0	+1/2
8	4	2	2	1	0	+1/2
9	4	0	2	1	0	+1/2
10	1	0	1	0	1	−1/2
11	2	1	1	0	1	−1/2
12	3	2	1	0	1	−1/2
13	3	0	3	1	1	−1/2
14	4	3	1	0	1	−1/2
15	1	1	3	1	1	−1/2

see Sect. 2.6). How should the infinite set of states be cut off? In principle there is no prescribed method, one should examine how the perturbation couples the basis states. A good choice is usually an energy cutoff, which in the present case corresponds to allowing the same number of quanta in all directions, or explicitly $n_z + N_\rho < N_{\text{cutoff}}$ (this may be a bad choice if the nucleus is strongly deformed).

Let us set up the basis for $\Omega = \frac{1}{2}$ and $N_{\text{cutoff}} = 5$. There are states contributing with $\mu = 0$ and $\mu = 1$. They are given in Table 7.2.

Of course this list can easily be generated through computer loops. The 15×15 matrix can then be filled with matrix elements; the energy along the diagonal and the spin–orbit matrix elements calculated according to the formula derived in this exercise with the quantum numbers of the left and right state. Finally the matrix is handed to a diagonalization program which returns a list of numerical eigenenergies and, if desired, the numerical coefficients giving the expansion of the true eigenvector in terms of the 15 basis states. The same must then be repeated for the other positive values of Ω.

To be assured of the accuracy of the expansion, this calculation should be repeated with different cutoff limits until convergence is achieved; the value N_{cutoff} necessary depends on the deformation and on the desired accuracy.

EXERCISE

7.7 The Crossing of Energy Levels

Problem. Regard the case of two energy levels which intersect as functions of a parameter. How do the eigenstates change if an interaction of constant matrix element is introduced?

Exercise 7.7

Solution. Assume that the parameter is t and that the intersection is at $t = 0$, where both levels have energy a. To simplify the formulas without losing generality, let us take a symmetric situation with the same slope in both:

$$E_1^{(0)}(t) = a - bt \quad , \quad E_2^{(0)}(t) = a + bt \quad . \tag{1}$$

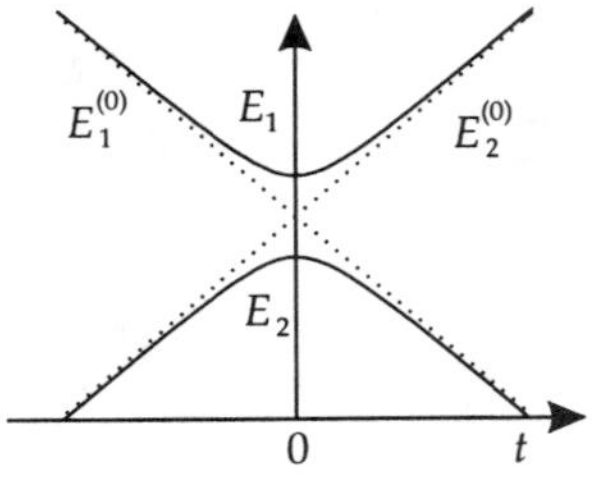

Fig. 7.9 The avoidance of crossed energy levels. Two noninteracting energy levels *(dashed)*, which intersect as functions of the parameter t, are repelled by an interaction to form the new levels shown as full curves.

The situation is indicated by the two dashed lines in Fig. 7.9. Adding an off-diagonal matrix element of magnitude d leads to the perturbed Hamiltonian

$$H' = \begin{pmatrix} a - bt & d \\ d & a + bt \end{pmatrix} \quad . \tag{2}$$

It is straightforward to calculate the eigenvalues from the characteristic equation. The two new eigenstates are given by

$$E_1(t) = a + \sqrt{b^2t^2 + d^2} \quad , \quad E_2(t) = a - \sqrt{b^2t^2 + d^2} \quad . \tag{3}$$

They are shown by the full lines in Fig. 7.9 and show the typical features of an avoided energy-level crossing. The perturbed states approach the unperturbed ones asymptotically far off from the original crossing, and the distance of closest approach is $2d$, so it is directly determined by the interaction matrix element. In the Nilsson diagrams of Figs. 7.6–7 many structures of this type can be seen whenever two levels that have the same exact quantum numbers (and thus have an interaction matrix element in the full Hamiltonian) come close to each other.

It is also interesting to examine the eigenvectors, which can also be computed easily. For the normalized eigenvector corresponding to the lower eigenvalue one finds

$$v_1 = \frac{|d|}{\left[\left(bt - \sqrt{b^2t^2 + d^2}\right)^2 + d^2\right]^{1/2}} \begin{pmatrix} bt - \sqrt{b^2t^2}/d \\ 1 \end{pmatrix} \quad . \tag{4}$$

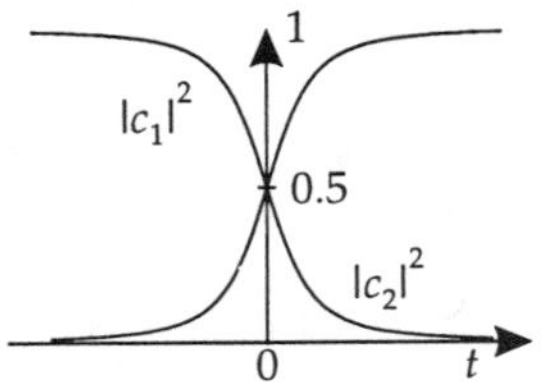

Fig. 7.10 Squared amplitudes $|c_i|^2$ of the eigenvector corresponding to the lower perturbed state in Fig. 7.9; as functions of the parameter t.

These results are illustrated in Figs. 7.9–10. For $t = 0$, i.e., at the former intersection point, the new state has equal amplitude in the unperturbed states. For $t \to -\infty$ it becomes identical to the state $E_1^{(0)}$, whose energy it also approaches, and for $t \to +\infty$ it becomes identical to the other one. An avoided energy-level crossing thus corresponds to an exchange of the two wave functions.

In nuclear dynamics this has very important consequences. Take the case of a quadrupole parameter, for example. If the nucleus executes a motion in the quadrupole direction, at each avoided energy-level crossing the single particle has to be shuffled into a new wave function with a possibly totally different spatial distribution (if motion is slow, so that the particle stays in the lowest available energy level). This corresponds to a large amount of motion associated with a small change in quadrupole moment and thus causes a large kinetic energy.

7.4 The Relativistic Mean-Field Model

7.4.1 Introduction

Conventional wisdom claims that relativistic effects are unimportant for low-energy nuclear structure problems. Indeed, considering that the largest kinetic energy of the nucleons in the nucleus is determined by the Fermi momentum $k_F \approx 1.4\,\mathrm{fm}^{-1}$ to be

$$T_{\max} = \frac{\hbar^2 k_F^2}{2m} \approx 38\,\mathrm{MeV} \quad , \tag{7.147}$$

this corresponds to $\gamma = (1 - v^2/c^2)^{-1/2} \approx 1.04$ or a velocity $v \approx 0.29c$. Thus one would expect only minor modifications due to relativistic kinematics. Relativistic versions of nuclear structure models have nevertheless become important in recent years for the following reasons.

- The relativistic mean-field model (discussed in the next sections) is as successful as the Skyrme-force Hartree–Fock models in describing the single-particle structure of nuclei *and provides a natural explanation of the spin–orbit force* [Du56, Mi72, Wa74].
- The relativistic theory of nuclear matter has had some success in clearing up the long-standing problems encountered in the nonrelativistic theory [Br84, Te86].
- One may hope to link the model of nucleons interacting through meson fields to a more fundamental field-theoretic description of nuclear interactions.
- It may serve as the basis for an extrapolation to dense and hot nuclear matter, where relativity certainly becomes important.

In the spirit of this book we will limit the discussion to the simplest version of the model, which will be seen to be something very similar to a relativistic version of the Skyrme-force Hartree–Fock theory, and not go into the advanced field-theoretic discussion, for which the reader is referred to the review literature [Ce86, Se86]. A more thorough introduction to the practical application of the model is given in [Re89].

7.4.2 Formulation of the Model

In a relativistic description of interacting particles the concept of an instantaneous force described, for example, by potentials, is not appropriate. Instead, the interaction must be mediated by fields which are independent degrees of freedom. The fields can be classified by their internal angular momenta, parity, and isospin. The fields usually considered are as follows.

1. The σ meson: a scalar and isoscalar meson which provides an attractive interaction.
2. The ω meson: an isoscalar vector meson leading to a repulsive interaction.
3. The ρ meson: an isovector vector meson needed for a better description of isospin-dependent effects in nuclei.

4. The photon γ describing the electromagnetic interaction.

The initial breakthrough of these models came with the demonstration that the two mesons σ and ω already suffice to describe the saturation of nuclear matter and the binding properties of nuclei quite well. In fact, choosing a larger mass and a stronger coupling for the ω meson than for the σ already produces the long-range attraction and short-range repulsion familiar from standard nucleon–nucleon interactions. Later refinements will be discussed as they appear below.

The starting point for a relativistic model is a Lagrange density $\mathcal{L}$. It should contain the free Lagrangians for the nucleon and the meson fields (this will always include the photon as well) as well as a term coupling the nucleon to the mesons. We can thus write

$$\mathcal{L} = \mathcal{L}_{\text{nucleon}} + \mathcal{L}_{\text{mesons}} + \mathcal{L}_{\text{couple}} \quad . \tag{7.148}$$

The free Lagrangian for the nucleon is standard:

$$\mathcal{L}_{\text{nucleon}} = \hat{\bar{\psi}}(\mathrm{i}\gamma^\mu \partial_\mu - m_{\mathrm{B}})\hat{\psi} \quad . \tag{7.149}$$

Note that the nucleon mass is denoted by m_{B} in this chapter to distinguish it more clearly from the various other masses occurring. Similarly the Lagrangians for the free meson fields are easily written down:

$$\begin{aligned}\mathcal{L}_{\text{mesons}} = {} & \tfrac{1}{2}\left(\partial^\mu \hat{\sigma} \partial_\mu \hat{\sigma} - m_\sigma^2 \hat{\sigma}^2\right) \\ & - \tfrac{1}{2}\left(\overline{\partial^\nu \hat{\omega}^\mu} \partial_\mu \hat{\omega}_\nu - m_\omega^2 \hat{\omega}^\mu \hat{\omega}_\mu\right) \\ & - \tfrac{1}{2}\left(\overline{\partial^\nu \hat{\boldsymbol{R}}^\mu} \cdot \partial_\mu \hat{\boldsymbol{R}}_\nu - m_\rho^2 \hat{\boldsymbol{R}}^\mu \cdot \hat{\boldsymbol{R}}_\mu\right) \\ & - \tfrac{1}{2}\overline{\partial^\nu \hat{A}^\mu} \partial_\mu \hat{A}_\nu \quad .\end{aligned} \tag{7.150}$$

Note that the usual relativistic units of $\hbar = c = 1$ are used for the discussion of this model. The fields $\hat{\sigma}$ and $\hat{\omega}_\mu$ describe the corresponding mesons; for the ρ meson the notation $\hat{\boldsymbol{R}}_\mu$ is used to distinguish it from the densities, and $\hat{A}_\mu$ stands for the photon field. Also an antisymmetrized derivative was defined via

$$\overline{\partial^\nu A^\mu} = \partial^\nu A^\mu - \partial^\mu A^\nu \quad . \tag{7.151}$$

This expression corresponds to the field-strength tensor which for the electromagnetic case is usually denoted by $F_{\mu\nu}$, and the Lagrangian is given by

$$\mathcal{L} = -\tfrac{1}{4} F_{\mu\nu} F^{\mu\nu} = \tfrac{1}{2}\overline{\partial^\nu A^\mu} \partial_\mu A_\nu \quad . \tag{7.152}$$

The present notation avoids having to introduce different letters for the field strengths associated with the various meson fields. For the interaction the natural choice is to use *minimal coupling*, but this is supplemented by a nonlinear self-coupling of the σ meson,

$$U(\hat{\sigma}) = \tfrac{1}{3} b_2 \hat{\sigma}^3 + \tfrac{1}{4} b_3 \hat{\sigma}^4 \quad , \tag{7.153}$$

first introduced by Boguta and Bodmer [Bo77] and since widely accepted. The purpose of this term is to improve the compressibility of nuclear matter in the model. The Lagrangian usually adopted is thus

$$\mathcal{L}_{\text{coupling}} = -g_\sigma \hat{\sigma} \hat{\rho}_s - g_\omega \hat{\omega}^\mu \hat{\rho}_\mu - \tfrac{1}{2} g_\rho \boldsymbol{R}^\mu \cdot \hat{\boldsymbol{\rho}}_\mu \\ - A^\mu \hat{\rho}_\mu^{\mathrm{C}} - U(\hat{\sigma}) \quad . \tag{7.154}$$

There are four different densities in the coupling terms:

- the *scalar density* $\hat{\rho}_s = \hat{\bar{\psi}}\hat{\psi}$, which describes the difference of the densities of the positive and negative energy components in the wave functions,
- the *vector density* $\hat{\rho}_\mu = \hat{\bar{\psi}}\gamma_\mu\hat{\psi}$ describing the sum of these,
- the *isovector density* $\hat{\boldsymbol{\rho}}_\mu = \hat{\bar{\psi}}\boldsymbol{\tau}\gamma_\mu\hat{\psi}$, and
- the charge density $\hat{\rho}_\mu^{\mathrm{C}} = \frac{1}{2}e\hat{\bar{\psi}}(1+\tau_0)\gamma_\mu\hat{\psi}$.

The model as it stands contains as free parameters the meson masses m_σ, m_ω, and m_ρ, as well as the coupling constants g_σ, g_ω, g_ρ, b_2, and b_3. For the nucleon mass m_{B} the free value is usually employed.

For practical applications a number of approximations are necessary, because the solution of the full field-theoretic problem is clearly impossible. The first step is to replace the field operators for the mesons and the photon by their expectation values (the *mean field approximation*), formally $\hat{\sigma} \rightarrow \sigma = \langle\hat{\sigma}\rangle$ etc. The role of the meson fields then reduces to that of potentials generated by the appropriate nucleon densities, and the nucleons behave as noninteracting particles moving in these mean fields. This implies that the nucleon field operator can be expanded in single-particle states $\phi_\alpha(x^\mu)$,

$$\hat{\psi} = \sum_\alpha \phi_\alpha(x^\mu)\, \hat{a}_\alpha \quad , \tag{7.155}$$

while the densities reduce to sums over densities of the single-particle states,

$$\hat{\rho}_{\mathrm{s}} \longrightarrow \sum_{\alpha<F} \bar{\phi}_\alpha \phi_\alpha - \rho_{\mathrm{s}}^{\mathrm{vacuum}} \quad , \tag{7.156}$$

and similarly for the other densities. Here $\alpha < F$ as usual denotes summation over all states below the Fermi level, and $\rho_{\mathrm{s}}^{\mathrm{vacuum}}$ is the corresponding density for free particles at baryon number zero. The presence of this vacuum term and the need to sum over *all* levels below the Fermi level, including all negative energy states, is unavoidable in the relativistic theory but makes the problem very hard to solve. For this reason another approximation has to be introduced, the *no-sea approximation* (i.e., exclusion of the filled Dirac sea of negative energy states) [Re89], which assumes that the sum over all negative energy states in the first term of (7.156) cancels the vacuum contribution. It corresponds to the neglection of vacuum polarization.

The net result of these approximations is that we will have a number of occupied single-particle orbitals ϕ_α, $\alpha = 1, \ldots, \Omega$, which determine the densities such as

$$\rho_{\mathrm{s}} = \sum_{\alpha=1}^{\Omega} w_\alpha \bar{\phi}_\alpha \phi_\alpha \tag{7.157}$$

and analogously for the vector density etc. The weight factors w_α were introduced to allow for a phenomenologically introduced pairing (see Sect. 7.5), and the number of states Ω then is larger than the number of nucleons A. The meson fields are classical fields with these densities as sources and in turn appear in the Dirac equations for the single-particle states, so that a self-consistent problem such as in standard Hartree–Fock results. We will now derive the field equations for the static case with a few additional restrictions.

In nuclear structure calculations we are usually interested in stationary states. Also it is generally true that proton and neutron states do not mix, i.e., that the single-particle states are eigenstates of τ_0. Thus only the components of the ρ field and the isovector vector density with isospin projection zero will appear, viz. $R_{0\mu}$ and $\rho_{0\mu}$. Stationarity implies that all time derivatives and also the spatial components of the densities, which describe current densities, vanish. The latter fact also leads to the vanishing of the spatial components of the fields themselves, since their source terms are just these densities. The only fields left are thus σ, ω_0, R_{00}, and A_0. Furthermore the time dependence of the wave functions may be separated off according to

$$\phi_\alpha(x^\mu) = \phi_\alpha(\boldsymbol{r})\, e^{i\varepsilon_\alpha t} \quad , \tag{7.158}$$

with ε_α the single-particle energy.

The equations of motion can now be obtained by varying the action $S = \int d^4x\ \mathcal{L}$ with respect to the fields. The trivial calculation leads to the Dirac equation for the nucleonic states in the presence of the fields,

$$\begin{aligned}\varepsilon_\alpha\gamma_0\phi_\alpha = \Big[&-i\boldsymbol{\gamma}\cdot\nabla + m_B + g_\sigma\sigma + g_\omega\omega_0\gamma_0 \\ &+ \tfrac{1}{2}g_\rho R_{00}\gamma_0\tau_0 + \tfrac{1}{2}eA_0\gamma_0(1+\tau_0)\Big]\phi_\alpha \quad ,\end{aligned} \tag{7.159}$$

as well as to the equations describing the fields

$$\begin{aligned}\left(-\Delta + m_\sigma^2\right)\sigma + U'(\sigma) &= -g_\sigma\rho_s \quad , \\ \left(-\Delta + m_\omega^2\right)\omega_0 &= g_\omega\rho_0 \quad , \\ \left(-\Delta + m_\rho^2\right)R_{00} &= \tfrac{1}{2}g_\rho\rho_{00} \quad , \\ -\Delta A_0 &= \rho_0^C \quad .\end{aligned} \tag{7.160}$$

This set of equations, together with the definition of the densities, constitutes a self-consistent field problem very similar to nonrelativistic Hartree–Fock. The approximations made to arrive there from the very general starting point were, however, quite severe, and it is not clear, and the subject of ongoing efforts to find out, how well they can be justified. Nevertheless, one may always regard these equations as defining an *effective model* of the same spirit as the Skyrme force approach and it has certainly been reasonably successful in practical applications.

A very interesting feature of the model becomes apparent from Fig. 7.11 in which the meson fields for ^{208}Pb are plotted. The σ and ω fields are quite large, being comparable in magnitude to the nucleon mass (the ρ meson in contrast is a relatively minor correction), but almost cancel, so that the result is the standard nuclear potential. This implies that wherever these potentials add up in a constructive way, the relativistic effects are nonnegligible, which is responsible, for example, for the fact that the spin–orbit coupling has the right magnitude naturally in this model with requiring additional fit parameters.

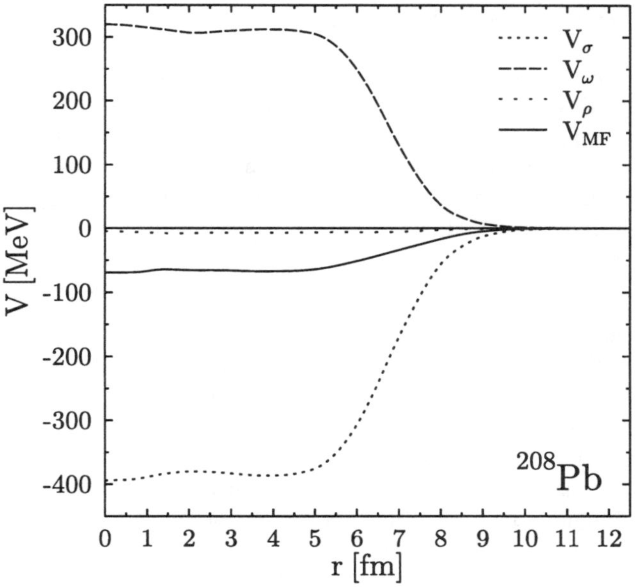

Fig. 7.11 Plot of the individual mesonic potentials in the nucleus ^{208}Pb for a typical parameter set. V_{MF} is the total potential resulting from the near-cancellation of the σ and ω terms.

The mean-field model is an approximation that neglects many-particle interactions as well as antiparticle contributions. There have been many attempts to take several kinds of higher-order corrections into account both for nuclear matter and finite nuclei. For example in the relativistic Hartree approximation (Dirac–Hartree theory) vacuum fluctuation effects are included that turn out to be considerable. After a refit of the parameters, though, only minor differences remain to mean-field theory both for nuclear matter [Ch76b, Ch77b] and for finite nuclei [Pe86]. In other calculations the influence of many-particle correlations were considered in various approximations (Dirac–Hartree–Fock [Se86, Bl87], relativistic Bethe–Brueckner–Goldstone [An83]) or even with a consideration of the effects caused by the Dirac sea [Ho87]. These corrections are also large, but again after a refitting of the parameters the final results are affected only slightly, so that one may assume that they are really included implicitly in the effective Lagrangian of the mean-field model. Aside from these "one-loop" contributions even "two-loop" contributions were considered. This does not, however, imply a perturbation expansion, since each approximation contains all orders of the coupling constants. Both the magnitude and the character of the two-loop corrections lead to physically unsatisfactory results in this approximation [Se89]. Even expensive lattice calculations were employed to search for corrections to the mean-field model [Br92]. An extensive discussion of the various approximations studied can be found in [Re89, Se92].

7.4.3 Applications

It is instructive to examine the limiting case of symmetric nuclear matter as for the Skyrme forces (Sect. 7.2.9). In this case the isospin density will vanish, so that the ρ meson plays no role. The Coulomb force is never considered in infinite symmetric nuclear matter because in principle it should be infinite. The wave functions are plane wave states labelled by the momentum $\boldsymbol{k}$ and the spin direction. An important difference is that in (7.159) the *effective mass*

$$m_{\mathrm{B}}^* = m_{\mathrm{B}} + g_\sigma \sigma \tag{7.161}$$

appears in the same mathematical position as the free mass m_{B} in the free Dirac equation (i.e., the term not multiplied by any γ matrix). In the solutions we thus have to use the effective mass everywhere the free mass appears in the free Dirac spinors.

The spinors must be normalized such that the baryon number, which corresponds to the density ρ_0, gets a contribution of 1 from each single-particle wave function; the density itself is then given by the familiar calculation for a Fermi gas,

$$\rho_0 = \sum_{\text{isospin}} \sum_{\text{spin}} \int_0^{k_{\mathrm{F}}} \frac{\mathrm{d}^3 k}{(2\pi)^3} = \frac{2}{3\pi^2} k_{\mathrm{F}}^3 \quad , \tag{7.162}$$

while the scalar density turns out to be

$$\begin{aligned} \rho_{\mathrm{s}} &= \sum_{\text{isospin}} \sum_{\text{spin}} \int_0^{k_{\mathrm{F}}} \frac{\mathrm{d}^3 k}{(2\pi)^3} \frac{m_{\mathrm{B}}^*}{\sqrt{k^2 + m_{\mathrm{B}}^{*2}}} \\ &= \frac{1}{\pi^2} m_{\mathrm{B}}^* \left(k_{\mathrm{F}} \sqrt{k_{\mathrm{F}}^2 + m_{\mathrm{B}}^{*2}} - m_{\mathrm{B}}^{*2} \log \frac{k_{\mathrm{F}} + \sqrt{k_{\mathrm{F}}^2 + m_{\mathrm{B}}^{*2}}}{m_{\mathrm{B}}^*} \right) \quad . \end{aligned} \tag{7.163}$$

The two remaining meson fields must be uniform and are determined from the equations

$$\omega_0 = \frac{g_\omega}{m_\omega^2} \rho_0 \quad , \quad m_\sigma^2 \sigma + U'(\sigma) = -g_\sigma \rho_{\mathrm{s}} \quad . \tag{7.164}$$

While ω_0 is directly a function of the nucleon density, σ must be determined numerically from this equation, because not only is it third order itself, but the scalar density according to (7.163) inturn depends on σ via m_{B}^* in a complicated way (k_{F} is determined from the nucleon density via (7.162).

The energy per nucleon of the system can be calculated by starting from the Lagrangian, determining the corresponding Hamiltonian density and then inserting the approximations made up to here. It is a straightforward calculation yielding the following result:

$$
\begin{aligned}
E/A = \frac{1}{\rho_0}\Bigg[\sum_{\substack{\text{spin}\\\text{isospin}}} \int_0^{k_\mathrm{F}} \frac{\mathrm{d}^3k}{(2\pi)^3}\sqrt{k^2+m_\mathrm{B}^{*2}} \\
-\tfrac{1}{2}(g_\sigma \sigma \rho_\mathrm{s} + g_\omega \omega_0 \rho_0) + U(\sigma) - \tfrac{1}{2}\sigma U'(\sigma)\Bigg] \\
= \frac{3}{4k_\mathrm{F}^3}\left[k_\mathrm{F}\sqrt{k_\mathrm{F}^2+m_\mathrm{B}^{*2}}(2k_\mathrm{F}^2+m_\mathrm{B}^{*2}) - m_\mathrm{B}^{*4}\log\frac{k_\mathrm{F}+\sqrt{k_\mathrm{F}^2+m_\mathrm{B}^{*2}}}{m_\mathrm{B}^*}\right] \\
-\frac{1}{2}g_\sigma \sigma \frac{\rho_\mathrm{s}}{\rho_0} - \frac{1}{2}\frac{g_\omega^2}{m_\omega^2}\rho_0 + \frac{U(\sigma)-\frac{1}{2}\sigma U'(\sigma)}{\rho_0} \quad . \qquad (7.165)
\end{aligned}
$$

Assuming any value for the density ρ_0, one can determine k_F and m_B^* using (7.162) and (7.164) and in this way construct the equation of state of nuclear matter (at zero temperature). As for the Skyrme forces, the ground-state density, the binding energy, and the incompressibility of nuclear matter may then be related to the parameters of the model.

Among the many sets of parameters in the literature one of the most carefully obtained is that of Reinhard et al. [Re86], who used a calculation of the properties of a number of spherical nuclei in which the parameters of the model were then adjusted automatically to provide the best agreement with experimental data in the sense of a least-squares fit. The quantities considered were the binding energies, radii, and surface diffusenesses. Table 7.3 compares these parameters with the original set of Walecka [Wa74] and also shows the associated nuclear matter properties. Note that the data depend too little on the mass of the ρ meson for that value to be fitted; instead the experimental value was used.

Table 7.3 Typical parameter sets for the relativistic mean-field model.

	[Wa74]	[Re86]
g_σ	9.57371	10.1377
g_ω	11.6724	13.2846
g_ρ	0.0	9.95145
b_2	0.0	−12.1724
b_3	0.0	−36.2646
m_σ	550.00	492.25
m_ω	783.00	795.36
m_ρ	763	763
E/A	−15.75	−16.42
$\rho_{\mathrm{n.m.}}$	0.194	0.152
K	544.6	211.7
$m_\mathrm{B}^*/m_\mathrm{B}$	0.56	0.57

The nuclear-matter properties show clearly the basic problem of the linear version of the model: the incompressibility is much too large, whereas the nonlinear fit easily comes close to the value estimated from experiment, although this property was not directly used in the fit. The problem in the linear model seems to be unavoidable.

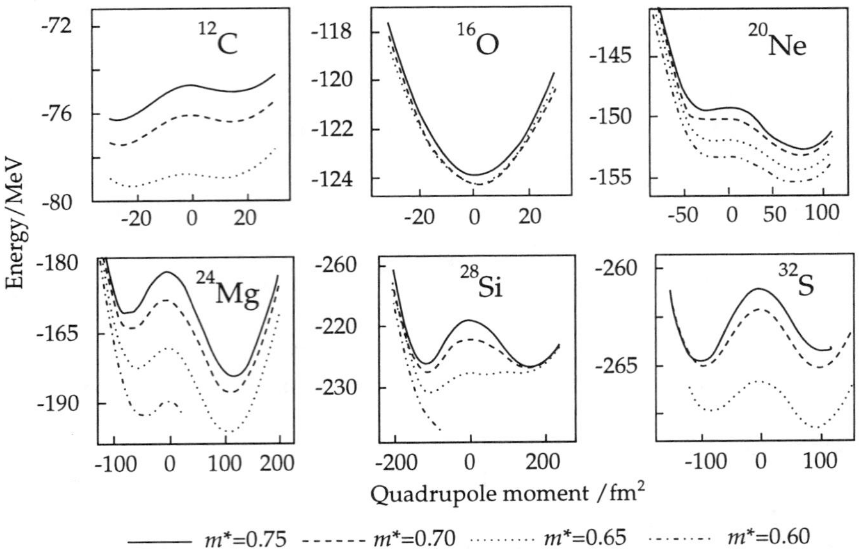

Fig. 7.12 Potential-energy curves for various light nuclei in the relativistic mean-field model, calculated on an axial grid with quadrupole constraint by Fink et al. [Fi89]. The different curves for each nuclide correspond to parameter sets providing an optimal fit of spherical nuclei, but with the effective mass constrained to a prescribed value. This value is $m_B^*/m_B = 0.75$ *(full curve)*, 0.7 *(dashed line)*, 0.65 *(dotted line)*, and 0.6 *(dash-dotted line)*. Some of the curves end for $m_B^* = 0.6$ because the model becomes unstable.

The effective mass in nuclear matter seems small compared to values in other models. It is not yet clearly understood what the reason for this is; it has the additional disadvantage of causing instabilities in the model [Wa88].

By now the relativistic mean-field model has been used in a number of applications comparable to the Skyrme-force Hartree–Fock. Perhaps most interesting in our context are the calculations of quadrupole-deformed nuclei using a constraint [Fi89, Bl89, Ga88]. Figure 7.12 shows potential-energy curves for a number of light nuclei with various parametrizations for the model. Clearly the general structure of the surfaces seems to be relatively independent of the details of the parameters; the position of the minimum agrees quite well with the experimental information where available. The results also demonstrate a problem of the model: in the fits the nonlinear potential in the σ meson tends to negative values, making the model unstable for large values of σ, which occurs at high densities, and for the smallest values of the effective mass, even happens inside the nucleus due to shell oscillations of the density. One may argue that the particular form of this potential was selected with a view to renormalizability of the model and that if the model is not regarded as a field-theoretical model but rather as an effective one, other functional forms may be appropriate.

Figure 7.13 shows results for some Gadolinium isotopes. There is a systematic dependence of the deformations on the parameter set used, and the experimental values seem to prefer an effective mass close to 0.65 m_B. Such studies may thus help in a more refined determination of the parameters compared to those based on spherical nuclei only.

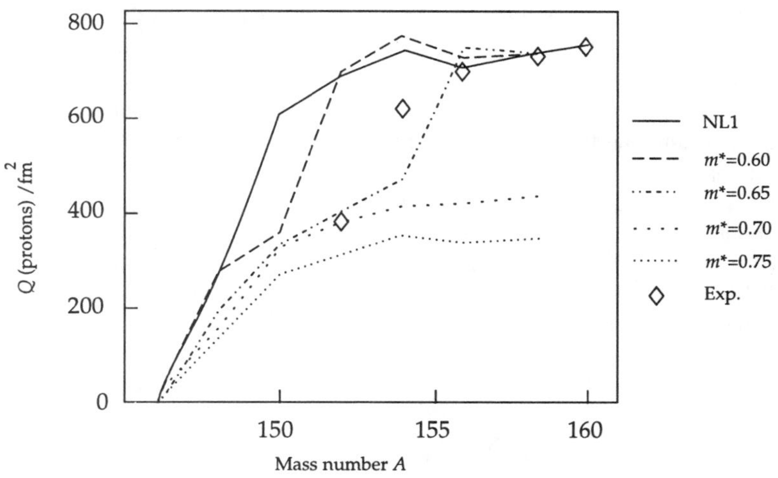

Fig. 7.13 Ground-state deformations for some even Gad- olinium isotopes in the mean-field model with different parametrization, calculated by Blum et al. [Bl89]. The meaning of the curves is indicated, NL1 denotes the unconstrained best fit of spherical nuclei, the other NLx correspond to $m_B^*/m_B = x$. The experimental data are indicated by squares.

7.5 Pairing

7.5.1 Motivation

The single-particle model as discussed up to now has some serious shortcomings. One of the problems becomes apparent when one considers the total nuclear angular momentum in a spherical nucleus. In this case the single-particle states will have good angular-momentum quantum numbers, and we may denote them, for example, by ϕ_{njm} with n standing for all nonangular-momentum quantum numbers. Let us call the set of states with $m = -j \dots j$ a j shell. For each j shell one of two cases is fulfilled:

- if all the substates with different m are filled, the total angular momentum of the nucleons concerned is always zero, because the space of occupied m states is invariant under any rotation;
- if only a part of the j shell is occupied, the angular momenta of the nucleons may result in different total angular momenta depending on the coupling. The energy, however, does not depend on the coupling, because in the single-particle model it is simply given by the sum of the single-particle energies of the occupied states. Thus we get degeneracy of the various total angular momenta. Any lifting of this degeneracy must be due to residual interactions.

Experimentally, a degeneracy of the ground state with different angular momenta does not occur; instead experiment establishes the following facts:

- for even–even nuclei the ground state always has zero angular momentum, i.e., the residual interaction lowers this particular state with respect to the other angular-momentum combinations;
- even–even nuclei are bound more tightly than odd nuclei.
- in even–even nuclei there is an *energy gap* of 1–2 MeV between the ground state and the lowest *single-particle* excitation.

Note that the last property does not refer to the many collective states, of which the ground-state rotational band, for example, is usually quite low in energy especially in heavy nuclei.

The residual interaction thus must strongly favor the coupling to zero angular momentum, and the pairing formalism provides a simple model interaction that has this property.

To get an idea of how to formulate the interaction, we investigate what is special about a pair of nucleons in the same j shell coupled to zero angular momentum (this, as is the rest of this section, is based on the discussion given by Mottelson [Mo59]). To see the principal argument, it is not necessary to include spin or antisymmetrization, so that

$$\begin{aligned} \Psi(\boldsymbol{r}_1, \boldsymbol{r}_2) &= \sum_{m_1 m_2} (jj0|m_1 m_2 0)\, \psi_{m_1}(\boldsymbol{r}_1)\, \psi_{m_2}(\boldsymbol{r}_2) \\ &= \frac{1}{\sqrt{2j+1}} \sum_m (-1)^{j-m}\, \psi_m(\boldsymbol{r}_1)\, \psi_{-m}(\boldsymbol{r}_2) \quad , \end{aligned} \tag{7.166}$$

and since without spin $\psi_m(\boldsymbol{r}) = f(r)\, Y_{jm}(\Omega)$ (note that here as in the following discussions the index j is suppressed, because we will always be dealing with one fixed value of j),

$$\begin{aligned} \Psi(\boldsymbol{r}_1, \boldsymbol{r}_2) &= \frac{1}{\sqrt{2j+1}} f(r_1) f(r_2) \sum_m (-1)^{j-m}\, Y_{jm}(\Omega_1)\, Y_{j-m}(\Omega_2) \\ &= \frac{(-1)^j}{4\pi} \sqrt{2j+1}\, f(r_1) f(r_2)\, P_j(\cos\theta_{12}) \quad , \end{aligned} \tag{7.167}$$

where θ_{12} is the angle between the directions Ω_1 and Ω_2. The Legendre polynomials are strongly peaked near the argument value 1, so that the two-particle probability distribution shows a preference for having the two nucleons close to each other. This result is not changed fundamentally when spin and antisymmetry are included. Note that in the wave function the single-particle states with projection m and $-m$ are paired, which are time-reversed with respect to each other and have a similar probability distribution, again supporting the argument of geometrical correlation.

The residual interaction responsible for coupling to angular momentum zero must thus be attractive and of short range to provide the desired correlation. Using, for example, an attractive δ-function interaction in fact produces a spectrum in which the energy depends on the total angular momentum J and the state with $J = 0$ is the lowest, but such an interaction is too complicated for most applications. A more practical method is to replace it by an *idealized pairing potential* that is constructed explicitly in such a way so as to lower *only* the $J = 0$ state. For the two-particle case, this state is given by

$$|j = 0\rangle = \frac{1}{\sqrt{2j+1}} \sum_{m=-j}^{j} (-1)^{j-m} |jm\rangle |j - m\rangle \quad , \tag{7.168}$$

and the second-quantization operator "creating" it is

$$\hat{A}^\dagger = \tfrac{1}{2} \sum_{m=-j}^{j} (-1)^{j-m}\, \hat{a}_m^\dagger \hat{a}_{-m}^\dagger \quad , \tag{7.169}$$

where the index j is suppressed in the operators (for the time being we will stay within one j shell). The state produced by this operator acting on the vacuum is not normalized; the factor $1/2$ in front of the operator is chosen with a view to its commutation properties.

For the further developments it is convenient to eliminate the factor $(-1)^{j-m}$ by redefining the phases of the wave functions. Denoting the usual angular-momentum eigenstates with the phases according to Condon and Shortley (see Sect. 2.3.2) by $|m\rangle_{\mathrm{CS}}$, we can introduce new states $|m\rangle_{\mathrm{BCS}}$ via [Ni61]

$$|m\rangle_{\mathrm{BCS}} = \begin{cases} |m\rangle_{\mathrm{CS}} & \text{if } m > 0 \quad , \\ (-1)^{j+m}|m\rangle_{\mathrm{CS}} & \text{if } m < 0 \quad . \end{cases} \tag{7.170}$$

The name "BCS" refers to the Bardeen–Cooper–Schrieffer theory of superconductivity [Ba57], which was first developed in the context of condensed-matter physics and then applied to nuclei by Belyaev [Be59].

The operator $\hat{A}^\dagger$ now takes the form

$$\begin{aligned} \hat{A}^\dagger &= \frac{1}{2}\sum_{m>0}(-1)^{j-m}\left(\hat{a}_m^\dagger \hat{a}_{-m}^\dagger\right)_{\mathrm{CS}} + \frac{1}{2}\sum_{m<0}(-1)^{j-m}\left(\hat{a}_m^\dagger \hat{a}_{-m}^\dagger\right)_{\mathrm{CS}} \\ &= \frac{1}{2}\left(\sum_{m>0}\hat{a}_m^\dagger \hat{a}_{-m}^\dagger + (-1)^{2m}\sum_{m<0}\hat{a}_m^\dagger \hat{a}_{-m}^\dagger\right)_{\mathrm{BCS}} \\ &= \sum_{m>0}\left(\hat{a}_m^\dagger \hat{a}_{-m}^\dagger\right)_{\mathrm{BCS}} \quad . \end{aligned} \tag{7.171}$$

Note that $(-1)^{2m} = -1$ as m is half-integer. For the rest of the section on pairing the BCS phases will be used.

The pairing potential is now tentatively constructed as

$$\hat{V}_{\mathrm{P}} = -G\hat{A}^\dagger\hat{A} = -G\sum_{m,m'>0}\hat{a}_{m'}^\dagger \hat{a}_{-m'}^\dagger \hat{a}_{-m}\hat{a}_m \quad , \tag{7.172}$$

with $G > 0$ giving the strength. The summation is over positive m values only; other definitions summing over the whole range have a factor of $\frac{1}{4}$ in front. The definition of $\hat{V}_{\mathrm{P}}$ is inspired by the particle-number operator and is supposed to "count" the pairs coupled to zero angular momentum. Since $\hat{A}^\dagger$ is not a simple creation operator, it has to be shown whether this operator really performs as desired. To see that, calculate the spectrum of $\hat{V}_{\mathrm{Pair}}$ in the space of antisymmetrized two-particle states $|m_1 m_2\rangle$. Its matrix elements are

$$\langle m_1' m_2'|\hat{V}_{\mathrm{P}}|m_1 m_2\rangle = -G\,\delta_{m_1',-m_2'}\,\delta_{m_1,-m_2} \tag{7.173}$$

(actually the sign of this result is true only if the ordering of the positive and negative projection states is the same on both, for example, m_1, $m_1' > 0$ and m_2, $m_2' < 0$). The number of states $|m_1 m_2\rangle$ is $N = \binom{2j+1}{2}$, which just corresponds to the number of ways that two occupied states can be selected among the $2j+1$ available. Amongst these there are $\Omega = (2j+1)/2$ states of the form $|m\,-m\rangle$ with $m > 0$, and $\hat{V}_{\mathrm{P}}$ has a matrix element of $-G$ between all of these and a zero matrix element otherwise. Assume that we arrange the two-particle basis such that the first

Ω states are those of the form $|m\,-m\rangle$. A vector in this space with components $(c_1, \ldots, c_\Omega, c_{\Omega+1}, \ldots, c_N)$ is transformed under the action of $\hat{V}_\mathrm{P}$ according to

$$\hat{V}_\mathrm{P} \begin{pmatrix} c_1 \\ \vdots \\ c_\Omega \\ c_{\Omega+1} \\ \vdots \\ c_N \end{pmatrix} = -G \begin{pmatrix} C \\ \vdots \\ C \\ 0 \\ \vdots \\ 0 \end{pmatrix} \quad , \tag{7.174}$$

with $C = \sum_{j=1}^{\Omega} c_j$. Clearly $\hat{V}_\mathrm{P}$ maps any vector into the vector with the Ω first components equal and zero otherwise, so that it is a projector into that one-dimensional subspace. This implies that it has a zero eigenvalue for all vectors outside that space and the only eigenvector with a nonvanishing eigenvalue is given by $c_j = 1$, $j = 1, \ldots, \Omega$ and $c_j = 0, j > \Omega$ with eigenvalue $-G\Omega$. We have thus constructed the desired pairing potential which leaves all states unchanged except for the one state given by $\sum_m |m\,-m\rangle$, which has zero total angular momentum and is lowered by $-G\Omega$.

The next step will be to examine what happens if there are more than two particles in one j shell.

7.5.2 The Seniority Model

Let us try to extend the considerations of the previous section to the case of N particles in a j shell. This problem was first studied by Racah [Ra43]. Again we will use the operator

$$\hat{A}^\dagger = \sum_{m>0} \hat{a}_m^\dagger \hat{a}_{-m}^\dagger \quad , \tag{7.175}$$

which creates two particles coupled to zero total angular momentum. Applying this operator repeatedly does not simply create many such pairs; for example the operator $(\hat{A}^\dagger)^2$ will contain contributions proportional to $(\hat{a}^\dagger)^2$ which vanish because of the Pauli principle.

For the two-particle case, $N = 2$, we already know that the state $\hat{A}^\dagger|0\rangle$ has lower energy than all other two-particle states, which are degenerate. The construction of states with higher numbers of N is a good example of operator algebra. Aside from $\hat{A}^\dagger$ the other operators that may play a role are the particle-number operator

$$\hat{N} = \sum_{m=-j}^{j} \hat{a}_m^\dagger \hat{a}_m \tag{7.176}$$

(note that for this operator the summation over m extends to both signs), and the pairing potential

$$\hat{V}_\mathrm{P} = -G\hat{A}^\dagger\hat{A} = -G \sum_{m,m'>0} \hat{a}_{m'}^\dagger \hat{a}_{-m'}^\dagger \hat{a}_{-m} \hat{a}_m \quad . \tag{7.177}$$

We will use the commutation relations between these operators to construct the eigenstates of $\hat{V}_P$. Because $\hat{V}_P$ does not change the particle number, it is clear that $[\hat{N}, \hat{V}_P] = 0$. The other commutation relations can also be evaluated easily:

$$\begin{aligned}
[\hat{A}^\dagger, \hat{N}] &= \sum_{m>0,m'} [\hat{a}_m^\dagger \hat{a}_{-m}^\dagger, \hat{a}_{m'}^\dagger \hat{a}_{m'}] \\
&= \sum_{m>0,m'} (\hat{a}_m^\dagger \hat{a}_{-m}^\dagger \hat{a}_{m'}^\dagger \hat{a}_{m'} - \hat{a}_{m'}^\dagger \hat{a}_{m'} \hat{a}_m^\dagger \hat{a}_{-m}^\dagger) \\
&= \sum_{m>0,m'} (\hat{a}_{m'}^\dagger \hat{a}_m^\dagger \hat{a}_{-m}^\dagger \hat{a}_{m'} - \hat{a}_{m'}^\dagger \hat{a}_{m'} \hat{a}_m^\dagger \hat{a}_{-m}^\dagger) \\
&= \sum_{m>0,m'} (\delta_{m',-m}\ \hat{a}_{m'}^\dagger \hat{a}_m^\dagger - \hat{a}_{m'}^\dagger \hat{a}_m^\dagger \hat{a}_{m'} \hat{a}_{-m}^\dagger - \hat{a}_{m'}^\dagger \hat{a}_{m'} \hat{a}_m^\dagger \hat{a}_{-m}^\dagger) \\
&= \sum_{m>0,m'} (\delta_{m',-m}\ \hat{a}_{m'}^\dagger \hat{a}_m^\dagger - \delta_{mm'}\ \hat{a}_{m'}^\dagger \hat{a}_{-m}^\dagger) \\
&= \sum_{m>0} \hat{a}_{-m}^\dagger \hat{a}_m^\dagger - \sum_{m>0} \hat{a}_m^\dagger \hat{a}_{-m}^\dagger \\
&= -2\hat{A}^\dagger \quad .
\end{aligned} \tag{7.178}$$

This result expresses the fact that $\hat{A}^\dagger$ raises the particle number by 2:

$$\hat{N}\hat{A}^\dagger = \hat{A}^\dagger(\hat{N} + 2) \quad . \tag{7.179}$$

For the second commutation relation we can proceed similarly with successive permutation, noting that now the indices are always positive, so that, for example, $\hat{a}_m^\dagger$ anticommutes with $\hat{a}_{-n}$:

$$\begin{aligned}
[\hat{V}_P, \hat{A}^\dagger] &= -G \sum_{mm'n>0} [\hat{a}_m^\dagger \hat{a}_{-m}^\dagger \hat{a}_{-m'} \hat{a}_{m'}, \hat{a}_n^\dagger \hat{a}_{-n}^\dagger] \\
&= -G \sum_{mm'n>0} (\hat{a}_m^\dagger \hat{a}_{-m}^\dagger \hat{a}_{-m'} (\delta_{m'n} - \hat{a}_{-n}^\dagger - \hat{a}_m^\dagger \hat{a}_{-m}^\dagger \hat{a}_n^\dagger (\delta_{m'n} - \hat{a}_{-m'}) \hat{a}_{m'}) \\
&= -G \left(- \sum_{m>0} \hat{a}_m^\dagger \hat{a}_{-m}^\dagger \sum_n \hat{a}_n^\dagger \hat{a}_n + \sum_{m>0} \hat{a}_m^\dagger \hat{a}_{-m}^\dagger \sum_{m'n>0} \delta_{m'n} \right) \\
&= -G(- \hat{A}^\dagger \hat{N} + \Omega \hat{A}^\dagger) \\
&= -G(\Omega + 2 - \hat{N})\ \hat{A}^\dagger \quad ,
\end{aligned} \tag{7.180}$$

where Ω again is the number of positive angular-momentum projections. This result implies that $\hat{A}^\dagger$ shifts the eigenvalues of $\hat{V}_P$ by amount that is dependent on the particle number. We can now also evaluate the commutation of $\hat{V}_P$ with powers of $\hat{A}^\dagger$:

$$\begin{aligned}
[\hat{V}_P, (\hat{A}^\dagger)^\nu] &= \hat{V}_P (\hat{A}^\dagger)^\nu - (\hat{A}^\dagger)^\nu \hat{V}_P \\
&= [\hat{V}_P, \hat{A}^\dagger] (\hat{A}^\dagger)^{\nu-1} + \hat{A}^\dagger [\hat{V}_P, \hat{A}^\dagger] (\hat{A}^\dagger)^{\nu-2} \\
&\quad + \cdots + (\hat{A}^\dagger)^{\nu-1} [\hat{V}_P, \hat{A}^\dagger] \\
&= -G \sum_{i=0}^{\nu-1} (\hat{A}^\dagger)^i (\Omega + 2 - \hat{N})\ (\hat{A}^\dagger)^{\nu-i} \quad .
\end{aligned} \tag{7.181}$$

Now commute the leading operator to the right using $(\hat{A}^\dagger)^i \hat{N} = (\hat{N} - 2i)(\hat{A}^\dagger)^i$ to get

$$\begin{aligned}
\left[\hat{V}_{\rm P}, (\hat{A}^\dagger)^\nu\right] &= -G \sum_{i=0}^{\nu-1} (\Omega + 2 - \hat{N} + 2i)(\hat{A}^\dagger)^\nu \\
&= -G\left[\nu(\Omega + 2 - \hat{N}) + \nu(\nu - 1)\right](\hat{A}^\dagger)^\nu \quad . \qquad (7.182)
\end{aligned}$$

We can now proceed to construct the N-particle states. A basis for these states is obtained by distributing the particles over the $2j + 1 = 2\Omega$ single-particle levels; there are $d_N = \binom{2\Omega}{N}$ such configurations. Note that, for example, for $N = 2\Omega$ there is only one such state, since all levels are filled. The number of configurations in which all particles are in $(m, -m)$ combinations is $p_N = \binom{\Omega}{N/2}$, i.e., the number of ways to distribute the pairs over the positive-projection single-particle states.

For $N = 2$ the paired state is $\hat{A}^\dagger|0\rangle$, and its energy can be determined from

$$\hat{V}_{\rm P}\hat{A}^\dagger|0\rangle = \left[\hat{V}_{\rm P}, \hat{A}^\dagger\right]|0\rangle = -G\left(\Omega + 2 - \hat{N}\right)\hat{A}^\dagger|0\rangle = -G\Omega\hat{A}^\dagger|0\rangle \quad , \qquad (7.183)$$

so that the old result is recovered. For $N = 4$ we expect the lowest state to be $(\hat{A}^\dagger)^2|0\rangle$. Being built out of two states with $J = 0$, its angular momentum is zero and its energy again is easily calculated from the commutation relation:

$$\begin{aligned}
\hat{V}_{\rm P}(\hat{A}^\dagger)^2|0\rangle &= \left[\hat{V}_{\rm P}, (\hat{A}^\dagger)^2\right]|0\rangle \\
&= -G\left[2(\Omega + 2 - \hat{N}) + 2\right](\hat{A}^\dagger)^2|0\rangle \\
&= -2G(\Omega - 1)(\hat{A}^\dagger)^2|0\rangle \quad , \qquad (7.184)
\end{aligned}$$

so that the energy of this state is $E = -2G(\Omega - 1)$. The effect of the Pauli principle already reduces the energy gained to less than twice that of a single pair.

What about the other possible states? For the two-particle case we have already seen that the remaining $d_N - 1$ states have zero energy. Let us denote them by $\hat{B}_i^{\dagger(2)}|0\rangle$, $i = 1, \ldots, d_N - 1$, so that

$$\hat{V}_{\rm P}\hat{B}_i^{\dagger(2)}|0\rangle = 0 \quad . \qquad (7.185)$$

These allow the construction of a second group of four-particle states $\hat{A}^\dagger\hat{B}_i^{\dagger(2)}|0\rangle$ whose energy is given by

$$\begin{aligned}
\hat{V}_{\rm P}\hat{A}^\dagger\,\hat{B}_i^{\dagger(2)}|0\rangle &= \left[\hat{V}_{\rm P}, \hat{A}^\dagger\right]\,\hat{B}_i^{\dagger(2)}|0\rangle \\
&= -G\left(\Omega + 2 - \hat{N}\right)\,\hat{A}^\dagger\hat{B}_i^{\dagger(2)}|0\rangle \\
&= -G(\Omega - 2)\,\hat{A}^\dagger\hat{B}_i^{\dagger(2)}|0\rangle \quad , \qquad (7.186)
\end{aligned}$$

where the preceding equation was utilized. By now we have constructed d_2 special four-particle states. To derive the energy of the others, first note that the potential $\hat{V}_{\rm P}$ is negative-definite, since for any $|\Psi\rangle$ we have

$$\langle\Psi|\hat{V}_{\rm P}|\Psi\rangle = -G\langle\Psi|\hat{A}^\dagger\hat{A}|\Psi\rangle = -G\left|\hat{A}|\Psi\rangle\right|^2 \leq 0 \quad . \qquad (7.187)$$

Thus all of its eigenvalues must be negative, and we will show that the sum of the eigenvalues derived up to now exhausts the trace of $\hat{V}_P$, leaving only zero for the remaining eigenvalues. This sum is

$$-2G(\Omega-1) - G(d_2-1)(\Omega-2) = -G\Omega\left(2\Omega^2 - 5\Omega + 3\right) \quad . \tag{7.188}$$

The trace of $\hat{V}_P$ on the other hand is

$$\begin{aligned}\mathrm{Tr}\{\hat{V}_P\} &= -G \sum_{mm'>0} \sum_i \langle \Psi_i | \hat{a}^\dagger_m \hat{a}^\dagger_{-m} \hat{a}_{m'} \hat{a}_{-m'} | \Psi_i \rangle \\ &= -G \sum_{m>0} \sum_i \langle \Psi_i | \hat{a}^\dagger_m \hat{a}^\dagger_{-m} \hat{a}_m \hat{a}_{-m} | \Psi_i \rangle \quad . \end{aligned} \tag{7.189}$$

The states $|\Psi_i\rangle$, $i = 1, \ldots, d_4$, span the four-particle space. The matrix element counts how many such states there are with both m and $-m$ filled. If these are occupied, the remaining two particles can be distributed over the remaining $2\Omega - 2$ states, so that there are $\binom{2\Omega-2}{2}$ possibilities. The trace therefore is

$$\begin{aligned}\mathrm{Tr}\{\hat{V}_P\} &= -G \sum_{m>0} \binom{2\Omega-2}{2} \\ &= -G\Omega \binom{2\Omega-2}{2} \\ &= -G\Omega\left(2\Omega^2 - 5\Omega + 3\right) \end{aligned} \tag{7.190}$$

in agreement with (7.188). Thus all other states must have eigenvalues of zero, and we can denote them by $B_i^{\dagger(4)}|0\rangle$, $i = 1, \ldots, d_4 - d_2$.

Now the general case can be tackled. It appears that the number of "pairs" created by $\hat{A}^\dagger$ plays a special role; therefore it is advantageous to introduce a new quantum number s, the *seniority,* to keep track of them. It counts the number of nucleons *not in pairs*. For the cases discussed up to now we have the properties and construction of the states summarized in Table 7.4.

Table 7.4 Particle number N and seniority s for the different states up to $N = 4$.

N	s	state(s)
2	0	$\hat{A}^\dagger\|0\rangle$
2	2	$\hat{B}_i^{\dagger(2)}\|0\rangle$
4	0	$\left(\hat{A}^\dagger\right)^2\|0\rangle$
4	2	$\hat{A}^\dagger\hat{B}_i^{\dagger(2)}\|0\rangle$
4	4	$\hat{B}_i^{\dagger(4)}\|0\rangle$

This can be generalized first to *even* values of N by a recursive application of the preceding construction. For a given value of s we expect states of the following form:

$$\begin{aligned} s = 0: &\quad (\hat{A}^\dagger)^{N/2}|0\rangle \\ s = 2: &\quad (\hat{A}^\dagger)^{(N-2)/2}\hat{B}_i^{(2)}|0\rangle \\ \vdots \quad &\quad \vdots \\ s \text{ general}: &\quad (\hat{A}^\dagger)^{(N-s)/2}\hat{B}_i^{(s)}|0\rangle \quad . \end{aligned} \tag{7.191}$$

In this way the operators $\hat{B}_i^{(s)}$ are recursively defined by the states of the maximum allowed value of s. Since in general we require $\hat{V}_P\hat{B}_i^{(s)}|0\rangle = 0$, the energy of these states can again easily be computed via

$$\begin{aligned} &\hat{V}_P\left(\hat{A}^\dagger\right)^{(N-s)/2}\hat{B}_i^{(s)}|0\rangle \\ &= \left[\hat{V}_P, \left(\hat{A}^\dagger\right)^{(N-s)/2}\right]\hat{B}_i^{(s)}|0\rangle \\ &= -G\left[\frac{N-s}{2}(\Omega + 2 - \hat{N}) + \frac{N-s}{4}(N-s-2)\right]\left(\hat{A}^\dagger\right)^{(N-s)/2}\hat{B}_i^{(s)}|0\rangle \\ &= -\frac{G}{4}(N-s)(2\Omega + 2 - \hat{N} - s)\left(\hat{A}^\dagger\right)^{(N-s)/2}\hat{B}_i^{(s)}|0\rangle \quad , \end{aligned} \tag{7.192}$$

so that the energy formula will be

$$E_s^N = -\frac{G}{4}(N-s)(2\Omega + 2 - N - s) \quad . \tag{7.193}$$

For *odd* values of N the odd particle has to be created explicitly before the above construction commences. For $s = 1$, for example, the states should be

$$\left(\hat{A}^\dagger\right)^{(N-1)/2}\hat{a}_m^\dagger|0\rangle \quad , \quad m = -j, \ldots, +j \quad . \tag{7.194}$$

Then we can proceed as before and simply add an additional operator $\hat{a}_m^\dagger$ in front of the vacuum everywhere; because clearly $\hat{V}_P\hat{a}_m^\dagger|0\rangle = 0$, calculating the energies via the commutation relations produces the same result as (7.193) with only N and s reflecting the correct odd numbers.

Up to now the problem of what happens at larger values of N has been tacitly ignored. The recursive construction must fail as soon as the number of states does not increase anymore with N; this happens at $N = \Omega/2$. Owing to the Pauli principle, the states created then by continued application of $\hat{A}^\dagger$ will no longer be independent. We can, however, apply the same ideas if we start from the completely filled shell and then add *holes*. If we define creation and annihilation operators for holes via

$$\hat{\beta}_m^\dagger = \hat{a}_{-m} \quad , \quad \hat{\beta}_m = \hat{a}_{-m}^\dagger \quad , \tag{7.195}$$

the pairing potential can be rewritten by commutation

$$\begin{aligned}\hat{V}_{\mathrm{P}} &= -G \sum_{mm'>0} \hat{\beta}_{-m}\hat{\beta}_{m}\hat{\beta}^{\dagger}_{m'}\hat{\beta}^{\dagger}_{-m'} \\ &= -G \sum_{mm'>0} [\hat{\beta}^{\dagger}_{m'}\hat{\beta}^{\dagger}_{-m'}\hat{\beta}_{-m}\hat{\beta}_{m} + \delta_{mm'}(\hat{\beta}_{-m}\hat{\beta}^{\dagger}_{-m'} - \hat{\beta}^{\dagger}_{m'}\hat{\beta}_{m})] \\ &= -G \sum_{mm'>0} \hat{\beta}^{\dagger}_{m'}\hat{\beta}^{\dagger}_{-m'}\hat{\beta}_{-m}\hat{\beta}_{m} - G(\Omega - N) \quad . \end{aligned} \tag{7.196}$$

The pairing potential for holes thus has the same form except for a contribution depending only on N; since we cannot describe the relative position of the ground states of neighboring nuclei in this model anyway, this part can be ignored, and the level scheme will be symmetric with respect to $N = \Omega/2$. Note, however, that in the energy formula N should be replaced by the number of holes $2\Omega - N$, and that s in the formulas then refers to unpaired *holes*, and must be $\leq 2\Omega - N$.

We now summarize the results of this section. The levels of the nucleus, which were completely degenerate without regard to angular-momentum coupling in the single-particle model, are now split up owing to the pairing interaction, and the spectrum is described by the seniority quantum number. The lowest state is that with the smallest value of s, i.e., $s = 0$ or $s = 1$. The distance to the next-higher state is found to be

$$E^{N}_{s+2} - E^{N}_{s} = G(\Omega - s) \quad , \tag{7.197}$$

so that in particular an energy gap of $G\Omega$ results between the ground state ($s = 0$) and the first excited state in even–even nuclei. Also, since this state has seniority zero, it is clear that the nuclear spin must vanish. For odd nuclei, the gap has the smaller value $G(\Omega - 1)$.

In the limiting case of few particles in a large shell (i.e. for $\Omega \gg N, s, 1$) the energy may be approximated by

$$E^{N}_{s} \approx -\frac{G}{2}(N - s)\Omega \quad , \tag{7.198}$$

so that each pair contributes $G\Omega$. For larger numbers of particles the Pauli principle reduces the effect of pairing.

EXERCISE

7.8 Pairing in a $j = \frac{7}{2}$ Shell

Problem. Calculate the effect of pairing in the spectrum for nuclei in a $j = 7/2$ shell.

Solution. As mentioned above, we are not interested in the relative energy of nuclei with different N, but regard only the lowering of the states due to pairing. Writing the energy formula (7.193) for the case $\Omega = 4$, we get

$$E_s^N = -\frac{G}{4}(N - s)(10 - N - s) \quad , \tag{1}$$

and the spectrum is plotted in Fig. 7.14.

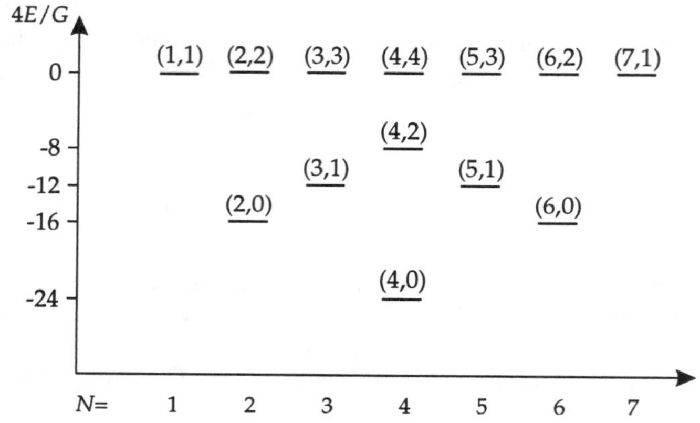

Fig. 7.14 The spectrum of the seniority model in a $j = 7/2$ shell. The individual states are labelled with (N, s), and the highest-lying states for each N are lined up at an energy of zero.

7.5.3 The Quasispin Model

There is a more elegant method to derive the results of the preceding section [Ke61]; although the results do not go beyond those presented there, the mathematical ideas are interesting enough to warrant a short discussion. The basic idea is to construct operators fulfilling angular-momentum commutation rules. Define the operators

$$\begin{aligned}
\hat{S}_+ &= \hat{A}^\dagger = \sum_{m>0} \hat{a}_m^\dagger \hat{a}_{-m}^\dagger \quad , \\
\hat{S}_- &= \hat{A} = \sum_{m>0} \hat{a}_{-m} \hat{a}_m \quad , \\
\hat{S}_0 &= \tfrac{1}{2} \sum_{m>0} \left(\hat{a}_m^\dagger \hat{a}_m + \hat{a}_{-m}^\dagger \hat{a}_{-m} - 1 \right) \quad ,
\end{aligned} \tag{7.199}$$

for m running from $-j$ to j. A simple check shows that these fulfil

$$[\hat{S}_+, \hat{S}_-] = 2\hat{S}_0 \quad , \quad [\hat{S}_0, \hat{S}_+] = \hat{S}_+ \quad , \quad [\hat{S}_0, \hat{S}_-] = -\hat{S}_- \quad , \tag{7.200}$$

i.e., exactly the same commutation relations as the angular-momentum operators $\hat{J}_+$, $\hat{J}_-$, and $\hat{J}_0$. This justifies introducing the name of "quasispin". Now the pairing potential can be expressed as

$$\hat{V}_{\mathrm{P}} = -G\,\hat{A}^\dagger \hat{A} = -G\,\hat{S}_+\hat{S}_- \quad , \tag{7.201}$$

while the operator $\hat{S}_0$ can be rewritten as

$$\hat{S}_0 = \tfrac{1}{2}\hat{N} - \tfrac{1}{2}\sum_{m>0} 1 = \tfrac{1}{2}(\hat{N} - \Omega) \quad . \tag{7.202}$$

The eigenvalues of $\hat{V}_{\mathrm{P}}$ can be related to those of the quasispin by noting that the square of the "angular momentum" operator must be defined analogously as

$$\hat{S}^2 = \hat{S}_+\hat{S}_- - \hat{S}_0 + \hat{S}_0^2 \quad , \tag{7.203}$$

so that

$$\hat{V}_{\mathrm{P}} = -G\left(\hat{S}^2 - \hat{S}_0^2 + \hat{S}_0\right) \quad . \tag{7.204}$$

The problem is thus solved by the well-known angular momentum eigenstates. Denoting the eigenvalue of $\hat{S}^2$ by $\sigma(\sigma + 1)$ and using (7.202), we find

$$E_\sigma^N = -G\left[\sigma(\sigma+1) - \tfrac{1}{4}(N - \Omega)^2 + \tfrac{1}{2}(N - \Omega)\right] \quad . \tag{7.205}$$

Formally the problem is solved, but the relation of the new quantum number σ to the seniority s, which it replaces, has to be found. Here a comparison with our previous construction helps. The states of fixed σ are distinguished by different eigenvalues of the "projection" $\hat{S}_0$, corresponding to a difference of 2 in $\hat{N}$ between neighboring projections, but because the operators $\hat{S}_\pm$ create and annihilate pairs, they must all have the same seniority. Thus σ must be a function of s and Ω. It can be derived by examining the state with the lowest projection $-\sigma = \frac{1}{2}(N - \Omega)$ (from the definition of $\hat{S}_0$), and the corresponding state has the property that

$$\hat{S}_-|-\sigma\rangle = \hat{A}|-\sigma\rangle = 0 \quad . \tag{7.206}$$

But physically this means that there are no pairs in the state, i.e., $s = N$. So σ must be given by

$$\sigma = \tfrac{1}{2}(\Omega - s) \quad . \tag{7.207}$$

Inserting this result into (7.204) produces (7.193), and by arguments similar to that for σ one may check that the allowed range for the quantum numbers is also identical.

7.5.4 The BCS Model

The seniority and quasispin models illustrate the action of the pairing force quite well, but are severely restricted by the assumption of a partially filled j shell, which is applicable only to spherical nuclei. The formalism was based on the coupling of two nucleons to zero angular momentum through a sum over pairs of states $(m, -m)$. For deformed nuclei, there is no longer any degeneracy of different projections within a j shell; but Kramers degeneracy still ensures the existence of pairs of degenerate, mutually time-reversal conjugate states, which should be coupled strongly by a short-range force. In a solid, for which case the theory was originally developed by Bardeen, Cooper, and Schrieffer [Ba57], these are momentum eigenstates $(\boldsymbol{k}, -\boldsymbol{k})$, whereas for a nucleus the angular-momentum projection onto the intrinsic axis will be the crucial quantum number. The application to nuclei was pioneered by Belyaev [Be59]. An extended version of the method is the so-called *Lipkin–Nogami* pairing model, which is discussed in [Li60, No64, Pr73] and can be understood easily once the techniques presented in this chapter are known.

It is still customary to denote the two states by k and $-k$, even when k refers to the angular-momentum projection Ω. In the following k will be used to represent all quantum numbers of the single-particle states; the only important property we need is that there are always two states k and $-k$ related to each other by time reversal and coupled preferentially by the pairing force.

We may start with a Hamiltonian that contains a pure single-particle part plus a residual interaction acting only on such pairs:

$$\hat{H} = \sum_k \varepsilon_k^0 \, \hat{a}_k^\dagger \hat{a}_k + \sum_{kk'>0} \langle k, -k|v|k', -k'\rangle \, \hat{a}_k^\dagger \hat{a}_{-k}^\dagger \hat{a}_{-k'} \hat{a}_{k'} \quad . \tag{7.208}$$

In spirit this is similar to the pairing force used up to now, except that the coupling to zero angular momentum is no longer implied and that we allow nonconstant matrix elements of the pairing potential. The restriction $k > 0$ in the second sum again means that only positive projections are being summed over. In the simplest case one may assume a constant matrix element $-G$ (we will come back to the more general case, though):

$$\hat{H} = \sum_k \varepsilon_k^0 \, \hat{a}_k^\dagger \hat{a}_k - G \sum_{kk'>0} \hat{a}_k^\dagger \hat{a}_{-k}^\dagger \hat{a}_{-k'} \hat{a}_{k'} \quad . \tag{7.209}$$

An analytic solution cannot be found in this case, but there is an approximate solution based on the *BCS state*

$$|\mathrm{BCS}\rangle = \prod_{k>0}^{\infty} \left(u_k + v_k \hat{a}_k^\dagger \hat{a}_{-k}^\dagger \right) |0\rangle \quad . \tag{7.210}$$

In this state each pair of single-particle levels $(k, -k)$ is occupied with a probability $|v_k|^2$ and remains empty with probability $|u_k|^2$. The parameters u_k and v_k will be determined through the variational principle. We will assume that they are real numbers; this will prove sufficiently general for time-independent problems.

We first examine a few properties of the BCS state.

- Normalization: the norm is given by

$$\langle \mathrm{BCS}|\mathrm{BCS}\rangle = \langle 0|\prod_{k>0}^{\infty}\left(u_k + v_k \hat{a}_{-k}\hat{a}_k\right)\prod_{k'>0}^{\infty}\left(u_{k'} + v_{k'}\hat{a}^\dagger_{k'}\hat{a}^\dagger_{-k'}\right)|0\rangle \quad . \tag{7.211}$$

  The terms in parentheses all commute for different indices, so only the product of two such terms with the same index needs to be considered:

$$\begin{aligned}&\left(u_k + v_k\hat{a}_{-k}\hat{a}_k\right)\left(u_k + v_k\hat{a}^\dagger_k\hat{a}^\dagger_{-k}\right)\\ &\quad = u_k^2 + u_k v_k\left(\hat{a}^\dagger_k\hat{a}^\dagger_{-k} + \hat{a}_{-k}\hat{a}_k\right) + v_k^2\,\hat{a}_{-k}\hat{a}_k\hat{a}^\dagger_k\hat{a}^\dagger_{-k} \quad .\end{aligned} \tag{7.212}$$

  Since the other terms in the product do not affect the states k and $-k$, this expression is effectively enclosed by the vacuum in the matrix element, and only the first and last terms will contribute. Thus the norm is

$$\langle \mathrm{BCS}|\mathrm{BCS}\rangle = \prod_{k>0}^{\infty}\left(u_k^2 + v_k^2\right) \quad , \tag{7.213}$$

  and for normalization we must require

$$u_k^2 + v_k^2 = 1 \quad . \tag{7.214}$$

- Particle number: clearly this is not a good quantum number for the BCS state. Its expectation value is

$$N = \langle \mathrm{BCS}|\hat{N}|\mathrm{BCS}\rangle = \langle \mathrm{BCS}|\sum_{k>0}\left(\hat{a}^\dagger_k\hat{a}_k + \hat{a}^\dagger_{-k}\hat{a}_{-k}\right)|\mathrm{BCS}\rangle \quad , \tag{7.215}$$

  and its value can be calculated by simply noting that for each value of k the operator projects the component proportional to v_k out of the BCS state and multiplies it by its particle number of 2, and then the projection onto the left-hand state will again only contribute a factor of v_k. The result is

$$N = \sum_{k>0} 2v_k^2 \quad . \tag{7.216}$$

  This naturally fits the interpretation of v_k^2 as the probability for having the pair $(k, -k)$ occupied.

- Particle-number uncertainty: the mean square deviation of the particle number is given by

$$\Delta N^2 = \langle \mathrm{BCS}|\hat{N}^2|\mathrm{BCS}\rangle - \langle \mathrm{BCS}|\hat{N}|\mathrm{BCS}\rangle^2 \quad . \tag{7.217}$$

  The matrix element of $\hat{N}^2$ follows simply from its expansion:

$$\hat{N}^2 = \sum_{kk'>0}\left(\hat{a}^\dagger_k\hat{a}_k + \hat{a}^\dagger_{-k}\hat{a}_{-k}\right)\left(\hat{a}^\dagger_{k'}\hat{a}_{k'} + \hat{a}^\dagger_{-k'}\hat{a}_{-k'}\right) \quad . \tag{7.218}$$

  Each term produces a factor of $2v_k^2 \times 2v_{k'}^2$ by counting the number of nucleons in the corresponding single-particle states. This result is not true, however, for the diagonal terms $k = k'$, where only one factor of $4v_k^2$ can be obtained from the wave function. Thus we have

$$\langle \mathrm{BCS}|\hat{N}^2|\mathrm{BCS}\rangle = 4 \sum_{\substack{k \neq k' \\ kk' > 0}} v_k^2 v_{k'}^2 + 4 \sum_{k>0} v_k^2 \quad . \tag{7.219}$$

Combining this with the previous result for the norm and using $u_k^2 = 1 - v_k^2$ yields

$$\Delta N^2 = 4 \sum_{k>0} u_k^2 v_k^2 \quad . \tag{7.220}$$

This last result shows that the uncertainty in the particle number is caused by those single-particle states that are fractionally occupied, i.e., for which neither u_k^2 nor v_k^2 is equal to unity. Since allowing only values of 0 or 1 for the occupation probabilities would mean reverting to the pure single-particle model, the effect of pairing must come in through this fractional occupation. The uncertainty in the particle number, while strictly speaking incorrect, may not be important as long as $\Delta N \ll N$. This has to be checked in the practical results.

As the trial wave function does not conserve the particle number, the desired expectation value has to be achieved through a constraint with a Lagrange multiplier. We are thus led to consider the variational condition

$$\delta \langle \mathrm{BCS}|\hat{H} - \lambda \hat{N}|\mathrm{BCS}\rangle = 0 \quad . \tag{7.221}$$

With the Hamiltonian of (7.209) and considering that the free parameters are the v_k, this is written more fully as

$$\frac{\partial}{\partial v_k} \langle \mathrm{BCS}| \sum_k (\varepsilon_k^0 - \lambda) \hat{a}_k^\dagger \hat{a}_k - G \sum_{kk'>0} \hat{a}_k^\dagger \hat{a}_{-k}^\dagger \hat{a}_{-k'} \hat{a}_{k'} |\mathrm{BCS}\rangle = 0 \quad . \tag{7.222}$$

The u_k depend on the v_k via the normalization $u_k^2 + v_k^2 = 1$, which yields $u_k \mathrm{d}u_k + v_k \mathrm{d}v_k = 0$ or

$$\frac{\partial}{\partial v_k} = \left.\frac{\partial}{\partial v_k}\right|_{u_k} - \frac{v_k}{u_k} \left.\frac{\partial}{\partial u_k}\right|_{v_k} \quad . \tag{7.223}$$

The evaluation of the matrix element is quite easy with the experience gained up to now. Examining which parts of the wave function are projected out by the operator to the left and right, we find

$$\begin{aligned} \langle \mathrm{BCS}|\hat{a}_k^\dagger \hat{a}_k|\mathrm{BCS}\rangle &= v_k^2 \quad , \\ \langle \mathrm{BCS}|\hat{a}_k^\dagger \hat{a}_{-k}^\dagger \hat{a}_{-k'} \hat{a}_{k'}|\mathrm{BCS}\rangle &= \begin{cases} u_k v_k u_{k'} v_{k'} & \text{for } k \neq k' \\ v_k^2 & \text{for } k = k' \end{cases} \quad . \end{aligned} \tag{7.224}$$

The pairing matrix element now reads

$$
\begin{aligned}
\langle \mathrm{BCS}| - G \sum_{kk'>0} \hat{a}_k^\dagger \hat{a}_{-k}^\dagger \hat{a}_{-k'} \hat{a}_{k'} |\mathrm{BCS}\rangle
&= -G \sum_{\substack{kk'>0 \\ k \neq k'}} u_k v_k u_{k'} v_{k'} - G \sum_{k>0} v_k^2 \\
&= -G \left(\sum_{k>0} u_k v_k \right)^2 - G \sum_{k>0} v_k^4 \quad ,
\end{aligned}
\tag{7.225}
$$

and the expectation value of the Hamiltonian becomes

$$
\langle \mathrm{BCS}|\hat{H} - \lambda \hat{N}|\mathrm{BCS}\rangle = 2 \sum_{k>0} (\varepsilon_k^0 - \lambda)\, v_k^2 - G \Big(\sum_{k>0} u_k v_k \Big)^2 - G \sum_{k>0} v_k^4 \, . \tag{7.226}
$$

This has to be differentiated according to (7.223), yielding

$$
\begin{aligned}
& 4(\varepsilon_k^0 - \lambda) v_k - 2G \left(\sum_{k'>0} u_{k'} v_{k'} \right) u_k - 4G v_k^3 \\
& \quad - \frac{v_k}{u_k} \Big[- 2G \Big(\sum_{k'>0} u_{k'} v_{k'} \Big) \Big] = 0 \quad .
\end{aligned}
\tag{7.227}
$$

All the equations for the different values of k are coupled through the term

$$
\Delta = G \sum_{k'>0} u_{k'} v_{k'} \quad . \tag{7.228}
$$

We proceed by assuming for the moment that Δ is known, deriving an explicit form for v_k and u_k, and then using the definition of Δ as a supplementary condition. If we abbreviate to

$$
\varepsilon_k = \varepsilon_k^0 - \lambda - G v_k^2 \quad , \tag{7.229}
$$

(7.227) reduces to

$$
2 \varepsilon_k v_k u_k + \Delta \big(v_k^2 - u_k^2 \big) = 0 \quad . \tag{7.230}
$$

Squaring this equation allows us to replace u_k^2 by v_k^2, and then we may solve for the latter:

$$
v_k^2 = \frac{1}{2} \left(1 \pm \sqrt{1 - \frac{\Delta^2}{\varepsilon_k^2 + \Delta^2}} \right) = \frac{1}{2} \left(1 \pm \frac{\varepsilon_k}{\sqrt{\varepsilon_k^2 + \Delta^2}} \right) \quad . \tag{7.231}
$$

The ambiguous sign results from the fourth-order equation appearing during the calculation; taking the square root of ε_k^2 contributed no further ambiguity. The correct sign can be selected by noting that for very large single-particle energies $\varepsilon_k \to \infty$ the occupation probabilities must go to zero; this is achieved by taking the negative sign. The final result is thus

$$
v_k^2 = \frac{1}{2} \left(1 - \frac{\varepsilon_k}{\sqrt{\varepsilon_k^2 + \Delta^2}} \right) \quad , \quad u_k^2 = \frac{1}{2} \left(1 + \frac{\varepsilon_k}{\sqrt{\varepsilon_k^2 + \Delta^2}} \right) \quad . \tag{7.232}
$$

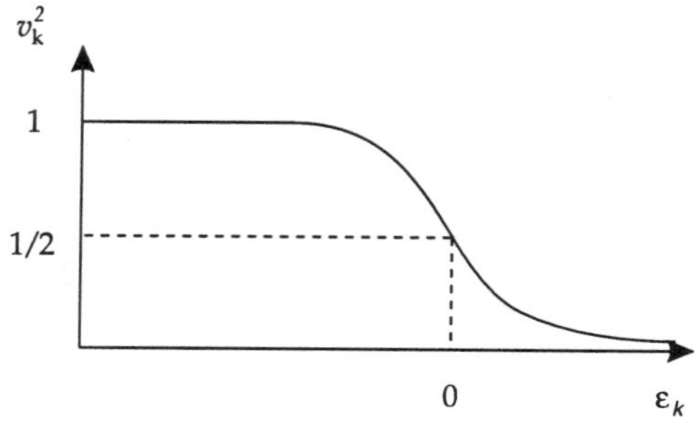

Fig. 7.15 The dependence of the occupation probabilities on the single-particle energy ε_k.

If we assume that λ and Δ have been determined, the behavior of these expressions is easily seen. For $\varepsilon_k = 0$, i.e. when $\varepsilon_k^0 - Gv_k^2 = \lambda$, both u_k^2 and v_k^2 are equal to $\frac{1}{2}$. For large negative values of ε_k we will have $u_k^2 \approx 0$ and $v_k^2 \approx 1$ and the reverse is true for large positive values. The width of the transition is governed by Δ. This behavior is illustrated in Fig. 7.15. We note that λ obviously plays the role of a generalized Fermi energy.

The unknown parameter can now be determined by inserting the explicit forms for u_k and v_k into its definition, yielding

$$\begin{aligned}\Delta &= G \sum_{k>0} u_k v_k \\ &= \sum_{k>0} \frac{G}{2} \sqrt{1 - \frac{\varepsilon_k^2}{\varepsilon_k^2 + \Delta^2}} \\ &= \frac{G}{2} \sum_{k>0} \frac{\Delta}{\sqrt{\varepsilon_k^2 + \Delta^2}} \quad ,\end{aligned} \tag{7.233}$$

i.e., the so-called *gap equation*

$$\Delta = \frac{G}{2} \sum_{k>0} \frac{\Delta}{\sqrt{\varepsilon_k^2 + \Delta^2}} \quad . \tag{7.234}$$

It can be solved iteratively using the known values of G and the single-particle energies ε_k^0. The other parameter λ then follows from simultaneously fulfilling the condition for the total particle number,

$$\sum_{k>0} 2v_k^2 = N \quad . \tag{7.235}$$

To do this the term $-Gv_k^2$ in the definition of the ε_k of (7.234) has to be neglected. This is usually done with the argument that it corresponds only to a renormalization of the single-particle energies.

Here are some hints for the practical application of the BCS model. The levels with $\varepsilon_k \approx 0$, i.e., those near the Fermi energy, will contribute most in the gap equation. This motivates us to treat protons and neutrons separately, since in all but the very light nuclei proton and neutron Fermi energies are quite different, which also leads to small matrix elements of a short-range pairing-type force between the respective wave functions. Thus the gap equation is written separately for the proton and neutron energy-level schemes and there will also be separate strengths G_p and G_n, gap parameters Δ_p and Δ_n, and Fermi energies λ_p and λ_n. Guidance values for the strength parameters are

$$G_\mathrm{p} \approx 17\,\mathrm{MeV}/A \quad , \quad G_\mathrm{n} \approx 25\,\mathrm{MeV}/A \quad , \tag{7.236}$$

but many other prescriptions can be found in the literature. Many authors use the pairing gap as the prescribed parameter, which simplifies the calculations considerably. It is also still a controversial question, whether for a deformed nucleus the pairing strength or gap depend on deformation.

The gap equation, (7.234), always has the trivial solution $\Delta = 0$. In most cases there is also the nontrivial paired solution with nonvanishing gap and lower energy. The two solutions cannot be linked by continuously increasing the pairing strength, so that one may talk of a *phase transition* between the paired and unpaired states of the nucleus.

7.5.5 The Bogolyubov Transformation

The BCS model may be formulated in a more elegant way by a transformation to new *quasiparticle* operators, the so-called Bogolyubov transformation developed by Bogolyubov and Valatin [Bo58, Va58, Bo59]. A welcomed by-product is a simple method of constructing the excited states of the nucleus as quasiparticle excitations.

The basic idea is to look for operators $\hat{\alpha}_k$ for which the BCS ground state is the vacuum state, i.e.,

$$\hat{\alpha}_k |\mathrm{BCS}\rangle = 0 \quad . \tag{7.237}$$

This is an analogous, but more general, problem to that for the Hartree–Fock state, for which the transformation to particles and holes leads to the desired formulation

$$\begin{aligned} \hat{\alpha}_k^\dagger &= \hat{a}_k^\dagger \quad , \quad \hat{\alpha}_k = \hat{a}_k \quad , \quad k > F \quad , \\ \hat{\alpha}_k^\dagger &= -\hat{a}_{-k} \quad , \quad \hat{\alpha}_k = -\hat{a}_{-k}^\dagger \quad , \quad k < F \quad , \end{aligned} \tag{7.238}$$

where $k < F$ and $k > F$ again refer to states occupied and unoccupied in the Hartree–Fock state, respectively. Note that for the holes ($k < F$) the creation of a hole k implies the destruction of a particle with angular-momentum projection k, so that its index should be denoted as $-k$.

In the case of the BCS state, the particle number is no longer conserved and it appears reasonable to try the more general transformation

$$\hat{\alpha}_k = p\hat{a}_k + q\hat{a}_{-k}^\dagger \quad , \tag{7.239}$$

where from the above arguments it is clear why the index has a different sign in the two terms. Applying this operator to the BCS state and noting that only the index k in the operator product can be important, we get the condition

$$\begin{aligned} 0 &= \left(p\hat{a}_k + q\hat{a}_{-k}^\dagger\right)\left(u_k + v_k \hat{a}_k^\dagger \hat{a}_{-k}^\dagger\right)|0\rangle \\ &= \left(qu_k + pv_k\right)\hat{a}_{-k}^\dagger |0\rangle \quad , \end{aligned} \tag{7.240}$$

yielding $qu_k + pv_k = 0$, which can be solved by $p = su_k$, $q = -sv_k$ with an as yet arbitrary real factor s. Doing the same calculation for $\hat{\alpha}_{-k}$, we get the two definitions

$$\begin{aligned} \hat{\alpha}_k &= s\, u_k\, \hat{a}_k - s\, v_k\, \hat{a}^\dagger_{-k} \quad , \\ \hat{\alpha}_{-k} &= t\, u_k\, \hat{a}_{-k} + t\, v_k\, \hat{a}^\dagger_k \quad , \end{aligned} \tag{7.241}$$

as well as the corresponding creation operators by Hermitian conjugation

$$\begin{aligned} \hat{\alpha}^\dagger_k &= s\, u_k\, \hat{a}^\dagger_k - s\, v_k\, \hat{a}_{-k} \quad , \\ \hat{\alpha}^\dagger_{-k} &= t\, u_k\, \hat{a}^\dagger_{-k} + t\, v_k\, \hat{a}_k \quad . \end{aligned} \tag{7.242}$$

It was assumed that u_k and v_k can be chosen to be real as before. The unknown factors s and t can be determined by requiring the usual fermion commutation rules, for example,

$$\delta_{kk'} = \left\{\hat{\alpha}_k, \hat{\alpha}^\dagger_{k'}\right\} = \delta_{kk'}\, s^2\left(u_k^2 + v_k^2\right) \quad . \tag{7.243}$$

All of them can be fulfilled by setting $s = t = 1$ and demanding that $u_k^2 + v_k^2 = 1$ as before.

The final version of the *Bogolyubov transformation* thus is the following definition of the *quasiparticle operators*:

$$\begin{aligned} \hat{\alpha}_k &= u_k \hat{a}_k - v_k \hat{a}^\dagger_{-k} \quad , & \hat{\alpha}^\dagger_k &= u_k \hat{a}^\dagger_k - v_k \hat{a}_{-k} \quad , \\ \hat{\alpha}_{-k} &= u_k \hat{a}_{-k} + v_k \hat{a}^\dagger_k \quad , & \hat{\alpha}^\dagger_{-k} &= u_k \hat{a}^\dagger_{-k} + v_k \hat{a}_k \quad , \end{aligned} \tag{7.244}$$

The inverse transformation is given by

$$\hat{a}_k = u_k \hat{\alpha}_k + v_k \hat{\alpha}^\dagger_{-k} \quad , \quad \hat{a}_{-k} = u_k \hat{\alpha}_{-k} - v_k \hat{\alpha}^\dagger_k \quad , \tag{7.245}$$

and the Hermitian conjugate for the creation operators.

The next task is to transform the Hamiltonian consisting of kinetic energy plus a two-body interaction

$$\hat{H} = \sum_{k_1 k_2} t_{k_1 k_2}\, \hat{a}^\dagger_{k_1} \hat{a}_{k_2} + \tfrac{1}{2} \sum_{k_1 k_2 k_3 k_4} \bar{v}_{k_1 k_2 k_3 k_4}\, \hat{a}^\dagger_{k_1} \hat{a}^\dagger_{k_2} \hat{a}_{k_4} \hat{a}_{k_3} \quad . \tag{7.246}$$

Here $\bar{v}_{k_1 k_2 k_3 k_4}$ refers to the antisymmetrized matrix element as in Sect. 7.2.4. Replacing the operators by the quasiparticle operators via the inverse Bogolyubov transformation as given above leads to a plethora of terms; we only discuss some general features for the simpler kinetic-energy terms (note that the matrix element $t_{k_1 k_2}$ does not depend on the signs of k_1 and k_2):

$$\begin{aligned} \sum_{k_1 k_2} & t_{k_1 k_2}\, \hat{a}^\dagger_{k_1} \hat{a}_{k_2} \\ &= \sum_{k_1 k_2 > 0} t_{k_1 k_2} \left(u_{k_1} \hat{\alpha}^\dagger_{k_1} + v_{k_1} \hat{\alpha}_{-k_1}\right)\left(u_{k_2} \hat{\alpha}_{k_2} + v_{k_2} \hat{\alpha}^\dagger_{-k_2}\right) \\ &\quad + \sum_{k_1 k_2 > 0} t_{k_1 k_2} \left(u_{k_1} \hat{\alpha}^\dagger_{-k_1} - v_{k_1} \hat{\alpha}_{k_1}\right)\left(u_{k_2} \hat{\alpha}_{k_2} + v_{k_2} \hat{\alpha}^\dagger_{-k_2}\right) \\ &\quad + \sum_{k_1 k_2 > 0} t_{k_1 k_2} \left(u_{k_1} \hat{\alpha}^\dagger_{k_1} + v_{k_1} \hat{\alpha}_{-k_1}\right)\left(u_{k_2} \hat{\alpha}_{-k_2} - v_{k_2} \hat{\alpha}^\dagger_{k_2}\right) \\ &\quad + \sum_{k_1 k_2 > 0} t_{k_1 k_2} \left(u_{k_1} \hat{\alpha}^\dagger_{-k_1} - v_{k_1} \hat{\alpha}_{k_1}\right)\left(u_{k_2} \hat{\alpha}_{-k_2} - v_{k_2} \hat{\alpha}^\dagger_{k_2}\right) \quad . \end{aligned} \tag{7.247}$$

Regard, for example, the term in $\hat{\alpha}_{k_1}^\dagger \hat{\alpha}_{k_2}$, which is

$$t_{k_1 k_2} \, u_{k_1} \, u_{k_2} \, \hat{\alpha}_{k_1}^\dagger \hat{\alpha}_{k_2} \quad , \tag{7.248}$$

whereas for the combination $\hat{\alpha}_{k_1} \hat{\alpha}_{k_2}^\dagger$ we have

$$t_{k_1 k_2} \, v_{k_1} \, v_{k_2} \, \hat{\alpha}_{k_1} \hat{\alpha}_{k_2}^\dagger \quad . \tag{7.249}$$

In the sum over k_1 and k_2 these may be combined to give

$$t_{k_1 k_2} \left[(u_{k_1} u_{k_2} - v_{k_1} v_{k_2}) \, \hat{\alpha}_{k_1}^\dagger \hat{\alpha}_{k_2} + v_{k_1}^2 \, \delta_{k_1 k_2} \right] \quad . \tag{7.250}$$

This simplest example illustrates the types of terms we expect: the operator products should be brought into *normal order*, i.e., all creation operators to the left of all annihilation operators, since in this case they will not contribute in the BCS ground state. Doing the commutation also generates terms with fewer operators like, as in the preceding example, one with no operators at all.

Treating all terms in this manner finally leads to a natural decomposition of the Hamiltonian according to the number of operators in the terms. Subtracting the term used to constrain the particle number, we may write it generically as

$$\hat{H} - \lambda \hat{N} = U + \hat{H}_{11} + \hat{H}_{20} + \hat{H}_{40} + \hat{H}_{31} + \hat{H}_{22} \quad , \tag{7.251}$$

where the two indices denote the number of creation and annihilation operators in the terms making up this particular part of the total $\hat{H}$. Before going into the details, let us discuss what the goals should be now. The term U is the energy of the BCS ground state with zero quasiparticles. $\hat{H}_{11}$ indicates the dependence of the energy of quasiparticle–quasihole excitations, and $\hat{H}_{20}$ violates quasiparticle number conservation and even implies that the BCS state will not be the true ground state. The other terms contain higher-order couplings and may be ignored for the moment. A reasonable interpretation of a BCS ground state with quasiparticle excitations requires $\hat{H}_{20} = 0$, and we can use this as the condition for determining the v_k (and, by implication, the u_k), which has so far been arbitrary. $\hat{H}_{20}$ turns out to be a sum of terms in $(\hat{\alpha}_k^\dagger \hat{\alpha}_{-k'}^\dagger + \hat{\alpha}_{-k} \hat{\alpha}_{k'})$ and requiring the coefficients to vanish leads to

$$\begin{aligned} 0 = & \left[t_{kk'} - \lambda \delta_{kk'} + \sum_{k''>0} \left(\bar{v}_{k-k''k'-k''} + \bar{v}_{kk''k'k''} \right) v_{k''}^2 \right] (u_k v_{k'} - u_{k'} v_k) \\ & + \sum_{k''>0} \bar{v}_{k-k'k''-k''} \, u_{k''} \, v_{k''} (u_k u_{k'} + v_k v_{k'}) \quad . \end{aligned} \tag{7.252}$$

This set of equations is a generalization of the Hartree–Fock equations (7.64), to which they reduce if the occupation numbers are restricted to 1 or 0 (the second term vanishes because $u_{k''} v_{k''} = 0$, the factor $(u_k v_{k'} - u_{k'} v_k)$ demands that k and k' differ in occupation, and the factor $v_{k''}^2$ in the sum restricts it to occupied levels). The second term thus may rightfully be regarded as the pairing term. It is convenient to introduce abbreviations for the *Hartree–Fock–Bogolyubov potential*,

$$\Gamma_{kk'}^{\mathrm{BCS}} = \sum_{k''} \bar{v}_{kk''k'k''} \, v_{k''}^2 \tag{7.253}$$

(note that the sum is now over both positive and negative values of k'', allowing the combination of the two terms in parentheses), and for the *pairing potential*

$$\Delta_{k-k'} = -\sum_{k''>0} \bar{v}_{k-k'k''-k''} u_{k''} v_{k''} \quad , \tag{7.254}$$

in terms of which the Hartree–Fock–Bogolyubov equations read

$$\begin{aligned} 0 = &\left(t_{kk'} - \lambda\delta_{kk'} + \Gamma_{kk'}^{\mathrm{BCS}}\right)(u_k v_{k'} + u_{k'} v_k) \\ &- \Delta_{k-k'}(u_k u_{k'} - v_k v_{k'}) \quad . \end{aligned} \tag{7.255}$$

It remains for us to indicate the other parts of the Hamiltonian. Using the same abbreviations we have

$$\begin{aligned} U &= \sum_{k>0}\left[\left(t_{kk} - \lambda + \tfrac{1}{2}\Gamma_{kk}^{\mathrm{BCS}}\right)2v_k^2 - \Delta_{k-k}u_k v_k\right] \quad , \\ \hat{H}_{11} &= \sum_{kk'>0}\left[\left(t_{kk'} - \lambda\delta_{kk'} + \Gamma_{kk'}^{\mathrm{BCS}}\right)(u_k u_{k'} - v_k v_{k'})\right. \\ &\quad \left. + \Delta_{k-k'}(u_k v_{k'} + v_k u_{k'})\right]\left(\hat{\alpha}_k^\dagger \hat{\alpha}_{k'} + \hat{\alpha}_{-k'}^\dagger \hat{\alpha}_{-k}\right) \quad . \end{aligned} \tag{7.256}$$

The terms with four operators are usually neglected, although it cannot of course be shown in general whether this is adequate.

The equations obtained from the Hartree–Fock–Bogolyubov transformation have yielded results that are somewhat more general than the simple pairing interaction used in the previous sections. In fact, we can now compute the pairing interaction from the underlying two-body interaction, and we should expect to recover the pairing theory only if the matrix elements are close to those of a simple pairing force. Since the case of a simple pairing force is also most often used in practical applications, we now reduce the above equations for the assumption of a diagonal pairing potential

$$\Delta_{k-k'} = \Delta_k \, \delta_{kk'} \quad . \tag{7.257}$$

In the same way as for the Hartree–Fock equations,the problem may be simplified by choosing the single-particle states as eigenstates of a suitably selected single-particle Hamiltonian $\hat{h}$. In this case the natural choice is

$$h_{kk'} = t_{kk'} - \lambda\delta_{kk'} + \Gamma_{kk'}^{\mathrm{BCS}} = \varepsilon_k \, \delta_{kk'} \quad , \tag{7.258}$$

and inserting this leads to the simplified form of the Hartree–Fock–Bogolyubov equations

$$2\varepsilon_k u_k v_k - \Delta_k\left(u_k^2 - v_k^2\right) = 0 \quad \text{for all } k \quad . \tag{7.259}$$

If the pairing matrix element is not diagonal, one may still select eigenfunctions according to (7.258), but (7.259) remains a matrix equation.

To see the connection with the BCS theory developed in the previous section, compare (7.259) with the BCS equation

$$2\varepsilon_k u_k v_k + \left(v_k^2 - u_k^2\right) G \sum_{k>0} u_k v_k = 0 \quad . \tag{7.260}$$

This is identical with (7.259) if

$$\Delta_k = G \sum_{k>0} u_k v_k \quad , \tag{7.261}$$

and this implies that the matrix element is constant:

$$\bar{v}_{k-k'k''-k''} = -G \; \delta_{kk'} \quad . \tag{7.262}$$

Solving (7.259) for u_k and v_k yields similar formulas as for the BCS theory, with only the constant gap replaced by Δ_k:

$$u_k^2 = \frac{1}{2}\left(1 + \frac{\varepsilon_k}{\sqrt{\varepsilon_k^2 + \Delta_k^2}}\right) \quad , \quad v_k^2 = \frac{1}{2}\left(1 - \frac{\varepsilon_k}{\sqrt{\varepsilon_k^2 + \Delta_k^2}}\right) \quad . \tag{7.263}$$

An analog to the gap equation is obtained by inserting these expressions into the version of (7.234) for diagonal pairing potential, yielding

$$\Delta_k = -\frac{1}{2} \sum_{k''>0} \frac{\bar{v}_{k-kk''-k''}}{\sqrt{\varepsilon_{k''}^2 + \Delta_{k''}^2}} \Delta_{k''} \quad . \tag{7.264}$$

The main new feature present in this formulation compared with simple BCS theory is the intricate coupling of the occupation numbers and the self-consistency problem. The single-particle Hamiltonian (7.258) depends on the occupation numbers v_k^2, which have to be determined by solving the gap equation (7.254) simultaneously with the iterations of the self-consistent field.

Finally we can insert the results for the pure pairing force into the other parts of the Hamiltonian. The ground-state energy becomes

$$U = \sum_{k>0} \left[\left(t_{kk} + \tfrac{1}{2} \Gamma_{kk}^{\mathrm{BCS}} - \lambda \right) 2 v_k^2 - \Delta_k u_k v_k \right] \quad , \tag{7.265}$$

and for the quasiparticle–quasihole part we get

$$\hat{H}_{11} = \sum_{k>0} \left[\varepsilon_k \left(u_k^2 - v_k^2 \right) + 2 \Delta_k u_k v_k \right] \left(\hat{\alpha}_k^\dagger \hat{\alpha}_k + \hat{\alpha}_{-k}^\dagger \hat{\alpha}_{-k} \right) \quad , \tag{7.266}$$

which may be simplified further using

$$u_k^2 - v_k^2 = \frac{\varepsilon_k}{\sqrt{\varepsilon_k^2 + \Delta_k^2}} \quad , \quad u_k v_k = \frac{1}{2} \frac{\Delta_k}{\sqrt{\varepsilon_k^2 + \Delta_k^2}} \quad . \tag{7.267}$$

The final result is

$$\hat{H}_{11} = \sum_{k>0} e_k \left(\hat{\alpha}_k^\dagger \hat{\alpha}_k + \hat{\alpha}_{-k}^\dagger \hat{\alpha}_{-k} \right) \tag{7.268}$$

with the *quasiparticle energy*

$$e_k = \sqrt{\varepsilon_k^2 + \Delta_k^2} \quad . \tag{7.269}$$

This has the form of a Hamiltonian of noninteracting quasiparticles. In this sense the problem of pairing correlations has been simplified considerably: the ground state now contains correlations between the nucleons via fractional occupation numbers and the excited states can be approximated as consisting of noninteracting quasiparticles with their energies related to the underlying single-particle Hartree–Fock eigenenergies via (7.269).

It is now easy to construct the excited states, although some more details related to the particle number are often ignored. The ground state contains contributions of various, but always even, total particle numbers, and will thus always describe an even–even nucleus. Acting on it with one quasiparticle creation operator will change the total numbers to odd ones, so it should describe an odd nucleus (we will come back to that in a moment). Adding two quasiparticle operators keeps an even total number, but how can we guarantee that the expectation number of $\hat{N}$ is not changed? In the formalism we demanded a fixed expectation value of the nucleon number only for the ground state, not for excited states. In fact, if $\hat{\alpha}_k^\dagger$ refers to a state far above the Fermi energy, it will be almost equal to $\hat{a}_k^\dagger$ (because $v_k \approx 1$ and $u_k \approx 0$), and the state with two quasiparticles will on average contain two nucleons more than the ground state. Far below the Fermi surface the quasiparticles almost coincide with usual holes. Only for states close to the Fermi surface the operators $\hat{\alpha}_k^\dagger$ create and annihilate nucleons with almost equal probability, so that the average particle number is not changed much and we stay in the same nucleus as was described by the ground state!

The lowest noncollective excited states of an even nucleus should thus take the form

$$\hat{\alpha}_k^\dagger \hat{\alpha}_{k'}^\dagger |\mathrm{BCS}\rangle \tag{7.270}$$

with excitation energy

$$e_k + e_{k'} > \Delta_k + \Delta_{k'} \quad . \tag{7.271}$$

There is thus again a gap determined by the Δ_k. For odd nuclei, as discussed above, the ground state will be

$$\hat{\alpha}_k^\dagger |\mathrm{BCS}\rangle \quad , \tag{7.272}$$

where the state with the lowest quasiparticle energy e_k is selected. An excited state can be created by simply shifting the quasiparticle to another state, $\hat{\alpha}_{k'}^\dagger |\mathrm{BCS}\rangle$, with excitation energy $e_{k'} - e_k$. Obviously this expression can be comparable to the distance between single-particle levels ε_k and there is no gap.

7.5.6 Generalized Density Matrices

The Hartree–Fock–Bogolyubov equations can be cast into a similarly simple form as the Hartree–Fock equations themselves by using density matrices [Bo59, Va61]. The density matrix for the BCS ground state can easily be evaluated:

$$\rho_{lk} = \langle \mathrm{BCS} | \hat{a}_k^\dagger \hat{a}_l | \mathrm{BCS} \rangle = v_k^2 \, \delta_{lk} \quad . \tag{7.273}$$

The Hartree–Fock–Bogolyubov equations, however, also contain the combination $u_k v_k$, which cannot be produced in this way. We thus need an additional "anomalous" density matrix

$$\kappa_{lk} = \langle \mathrm{BCS}|\hat{a}_k \hat{a}_l|\mathrm{BCS}\rangle \quad . \tag{7.274}$$

If $k > 0$, the matrix element will be nonzero only if $l = -k$, and the combination annihilates the pair created in the right-hand BCS state with amplitude v_k, picking up a factor u_k through the amplitude of the pair not present in the left-hand BCS state. The sign is positive, because in the BCS state the pair is created using $\hat{a}_k^\dagger \hat{a}_{-k}^\dagger$. If $k < 0$, we must have $l = -k$ again, but now there is a minus sign because the operators destroy the two nucleons in the wrong order. The general result is

$$\kappa_{lk} = \begin{cases} u_k v_k \delta_{k-l} & \text{for } k > 0 \\ -u_k v_k \delta_{k-l} & \text{for } k < 0 \end{cases} \quad . \tag{7.275}$$

Clearly κ is antisymmetric,

$$\kappa^{\mathrm{T}} = -\kappa \quad . \tag{7.276}$$

Our discussion has been restricted to the case of real coefficients u_k and v_k. The density-matrix treatment is usually cited in the literature for the case of complex coefficients (which is necessary, for example, in time-dependent problems), and we now give the formulas for this more general case. The above equation then has to be replaced by

$$\kappa^\dagger = -\kappa \tag{7.277}$$

which coincides with the old result for real matrices. We will ignore the complex case and use the simpler definitions below; this corresponds to leaving out some complex conjugations and using the transpose instead of the Hermitian conjugate throughout.

The density matrix will no longer be idempotent; instead

$$(\rho^2 - \rho)_{lk} = \left(v_k^4 - v_k^2\right) \delta_{lk} = -v_k^2 u_k^2 \, \delta_{lk} = -\sum_m \kappa_{lm} \kappa_{km} \tag{7.278}$$

or in pure matrix notation

$$\rho^2 - \rho = -\kappa \kappa^{\mathrm{T}} \quad . \tag{7.279}$$

It is trivial to show that

$$\rho \kappa = \kappa \rho \quad . \tag{7.280}$$

The matrices ρ and κ may be combined into a *generalized density matrix*

$$\mathcal{R} = \begin{pmatrix} \rho & \kappa \\ -\kappa & 1 - \rho \end{pmatrix} \quad . \tag{7.281}$$

The matrix $\mathcal{R}$ is constructed such that it is symmetric (Hermitian for complex values) and idempotent. The first of these properties follows from the symmetric nature of ρ and (7.276), whereas idempotence can be checked explicitly:

$$\mathcal{R}^2 = \begin{pmatrix} \rho^2 - \kappa^2 & \rho\kappa + \kappa - \kappa\rho \\ -\kappa\rho + \rho - \rho^2 & -\kappa^2 + (1-\rho)^2 \end{pmatrix} \quad . \tag{7.282}$$

In the off-diagonal elements the idempotence of ρ and (7.280) lead to the desired result; in the diagonal parts use (7.279) together with (7.276).

It is now a matter of simple calculation to show that defining a generalized Hamiltonian according to

$$\mathcal{H} = \begin{pmatrix} h & \Delta \\ -\Delta & -h \end{pmatrix} \tag{7.283}$$

allows one to formulate the Hartree–Fock–Bogolyubov equations in the concise form

$$[\mathcal{H}, \mathcal{R}] = 0 \quad , \tag{7.284}$$

reminiscent of the density-matrix formulation of standard Hartree–Fock theory and as useful for formal manipulations.

8. Interplay of Collective and Single-Particle Motion

8.1 The Core-plus-Particle Models

8.1.1 Basic Considerations

The collective model as treated up to now always deals with nuclei containing and even number of protons and neutrons. The next step of complication arises if one of these numbers is odd, i.e., if a single proton or neutron is added to such a collective nucleus. It must then be expected that excitations of both collective and single-particle character will be possible and will, in general be coupled. As a first approximation, it appears to be reasonable to regard the even–even nucleus as a *collective core* whose internal structure is not affected by the particle moving on its surface.

To be more specific, let us examine the case of a collective core described by a geometric collective model. The concept is thus that of a single particle moving in a potential generated by the time-dependent core. The particle "feels" an instantaneous potential of the Nilsson type (Sect. 7.4.2), and in turn its presence generates a force onto the core which distorts its collective motion. This is the *adiabatic approximation* which guided the initial development of this model [Bo53a, Ke56]. The situation is illustrated in Fig. 8.1, which also illustrates the angular-momentum coupling discussed later.

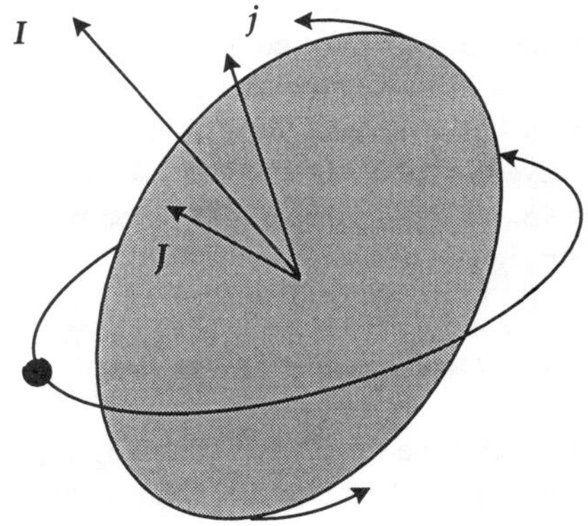

Fig. 8.1 Schematic illustration of the core-plus-particle model, illustrating also the coupling of the angular momentum of the core, J, with that of the odd particle, j, to a resulting total angular momentum I.

A useful approximation results from a consideration of the time scales involved. Typical single-particle energies near the Fermi level are of the order of 40 MeV, of which about half is kinetic energy, while the collective vibrations have an energy-level spacing of the order of 0.5 MeV. This shows that the collective motion is very much slower than that of the single particles, so that the latter can adapt to the core motions and the average potential follows the nuclear deformation. If the nuclear surface is defined in the usual way as

$$R(\theta,\phi) = R_0\Big(1 + \sum_\mu \alpha^*_{2\mu} Y_{2\mu}(\theta,\phi)\Big) \quad , \tag{8.1}$$

the nuclear surface should correspond to an equipotential surface of the potential. A potential possessing the correct equipotential surface can be set up in the form

$$V(\boldsymbol{r};\alpha_{2\mu}) = V_0\Big[r\Big/\Big(1 + \sum_\mu \alpha^*_{2\mu} Y_{2\mu}(\theta,\phi)\Big)\Big] \quad , \tag{8.2}$$

and this can be expanded to first order in the $\alpha_{2\mu}$:

$$V(\boldsymbol{r};\alpha_{2\mu}) \approx V(\boldsymbol{r},\alpha_{2\mu}=0) - r\frac{\mathrm{d}V_0}{\mathrm{d}r}\sum_\mu \alpha^*_{2\mu} Y_{2\mu}(\theta,\phi) \quad . \tag{8.3}$$

If V_0 is an oscillator potential $V_0(r) = \frac{1}{2}m\omega^2 r^2$, this becomes

$$V(\boldsymbol{r};\alpha_{2\mu}) = \tfrac{1}{2}m\omega^2 r^2 - m r^2\omega^2 \sum_\mu \alpha^*_{2\mu} Y_{2\mu}(\theta,\phi) \tag{8.4}$$

and agrees with the Nilsson-model potential. Note that the equipotential surfaces in this approximation are of the pure ellipsoid type and agree with the quadrupole shapes only to first order in α.

The best way to determine eigenstates for such a coupling potential depends on its strength. If the core is spherical, it will be relatively small, being generated only by the small oscillations around sphericity. In this case it appears sufficient to construct the solutions of the combined systems out of the wave functions of the spherical oscillator coupled with the single-particle functions of the spherical-shell model. This is the *weak-coupling limit.* If the ground state is strongly deformed, it is better to take the ground state-deformation directly into account in the core wave functions by using the rotation–vibration model; the single particle will then be expressed in terms of Nilsson model wave functions and this corresponds to the *strong-coupling limit.*

8.1.2 The Weak-Coupling Limit

As mentioned, the weak-coupling limit is built on a spherical vibrator and spherical single-particle model. Its Hamiltonian may thus be decomposed into

$$\hat{H} = \hat{H}_{\mathrm{coll}} + \hat{H}_{\mathrm{sp}} + \hat{H}_{\mathrm{coupling}} \tag{8.5}$$

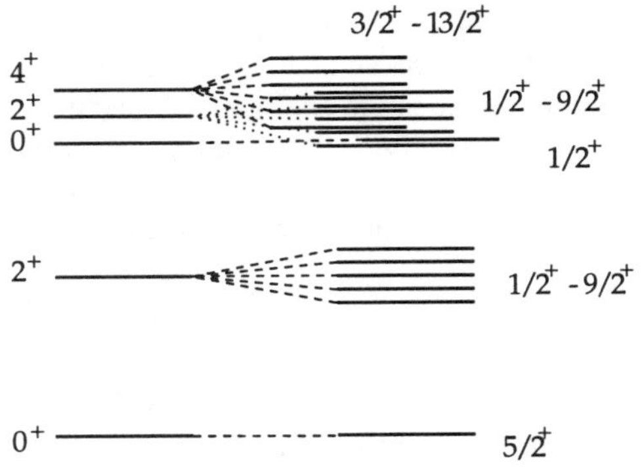

Fig. 8.2 Splitting of the collective vibrational states under the influence of the core–particle coupling.

with

$$
\begin{aligned}
\hat{H}_{\text{coll}} &= \frac{\sqrt{5}}{2B}\,[\pi \times \pi]^0 + \frac{\sqrt{5}C}{2}\,[\alpha \times \alpha]^0 \quad , \\
\hat{H}_{\text{sp}} &= -\frac{\hbar^2}{2m}\nabla^2 + \frac{m\omega^2}{2} r^2 + C\,\hat{\boldsymbol{l}} \cdot \hat{\boldsymbol{s}} + D\,\hat{\boldsymbol{l}}^2 \quad , \\
\hat{H}_{\text{coupling}} &= -m\omega^2 r^2 \sum_{\mu} \alpha^*_{2\mu} Y_{2\mu}(\theta, \phi) \quad .
\end{aligned} \tag{8.6}
$$

Here the full single-particle Hamiltonian is given, including spin–orbit coupling.

If it is sufficiently weak, the coupling term may be treated in perturbation theory. If we use the collective spherical-oscillator states $|NJM_J\rangle$ and the spherical single-particle oscillator states $|nljm\rangle$ as basis states, the states of the odd nucleus are given by angular-momentum coupling via

$$
|NJnlj;IM\rangle = \sum_{mM_J} (JlI\,|mM_J M)|NJM_J\rangle|nljm\rangle \quad . \tag{8.7}
$$

In the ground state the core must be in the phonon vacuum state $|0\rangle$ with $N = 0$, $J = 0$, and the particle in the lowest single-particle state allowed by the Pauli principle. If this is $|nljm\rangle$, the ground state of the odd nucleus will be

$$
|00nlj;jm\rangle = |0\rangle|nljm\rangle \quad . \tag{8.8}
$$

So obviously this ground state just shows the angular -omentum properties of the single-particle state. Since the spacing between single-particle levels is usually larger than that of collective oscillations, the lowest excited state will be built upon the one-phonon state $|12M_J\rangle = \hat{\beta}^\dagger_{M_J}|0\rangle$, leading to the odd-nucleus state

$$
|12nlj;IM\rangle = \sum_{mM_J} (2jI\,|mM_J M)\hat{\beta}^\dagger_{M_J}|0\rangle|nljm\rangle \quad . \tag{8.9}
$$

The total angular momentum ranges from $|j-2|$ through $j+2$. This construction can be continued to higher phonon multiplets analogously. It is illustrated in Fig. 8.2. These states, however, are as yet degenerate and will split up only under the influence of the core–particle coupling, which we treat now.

The coupling term $\hat{H}_{\text{coupling}}$ of (8.6) obviously couples collective states differing by 1 in the phonon number, and single-particle states with up to two units difference in angular momentum and in radial quantum number. For example, the ground state of the nucleus $|00nlj;jm\rangle$ couples to all states of the form $|12n'l'j';jm\rangle$, with the same resulting angular momentum j and projection m. Decomposing the matrix element into the collective and single-particle parts yields

$$\langle 12n'l'j';jm|\hat{H}_{\text{coupling}}|00nlj;jm\rangle = \sqrt{\frac{\hbar}{2B\omega_2}}\langle j'||Y_2||j\rangle\langle n'l'|m\omega^2 r^2|nl\rangle \quad .(8.10)$$

Here the matrix element of the creation operator contained in $\alpha_{2\mu}$ was inserted, the single-particle matrix element decomposed into angular and radial parts, and the latter expressed through the reduced matrix element. The radial ones are nontrivial and can be calculated to be given by

$$\langle n'l'|m\omega^2 r^2|nl\rangle = \hbar\omega \times \begin{cases} l + 2n + \frac{3}{2} & \text{for } n' = n,\ l' = l \\ \frac{1}{2}\sqrt{2n(2n+2l+1)} & \text{for } n' = n-1,\ l' = l \\ \frac{1}{2}\sqrt{(2n+2l+1)(2n+2l-1)} & \text{for } n' = n,\ l' = l-2 \\ \sqrt{(2n+2)(2n+2l+1)} & \text{for } n' = n+1,\ l' = l-2 \\ \frac{1}{2}\sqrt{2n(2n-2)} & \text{for } n' = n+1,\ l' = l+2 \end{cases} \quad (8.11)$$

All other matrix elements vanish. The true ground state of the system will thus be a superposition of the unperturbed ground state, mixed with a combination of the one-phonon state coupled to the low-lying single-particle states allowed by angular-momentum rules. Perturbation theory can be used to find a relatively simple treatment of the coupling.

In experiment, unfortunately, this model in its simplest version is not too successful. Although the angular momenta and parities of the first few excited states often correspond to the values predicted, their ordering and the energy spacings cannot be described quantitatively, and the coupling to other phonon or single-particle states has to be taken into account.

8.1.3 The Strong-Coupling Approximation

In the weak-coupling model it was assumed that the collective core is spherical and can be described by the spherical-vibrator wave functions. For a deformed core, one could in principle replace the spherical vibrator model by a generalized Hamiltonian such as discussed in Sect. 6.6.1 and expand the wave functions of the odd nucleus in a basis consisting of products of collective wave functions coupled with spherical single-particle states. There are two obstacles, however, to proceeding in this way:

1. The interaction between the particle and the core is expected to be quite strong in this case, since we know from the deformed-shell model (Sect. 7.3.2) that single-particle wave functions are strongly affected by the core deformation. Conceptually the deformed single-particle wave function should be "dragged along" by the core, following its rotations or vibrations closely. Expanding this

in a spherical basis which is unaware of the present orientation of the core may lead to a very bad first approximation.

2. The single-particle configuration changes with deformation: different states may be occupied depending on the instantaneous shape of the core. Since the core itself represents the occupied levels (except for the odd particle), care has to be taken not to introduce spurious double occupations, and the formalism becomes quite complicated.

In the strong-coupling Hamiltonian, on the other hand, the effect of the deformation of the core is taken into account directly in the basis states by using the wave functions of the rotation–vibration model coupled to those of the deformed shell model. The new problem arising in this formulation is how to deal with the coupling of the angular momenta, since the single-particle wave functions are given in the intrinsic frame of the collective model, which is not rotationally invariant. Below we will see that this can be done quite simply, and it is interesting to note that similar concepts have also been applied to other problems, such as octupole vibrations of a deformed nucleus. Although the coupling of the angular momenta of particle and core was first treated by Bohr and Mottleson [Bo53a] and Kerman [Ke56], the full treatment of the coupling to the vibrations was first given by Faessler [Fa64b].

The basic idea is thus to write the Hamiltonian and wave functions of the odd particle in the intrinsic frame of the deformed core. The single-particle angular momentum $\hat{\jmath}$ has to be coupled to that of the core, $\hat{\mathcal{M}}$, to give the total angular momentum of the nucleus

$$\hat{I} = \hat{\mathcal{M}} + \hat{\jmath} \quad . \tag{8.12}$$

The total Hamiltonian of the nucleus in the strong-coupling limit is split up into

$$\hat{H}_{\mathrm{sc}} = \hat{H}_{\mathrm{coll}} + \hat{H}_{\mathrm{sp}} \quad , \tag{8.13}$$

where it is assumed that there is no need for an additional coupling term, at least in the lowest order, since it is already included in the strong-coupling approximation itself. The collective part consists of

$$\hat{H}_{\mathrm{coll}} = \hat{H}_{\mathrm{rot}} + \hat{H}_{\mathrm{vib}} \tag{8.14}$$

(as before we ignore the rotation–vibration interaction) and can be written down immediately, by simply noting that the angular momentum of the core is now given by $\hat{\mathcal{M}} = \hat{I} - \hat{\jmath}$:

$$\begin{aligned} \hat{H}_{\mathrm{rot}} &= \frac{(\hat{I}' - \hat{\jmath}')^2 - (\hat{I}'_z - \hat{\jmath}'_z)^2}{2\mathcal{J}} + \frac{(\hat{I}'_z - \hat{\jmath}'_z)^2}{16B\eta^2} \quad , \\ \hat{H}_{\mathrm{vib}} &= -\frac{\hbar^2}{2B}\left(\frac{\partial^2}{\partial\xi^2} + \frac{1}{2}\frac{\partial^2}{\partial\eta^2}\right) + \frac{1}{2}C_0\xi^2 + C_2\eta^2 - \frac{\hbar^2}{16B\eta^2} \quad . \end{aligned} \tag{8.15}$$

The single-particle Hamiltonian is just that of the deformed-shell model written in intrinsic coordinates,

$$\begin{aligned} \hat{H}_{\mathrm{sp}} &= \frac{\hat{p}'^2}{2m} + \frac{m\omega^2}{2}r'^2 + C\,\hat{l}'\cdot\hat{s}' + D\,\hat{l}'^2 \\ &\quad - m\omega^2 r'^2\left[a_0 Y_{20}(\theta',\phi') + a_2\left(Y_{22}(\theta',\phi') + Y_{2-2}(\theta',\phi')\right)\right] \quad . \end{aligned} \tag{8.16}$$

The deformed potential was here expressed in terms of the intrinsic deformations $a_0 = \beta_0 + \xi$ and $a_2 = \eta$.

We now regroup the various parts of the Hamiltonian $\hat{H}_{\rm sc}$ by separating the terms that are diagonal in the basis functions from those that describe an interaction. In the single-particle Hamiltonian the only non-diagonal terms are those involving ξ and η, while from the collective part the single-particle angular-momentum terms have to be eliminated except for the intrinsic z projection, which is still a good quantum number in the deformed-shell model. The decomposition thus looks like this:

$$\hat{H}_{\rm sc} = \hat{H}_{\rm 0coll} + \hat{H}_{\rm 0sp} + \hat{H}' \tag{8.17}$$

with

$$\begin{aligned} \hat{H}_{\rm 0coll} &= \frac{\hat{I}'^2 - (\hat{I}'_z - \hat{\jmath}'_z)^2}{2\mathcal{J}} + \frac{(\hat{I}'_z - \hat{\jmath}'_z)^2 - 1}{16B\eta^2} \\ &\quad - \frac{\hbar^2}{2B}\left(\frac{\partial^2}{\partial\xi^2} + \frac{1}{2}\frac{\partial^2}{\partial\eta^2}\right) + \tfrac{1}{2}C_0\xi^2 + C_2\eta^2 \quad , \\ \hat{H}_{\rm sp} &= \frac{\hat{p}'^2}{2m} + \frac{m\omega^2}{2}r'^2 + C\,\hat{\boldsymbol{l}}'\cdot\hat{\boldsymbol{s}}' + D\,\hat{\boldsymbol{l}}'^2 \\ &\quad - m\omega^2 r'^2 \beta_0 Y_{20}(\theta',\phi') \quad , \end{aligned} \tag{8.18}$$

and

$$\begin{aligned} \hat{H}' &= \frac{\hat{\jmath}'^2}{2\mathcal{J}} - \frac{\hat{I}'_+\hat{\jmath}'_- + \hat{I}'_-\hat{\jmath}'_+ + 2\hat{I}'_z\hat{\jmath}'_z}{2\mathcal{J}} \\ &\quad - m\omega^2 r'^2\left[\xi Y'_{20} + \eta\left(Y'_{22} + Y'_{2-2}\right)\right] \quad . \end{aligned} \tag{8.19}$$

Since the rotation–vibration interaction has already been neglected, the term $\hat{H}'$ will also not be considered further; we only note that its first part describes the Coriolis force acting on the particle in the rotating frame, while the second part contains the interaction between the core vibrations and the odd particle.

The solutions for the Hamiltonian without interaction can be written down quite simply by just forming a product of the wave functions of the rotation–vibration model with those of the deformed-shell model; however, symmetrization because of the ambiguities of the choice of the intrinsic system (see Sect. 6.1.4) has to be considered separately. For the collective part the unsymmetrized wave functions are given by (see Sect. 6.4.3)

$$\psi_{\rm coll}(\xi,\eta,\boldsymbol{\theta}) = N\ \mathcal{D}^{(I)*}_{MK}(\boldsymbol{\theta})\ \chi_{K-\Omega\ n_\gamma}(\eta)\ \langle\beta|n_\beta\rangle \quad , \tag{8.20}$$

with an as yet unspecified normalization factor N. Here Ω is the eigenvalue of $\hat{\jmath}'_z$, i.e., the projection of the odd particle angular momentum onto the intrinsic z axis, and K is the corresponding projection of the total angular momentum. It is clear that K is the correct index in the rotational eigenfunction, but why is it replaced by $K - \Omega$ in the η-vibrational function? The reason is that the "centrifugal" potential in $1/\eta^2$ contains $\hat{I}'_z - \hat{\jmath}'_z$, so that for the η vibrations the effective potential contains $K - \Omega$ and must be taken over into the wave function.

The single-particle wave functions are those of the deformed-shell model, for which the only exact quantum number is the projection Ω. Distinguishing different states of the same Ω by an enumerative index κ, they solve

$$\hat{H}_{0\mathrm{sp}}\,\phi_{\kappa\Omega} = E_{\kappa\Omega}\,\phi_{\kappa\Omega} \quad . \tag{8.21}$$

The eigenfunctions of the total Hamiltonian $\hat{H}_{\mathrm{sc}}$ now become

$$\Psi_{IK\Omega n_\gamma n_\beta\kappa} = N\,\mathcal{D}^{(I)*}_{MK}(\boldsymbol{\theta})\,\chi_{K-\Omega\,n_\gamma}(\eta)\,\langle\beta|n_\beta\rangle\,\phi_{\kappa\Omega} \tag{8.22}$$

with the eigenenergies a sum of the two parts:

$$\begin{aligned} E_{IK\Omega n_\gamma n_\beta\kappa} &= E_{\kappa\Omega} + \frac{\hbar^2}{2\mathcal{J}}\left[I(I+1)-(K-\Omega)^2\right] \\ &\quad + \left(\tfrac{1}{2}|K-\Omega| + 2n_\gamma + 1\right)\hbar\omega_\gamma + \left(n_\beta + \tfrac{1}{2}\right)\hbar\omega_\beta \quad . \end{aligned} \tag{8.23}$$

Again the only adjustment was to replace K by $K-\Omega$ in the term arising from the η-vibrational contribution.

The allowed quantum numbers are as usual influenced by the symmetry requirements. Proceeding similarly as in Sect. 6.4.3, we note that $\hat{R}_3$ does not apply because the z' axis is special. Of the others, the simpler one is $\hat{R}_2$, which inverts the sign of η and adds $\pi/2$ to θ_3. As in the pure rotation–vibration model, the latter operation picks up a factor of $\exp(\mathrm{i}\pi K/2)$ from the rotation matrix, while inverting the sign of η now produces a factor of $(-1)^{(K-\Omega)/2}$ instead of $(-1)^{K/2}$ from the η-vibrational wave function. There is, however, an additional contribution $(-1)^{-\Omega/2}$ coming from the single-particle wave function, which as an eigenfunction of $\hat{\mathcal{J}}'_3$ contains a factor $\exp(-\mathrm{i}\Omega\theta_3)$. The total effect of $\hat{R}_2$ is thus

$$\begin{aligned} \hat{R}_2\Psi_{IK\Omega n_\gamma n_\beta\kappa} &= (-1)^{K/2}(-1)^{(K-\Omega)/2}(-1)^{-\Omega/2}\,\Psi_{IK\Omega n_\gamma n_\beta\kappa} \\ &= (-1)^{K-\Omega}\,\Psi_{IK\Omega n_\gamma n_\beta\kappa} \quad . \end{aligned} \tag{8.24}$$

The wave function is invariant only if $K-\Omega$ is even.

For $\hat{R}_1$ remember that this corresponds to replacing the choice of axes (x',y',z') by $(x',-y',-z')$. The effect on the rotation matrix has already been given:

$$\hat{R}_1\,\mathcal{D}^{(I)*}_{MK}(\boldsymbol{\theta}) = (-1)^{I-2K}\,\mathcal{D}^{(I)*}_{M-K}(\boldsymbol{\theta}) \quad . \tag{8.25}$$

We cannot omit $2K$ in the exponent, because it is now half-integer. Unfortunately the effect of $\hat{R}_1$ on the single-particle wave function cannot be formulated as simply. Since it does not lead to additional selection rules in the general case, we do not give any details and simply write down the symmetrized wave function as

$$\begin{aligned} \Psi_{IK\Omega n_\gamma n_\beta\kappa} &= \sqrt{\frac{2I+1}{16\pi^2}}\left[\mathcal{D}^{(I)*}_{MK}\,\phi_{\kappa\Omega} + (-1)^{I-2K}\,\mathcal{D}^{(I)*}_{M-K}\,\hat{R}_1\,\phi_{\kappa\Omega}\right] \\ &\quad \times \chi_{|K-\Omega|n_\gamma}\,\langle\beta|n_\beta\rangle \quad . \end{aligned} \tag{8.26}$$

The final task now is to construct the spectrum of the model. This will of necessity be much more complex than that of the pure rotation–vibration model. The crucial starting point is a knowledge of the ground-state deformation β_0. Examining the spectrum of the deformed-shell model in Figs. 7.6–7, one has to fill the states until all except the odd particle have been inserted. The single-particle level containing the odd particle then determines the lowest accessible function amidst the $\phi_{K_0\Omega_0}$, and the ones above this, denoted by $\phi_{K_i\Omega_i}$, give rise to single-particle excitations. There is thus a spectrum of such excitations with energies given by $E_{K_i\Omega_i} - E_{K_0\Omega_0}$.

On each of these single-particle excitations a rotation-vibration band is built, whose structure can be seen from (8.23). For example the ground-state band has

$$K = \Omega_0 \quad , \quad n_\gamma = n_\beta = 0 \quad , \tag{8.27}$$

and its sequence of angular momenta must be given by

$$I = \Omega_0,\ \Omega_0 + 1, \dots \quad . \tag{8.28}$$

For the bands with different K or built on another single-particle state the situation is similar. The principal restriction is the one from $\hat{R}_2$, namely that $K - \Omega_i$ be even. The quantum numbers thus run over

$$\begin{aligned} K &= \Omega_i,\ \Omega_i + 2, \dots \quad , \\ I &= |\Omega_i|,\ |\Omega_i| + 1, \dots \quad . \end{aligned} \tag{8.29}$$

In addition, β and γ vibrations may be excited. For β vibrations only n_β has to be varied, while γ vibrations are of course already present if $K > \Omega_i$ as in the rotation–vibration model, where the standard γ band corresponds to $K = 2$. Having $n_\gamma > 0$ is possible in principle, also.

In practice, there are often several single-particle levels close to the Fermi energy, and the Nilsson model is not precise enough to rely on the exact order. In these cases, usually all the angular momenta expected from the level scheme and the bands built thereupon occur in the experimental spectrum.

At the end of this brief description of the strong-coupling model, we note that the Coriolis coupling is often quite important and can change even the sequence of levels considerably. A thorough treatment unfortunately requires more extensive calculations so that it is omitted here. However, it is worthwhile to mention the original form of the Hamiltonian as developped in [Bo53a, Ke56], in which the vibrations where not considered, so that the coupling of the angular momenta becomes the crucial feature. Added to the standard Nilsson and rotation–vibration Hamiltonians we now find the interaction of the form

$$\hat{H}' = \frac{1}{2\mathcal{J}} \left(\hat{I}'_- \hat{\jmath}'_+ + \hat{I}'_+ \hat{\jmath}'_- \right) \quad . \tag{8.30}$$

Now an interesting feature of this coupling is that it has diagonal matrix elements only in basis states with $K = \Omega = \frac{1}{2}$ because of the way the components of opposite projection are mixed in (8.26). These diagonal elements can be evaluated to yield

$$-\frac{1}{2\mathcal{J}} \langle IK\Omega n_\gamma n_\beta \kappa | \hat{H}' | IK\Omega n_\gamma n_\beta \kappa \rangle = a(-1)^{I+1/2} \left(I + \tfrac{1}{2} \right) \quad , \tag{8.31}$$

where a, which contains the single-particle part of the matrix element, is called the *decoupling parameter*. It has the intersting consequence of a systematic modification of the $I(I+1)$ rule for $K = \frac{1}{2}$ bands:

$$E_K(I) = E_K^{(0)} + \frac{\hbar^2}{2\mathcal{J}} \left[I(I+1) + a(-1)^{I+1/2} \left(I + \tfrac{1}{2}\right) \delta_{K,1/2} \right] \quad . \tag{8.32}$$

The effect is seen in experiment, but theory does not do well in the quantitative description of the decoupling parameter, showing the limitations of this very simple Hamiltonian.

EXERCISE

8.1 The Spectrum of ^{183}W in the Stong-Coupling Model

Problem. Interpret the spectrum of the nucleus ^{183}W in terms of the strong-coupling model.

Solution. The lowest levels of this nucleus are given in Table 8.1.

Table 8.1 The lowest experimental levels of ^{183}W

I^π	Energy / keV
$1/2^-$	0.0
$3/2^-$	46.5
$5/2^-$	99.1
$7/2^-$	207.0
$3/2^-$	208.8
$5/2^-$	291.7
$9/2^-$	308.9
$9/2^+$	309.5
$7/2^-$	412.1
$7/2^-$	453.1
$9/2^-$	554.2

Another necessary piece of information not in this table is that the nucleus has a deformation of about $\beta_0 \approx 0.21$.

^{183}W has 74 protons and 109 neutrons. The odd particle is thus a neutron and we have to find its level in the Nilsson scheme of Fig. 7.8. One could in principle count all the filled levels with their two-fold Kramers degeneracies, but in practice it is easier to note which filled shells stay completely below the Fermi energy. In this case the shell at 82 stays filled so that, ignoring the $h_{11/2}$ state coming from this lower shell, we count up by two for each level we cross, getting to 108 with the $9/2^+$ state. The next levels available for the odd particle are a $1/2^-$, a $3/2^-$, and a $7/2^-$, with the differences too small to allow a reliable prediction. In this case, however, the ordering is right and we can split up the spectrum into bands based on the appropriate angular momenta. In addition there is a $9/2^+$ state, which can be understood as a hole state: a neutron jumps up to join the odd particle, so

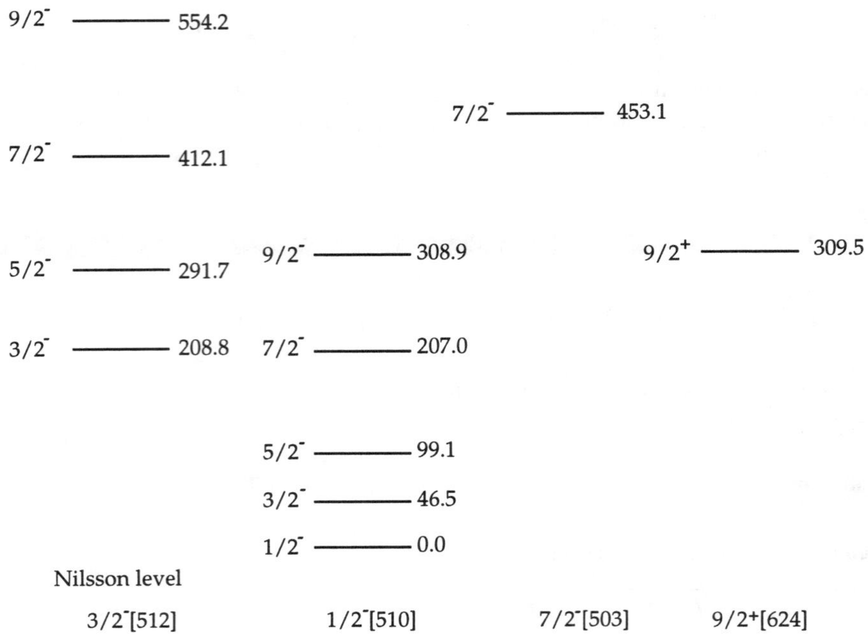

Fig. 8.3 Low-lying levels of ^{183}W assigned to bands based on different single-particle levels.

that the $9/2^+$ state now is singly occupied and determines the angular momentum and parity.

The interpretation of the spectrum gained in this way is given in Fig. 8.3, which decomposes it into the various bands based on the single-particle levels.

8.1.4 The Interacting Boson–Fermion Model

The IBM of Sect. 6.8 also lends itself to a coupling with an additional odd particle as the geometric collective model does. This combined model is called the *interacting boson–fermion model (IBFM)* and was introduced by Iachello and Scholten [Ia79, Ia80].

The principal ingredients of this model are quite straightforward: in addition to the s and d bosons of the IBM one introduces the odd-particle creation and annihilation operators $\hat{a}^\dagger_{jm}$ and $\hat{a}_{jm}$. The angular momenta of the odd particle is determined by the available orbitals, so that in general the wave function of the odd particle has contributions from a number of different single-particle states above the Fermi level. This treatment is closer in spirit to the weak-coupling model where the angular momenta of core and particle both used. The Hamiltonian is decomposed into boson, fermion, and coupling terms as

$$\hat{H} = \hat{H}_{\text{IBM}} + \hat{H}_{\text{fermion}} + \hat{H}_{\text{coupling}} \quad . \tag{8.33}$$

The first part is the standard IBM or IBM-2 Hamiltonian, and for the fermion part a general expansion in the same spirit is used:

$$\hat{H}_{\text{fermion}} = \sum_j n_j \left[\hat{a}_j^\dagger \times \hat{a}_j\right]^0$$
$$+ \sum_\lambda \sum_{j_1 j_2 j_3 j_4} v^{(\lambda)}_{j_1 j_2 j_3 j_4} \left[\left[\hat{a}_{j_1}^\dagger \times \hat{a}_{j_2}\right]^\lambda \times \left[\hat{a}_{j_3}^\dagger \times \hat{a}_{j_4}\right]^\lambda\right]^0 \quad . \qquad (8.34)$$

Finally the coupling part is set up in the same way as the sum over all products of $[\hat{a}_j^\dagger \times \hat{a}_{j'}]^\lambda$ with combinations of one boson creation and one annihilation operator coupled to the same λ. All of the terms conserve boson and fermion numbers separately, as it should be.

The actual construction of the eigenstates of this model goes too deeply into group theory to be presented here; let us instead mention some of the ideas employed and the successes gained. One problem is the large number of parameters appearing because of the general mathematical form of the Hamiltonian. In the geometric model the situation was simpler because of the clear-cut picture of motion in a deformed potential, which ideally determines the interaction completely. In the IBFM one can also try to determine the coefficients from the underlying physics, such calculations are presented in, for example, [Sc82, Br81, Bi82, Cu82].

An alternative method that is of independent interest is to use the group-theoretical decomposition of the combined symmetry groups for the boson and fermion part, which can best be done in the framework of supersymmetry [Ba83, Is84]. It has some success in the description of spectra and transition probabilities in isotope chains, similar to the way the IBM itself can be used to investigate the systematic changes of the collective structure through such a chain. Since the model links states of even–even nuclei with those of odd ones, it may well fulfil the claim of providing the first application of supersymmetry in nature.

8.2 Collective Vibrations in Microscopic Models

8.2.1 The Tamm–Dancoff Approximation

In Hartree–Fock models the excited states of the nucleus are the particle-hole excitations. The lowest of these are the one-particle/one-hole excitations with an energy corresponding to the difference between the energy of the last filled state and that of the first empty one.

If the nucleus has magic proton and neutron numbers these excitations will mostly have negative parity, because the successive shells in a harmonic oscillator are of alternating parity; the spin–orbit coupling leads only to an occasional presence of *intruder states* of opposite parity. Thus one would expect a large number of negative-parity states with about the energy of the shell gap and it also appears clear that the residual interaction will lift the degeneracies. These states should be easy to excite through electromagnetic dipole fields, since the electromagnetic excitations are described by one-particle operators.

In ^{16}O, for example, the shell gap is about 11.5 MeV. Instead of a concentration of states there with about equal probability of excitation, however, one observes a concentration of excitation probability in two states, one 3^- state near 6 MeV and two giant-resonance 1^- states near 22 and 25 MeV, respectively. The other negative-parity states are only excited with small probability. Thus it appears that the residual interactions lead to the presence of collective states with much larger electromagnetic excitation probabilities. In the following sections we will see how this can be explained.

Using again the notational convention that the indices i and j refer to single-particle states *below* and m and n to states *above* the Fermi level, the particle–hole can be written as

$$|mi\rangle = \hat{a}_m^\dagger \hat{a}_i \, |\mathrm{HF}\rangle \quad . \tag{8.35}$$

In the presence of the residual interaction these are of course not eigenstates of the Hamiltonian, but they may serve as the basis for a variational procedure. We thus study the variational problem

$$\delta \langle \Psi | \hat{H} | \Psi \rangle = 0 \quad , \tag{8.36}$$

with the variation restricted to *normalized* wave functions of the form

$$|\Psi\rangle = \sum_{mi} c_{mi} \, |mi\rangle = \sum_{mi} c_{mi} \; \hat{a}_m^\dagger \hat{a}_i \, |\mathrm{HF}\rangle \tag{8.37}$$

and the general two-body Hamiltonian

$$\hat{H} = \sum_{k_1 k_2} t_{k_1 k_2} \, \hat{a}_{k_1}^\dagger \hat{a}_{k_2} + \tfrac{1}{2} \sum_{k_1 k_2 k_3 k_4} v_{k_1 k_2 k_4 k_3} \, \hat{a}_{k_1}^\dagger \hat{a}_{k_2}^\dagger \hat{a}_{k_3} \hat{a}_{k_4} - \langle \mathrm{HF} | \hat{H} | \mathrm{HF} \rangle \quad . \tag{8.38}$$

The Hartree–Fock ground-state energywas subtracted since we are interested in the excitation energy only. Taking the constraint for normalization into account in analogy to Sect. 7.2.2 and varying with respect to c_{mi} leads to

$$\sum_{nj} \Big(\langle \mathrm{HF} | \hat{a}_i^\dagger \hat{a}_m \hat{H} \, \hat{a}_n^\dagger \hat{a}_j \, |\mathrm{HF}\rangle - E \, \langle \mathrm{HF} | \hat{a}_i^\dagger \hat{a}_m \hat{a}_n^\dagger \hat{a}_j \, |\mathrm{HF}\rangle \Big) \, c_{nj} = 0 \quad . \tag{8.39}$$

The second matrix element is $\delta_{ij} \, \delta_{mn}$, and the matrix element of the Hamiltonian can be evaluated with the standard methods. The potential matrix elements behave somewhat differently in diagonal terms than off-diagonally:

$$\begin{aligned} \langle mi | \hat{H} | mi \rangle &= t_{mm} - t_{ii} + \sum_j \big(\langle mj | \bar{v} | mj \rangle - \langle ij | \bar{v} | ij \rangle \big) + \langle mi | \bar{v} | im \rangle \\ &= \varepsilon_m - \varepsilon_i + \langle mi | \bar{v} | im \rangle \quad , \\ \langle mi | \bar{v} | nj \rangle &= \langle mj | \bar{v} | in \rangle \quad . \end{aligned} \tag{8.40}$$

In these expressions the definition of the Hartree–Fock single-particle energies and that of the antisymmetrized matrix elements was taken over from Chap. 7.2.

Inserting this result into (8.39) yields the *Tamm–Dancoff equations*

$$\sum_{jn} \big[(\varepsilon_m - \varepsilon_i) \, \delta_{mn} \, \delta_{ij} + \langle mj | \bar{v} | in \rangle \big] c_{nj}^{\nu} = E_\nu \, c_{mi}^{\nu} \quad , \tag{8.41}$$

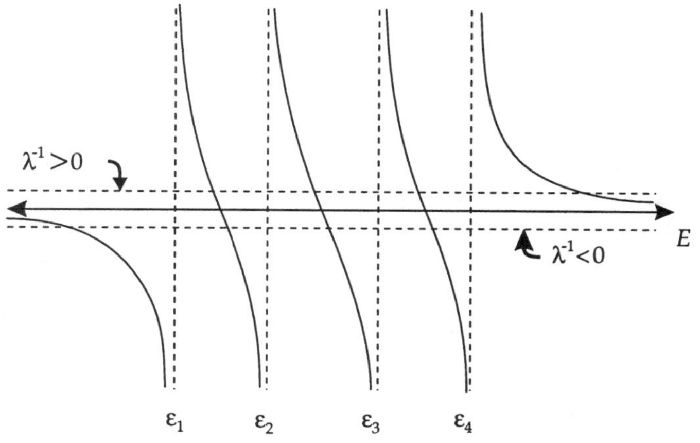

Fig. 8.4 Graphical illustration of the solution of (8.41). The function on the left-hand side of that equation is plotted vs. E. It has singularities at the particle–hole energies denoted, for this case, by ε_{1-4}. The intersection of the curve with the horizontal lines shown for both $\lambda < 0$ and $\lambda > 0$ produces the solutions marked by points, whose abscissas give the eigenenergies E_ν. Note that while most eigenvalues stay between the particle–hole energies, the lowest or highest, respectively, are pushed far down or up.

where the additional index ν was utilized to enumerate the different eigenstates. The energies E_ν are excitation energies above the ground state. The Tamm–Dancoff method was introduced into nuclear physics in [El57, Br59].

To understand how this equation can give rise to collective states, it is useful to examine the so-called *schematic model* [Br59, Br61]. This is based on a separable approximation to the potential matrix elements,

$$\langle mj|\bar{v}|in\rangle \approx \lambda D_{mi} D_{nj} \quad , \tag{8.42}$$

with λ a strength parameter. Note that the quantities D_{mi} correspond to one-particle matrix elements and may be assumed to be real. Before discussing the appropriateness of this approximation, we will first use it to derive properties of the solutions. Inserting (8.42) into (8.41) we get

$$\left(E_\nu - \varepsilon_m + \varepsilon_i\right) c^\nu_{mi} = \lambda D_{mi} \sum_{nj} D_{nj}\, c^\nu_{nj} \quad , \tag{8.43}$$

and this can be solved easily by noting that the sum on the right-hand side represents only an overall constant that can also be determined from the normalization of the wave functions, $\sum_{mi} |c^\nu_{mi}|^2 = 1$:

$$c^\nu_{mi} = \frac{D_{mi}}{E_\nu - \varepsilon_m + \varepsilon_i} \lambda \sum_{nj} D_{nj}\, c^\nu_{nj} \quad . \tag{8.44}$$

An equation for the energies results from regarding the expression

$$\sum_{mi} D_{mi} c^\nu_{mi} = \sum_{mi} \frac{D^2_{mi}}{E_\nu - \varepsilon_m + \varepsilon_i} \left(\lambda \sum_{nj} D_{nj}\, c^\nu_{nj}\right) \quad , \tag{8.45}$$

from which we get

$$\sum_{mi} \frac{D_{mi}^2}{E_\nu - \varepsilon_m + \varepsilon_i} = \frac{1}{\lambda} \quad . \tag{8.46}$$

As a function of E_ν the left-hand side of this equation has singularities at all the particle–hole energies $\varepsilon_{mi} = \varepsilon_m - \varepsilon_i$; it is positive above and negative below these. For $E_\nu \to \pm\infty$ it approaches ± 0. Its behavior is plotted in Fig. 8.4. The eigenenergies are obtained by searching for the intersections of this curve with the constant $\frac{1}{\lambda}$, which may have either sign. Clearly all the eigenenergies will be bracketed between the particle–hole energies except one which is shifted very much upward (for $\lambda > 0$) or downward (for $\lambda < 0$). We will now show that this state shows collective behavior.

To see that this special state is the collective one, examine the case of all particle–hole energies equal, $\varepsilon_m - \varepsilon_i = \varepsilon$. In this case the graphical construction shows that all eigenenergies will be equal to ε except for that one state leading to a nonvanishing denominator; its energy is given by

$$E = \varepsilon + \lambda \sum_{mi} D_{mi}^2 \tag{8.47}$$

and the expansion coefficients are

$$c_{mi} = \frac{D_{mi}}{\sqrt{\sum_{nj} D_{nj}^2}} \quad . \tag{8.48}$$

This clearly demonstrates that the collective state is a highly coherent excitation of single particles.

Whether a separable expansion for the residual interaction is meaningful, depends, of course, on the precise assumptions made for that interaction. Often the matrix elements D_{mi} are replaced by those of the dipole operator, because neighboring shells differ in parity, so that the dipole operator is the simplest one connecting these states.

In practical applications the particle–hole states have to be coupled to good angular momentum and, for light nuclei, isospin. One finds that for the usual interactions the states with unit isospin are pushed up in energy (i.e., have $\lambda > 0$), while those with $T = 0$ have $\lambda < 0$.

EXERCISE

8.2 Tamm–Dancoff Calculation for ^{16}O

Problem. Explain how to set up a Tamm–Dancoff calculation for the nucleus ^{16}O.

Exercise 8.2

Table 8.2. Particle-hole states in ^{16}O obtained by coupling a particle in the sd shell with a hole in the p shell. The specific states coupled are indicated in the first column and row and lead to the coupled angular momenta and parities listed.

	$1d_{3/2}$	$2s_{1/2}$	$1d_{5/2}$
$1p_{1/2}$	$1^-, 2^-$	$0^-, 1^-$	$2^-, 3^-$
$1p_{3/2}$	$0^-, 1^-, 2^-, 3^-$	$1^-, 2^-$	$1^-, 2^-, 3^-, 4^-$

Solution. The shell structure of ^{16}O is apparent from the level diagram of the single-particle model of Sect. 7.4.1. The last occupied shell is the negative-parity p shell with the states $1p_{3/2}$ and $1p_{1/2}$, while the first empty shell is the sd shell containing the states $1d_{5/2}$, $2s_{1/2}$, and $1d_{3/2}$ of positive parity. Coupling the angular momenta thus allows the combinations listed in Table 8.2.

This table is the same for both $T = 0$ and $T = 1$. The space in which the wave functions are expanded is of dimension 2 for 0^-, 5 for 1^-, 5 for 2^-, and 3 for 3^-, and there is only one state of type 4^-. The calculation therefore takes place in relatively small spaces (this increases rapidly, of course, for heavier systems). The particle-hole excitation energies are of the order of $\hbar\omega = 41\ \text{MeV} \times A^{-1/3}$, i.e., in this case, near 16 MeV. Experimentally one finds three prominent collective states, one 3^-, $T = 0$ at 6 MeV, and two 1^-, $T = 1$ states at 22.6 MeV and 25.2 MeV. Calculations with a realistic residual interaction can explain this structure and show that the collective states indeed correspond to mixtures of single-particle states with a broad distribution over the underlying particle–hole states.

There is another way of deriving the Tamm–Dancoff approximation, which will provide a convenient basis for future developments, especially since it is related to the collective-model way of treating the excited states as bosons. The collective state we are looking for, $|\nu\rangle$, solves the stationary Schrödinger equation

$$\hat{H}|\nu\rangle = E_\nu|\nu\rangle \quad . \tag{8.49}$$

We now define an operator $\hat{Q}_\nu^\dagger$, which "creates" this state out of the vacuum and is also similar to a creation operator in that the associated annihilation operator produces zero when applied to the vacuum:

$$\hat{Q}_\nu^\dagger|0\rangle = |\nu\rangle \quad , \quad \hat{Q}_\nu|0\rangle = 0 \quad . \tag{8.50}$$

There is no approximation in this requirement; in fact, such an operator can formally be given as $\hat{Q}_\nu^\dagger = |\nu\rangle\langle 0|$, which is easily seen to fulfill both conditions. The introduction of an approximation to the operator $\hat{Q}_\nu^\dagger$ will be based on a variational principle as usual,

$$0 = \delta\langle\nu|\hat{H} - E_\nu|\nu\rangle = \delta\langle 0|\hat{Q}_\nu\hat{H}\hat{Q}_\nu^\dagger - E_\nu\hat{Q}_\nu\hat{Q}_\nu^\dagger|0\rangle \quad . \tag{8.51}$$

We can proceed as for the standard version of the variational principle, noting that instead of the wave function and its complex conjugate, the operator $\hat{Q}_\nu$ and its adjoint may be varied independently. A variation of $\hat{Q}_\nu$ leads to

$$\langle 0|\delta\hat{Q}_\nu \hat{H}\hat{Q}_\nu^\dagger|0\rangle = E\langle 0|\delta\hat{Q}_\nu\hat{Q}_\nu^\dagger|0\rangle \quad . \tag{8.52}$$

Noting that

$$\hat{H}\hat{Q}_\nu^\dagger|0\rangle = \left[\hat{H},\hat{Q}_\nu^\dagger\right]|0\rangle + E_0\hat{Q}_\nu^\dagger|0\rangle \tag{8.53}$$

with E_0 the energy of the ground state, we can rewrite it as

$$\langle 0|\delta\hat{Q}_\nu\left[\hat{H},\hat{Q}_\nu^\dagger\right]|0\rangle = (E_\nu - E_0)\,\langle 0|\delta\hat{Q}_\nu\hat{Q}_\nu^\dagger|0\rangle \quad . \tag{8.54}$$

Finally everything may be written in terms of commutators:

$$\langle 0|\left[\delta\hat{Q}_\nu,\left[\hat{H},\hat{Q}_\nu^\dagger\right]\right]|0\rangle = (E_\nu - E_0)\,\langle 0|\left[\delta\hat{Q}_\nu,\hat{Q}_\nu^\dagger\right]|0\rangle \quad . \tag{8.55}$$

This can be proven using $\langle 0|\hat{Q}_\nu^\dagger = \langle 0|\hat{H}\hat{Q}_\nu^\dagger = 0$. Equation (8.55) will serve as the basis for deriving the random-phase approximation in the next section; before that, however, it will be checked in an exercise whether the Tamm–Dancoff approximation can be recovered from the new formalism.

EXERCISE

8.3 Derivation of the Tamm–Dancoff Equation

Problem. Derive the Tamm–Dancoff approximation from (8.55).

Solution. In the Tamm–Dancoff approximation the phonon vacuum is the Hartree–Fock ground state $|\mathrm{HF}\rangle$ and the ground state energy was set to zero. From the trial state as defined for the Tamm–Dancoff approximation, the operator $\hat{Q}_\nu^\dagger$ should be defined as

$$\hat{Q}_\nu^\dagger = \sum_{mi} c_{mi}^\nu\,\hat{a}_m^\dagger\hat{a}_i \tag{1}$$

with the c_{mi}^ν being the variation parameters. This operator clearly fulfills the desired requirement $\hat{Q}_\nu|0\rangle = 0$, so we can revert the last arguments leading to (8.55) and leave out the commutators, getting

$$\sum_{mi}\delta c_{mi}^{\nu *}\sum_{nj}\langle \mathrm{HF}|\hat{a}_i^\dagger\hat{a}_m(\hat{H} - E_\nu)\,\hat{a}_n^\dagger\hat{a}_j|\mathrm{HF}\rangle c_{nj}^\nu = 0 \quad , \tag{2}$$

which is identical to (8.39), since the variations in the c_{mi} are independent.

8.2.2 The Random-Phase Approximation (RPA)

The formalism developed at the end of the last section seemed needlessly complicated for deriving the Tamm–Dancoff approximation. It is, however, very well suited for the development of more advanced theories, as we will see presently.

The name of "random-phase approximation" referred to an approximation made in one of the original derivations; see for example the treatment in [La64].

A serious shortcoming of the Tamm–Dancoff approximation is the use of the unmodified Hartree–Fock ground state. The presence of the residual interaction should also modify the ground state itself, and to bring it in line with the excited states one needs only to admit particle–hole admixtures to construct a new ground state $|\mathrm{RPA}\rangle$. The crucial difference from the Hartree–Fock ground state is that excited states can be constructed not only with the usual particle-hole creation operators $\hat{a}_m^\dagger \hat{a}_i$, but also with combinations such as $\hat{a}_i^\dagger \hat{a}_m$, which take particle–hole excitations out of the ground state (note that the tacit assumption still holds that indices m, n refer to single-particle states above and i, j to states below the Fermi energy). Our operator for generating collective states can thus take the more general form

$$\hat{Q}_\nu^\dagger = \sum_{mi} x_{mi}^\nu \, \hat{a}_m^\dagger \hat{a}_i - \sum_{mi} y_{mi}^\nu \, \hat{a}_i^\dagger \hat{a}_m \tag{8.56}$$

and should fulfill the condition

$$\hat{Q}_\nu |\mathrm{RPA}\rangle = 0 \quad . \tag{8.57}$$

The variation of the operator involves either a variation of the coefficients x_{mi}^ν or of the y_{mi}^ν, and (8.55) leads to two equations multiplying δx_{mi}^ν and δy_{mi}^ν, respectively:

$$\begin{aligned}
\langle \mathrm{RPA}| \left[\hat{\mathrm{a}}_\mathrm{i}^\dagger \hat{\mathrm{a}}_\mathrm{m}, \left[\hat{\mathrm{H}}, \hat{\mathrm{Q}}_\nu^\dagger\right]\right] |\mathrm{RPA}\rangle &= (E_\nu - E_0)\, \langle \mathrm{RPA}| \left[\hat{\mathrm{a}}_\mathrm{i}^\dagger \hat{\mathrm{a}}_\mathrm{m}, \hat{\mathrm{Q}}_\nu^\dagger\right] |\mathrm{RPA}\rangle \quad , \\
\langle \mathrm{RPA}| \left[\hat{\mathrm{a}}_\mathrm{m}^\dagger \hat{\mathrm{a}}_\mathrm{i}, \left[\hat{\mathrm{H}}, \hat{\mathrm{Q}}_\nu^\dagger\right]\right] |\mathrm{RPA}\rangle &= (E_\nu - E_0)\, \langle \mathrm{RPA}| \left[\hat{\mathrm{a}}_\mathrm{m}^\dagger \hat{\mathrm{a}}_\mathrm{i}, \hat{\mathrm{Q}}_\nu^\dagger\right] |\mathrm{RPA}\rangle \quad .
\end{aligned} \tag{8.58}$$

To proceed further, some information about the state $|\mathrm{RPA}\rangle$ must be used in principle, but there is fortunately a way around this. On the right-hand side of this equation the commutators can be evaluated,

$$\left[\hat{a}_i^\dagger \hat{a}_m, \hat{a}_n^\dagger \hat{a}_j\right] = \delta_{mn}\, \delta_{ij} - \delta_{mn}\, \hat{a}_j \hat{a}_i^\dagger - \delta_{ij}\, \hat{a}_n^\dagger \hat{a}_m \quad , \tag{8.59}$$

and one needs the expectation value of this expression in the state $|\mathrm{RPA}\rangle$. Now the last term should be small as the number of particles above the Fermi energy. In the same way, the first term should be small, being produced by the presence of holes in the RPA state. We are thus led to the approximation

$$\begin{aligned}
&\langle \mathrm{RPA}| \left[\hat{\mathrm{a}}_\mathrm{i}^\dagger \hat{\mathrm{a}}_\mathrm{m}, \hat{\mathrm{a}}_\mathrm{n}^\dagger \hat{\mathrm{a}}_\mathrm{j}\right] |\mathrm{RPA}\rangle \\
&\quad = \delta_{ij}\, \delta_{mn} - \delta_{mn}\, \langle \mathrm{RPA}|\hat{\mathrm{a}}_\mathrm{j} \hat{\mathrm{a}}_\mathrm{i}^\dagger|\mathrm{RPA}\rangle - \delta_{\mathrm{ij}}\, \langle \mathrm{RPA}|\hat{\mathrm{a}}_\mathrm{n}^\dagger \hat{\mathrm{a}}_\mathrm{m}|\mathrm{RPA}\rangle \\
&\quad \approx \delta_{ij}\, \delta_{mn} \\
&\quad = \langle \mathrm{HF}| \left[\hat{a}_i^\dagger \hat{a}_m, \hat{a}_n^\dagger \hat{a}_j\right] |\mathrm{HF}\rangle \quad .
\end{aligned} \tag{8.60}$$

A consequence is that the particle–hole creation operators $\hat{a}_i^\dagger \hat{a}_m$ act as if they fulfilled the boson commutation relation $\left[\hat{a}_i^\dagger \hat{a}_m, \hat{a}_n^\dagger \hat{a}_j\right] = \delta_{ij}\delta_{mn}$, and for this reason this method is also known as the quasi-boson approximation. Note that this approximation strictly speaking violates the Pauli principle.

The approximation is now completed by replacing the RPA state also by the TDHF state on the left-hand side of (8.58). Note that this does not mean going back to the Tamm–Dancoff approximation, because we still have the sums over both the x_{nj}^ν and y_{nj}^ν. The matrix elements can then be easily evaluated and we define abbreviations for them:

$$\begin{aligned} A_{mi,nj} &= \langle \mathrm{HF}| \left[\hat{a}_i^\dagger \hat{a}_m, \left[\hat{H}, \hat{a}_n^\dagger \hat{a}_j\right]\right] |\mathrm{HF}\rangle = (\varepsilon_m - \varepsilon_i)\, \delta_{mn}\, \delta_{ij} + \bar{v}_{mjin} \quad , \\ B_{mi,nj} &= -\langle \mathrm{HF}| \left[\hat{a}_i^\dagger \hat{a}_m, \left[\hat{H}, \hat{a}_j^\dagger \hat{a}_n\right]\right] |\mathrm{HF}\rangle = \bar{v}_{mnij} \quad . \end{aligned} \tag{8.61}$$

The RPA equations then take the form

$$\begin{aligned} \sum_{nj} \left(A_{mi,nj}\, x_{nj}^\nu + B_{mi,nj}\, y_{nj}^\nu\right) &= (E_\nu - E_0)\, x_{mi}^\nu \quad , \\ \sum_{nj} \left(B_{mi,nj}^*\, x_{nj}^\nu + A_{mi,nj}^*\, y_{nj}^\nu\right) &= -(E_\nu - E_0)\, y_{mi}^\nu \quad . \end{aligned} \tag{8.62}$$

Often these are written in a matrix notation, where the indices are assigned in an obvious way,

$$\begin{pmatrix} A & B \\ B^* & A^* \end{pmatrix} \begin{pmatrix} X^\nu \\ Y^\nu \end{pmatrix} = (E_\nu - E_0) \begin{pmatrix} 1 & 0 \\ 0 & -1 \end{pmatrix} \begin{pmatrix} X^\nu \\ Y^\nu \end{pmatrix} \quad . \tag{8.63}$$

The RPA approximation was first developped in the context of the electron gas [Bo53b] and then soon applied in nuclear theory, for example in [Fe53, Fe57, Ba60].

That this indeed provides a generalization of the Tamm–Dancoff equations becomes apparent if it is considered what happens when the y are set equal to zero: the equations then simply reduce to the Tamm–Dancoff equations. Physically this is evident, as the coefficients Y measure the correlations in the ground state, which are neglected in Tamm–Dancoff.

The overall normalization of the excited states in the RPA requires that

$$\sum_{mi} \left(|x_{mi}^\nu|^2 + |y_{mi}^\nu|^2\right) = 1 \quad . \tag{8.64}$$

Now we come to some general features of the RPA. One problem that is apparent from the matrix form of the equations (8.63) is their non-Hermitian nature. Consequently, the eigenenergies may be complex; in practice, however, this occurs only under unusual circumstances. The ground state itself has not been given explicitly. In fact, its form is quite complicated; it must be determined from the condition that $\hat{Q}^\nu|\mathrm{RPA}\rangle = 0$ for all ν.

As in the Tamm–Dancoff approximation, one can study an extended version of the schematic model for separable matrix elements:

$$\bar{v}_{mnij} = \lambda\, D_{mi}\, D_{nj} \quad . \tag{8.65}$$

If we assume further that the matrices D_{mi} are real and symmetric, the RPA equations reduce to

$$\begin{aligned}
\lambda \sum_{nj} D_{jn}\, D_{mi}\, x_{nj}^{\nu} + \lambda \sum_{nj} D_{mi}\, D_{nj}\, y_{nj}^{\nu} &= (E_\nu - E_0 - \varepsilon_m + \varepsilon_i)\, x_{mi}^{\nu} \quad , \\
\lambda \sum_{nj} D_{mi}\, D_{jn}\, x_{nj}^{\nu} + \lambda \sum_{nj} D_{mi}\, D_{nj}\, y_{nj}^{\nu} &= -(E_\nu - E_0 + \varepsilon_m - \varepsilon_i)\, y_{mi}^{\nu} \quad .
\end{aligned} \tag{8.66}$$

Now the same trick works as for the simple schematic model. If we note that the same sum

$$S = \sum_{nj} D_{nj}\, x_{nj}^{\nu} + \sum_{nj} D_{nj}\, y_{nj}^{\nu} \tag{8.67}$$

appears in both equations, these may immediately be solved for the unknown coefficients:

$$\begin{aligned}
x_{mi}^{\nu} &= \frac{\lambda S D_{mi}}{E_\nu - E_0 - \varepsilon_m + \varepsilon_i} \quad , \\
y_{mi}^{\nu} &= \frac{\lambda S D_{mi}}{-E_\nu + E_0 - \varepsilon_m + \varepsilon_i} \quad .
\end{aligned} \tag{8.68}$$

Inserting these solutions into the definition of S leads to

$$\begin{aligned}
S &= \sum_{mi} \lambda\, D_{mi}^2\, S \left(\frac{1}{\varepsilon_i - \varepsilon_m + E_\nu - E_0} + \frac{1}{\varepsilon_i - \varepsilon_m - E_\nu + E_0} \right) \\
&= \sum_{mi} \frac{2\lambda\, D_{mi}^2\, S\, (\varepsilon_i - \varepsilon_m)}{(\varepsilon_i - \varepsilon_m)^2 - (E_\nu - E_0)^2} \quad ,
\end{aligned} \tag{8.69}$$

or

$$1 = \sum_{mi} \frac{2\lambda\, D_{mi}^2\, (\varepsilon_m - \varepsilon_i)}{(E_\nu - E_0)^2 - (\varepsilon_m - \varepsilon_i)^2} \quad . \tag{8.70}$$

EXERCISE

8.4 The Extended Schematic Model

Problem. Examine the properties of the extended schematic model in the case of degenerate particle–hole energies. Can the energy of the collective state fall below that of the ground state?

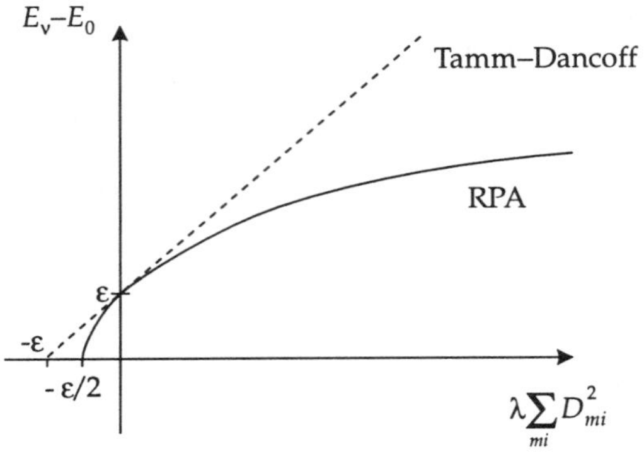

Fig. 8.5 The dependence of the collective excitation energy on the interaction strength, given by the parameter $\lambda \sum_{mi} D_{mi}^2$. Both curves refer to the case of completely degenerate particle–hole spectrum in the appropriate schematic model.

Solution. Inserting the degenerate case $\varepsilon_m - \varepsilon_i \equiv \varepsilon$ into (8.70) produces

$$1 = \frac{2\lambda \sum_{mi} D_{mi}^2 \, \varepsilon}{(E_\nu - E_0)^2 - \varepsilon^2} \tag{1}$$

with the solution

$$E_\nu - E_0 = \sqrt{\varepsilon^2 + 2\varepsilon\lambda \sum_{mi} D_{mi}^2} \tag{2}$$

for the collective state. Compare this with the schematic model associated with the Tamm–Dancoff approximation,

$$E_\nu - E_0 = \varepsilon + \lambda \sum_{mi} D_{mi}^2 \quad , \tag{3}$$

which is just the lowest-order expansion of the new result. Clearly the energy of the collective state depends only on the parameter $\lambda \sum_{mi} D_{mi}^2$, which characterizes the strength of the residual interaction. The functional behavior is plotted in Fig. 8.5, showing that the collective excitation energy may indeed become negative for a sufficiently strong negative interaction strength. The physical interpretation of such a collapse of the collective excitation is quite simple: the interaction leads to a deformed minimum in the potential energy surface, so that vibrational excitations around the spherical shape are no longer stable.

8.2.3 Time-Dependent Hartree–Fock and Linear Response

A totally different and quite enlightening derivation of the RPA approximation is the so-called linear-response theory, which also employs the time-dependent form of the Hartree–Fock equations [Eh59, Go59, Th61]. Instead of looking at the stationary states of the Hamiltonian in a restricted space of particle–hole excitations, the time-dependent response of the nucleus to an external stimulus is investigated. The derivation is also quite educational in its elegant use of density matrix methods.

Assume that the external perturbation is described by a harmonically time-dependent one-particle operator

$$\hat{F}(t) = \hat{F}\,\mathrm{e}^{-\mathrm{i}\omega t} + \hat{F}^\dagger\,\mathrm{e}^{\mathrm{i}\omega t} \quad . \tag{8.71}$$

This operator is Hermitian by construction. In second quantization it may be expressed as

$$\hat{F}(t) = \sum_{kl} f_{kl}(t)\,\hat{a}_k^\dagger \hat{a}_l \quad . \tag{8.72}$$

Later we will utilize perturbation theory, so that the perturbation is also assumed to be small. We expect to see strong resonances in the nuclear response whenever $\hbar\omega$ is close to the energy of a collective state. The dipole operator, which is linked to giant resonances as demonstrated in Sect. 6.8, may serve as an example for such a perturbation.

Under the influence of the external time-dependent perturbation the wave function $|\Phi(t)\rangle$ of the nucleus becomes time dependent, too, and this also holds for the associated one-particle density matrix

$$\rho_{kl} = \langle\Phi(t)|\hat{a}_l^\dagger \hat{a}_k|\Phi(t)\rangle \quad . \tag{8.73}$$

In addition to the smallness of the perturbation we will also assume that the state of the nucleus is always given by a Slater determinant, i.e., $\rho^2(t) = \rho(t)$. The equation of motion of the density matrix is easily obtained from the Schrödinger equation for $|\Phi(t)\rangle$:

$$\mathrm{i}\hbar\dot{\rho}_{kl}(t) = \langle\Phi(t)|\left[\hat{a}_l^\dagger \hat{a}_k, \hat{H}\right]|\Phi(t)\rangle \quad . \tag{8.74}$$

Inserting the standard two-body Hamiltonian

$$\hat{H} = \sum_{ij} t_{ij}\,\hat{a}_i^\dagger \hat{a}_j + \tfrac{1}{2}\sum_{ijkl} v_{ijkl}\,\hat{a}_i^\dagger \hat{a}_j^\dagger \hat{a}_l \hat{a}_k \tag{8.75}$$

(with suitably renamed indices) and evaluating the commutator yields

$$\mathrm{i}\hbar\dot{\rho}_{kl} = \sum_p \left(t_{kp}\,\rho_{pl} - \rho_{kp}\,t_{pl}\right) + \tfrac{1}{2}\sum_{prs}\left(\bar{v}_{kprs}\,\rho^{(2)}_{rslp} - \bar{v}_{rslp}\,\rho^{(2)}_{kprs}\right) \quad , \tag{8.76}$$

where $\rho^{(2)}$ is the two-body density matrix defined by

$$\rho^{(2)}_{klpq} = \langle\Phi(t)|\hat{a}_p^\dagger \hat{a}_q^\dagger \hat{a}_l \hat{a}_k|\Phi(t)\rangle \quad . \tag{8.77}$$

In principle the two-body density matrix describes two-body correlations, which appear in the equation for the one-particle density matrix because it contains the effects of two-body scatterings via the interaction v. Unfortunately it implies that this equation is not complete. Deriving the time dependence of the two-body density matrix, however, shows that it in turn is coupled to the three-body density matrix, and so on. One therefore has to cut off this chain at some point, and this is where the above assumption of $|\Phi(t)\rangle$ being a Slater determinant at all times comes in. For a Slater determinant we have

$$\langle \Phi(t)|\hat{a}_p^\dagger \hat{a}_q^\dagger \hat{a}_l \hat{a}_k|\Phi(t)\rangle = \rho_{kp}\,\rho_{ql} - \rho_{kq}\,\rho_{lp} \tag{8.78}$$

as can be seen immediately from the diagonal representation, and now the time development of the one-particle density matrix becomes self-contained:

$$\mathrm{i}\hbar\dot{\rho} = \left[t + \Gamma, \rho\right] = \left[h(\rho), \rho\right] \tag{8.79}$$

with a mean potential

$$\Gamma_{kl} = \sum_{pq} \bar{v}_{kqlp}\,\rho_{pq} \tag{8.80}$$

and the density-dependent single-particle Hamiltonian $\hat{h}$, which is represented by the corresponding matrix

$$h(\rho) = t + \Gamma \quad . \tag{8.81}$$

Actually this looks like a time-dependent version of the density-matrix formulation of Hartree–Fock, and for this reason the associated approximation is referred to as the *time-dependent Hartree–Fock approximation* (TDHF). Later, in Sect. 9.4, we will see that the wave functions in this case simply satisfy time-dependent Schrödinger equations where the mean field appears as the potential. Here, however, we are interested in the small-oscillation limit.

Adding the external perturbation to the single-particle Hamiltonian produces

$$\mathrm{i}\hbar\dot{\rho} = \left[h(\rho) + F(t), \rho\right] \quad , \tag{8.82}$$

where $F(t)$ denotes the matrix $\langle k|\hat{F}(t)|l\rangle$ of the perturbation operator in the single-particle states. If the unperturbed ground state of the nucleus has a density matrix ρ_0, the time-dependent one can be expanded as

$$\rho(t) = \rho_0 + \delta\rho(t) \quad . \tag{8.83}$$

The condition $\rho^2(t) = \rho(t)$ to first order in $\delta\rho(t)$ requires

$$\delta\rho(t) = \rho_0\,\delta\rho(t) + \delta\rho(t)\,\rho_0 \tag{8.84}$$

and multiplying on the left with ρ_0 immediately yields

$$\rho_0\,\delta\rho\,\rho_0 = 0 \quad . \tag{8.85}$$

This means that $\delta\rho(t)$ has vanishing matrix elements between the single-particle states occupied in the Hartree–Fock ground state. Similarly for the projector onto the unoccupied states $\sigma_0 = 1 - \rho_0$ we get

$$\sigma_0\,\delta\rho\,\sigma_0 = 0 \quad , \tag{8.86}$$

so that the corresponding matrix elements vanish also. In the notation of Sect. 7.2.4 one may also summarize it as $(\delta\rho)_{pp} = (\delta\rho)_{hh} = 0$.

To simplify the following discussions we will assume that the single-particle states are just the eigenstates of the single-particle Hamiltonian in the ground state, $h_0 = h(\rho_0)$, i.e.,

$$(\rho_0)_{ij} = \delta_{ij} , \quad (\rho_0)_{mn} = (\rho_0)_{mi} = (\rho_0)_{im} = 0 \quad . \tag{8.87}$$

Here again the convention was used that the indices i, j refer to occupied, m, n to unoccupied, and k, l to arbitrary single-particle states. The single-particle Hamiltonian is given by

$$(h_0)_{kl} = h(\rho_0)_{kl} = \varepsilon_k \delta_{kl} \quad . \tag{8.88}$$

Note that the single-particle Hamiltonian h deviates from its ground-state counterpart h_0 owing to of the time-dependent density caused by the perturbation.

In first order ("linear response") the equation of motion of the density matrix becomes

$$\mathrm{i}\hbar\delta\rho = \left[h_0, \delta\rho\right] + \left[\frac{\delta h}{\delta\rho}\delta\rho, \rho_0\right] + \left[F, \rho_0\right] \quad . \tag{8.89}$$

This requires some explanation: the density dependence of h was expanded to first order as

$$\rho|_{\rho=\rho_0+\delta\rho} \approx \rho_0 + \frac{\delta h}{\delta\rho}\delta\rho \tag{8.90}$$

and the matrix notation $\frac{\delta h}{\delta\rho}\delta\rho$ stands for

$$\left(\frac{\delta h}{\delta\rho}\delta\rho\right)_{kl} = \sum_{im}\left(\left.\frac{\partial h_{kl}}{\partial\rho_{mi}}\right|_{\rho=\rho_0}\delta\rho_{mi} + \left.\frac{\partial h_{kl}}{\partial\rho_{im}}\right|_{\rho=\rho_0}\delta\rho_{im}\right) \quad . \tag{8.91}$$

The summation is written such that only the non-vanishing matrix elements of type $\delta\rho_{mi}$ and $\delta\rho_{im}$ appear.

An important property of (8.89) is that it has vanishing *pp*- and *hh*-matrix elements. For the second and last terms on the right-hand side this follows from

$$\rho_0 \left[A, \rho_0\right] \rho_0 = \rho_0\, A\, \rho_0 - \rho_0\, A\, \rho_0 = 0 \quad , \tag{8.92}$$

which holds for any matrix A. For the first term use $[h_0, \rho_0] = 0$ and (8.85); similarly on the left-hand side. For σ_0 instead of ρ_0 the arguments can be applied in the same way.

The perturbation of the density matrix should have the same time dependence as the perturbing field and we can express it as

$$\delta\rho(t) = \rho^{(1)}\mathrm{e}^{-\mathrm{i}\omega t} + \rho^{(1)^\dagger}\mathrm{e}^{\mathrm{i}\omega t} \quad , \tag{8.93}$$

taking hermiticity into account. Since the two contributions are linearly independent, they may be regarded separately. Examine the *ph* matrix elements (indices mi) containing $\exp(\mathrm{i}\omega t)$ separately for the various terms:

$$\begin{aligned} \mathrm{i}\hbar(\delta\rho)_{mi} &= \hbar\omega\rho^{(1)}_{mi}\,\mathrm{e}^{-\mathrm{i}\omega t} \quad , \\ \left[h_0, \delta\rho\right]_{mi} &= \sum_k \left(\varepsilon_m\,\delta_{mk}\,\rho^{(1)}_{ki} - \rho^{(1)}_{mk}\,\varepsilon_k\,\delta_{ki}\right)\mathrm{e}^{-\mathrm{i}\omega t} \\ &= \sum_{nj}(\varepsilon_m - \varepsilon_i)\,\delta_{mn}\,\delta_{ij}\,\rho^{(1)}_{nj}\,\mathrm{e}^{-\mathrm{i}\omega t} \quad . \end{aligned} \tag{8.94}$$

In the last step the summation indices were renamed to express the fact that the Kronecker symbols restrict the summation over k to above and that over l to below the Fermi energy. For the last term to be treated we note that

$$\left[\frac{\delta h}{\delta \rho}\,\delta\rho, \rho_0\right]_{mi} = \left[\frac{\delta h}{\delta \rho}\,\delta\rho\right]_{mi} \tag{8.95}$$

because the reversed term does not have ph matrix elements owing to the leading ρ_0. Furthermore the trailing ρ_0 drops out for this type of matrix element. We can split up the implied sum in this expression into ph and hp contributions:

$$\sum_{kl} \frac{\partial h_{mi}}{\partial \rho_{kl}}\delta\rho_{kl} = \sum_{nj}\left(\frac{\partial h_{mi}}{\partial \rho_{nj}}\rho^{(1)}_{nj} + \frac{\partial h_{mi}}{\partial \rho_{jn}}\rho^{(1)}_{jn}\right) \mathrm{e}^{-\mathrm{i}\omega t} \quad . \tag{8.96}$$

Finally, identical arguments can be applied to the perturbation term:

$$[f, \rho_0]_{mi} = f_{mi} \quad . \tag{8.97}$$

Putting all of these results together, we get the equation

$$\sum_{nj} A_{mi,nj}\;\rho^{(1)}_{nj} + B_{mi,jn}\;\rho^{(1)}_{jn} - \hbar\omega\rho^{(1)}_{mi} = -f_{mi} \tag{8.98}$$

with the definitions

$$\begin{aligned} A_{mi,nj} &= (\varepsilon_m - \varepsilon_i)\,\delta_{mn}\;\delta_{ij} + \frac{\partial h_{mi}}{\partial \rho_{nj}} \quad , \\ B_{mi,jn} &= \frac{\partial h_{mi}}{\partial \rho_{jn}} \quad . \end{aligned} \tag{8.99}$$

The left-hand side of this equation is identical to the RPA equations if we assume that

$$\frac{\partial h_{kl}}{\partial \rho_{k'l'}} = \bar{v}_{kl'lk'} \quad . \tag{8.100}$$

For standard Hartree–Fock this is indeed true as in this case

$$\hat{h} = \sum_{kl}\left(t_{kl} + \sum_i \bar{v}_{ikil}\right)\hat{a}^\dagger_k \hat{a}_l \quad , \tag{8.101}$$

and this can be written in terms of the density matrix and with summation over the whole index range as

$$h_{kl} = t_{kl} + \sum_{k'l'} \bar{v}_{k'kl'l}\;\rho_{k'l'} \quad . \tag{8.102}$$

The eigenmodes of the system are obtained by solving the homogeneous equation without perturbation, which is just the RPA equation itself.

The present equation is, however, slightly more general than RPA, because the density dependence of the single-particle Hamiltonian could be due to, for example, a three-body force such as is included in the Skyrme forces. /

9. Large-Amplitude Collective Motion

9.1 Introduction

The study of nuclear shapes far off from the ground state has attracted considerable interest with the advent of heavy ion accelerators. Before that the only process known to involve large deviations from the equilibrium state was nuclear fission, so heavy-ion reactions, which in a certain sense constitute the "inverse process" to fission, have opened the way to a much more detailed study. The theories discussed in this chapter are all devoted to large surface deformations in the sense of Chap. 6; other ways to excite the nucleus far from the ground state include high angular momenta or temperatures.

The starting point is the construction of a Lagrangian, which depends on some set of collective parameters denoted by a vector $\boldsymbol{\beta} = \{\beta_i\}$, $i = 1, \ldots, N$, and the associated velocities $\dot{\boldsymbol{\beta}}$. An example of such a surface parametrization is provided by two-center shell models (Sect. 9.2.3). We will assume the following form of the Lagrangian:

$$L(\boldsymbol{\beta}, \dot{\boldsymbol{\beta}}) = T(\boldsymbol{\beta}, \dot{\boldsymbol{\beta}}) - V(\boldsymbol{\beta}) \tag{9.1}$$

with

$$T(\boldsymbol{\beta}, \dot{\boldsymbol{\beta}}) = \tfrac{1}{2}\dot{\boldsymbol{\beta}} \cdot \boldsymbol{B} \cdot \dot{\boldsymbol{\beta}} \tag{9.2}$$

the kinetic energy and $V(\boldsymbol{\beta})$ the potential. The symbol $\boldsymbol{B}$ denotes a *symmetric* tensor of *mass parameters*, which may in turn depend on the coordinate, so that spelled out in full the kinetic energy is

$$T(\boldsymbol{\beta}, \dot{\boldsymbol{\beta}}) = \tfrac{1}{2} \sum_{i,j=1}^{N} \dot{\beta}_i \, B_{ij}(\boldsymbol{\beta}) \, \dot{\beta}_j \quad . \tag{9.3}$$

Assuming the very existence of such a Lagrangian corresponds to quite a severe approximation even within the classical treatment. It requires the following.

- The internal complicated single-particle structure should be determined uniquely by the collective parameters, i.e., that the state of the nucleus should be a function of its surface shape alone. Usually it is assumed that the lowest possible state for the given surface shape is realized, so that during the collective motion no internal excitation is possible; this is called the *adiabatic approximation* and will be discussed in connection with the cranking model. Without this approximation the internal state of the nucleus will certainly depend on its previous collective motion. While some degree of excitation can be accounted for by a classical friction force, this also requires a temperature-dependent model of nuclear structure to describe the change in internal energy through the dissipation. Using a friction force, however, makes eventual quantisation quite problematic.

- The kinetic energy must retain the simple quadratic form. Again memory effects are excluded, but also higher-order corrections which would lead to higher-order dependences on the velocities. Note that the mass-parameter tensor *must* be allowed to depend on deformation: this describes the different amount of energy required to vary the collective parameters dynamically at different deformation. In the same way, the masses for different coordinates must have different magnitudes to allow for the scales involved.

It is a dangerous but tempting and widely used practice to try to read off the dynamical behavior of the nucleus from the potential energy alone, ignoring the mass parameters. If we look at the classical equations of motion, it is clear what makes this risky:

$$\begin{aligned} 0 &= \frac{d}{dt}\frac{\partial L}{\partial \dot\beta_i} - \frac{\partial L}{\partial \beta_i} \\ &= \sum_j B_{ij}\ddot\beta_j + \sum_{jk}\left(\frac{\partial B_{ij}}{\partial \beta_k} - \frac{1}{2}\frac{\partial B_{jk}}{\partial \beta_i}\right)\dot\beta_j\dot\beta_k + \frac{\partial V}{\partial \beta_i} \quad . \end{aligned} \tag{9.4}$$

Clearly if the mass parameters depend strongly on the coordinates or if the nondiagonal ones are nonzero the additional terms in (9.4) may strongly distort the motion.

The following sections explain some of the approaches used today for calculating the potentials and mass parameters.

9.2 The Macroscopic-Microscopic Method

9.2.1 The Liquid-Drop Model

The liquid-drop model was used in Sect. 6.2.1 to study the change of energy of a nucleus for small deformations. Although it does not explain some of the more refined features such as deformed ground states and double-humped fission barriers, it has been highly successful in Bohr and Wheeler's original explanation of fission and still is the basic ingredient of many theories of larger deformations. With the Coulomb and surface energies of the spherical nucleus denotedby E_{C0} and E_{S0}, the energy of deformation relative to the ground state is

$$E_{\mathrm{LDM}} = E_{\mathrm{S}}(\beta) - E_{\mathrm{S0}} + E_{\mathrm{C}}(\beta) - E_{\mathrm{C0}} \quad . \tag{9.5}$$

For small deformations described by the $\alpha_{\lambda\mu}$ the results of Sect. 6.2.1 may be inserted to yield the deformation energy in units of the surface energy of the sphere

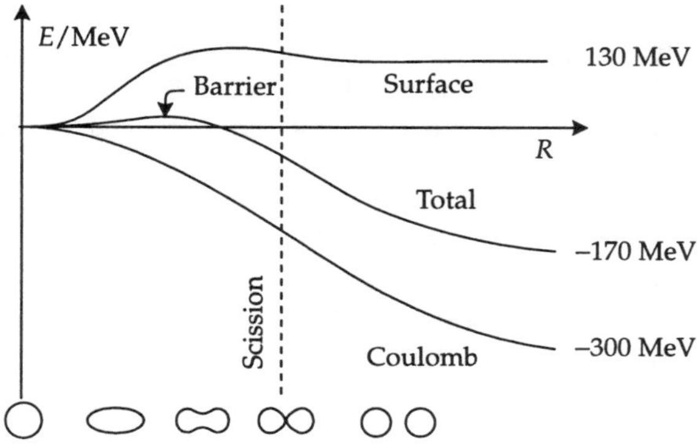

Fig. 9.1 Typical fission potential in the liquid-drop model, plotted here as a function of the two-center distance together with the surface and Coulomb energy contributions individually. The asymptotic numbers are indicative of the Uranium region, and the fission barrier is exaggerated to make it clearly visible. An indication of the associated shapes is given below the graph. Note that the surface energy exceeds that of the two separate fragments near scission because the nascent fragments are deformed.

$$\frac{E_{\rm LDM}}{E_{\rm S0}} = \frac{1}{2\pi}\sum_{\lambda\mu}\left[\frac{(\lambda-1)(\lambda+2)}{4} - x\frac{5(\lambda-1)}{2\lambda+1}\right]|\alpha_{\lambda\mu}|^2 \quad . \tag{9.6}$$

Here x is the *fissility* given by

$$x = \frac{E_{\rm C0}}{2E_{\rm S0}} = \frac{Z^2/A}{(Z^2/A)_{\rm C}} \tag{9.7}$$

with

$$(Z^2/A)_{\rm C} = \frac{40\pi\sigma r_0^3}{3e^2} \approx 50 \quad . \tag{9.8}$$

This result already shows the balance between surface and Coulomb effects: while the surface energy increases with deformation, the Coulomb energy decreases. For each multipolarity the balance of the two depends on the fissility; the drop is least stable for $\lambda = 2$ and becomes unstable against quadrupole deformations if $x = 1$, i.e. if Z^2/A exceeds the critical value of 50.

For larger deformations these expressions are no longer valid. If the nucleus splits up into two fragments, both terms approach the limiting values given by the corresponding expressions for the two separate fragments. While the surface energy does not change much any more once the nucleus has split up, the Coulomb energy still has the long-range contribution that decays only inversely with the distance between the fragments. For heavy nuclei the initial rise in energy is thus followed by a steep descent, and there is a barrier between them. For example in the Uranium isotopes the liquid-drop barrier has a height of close to 12 MeV while the energy of the fragments lies about 180 MeV below that. So one should bear in mind that the fission barrier is a relatively small effect compared to the huge energy gain. For this reason also the shell corrections discussed in the next section can play such an important role for the barrier in spite of their small size; for Uranium, for example, the barrier is lowered to about 6 MeV.

For more complicated shape parametrisations the liquid-drop energy can naturally only be evaluated numerically. Figure 9.1.1 shows the surface, Coulomb, and total deformation energies for a typical case.

Finally we note that there are more refined versions of the liquid-drop model. One of these already mentioned is the *droplet model*, which takes into account many effects associated with the finite width of the nuclear surface such as compression, polarization, curvature, etc. [My76]. There are also various energy functional formulations which determine the energy by folding the density distributions with some interaction potentials. One example is the *Yukawa-plus-exponential* formulation, which has also been used widely in fission and heavy-ion interaction potentials [Kr79].

9.2.2 The Shell-Correction Method

Although the liquid drop model explains the gross features of fission quite well, there are systematic deviations from the smooth mass and charge dependences predicted in this model. One of these is the approximately constant height of the barriers in the actinide region, contrasted to a predicted steep decline with increasing charge number; also the absolute values are correct only within a factor of two, roughly. This is caused by shell effects, as shall be seen below, which produce a deformed ground state with a lowered energy and also modify the barrier. The second problem is the preference for asymmetric mass splits in many fissioning systems, which have meanwhile been explained as a lowering of asymmetric barriers due to shell structure effects. Additionally, the observation of fission isomers [Po62] has led to the discovery of double-humped fission barriers, which could be understood as effects of shell closures at larger deformations.

These considerations might lead one to assume that the potential energy could be calculated easily in a deformed phenomenological shell model such as a two-center model (Sect. 9.2.3). The potential energy should be given by

$$V(\beta) = \tfrac{3}{4} \sum_{i \text{ occ}} \varepsilon_i(\beta) \quad , \tag{9.9}$$

where the virial theorem was used to express the potential energy, half of which has to be subtracted to avoid double counting, as one half of the single-particle energy (this trick is possible for a pure harmonic-oscillator potential). Calculating this sum in practice, however, yields disappointing results. The sum varies quite strongly with deformation with oscillations of an order of magnitude of 100 MeV, typically for heavy nuclei. This is clearly in disagreement with the experimental fission barriers and is not ameliorated by more sophisticated expressions for the potential energy.

The reason for this unphysical behavior lies in the deeply bound lower states of the level scheme, whose energy changes strongly with deformation. Self-consistent models such as Skyrme-force Hartree–Fock yield results much closer to the liquid-drop model and in good agreement with experiment. Even today, though, it is difficult to use these for all purposes, since in fission theory one would often like to study the dependence of the potential-energy surface on several collective parameters, and doing a Hartree–Fock calculation with more than one constraint still requires excessive computer resources.

The *shell correction prescription* proposed by Strutinsky [St67, St68] therefore meant tremendous progress and is still widely used today. Its basic idea is to use the liquid-drop model to describe the bulk part of the nuclear binding energy depending smoothly on mass and charge numbers, adding only a correction from the phenomenological single-particle model which describes the deviation from a smooth shell structure *near the Fermi surface*. Calculating such a correction requires the subtraction of the "smoothly varying" part of the sum of single-particle energies. The true distribution of single-particle energies is given by

$$g(\varepsilon) = \sum_{i=1}^{\infty} \delta(\varepsilon - \varepsilon_i) \quad , \tag{9.10}$$

and the total energy of the occupied levels by

$$U = \int_{-\infty}^{\varepsilon_F} \mathrm{d}\varepsilon \; g(\varepsilon)\varepsilon \quad . \tag{9.11}$$

Integrating up to the Fermi energy ε_F ensures that only the occupied levels are actually being summed. Similarly the total number of particles is given by

$$N = \int_{-\infty}^{\varepsilon_F} \mathrm{d}\varepsilon \; g(\varepsilon) \quad . \tag{9.12}$$

As is usual in the phenomenological single-particle models, protons and neutrons are treated quite separately, so that N denotes the number of particles of the species being treated. A "smoothed" sum of single-particle energies is now obtained by replacing each delta function by a smooth distribution:

$$\tilde{g}(\varepsilon) = \sum_{i=1}^{\infty} f\left(\frac{\varepsilon - \varepsilon_i}{\gamma}\right) \quad . \tag{9.13}$$

The function f should be similar to a Gaussian

$$f\left(\frac{\varepsilon - \varepsilon_i}{\gamma}\right) = \frac{1}{\sqrt{2\pi}\,\gamma} \exp\left(\frac{\varepsilon - \varepsilon_i}{\gamma}\right)^2 \quad , \tag{9.14}$$

with an integral normalized to 1. And the parameter γ should be chosen suitably such that the smoothed level distribution shows fluctuations only near major shell structures, and this suggests a value near $\hbar\omega$ for the oscillator model. Indeed, as we will see, a typical value from experience is $\gamma \approx 1.2\,\hbar\omega$.

One should note that in this procedure the desired smoothing over neighboring mass and charge numbers was actually replaced by one over the single-particle energies. This is reasonable given the relation between occupation and total particle number.

Given the smoothed distribution of levels, the first quantity to be determined is the Fermi energy for this case, $\tilde{\varepsilon}_{\mathrm{F}}$, which has to fulfil

$$N = \int_{-\infty}^{\tilde{\varepsilon}_{\mathrm{F}}} \mathrm{d}\varepsilon \, \tilde{g}(\varepsilon) \quad , \tag{9.15}$$

and then the smoothed total energy can be determined as

$$\tilde{U} = \int_{-\infty}^{\tilde{\varepsilon}_{\mathrm{F}}} \mathrm{d}\varepsilon \, \tilde{g}(\varepsilon) \, \varepsilon \quad . \tag{9.16}$$

Note that for all those levels that are sufficiently below the Fermi energy the integral will contribute just the single-particle energy,

$$\int_{-\infty}^{\varepsilon_{\mathrm{F}}} \mathrm{d}\varepsilon \, f\left(\frac{\varepsilon - \varepsilon_i}{\gamma}\right) \, \varepsilon \approx \varepsilon_i \quad , \tag{9.17}$$

while levels far above contribute zero, so that the *shell correction*

$$\delta U = U - \tilde{U} \tag{9.18}$$

has the desired property of depending only on the levels structure near the Fermi energy (levels far above again will contribute zero).

What kind of behavior should one expect for the shell correction? The results should depend on the distribution of levels within a typical smearing width γ near the Fermi surface. If there are many levels arranged practically uniformly, the smoothed level density will be almost a constant and δU should be close to zero. If there is a gap near ε_{F}, like for magic nuclei, the smoothed density will be have a minimum there and $\tilde{\varepsilon}_{\mathrm{F}}$ will fall into the gap, causing a larger contribution of higher energies in $\tilde{U}$ than in U, so that $\tilde{U} > U$ or $\delta U < 0$. Conversely, for a lower level density near the Fermi energy the shell correction will be positive. These considerations agree with the expectation that magic nuclei should have enhanced binding compared to the liquid drop.

From the above discussion the principle of the method is clear and it is also quite apparent how δU should be calculated. There are, however, a number of details that should not be omitted.

1. Taking a Gaussian is not quite sufficient. The reason is that a smoothing procedure should not change the result appreciably when applied a second time. In principle we should require

$$\tilde{g}(\varepsilon) = \frac{1}{\gamma} \int_{-\infty}^{\infty} \mathrm{d}\varepsilon' \, \tilde{g}(\varepsilon') \, f\left(\frac{\varepsilon' - \varepsilon}{\gamma}\right) \quad , \tag{9.19}$$

which is clearly not true for f of Gaussian form, but can be fulfilled exactly only with a δ function, which of course is unacceptable because it corresponds to no smoothing at all. The solution is that the formula need not be valid for an

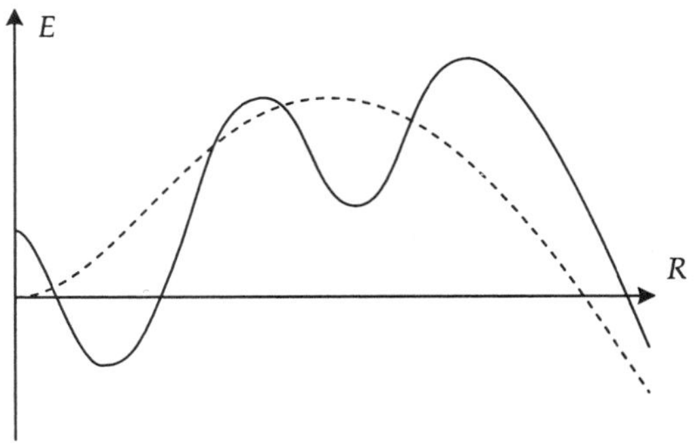

Fig. 9.2 Typical liquid-drop fission barrier *(dashed)* and total energy as functions of center separation in a two-center shell model for a nucleus in the actinide region. The oscillating shell correction produces a deformed minimum and a double-humped barrier.

arbitrary function $\tilde{g}(\varepsilon)$, but only for those which are already sufficently smooth on the scale of γ. The customary way to construct f is to use a product of a Gaussian and a polynomial. The polynomial must be even in order not to change the averages over a level. The result of the calculations is

$$f(x) = \frac{1}{\sqrt{\pi}} \mathrm{e}^{-x^2} \sum_{i=0}^{n} L_i^{1/2}(x^2) \tag{9.20}$$

with n a new parameter determining the degree of the polynomial; for example sixth order for $n = 3$.

2. It remains to determine the parameters n and γ. Ideally, the results should not depend on the precise values in a "broad" range of reasonable values. In practice one finds that $n = 3$ usually is sufficient, and for oscillator potentials there is indeed a broad range of γ values near $\gamma = \hbar\omega$ with approximately constant δU. For finite-well potentials this is often not the case and there are additional problems due to the presence of the continuum which contributes to the smoothed integral; many suggestions, not all of them convincing, have been made to solve these problems and the interested reader can find a thorough discussion and more background in the well-known "Funny Hills" review paper [Br72].

In any case, the theoretical reliability of the calculated shell corrections is estimated to be of the order of 0.5 MeV, with δU typically being less than 10 MeV. Despite this fact and its shaky theoretical foundations the method has enjoyed popularity because of its undeniable successes. The discrepancy between calculated and experimental binding energies is reduced from ± 10 MeV to ± 1 MeV. The ground state deformations and fission barriers are described quite well, and possibly the biggest success was the prediction of strong shell effects for deformed shapes, which led to the prediction of fission isomers. Since even a cursory examination of the level schemes as functions of deformation already shows the existence of quite pronounced gaps, it may appear surprising that it was widely believed that shell effects die out fast with increasing deformation. The present status is illustrated by Fig. 9.2, which shows the liquid-drop and total energies for a typical actinide nucleus as functions of center separation in a two-center shell model. Clearly the shell corrections produce both the deformed ground state and the second minimum right near the position of the liquid-drop barrier.

The shell correction is tiny compared to the liquid-drop surface and Coulomb energies, which for a heavy nucleus are both of the order of many hundreds of MeV. Its importance is due solely to the fact that for small deformations the two liquid-drop contributions together are almost independent of deformation, and that the small oscillations in the shell structure then become quite apparent in experiment because they determine such easily measurable quantities as barrier hights, lifetimes, etc. A calculation of fission lifetimes in this approach normally takes a model of two-center type with several shape parameters, evaluates the deformation energy, and finally evaluates the liftime using the WKB penetration probability well known from α decay, and given in its one-dimensional version by

$$P = \exp\left[- \int d\beta \, \sqrt{2m(V(\beta) - E)/\hbar^2} \right] \quad , \tag{9.21}$$

with the integral evaluated along that part of the trajectory where $E < V(\beta)$. For the case of collective parameters, there is first the need for a multidimensional generalization, and second for deformation-dependent mass parameters. The first is trivial if one looks only at the simplest approximation of calculating penetration along a known path and without consideration of neighboring ones. As conserns the mass parameter, in many fission calculations only a relative motion degree of freedom was used together with the reduced mass of the fragments replacing m, but this is clearly not correct. We will discuss the calculation of mass parameters in more detail later (Sect. 9.3).

9.2.3 Two-Center Shell Models

The expansion of the nuclear surface in spherical harmonics is not sufficient for going all the way to separate daughter nuclei in fission. Fundamentally there is the problem that the radius becomes a multivalued function of the angle if the nuclear neck is sufficiently constricted, but it is also inconvenient to have to deal with a large number of expansion parameters. Therefore many specalized models have been developped that allow nuclear shapes specially adapted to the fission process.

One of the more widely used models is the two-center shell model based on a double oscillator potential [Ma72]. We discuss it here as an example because of its simplicity, but remark that the model was specially designed to be computable with the computer resources of twenty years ago. Yet it is still sufficient to illustrate the basic ingredients of such an approach.

The general shapes considered in the model are illustrated in Fig. 9.3. The ellipsoidal shapes for the fragments allow individual deformations, and the overall size ratio makes for fragments of different mass, allowing for asymmetric fission. The center separation $z_2 - z_1$, of course, plays the role of a fission coordinate. Finally, one can vary the size of the neck between the two fragments. This yields a total of five parameters:

1. $\Delta z = z_2 - z_1$, the *center separation*;
2. $\xi = (A_1 - A_2)/(A_1 + A_2)$, the *asymmetry*, depending on the fragment masses (which are simply estimated from the volumes of the ellipsoids);
3. $\beta_1 = a_1/b_1$;

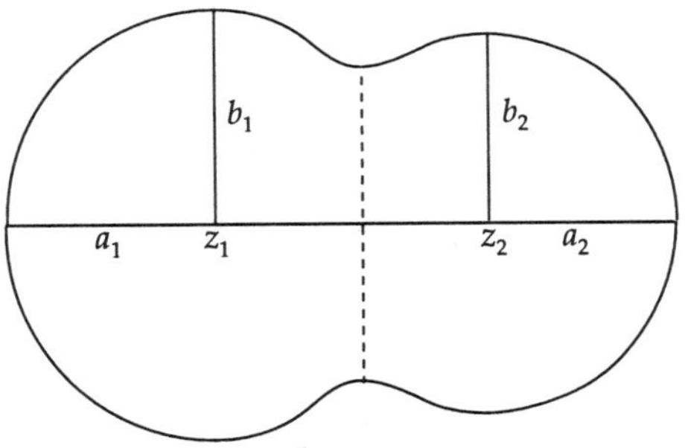

Fig. 9.3 Nuclear shape in the two-center shell model. The fragments are composed of ellipsoids with semiaxes $a_{1,2}$ and $b_{1,2}$. The centers of the ellipsoids are at positions $z_{1,2}$ on the z axis, which is the symmetry axis, and the surface is interpolated with a smooth curve near the neck.

4. $\beta_2 = a_2/b_2$, the *fragment deformations*; and
5. ε, a parameter controlling the neck, which appears in the potential definition.

An additional parameter is, of course, the total mass $A = A_1 + A_2$, which is reflected in the overall volume of the nuclear shape, $V = 4\pi r_0^3 A$.

The single-particle Hamiltonian appropriate to this shape is given mainly by the potential. It is decomposed according to

$$\hat{H} = -\frac{\hbar^2}{2m}\nabla^2 + V(\rho, z) + V_{ls} + V_{l^2} \quad . \tag{9.22}$$

The decisive geometric term is the oscillator potential, whose definition depends on the location along the z axis:

$$V(\rho, z) = \begin{cases} \frac{m}{2}\omega_{z1}^2 z'^2 + \frac{m}{2}\omega_{\rho 1}^2 \rho^2 & \text{for } z < z_1, \\ \frac{m}{2}\omega_{z1}^2 z'^2(1 + c_1 z' + d_1 z'^2) + \frac{m}{2}\omega_{\rho 1}^2(1 + g_1 z'^2)\rho^2 & \text{for } z_1 < z < 0, \\ \frac{m}{2}\omega_{z2}^2 z'^2(1 + c_2 z' + d_2 z'^2) + \frac{m}{2}\omega_{\rho 2}^2(1 + g_2 z'^2)\rho^2 & \text{for } 0 < z < z_2, \\ \frac{m}{2}\omega_{z2}^2 z'^2 + \frac{m}{2}\omega_{\rho 2}^2 \rho^2 & \text{for } z > z_2, \end{cases} \tag{9.23}$$

with

$$z' = \begin{cases} z - z_1 & \text{for } z < 0 \\ z - z_2 & \text{for } z > 0 \end{cases} \quad . \tag{9.24}$$

The coefficients c_i, d_i, and g_i are determined such that there is a smooth transition between the two oscillator potentials at $z = 0$; the details are unimportant for this discussion.

This potential was designed with a view to making its matrix elements analytically calculable. From the modern point of view, the spin–orbit potential is more interesting. It is defined as

$$V_{ls} = \begin{cases} -\left\{\frac{\hbar\kappa_1}{m\omega_1}, (\nabla V \times \hat{\boldsymbol{p}}) \cdot \hat{\boldsymbol{s}}\right\} & \text{for } z < 0 \quad , \\ -\left\{\frac{\hbar\kappa_2}{m\omega_2}, (\nabla V \times \hat{\boldsymbol{p}}) \cdot \hat{\boldsymbol{s}}\right\} & \text{for } z > 0 \quad . \end{cases} \tag{9.25}$$

There are several points to note: the orbital angular-momentum operator is replaced by an expression in $\nabla V \times \hat{\boldsymbol{p}}$, which reduces to $\boldsymbol{r} \times \hat{\boldsymbol{p}}$ for a spherical oscillator potential. This is necessary to make it go over into the correct spin–orbit potential

Fig. 9.4 Levels of the two-center shell model in their dependence on the center separation Δz. The mass asymmetry also increases during the separation, leading to spectra characterized by different oscillator spacings for the final fragments; the levels have been assigned to the heavy or light fragment with a superscript "H" or "L", respectively.

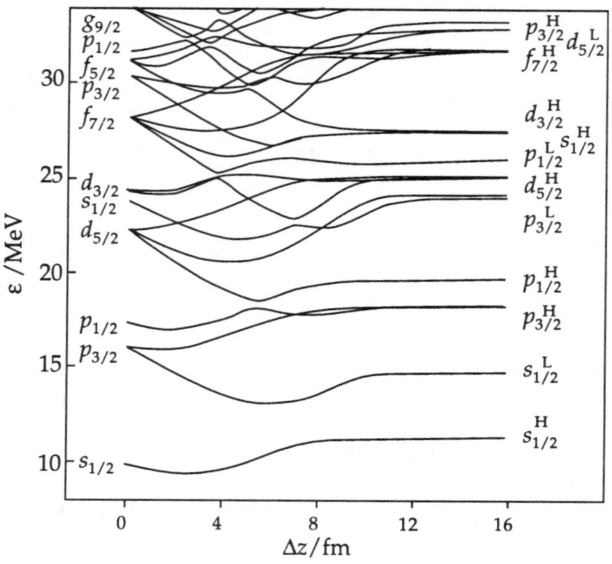

in the separated fragments. All phenomenological models use such an expression, while the self-consistent models include this automatically. A provision which is more characteristic for the oscillator model is that the spin–orbit strength is different for the fragments, which also has to be interpolated (the anticommutator in the expressions is needed to make the potential hermitian). The l^2 term is set up similarly.

The parameters in the potential can be determined from the geometric shape with the condition that the nuclear surface correspond to an equipotential $V(\rho, z) = \frac{1}{2}m\omega_0^2R^2$ with R the radius of the equivalent spherical nucleus. The neck parameter is determined as the ratio of the interpolated to the pure oscillator potential ($c_i = d_i = 0$) at ($\rho = 0, z = 0$) and is thus only indirectly related to the shape.

Figure 9.4 shows a characteristic level diagram for fission. The regular shell structure of the spherical nucleus splits up with increasing deformation (similarly to that from the Nilsson model) but then is sorted into the separate shell structures of the fragments for large center separation. In this case the fragments are spherical but of different mass, so that the asymptotic levels can easily be assigned to the light or heavy fragment. The holes in the level scheme along the way are responsible for the shell corrections, and their dependence on the asymmetry leads to a preference for asymmetric fission in many cases [Ma74].

Let us now examine some interesting applications of this model. One of the most important features of fission, which plays a crucial role in the determination of the lifetimes, is the preference for asymmetric shapes. This was first understood through the use of asymmetric two-center models as being due to the shell structure of the nascent fragments already influencing the potential energy well before separation [Mo71]. Since we will deal with superheavy nuclei shortly, this point is illustrated for the (still) hypothetical superheavy nucleus $^{298}_{114}X$. The dependence of the potential-energy surface on center separation and asymmetry is shown in

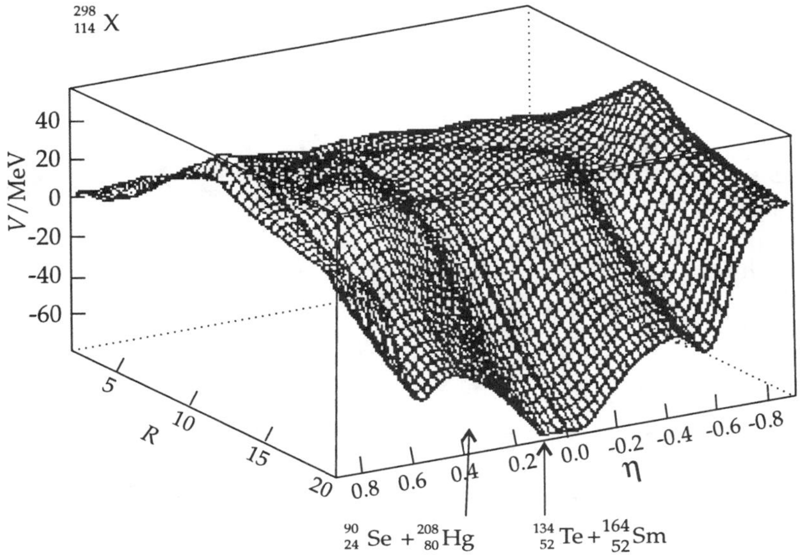

Fig. 9.5 Potential-energy surface for the hypothetical superheavy nucleus $^{298}_{114}$X in its dependence on the center separation and asymmetry. The valleys corresponding to the preferred asymmetries are clearly visible and also offer the opportunity for producing such nuclei using "cold" target-projectile combinations such as those listed. Of course the bombarding energy must be chosen such that the barrier within a cold valley is just overcome.

Fig. 9.5. Clearly as the nucleus proceeds towards fission, there are a number of valleys, indicating the preferred binary mass split-ups; these are usually characterized by the formation of one daughter nucleus close to magic numbers. Examining the most favorable shapes near scission one then usually finds this fragment spherical and the other one strongly deformed. There are, however, also cold valleys due to deformed shell structure, even though these valleys are less pronounced than for spherical fragments [Sa92].

In Fig. 9.5 there is a preference for slight asymmetry near $\eta \approx \pm 0.1$ and then a pronounced valley again near $\eta \approx \pm 0.4$. The latter has one fragment close to doubly-magic lead and is the valley responsible for the strong asymmetry in actinide fission (these nuclei are too light to allow the combinations present in the deepest valley here).

Note also that the valleys have almost constant structure as we increase Δz closer to scission, so that it appears that the decision for the mass distribution of the fragments is made already close to the barrier.

Such potential-energy surfaces can be used as a basis for calculating the mass distributions expected. For this purpose a collective Schrödinger equation can be set up for the asymmetry degree of freedom and then solved for known sets of potentials and masses [Li73, Ma74, Fi74]. If fission proceeds by tunneling and if the coupling of asymmetry and relative motion is unimportant (for a discussion see [Ma76b]), the asymmetry distribution should be given by the ground-state collective wave function. An example for such a calculation is given in Fig. 9.6 for the fission of ^{256}Fm [Lu80]. For spontaneous fission the experimental data are reproduced

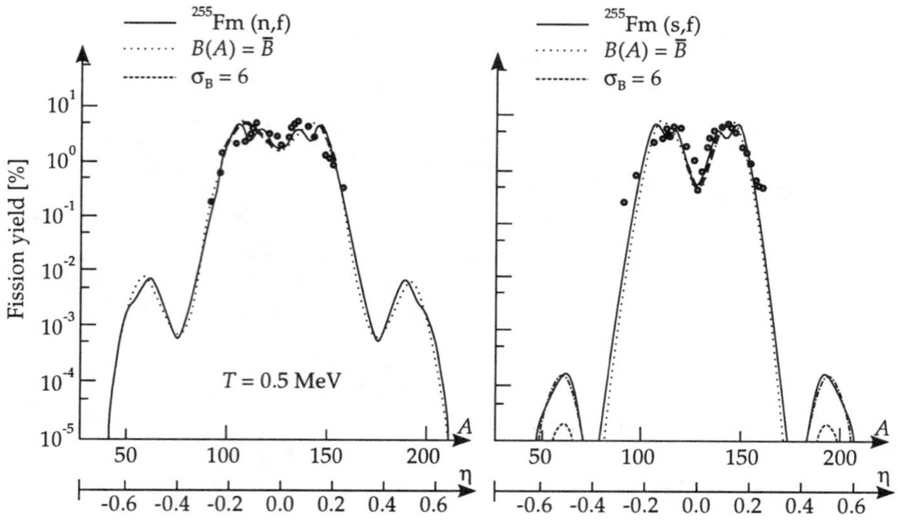

Fig. 9.6 Mass distribution for fission of ^{256}Fm. The full curves are calculated from the ground state of the collective Schrödinger equation, while the dash-dotted curve shows the result of replacing the deformation-dependent mass parameter by its average. The experimental values [Fl72, Fl75] are indicated by small circles. The right-hand side shows spontaneous fission, while the left corresponds to neutron-induced fission, which was treated in theory by assuming a temperature of $T = 0.5$ MeV populating excited asymmetry states. From [Lu80].

reasonably well, and adding a temperature-induced excitation of higher collective states also describes the broadening of the distribution for neutron-induced fission.

An alternative method for looking at tunneling lifetimes and mass distributions involves the penetration factors through the potential. Since tunnelling takes place in a multidimensional collective space, this should in principle be done as a sum over contributions from all possible paths through the collective space; practically this is still impossible, so that usually the path of maximum penetrability is regarded alone, with the assumption that the contributions from other paths fall off exponentially. This corresponds to the formula

$$G = \frac{2}{\hbar} \int_{t_i}^{t_f} \mathrm{d}t \, \sqrt{2\left[V\left(\boldsymbol{q}(t)\right) - E\right]} \, \sqrt{\sum_{ij} B_{ij}\left(\boldsymbol{q}(t)\right) \dot{q}_i(t) \dot{q}_j(t)} \tag{9.26}$$

for the Gamow factor, where $\boldsymbol{q}(t) = \{q_i(t)\}$ is the parametrization of the fission path and the B_{ij} are the mass parameters. An example of such a calculation, which also illustrates the role of the deformation-dependent mass parameters quite well, is given in Fig. 9.7. Note especially how the optimum paths do not follow the intuitive route through the potential that would be correct for constant mass parameters. Still it is interesting to see that the lifetimes of known nuclei can be reproduced quite well over many orders of magnitude in such a calculation (see Fig. 9.8).

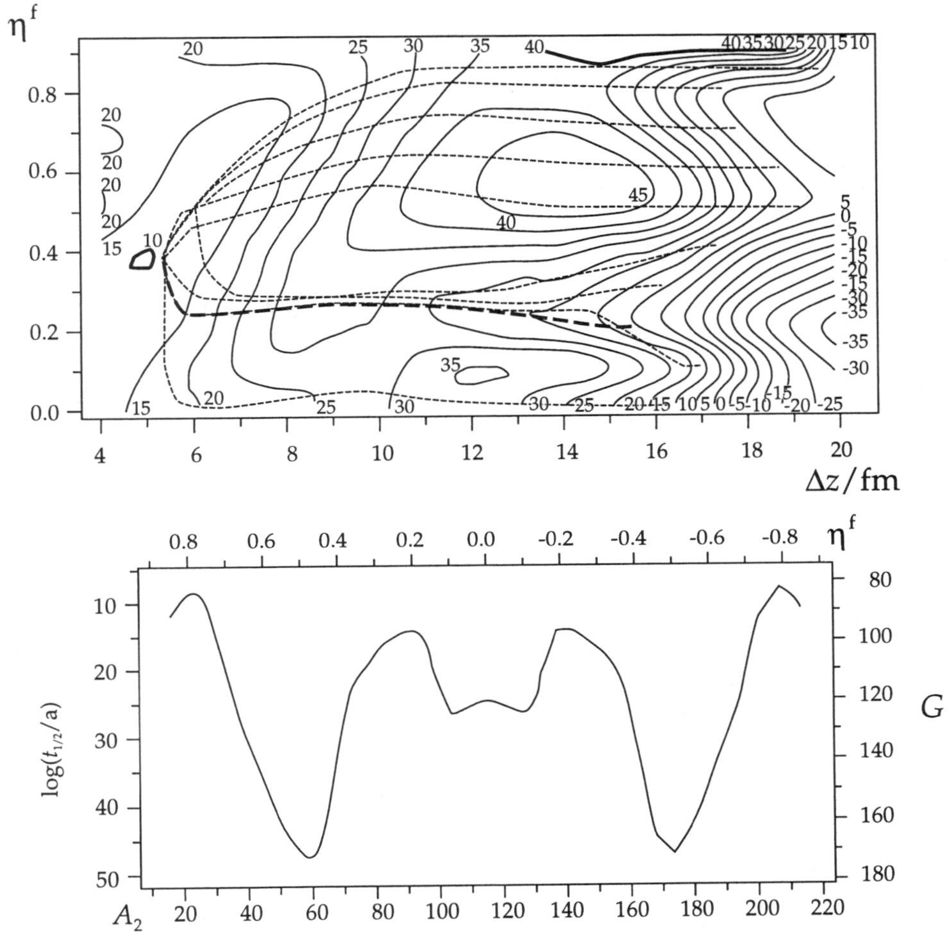

Fig. 9.7 Calculation of spontaneous fission properties for the nucleus ^{232}U. The upper graph shows the potential with the paths of maximum penetrability shown as dashed lines. They were obtained by fixing the initial point (near the ground state to the left) and a final point at the exit from the barrier and with a prescribed mass asymmetry η. The strong deformation dependence of the three mass parameters (one for η, one for Δz, and one coupling the two) is reflected in the way the fission paths do not follow the intuitive classical path. The lower graph gives the expected variation of the lifetime with mass asymmetry, which is qualitatively similar to the experimental mass distribution. In this lower graph the scale on the right shows the Gamow factor, while on the left the half-life time $t_{1/2}$ is given. From [Kl91].

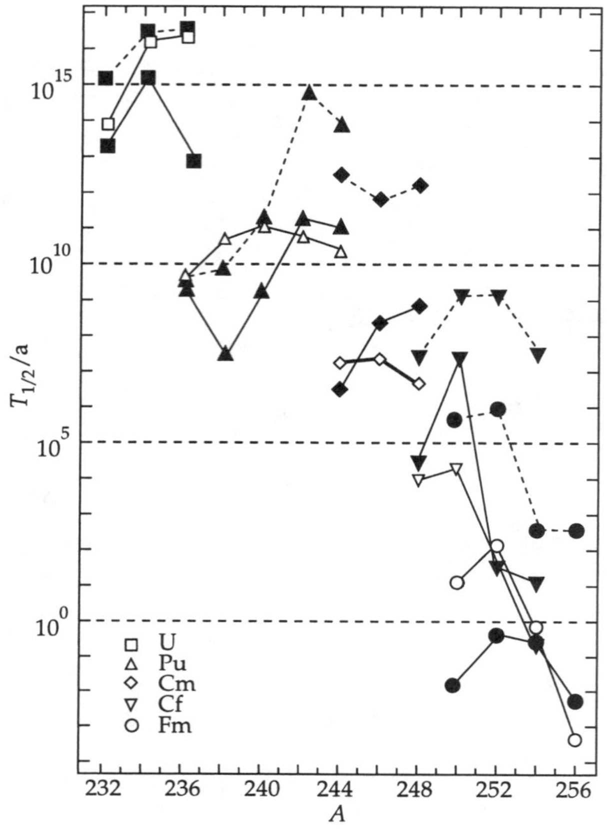

Fig. 9.8 Calculated and experimental fission half-life times for some actinide nuclei. Symbol shapes distinguish the elements, and the abscissa shows the mass number. The filled symbols correspond to theory and the open ones to measurements. The symbols linked by dashed lines are predictions for the most favorable cluster decay (see below).

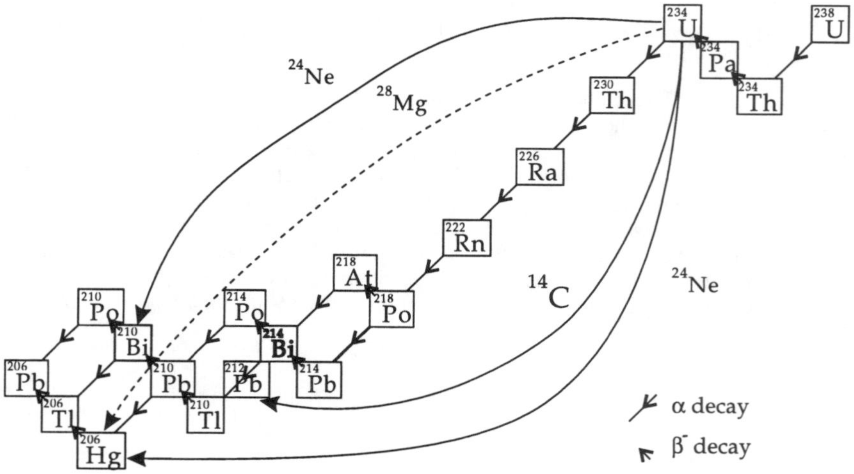

Fig. 9.9 Illustration of cluster decays. In the decay chain of ^{238}U, which normally consists of many α and β^- decays that change Z and N by at most two units, the cluster decays shown by the long arrows provide a giant jump down in both numbers.

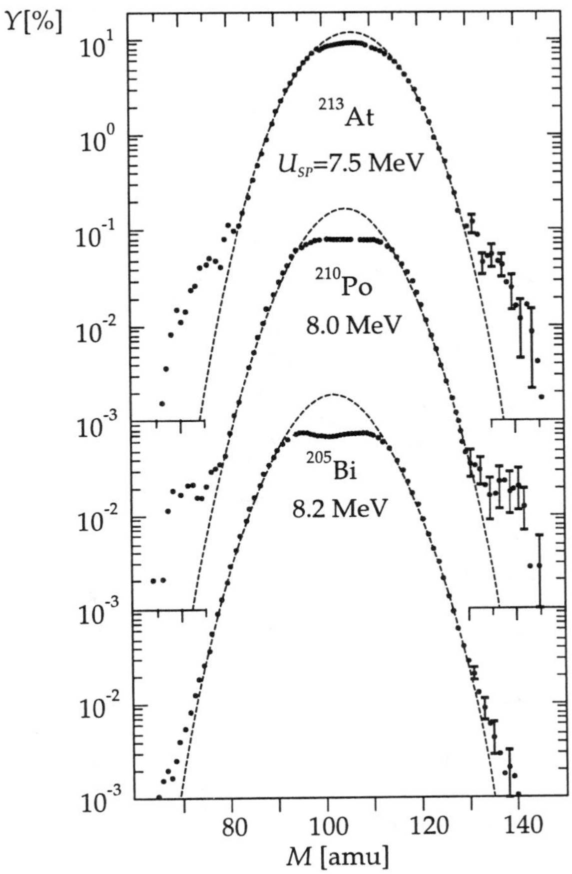

Fig. 9.10 Experimental data for the mass distributions of some fissioning nuclei with masses near 200. In this experiment the exponential tails of the distribution could be followed down to such low probabilities that the shoulders due to superasymmetric fission become clearly visible. From [It92].

All such calculations predict not only the standard asymmetric fission, but also additional valleys called "superasymmetric" fission – corresponding to the peaks near $\eta = \pm 0.5$ in Fig. 9.6 – and even *cluster decays*, the decay by emission of light nuclei such as carbon or neon or as a limiting process, α decay itself. Cluster decays are reviewed in [Gr89, Po89, Sa89], and Fig. 9.9 illustrates the dramatic shortcut taken in decay chains by cluster decays.

Although the predictions for cluster decay and highly asymmetric fission in general [Lu78, Sa80] were originally not taken seriously, the subsequent experimental discovery of cluster decays [Ro84] and some indications for superasymmetric fission (see below) have indeed demonstrated that fission shows a continuous mass distribution that only appeared to be divided into the two separate processes fission and α decay, because some intermediate mass splits are of very low probability. Fig. 9.10 shows first experimental evidence for additional " superasymmetric" shoulders in the mass distribution [It92].

Calculations of fission barriers are extremely important for the prediction of superheavy elements. Fig. 9.11 shows the region of known isotopes, amongst which the stable ones are only a small fraction, surrounded by a sea of instability. While

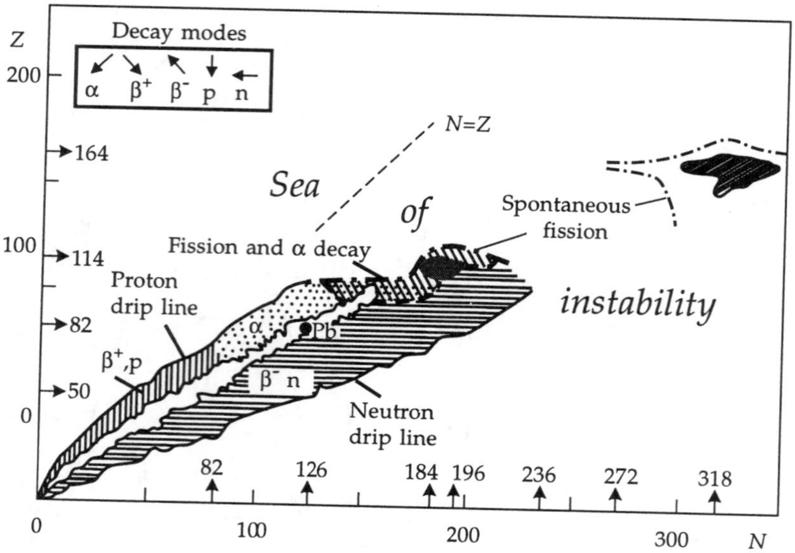

Fig. 9.11 Landscape of the isotopes, illustrating the possible islands of superheavy elements. For the known isotopes the prevalent decay modes are given, enclosing the island of stable nuclei which stretches from the origin through Pb and into the actinide region. It is separated from the first island of superheavy nuclei near magic numbers $(Z, N) \approx (114, 184)$ by a region of α- and fission-unstable nuclei. There may even be a second island near magic numbers $(Z, N) \approx (164, 318)$. In each region the dominant decay modes are indicated. The "drip lines" show where an additional proton or neutron can no longer be bound (their precise location is under intensive investigation at present).

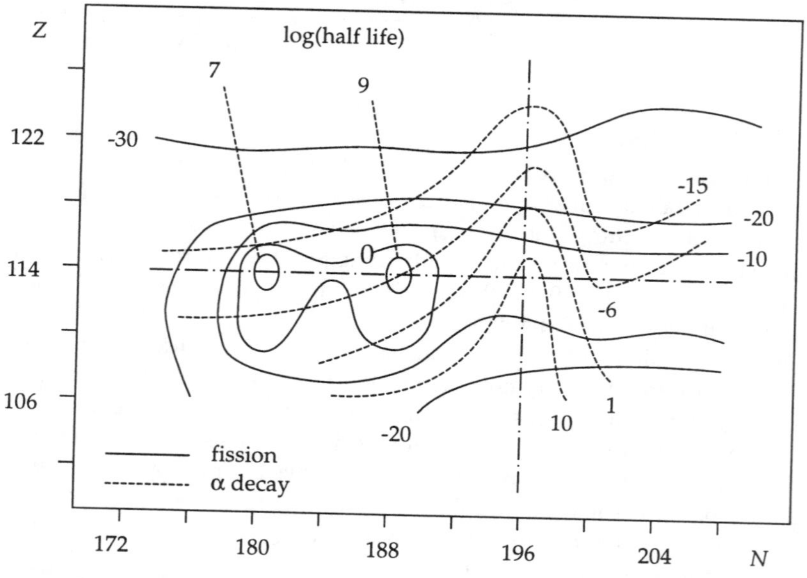

Fig. 9.12 Logarithm of the lifetimes of nuclei in the first superheavy island with respect to spontaneous fission and α decay. The numbers on the isolines correspond to $\log(T_{1/2}/1\text{a})$. From [Gr69].

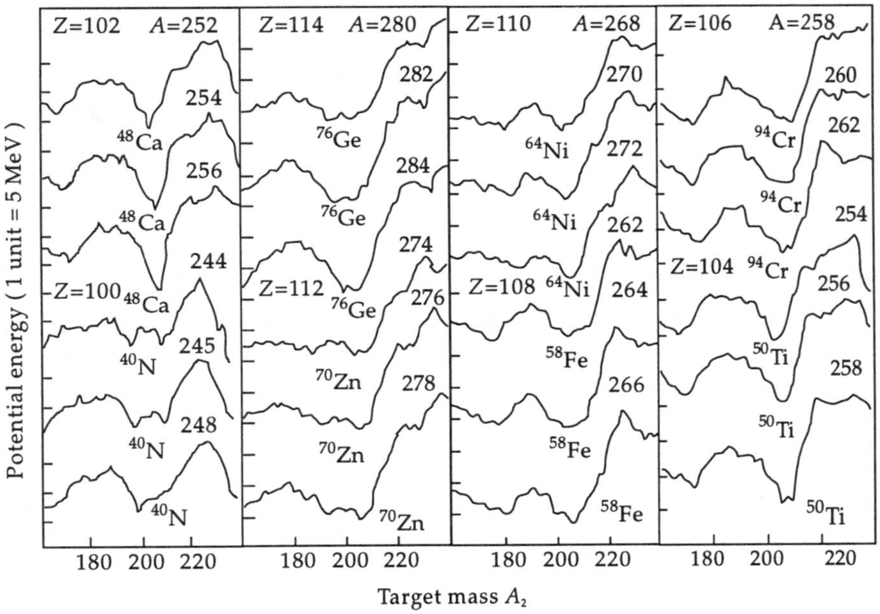

Fig. 9.13 Overview of the potential energies as functions of mass asymmetry for producing some of the elements at the upper end of the periodic table. In each of these curves, which correspond to the indicated (Z, N) combinations, for the deepest valley the projectile nucleus is indicated; the corresponding target must have just the missing mass and neutron numbers. From [Gu77].

at present the island is enlarged on both sides by the production of very neutron- or proton-rich nuclei, it is much more difficult to make progress in the direction of increasing total mass, because α decay and spontaneous fission reduce the lifetimes drastically in that direction. This trend can, however, be partly reversed by the presence of additional magic numbers. Most theories [Gr69, Ni69, Fi72, Ra74, Ir75] predict magic numbers near $(Z, N) = (114, 184)$ and possibly also for $(164, 318)$, but the details vary: the most favorable proton number may be 110 or 114, and the lifetimes predicted range from several months [Gr69] to billions of years [Fi72]. A detailed review can be found in [Ku89]. Details of the first of these calculations are shown in Fig. 9.12. One should keep in mind, though, that lifetimes for fission depend on tunneling probabilities which can change tremendously with small changes of the barrier height or width. The uncertainty of such predictions is therefore probably a few orders of magnitude.

There is, of course, the big practical problem of how to produce superheavy elements. Since the ratio of neutrons to protons increases with A, one cannot easily combine nuclei from the known region, because the compound system would be severely neutron-deficient. Producing a compound system with too many nucleons and hoping for the emission of superfluous ones has up to now not been successful, instead the heaviest elements have so far been produced by using very neutron-rich projectiles and targets in combinations corresponding to the asymmetric valleys in

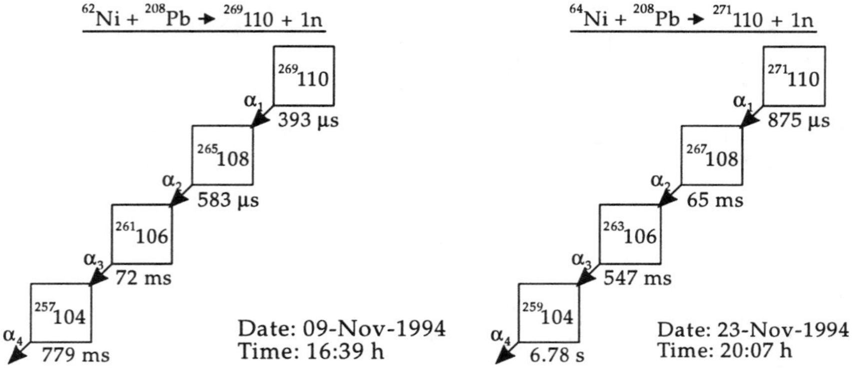

Fig. 9.14 Decay chains that led to the identification of element 110 through an experimental group at GSI at the dates and times indicated [Ho95a].

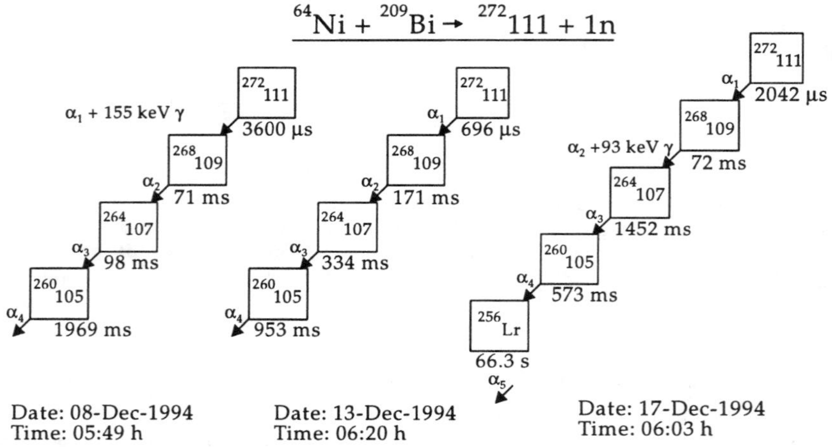

Fig. 9.15 Decay chains for element 111 [Ho95b].

the potential as suggested by theory [Sa76, Gu77]. Figure 9.5, for example, shows ${}^{90}_{24}$Se+${}^{208}_{80}$Hg and ${}^{134}_{52}$Te+${}^{164}_{52}$Sm as suitable combinations for producing ${}^{298}_{114}$X.

An overview of these "cold valleys" for producing elements at the upper end of the known periodic system is given in Fig. 9.13. Experimenters at the GSI Laboratory have indeed discovered elements 107–111 very recently by using just projectile–target combinations. They utilize the fast α emission of these neutron-poor nuclei to identify them through the observation of the α-decay chains (see Figs. 9.14–15). Although these nuclei are not as close to the island of superheavies as one would assume, because they are quite neutron deficient, this raises the hopes that the island itself will become accessible in the near future. Figures 9.14–15 show the exciting experimental findings that led to the discovery of the two new heaviest elements [Ho95a–b]. In each case a series of α particles (sometimes incomplete) of distinct energies and lifetimes could be measured, so that a small number or even a single event was sufficient to establish the new elements.

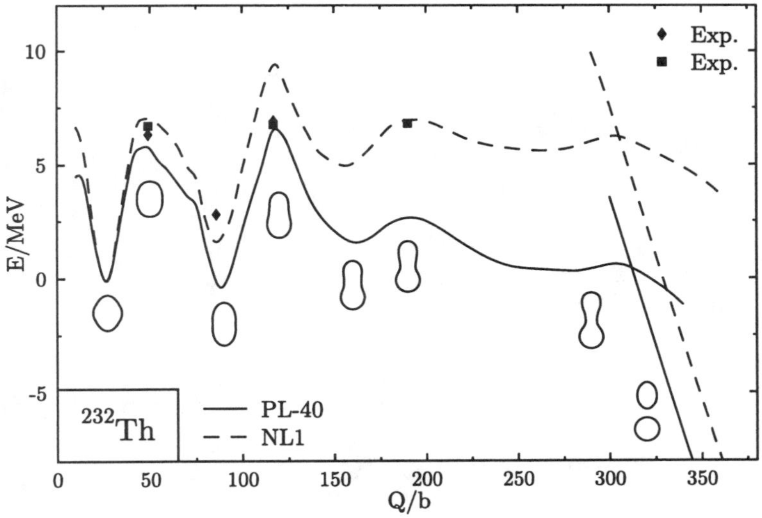

Fig. 9.16 Fission potential for the nucleus ^{232}Th calculated in the relativistic mean-field model (see Sect. 7.4) with the two different parametrizations "PL-40" and "NL1" by Rutz et al. [Ru95]. The constraint in this case is the mass quadrupole moment Q. Both calculations agree qualitatively, but show increaingly different behavior at large deformation. The deformed ground-state minimum and the fission isomer are clearly visible; in this case the first barrier is higher. There is also a third minimum corresponding to a very elongated shape. The experimental data [Lo70, Wa77, Bj80, Zh86] are indicated by diamonds and squares. The steep curves starting near $Q = 300\,\mathrm{b}$ belong to a second valley distinguished by separated fragments, as is shown by the small shape plots near the curves. Probably the fissioning nucleus will cross over to this valley, but where this happens depends on the barrier between the two valleys, which can be determined only with an additional constraint, possibly of hexadecupole nature. More detail for the shapes is given in the next figure.

9.2.4 Fission in Self-Consistent Models

An alternative method for looking at fission is provided by self-consistent models. With today's computers it is quite feasible to calculate potential-energy surfaces with one or even two constraints for heavy nuclei. Compared to the phenomenological models with prescribed shapes this has the following advantages.

1. There is no need for shell corrections. The total energies calculated are in good agreement with the liquid-drop plus shell correction results for the phenomenological models.
2. For fixed constraints one may hope that the shape degrees of freedom not frozen by constraints adjust in such a way as to produce the energetically most favorable configuration, so that there is not so much of a need to guess the optimum shape beforehand. If the numerical method allows asymmetric deformations, for example, the shapes along the fission path will automatically become asymmetric wherever this is favored.
3. The potential and the density distributions themselves can become more complicated; for example, the growing repulsion by the large nuclear charge can

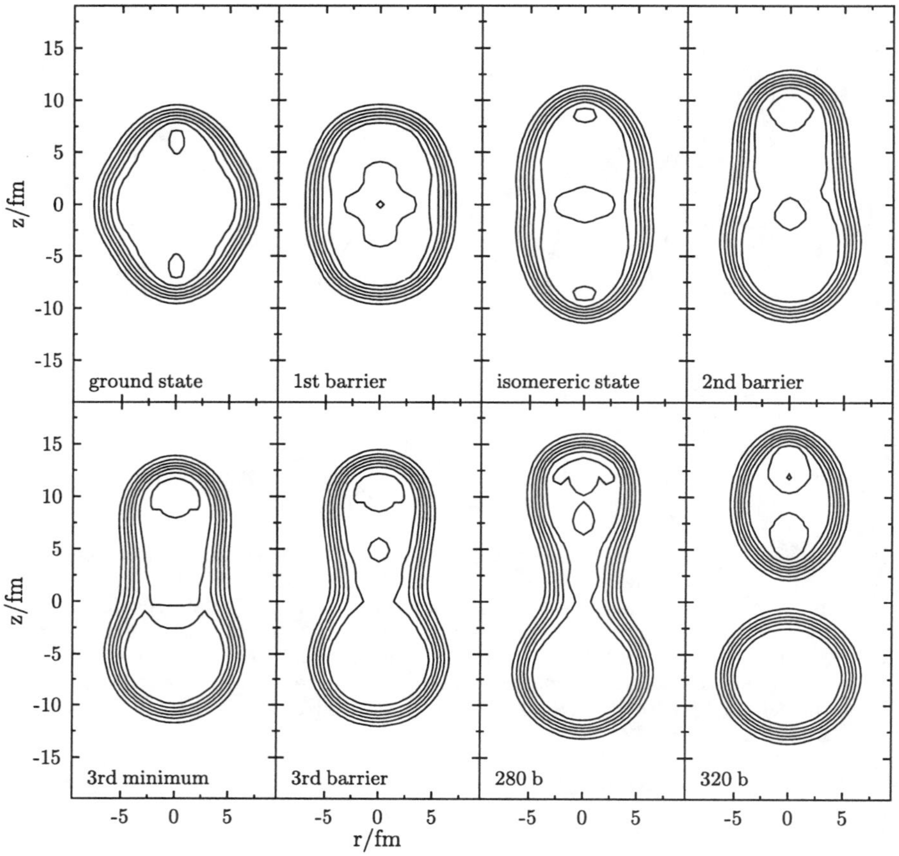

Fig. 9.17 Nuclear shapes corresponding to various points along the fission path of Fig. 9.16, plotted as density contour lines. The last one for $Q = 340\,\mathrm{b}$ is from the second valley and clearly has fissioned already for a smaller quadrupole moment than the elongated shape for $Q = 380\,\mathrm{b}$ from the first valley. Observe also that near the second barrier there is a decisive preference for an asymmetric configuration,leading to preferred asymmetric mass splits. The shape in the third minimum is extremely elongated and thus is a "superdeformed" configuration.

push the protons out of the center of the nucleus; in a phenomenological model this has to be allowed for by changing the potential by hand – usually this is not included in the models.

4. On the other hand, there is often a need for more constraints. The results presented below show that the minimization may lead to more than one minimum for the same set of constraints, showing that there are different valleys in the multidimensional potential-energy surface distinguished by some additional shape parameter. If these cross energetically, the transition between the valleys can be described only by using an additional constraint which allows one to also examine the barrier between the minima.

Figure 9.16 shows typical results for a fission barrier in the relativistic mean-field model with a quadrupole constraint. Of special interest in the light of the discussion above is the appearence of two distinct valleys at large deformation, one corresponding to very long and necked-in but still connected shapes, and the other corresponding to separate fragments. Since they have the same quadrupole moment, they appear as disjoint solutions with this one constraint, but could be linked continuously by crossing a barrier if a suitable second constraint is added. The corresponding shapes in Fig. 9.17 illustrate this effect and also the preference for asymmetric configurations near the second barrier.

9.3 Mass Parameters and the Cranking Model

9.3.1 Overview

As we will see, the problem of kinetic energies is far less settled than that of the potential energies. Although there are some theoretical methods for the calculation of mass parameters, none of these is universally accepted and it is very difficult to check their predictions in experiment. Large-scale collective motion occurs only in fission and heavy-ion reactions, which are sufficiently complicated that if the theory does not agree with experiment it is difficult to see whether this is caused by an insufficient number of collective coordinates, friction, or a real failure of the underlying models. Usually, however, effective parameters have to be introduced both in the mass parameters and the friction terms to get even close to the experimental data. We will now briefly discuss the two widely used approaches to mass-parameter calculations: the irrotational-flow method and the cranking model.

9.3.2 The Irrotational-Flow Model

The irrotational flow model was introduced for small deformations in Sect. 6.2.1. It can be generalized easily: a time dependence of the collective coordinates induces a local velocity of the nuclear surface, which can be used as a boundary condition to solve for the velocity field inside the nucleus under the assumption of irrotational flow. Integrating the kinetic energy of the fluid will then yield a kinetic energy which is of second order in the collective velocities, so that the mass parameters can be read off. Although this model is thus well defined, it is by no means easy to construct solutions for general surface shapes and motions. A well-known approximate method for cylindrically symmetric shapes is known as the Werner–Wheeler method [Ke64, Ni68]. Note that there is still no firm theoretical support for this model: it is widely used mainly because it is well defined, calculable, and expected at least to give the order-of-magnitude differences between the mass parameters for collective coordinates of various types; also the alternatives have problems of their own.

9.3.3 The Cranking Formula

Another method for the calculation of mass parameters is the *cranking model* invented by Inglis. The name reflects the idea of an external influence forcing the phenomenological single-particle potential and wave functions to follow a time-dependent change in the shape parameters. During the calculation of the deformation energy the crucial assumption was that of adiabasy. For consistency this has to be somehow extended to the dynamic case as well. We may imagine that the collective coordinates change very slowly, so that the nucleus can settle down into its adiabatic state at each point. On the other hand, the wave functions have to change during the motion and this time dependence is associated with a kinetic energy, which we want to investigate. The nucleus is thus not truly in the adiabatic state, but should not contain any real excitations aside from what is necessary to support the collective motion. How this distinction will work will become clear during the derivation.

We start from a Hamiltonian $\hat{H}(\boldsymbol{r}_1,\ldots,\boldsymbol{r}_A;\beta_1,\ldots,\beta_M)$ that depends parametrically on the deformation parameters $\beta_1,\ldots,\beta_M$. The associated adiabatic eigenfunctions depend on the parameters and fulfil

$$\begin{aligned}\hat{H}(\boldsymbol{r}_1,\ldots,\boldsymbol{r}_A;\beta_1,\ldots,\beta_M)\,\psi_i(\boldsymbol{r}_1,\ldots,\boldsymbol{r}_A;\beta_1,\ldots,\beta_M)&\\ = E_i(\beta_1,\ldots,\beta_M)\,\psi_i(\boldsymbol{r}_1,\ldots,\boldsymbol{r}_A;\beta_1,\ldots,\beta_M)&\quad. \end{aligned}\tag{9.27}$$

For brevity in the following developments we suppress the particle coordinates and use the vector notation $\boldsymbol{\beta}$, so that the equation reads

$$\hat{H}(\boldsymbol{\beta})\,\psi_i(\boldsymbol{\beta}) = E_i(\boldsymbol{\beta})\,\psi_i(\boldsymbol{\beta})\quad. \tag{9.28}$$

Now assume a prescribed time dependence $\boldsymbol{\beta} = \boldsymbol{\beta}(t)$, which is assumed to be so slow that no excitations are generated. This means that the characteristic time scale of the collective motion should be very long compared to the single-particle time scale. We now have a time-dependent Hamiltonian and basis wave functions. The solution of the time-dependent Schrödinger equation

$$\mathrm{i}\hbar\frac{\partial}{\partial t}\Psi = \hat{H}\left(\boldsymbol{\beta}(t)\right)\Psi \tag{9.29}$$

can be expanded in the adiabatic eigenstates:

$$\begin{aligned}&\Psi(\boldsymbol{r}_1,\ldots,\boldsymbol{r}_A,t)\\ &\quad=\sum_i c_i(t)\,\psi_i(\boldsymbol{r}_1,\ldots,\boldsymbol{r}_A;\boldsymbol{\beta}(t))\,\exp\left(-\frac{\mathrm{i}}{\hbar}\int^t \mathrm{d}t'\,E_i\left(\boldsymbol{\beta}(t')\right)\right)\quad.\end{aligned}\tag{9.30}$$

The phase factor is designed such as to cancel the phase due to the eigenenergies in the equation. Inserting this expression into the Schrödinger equation yields

$$\begin{aligned}&\sum_i\left(\mathrm{i}\hbar\dot{c}_i + \mathrm{i}\hbar c_i\dot{\boldsymbol{\beta}}\cdot\nabla_{\boldsymbol{\beta}} + E_i\,c_i\right)\psi_i(\boldsymbol{\beta})\,\exp\left(-\frac{\mathrm{i}}{\hbar}\int^t\mathrm{d}t'\,E_i\left(\boldsymbol{\beta}(t')\right)\right)\\ &\quad=\sum_i c_i\,E_i(\boldsymbol{\beta})\,\psi_i(\boldsymbol{\beta})\,\exp\left(-\frac{\mathrm{i}}{\hbar}\int^t\mathrm{d}t'\,E_i\left(\boldsymbol{\beta}(t')\right)\right)\quad.\end{aligned}\tag{9.31}$$

As is suggested, the vector notation for the $\boldsymbol{\beta}$ was extended to

$$\boldsymbol{\beta} \cdot \nabla_\beta = \sum_{k=1}^{M} \beta_k \frac{\partial}{\partial \beta_k} \quad . \tag{9.32}$$

Equation (9.31) may be converted into an equation for the expansion parameters by forming the matrix element using

$$\int \mathrm{d}^3 r_1 \cdots \int \mathrm{d}^3 r_A \psi_j^*(\boldsymbol{r}_1, \ldots, \boldsymbol{r}_A; \boldsymbol{\beta}) \times (9.31) \quad . \tag{9.33}$$

Cancelling also the phase factors as far as possible results in

$$\begin{aligned} \mathrm{i}\hbar c_j &= -\mathrm{i}\hbar \sum_i c_i \langle \psi_j | \boldsymbol{\beta} \cdot \nabla_\beta | \psi_i \rangle \\ &\quad \times \exp\left\{ -\frac{\mathrm{i}}{\hbar} \int^t \mathrm{d}t' \, \left[E_i(\boldsymbol{\beta}(t')) - E_j(\boldsymbol{\beta}(t')) \right] \right\} \quad . \end{aligned} \tag{9.34}$$

Now we have to use the assumption of different time scales and adiabasy. We are really interested in the solution only for a short interval of time around t, because in the adiabatic limit the state of the system at time t should be completely determined at that time, without any influence of the previous history of the system. During such an interval both the matrix element and the eigenenergies in the phase factor change very little, because they show only the slow dependence on the collective coordinates. On the other hand, the phase factor itself changes rapidly with time. Adiabasy implies that the coefficient of the ground state for fixed deformation $\boldsymbol{\beta}(t)$ is close to unity, $c_0 \approx 1$, while all the others are quite small. Therefore on the right-hand side only $i = 0$ contributes, and we get

$$\begin{aligned} c_j(t) &= \frac{\langle \psi_j | \boldsymbol{\beta} \cdot \nabla_\beta | \psi_0 \rangle}{E_j - E_0} \\ &\quad \times \exp\left\{ -\frac{\mathrm{i}}{\hbar} \int^t \mathrm{d}t' \, \left(E_0(\boldsymbol{\beta}(t')) - E_j(\boldsymbol{\beta}(t')) \right) \right\} \quad \text{for} \quad j \neq 0 \quad , \end{aligned} \tag{9.35}$$

whereas for $j = 0$ one may simply utilize normalization,

$$|c_0|^2 = 1 - \sum_{i \neq 0} |c_i|^2 \quad . \tag{9.36}$$

Actually there is a subtle point in (9.35). The constant of integration was set equal to zero, so that at $t = 0$ we do not get the unperturbed ground state $c_0 = 1$, $c_j = 0 (j > 0)$. If this were used as the initial condition, the solution would describe a strong variation of the amplitudes with time. In fact, the underlying picture was not that of a system in the adiabatic ground state at $t = 0$, suddenly starting a collective motion, but the collective motion has been going on already and the system has adapted to that. Equation (9.35) describes exactly that: while the complex coefficients oscillate rapidly, their amplitude is only slowly time dependent. This becomes apparent also from the total energy, which is the expectation value of the Hamiltonian:

$$\langle \Psi(t)|\hat{H}\left(\boldsymbol{\beta}(t)\right)|\Psi(t)\rangle = \sum_j |c_j(t)|^2 E_j(t) = E_0\left(\boldsymbol{\beta}(t)\right) + \tfrac{1}{2}\boldsymbol{\beta}\cdot\boldsymbol{B}\cdot\boldsymbol{\beta} \quad , \tag{9.37}$$

with the mass parameter

$$B_{kl} = 2\hbar^2 \sum_{j\neq 0} \frac{\langle\psi_0|\partial/\partial\beta_k|\psi_j\rangle\langle\psi_j|\partial/\partial\beta_l|\psi_0\rangle}{E_j - E_0} \quad . \tag{9.38}$$

This is the cranking formula as first given by Inglis [In54, In56]. Its evaluation requires a model for the many-body wave functions ψ_j, but at this point we may already make some general observations. The matrix elements in the numerator carry information about the change of the single-particle states with deformation, and there will be a peak in the mass if the wave functions have to be rearranged strongly to change the deformation. Another reason for large values of the mass parameters may be the presence of excited states close to the ground state, which make the denominator nearly vanish. If the latter is due to an avoided energy-level crossing, the wave functions will also change rapidly during the crossing and a strong peak should result.

9.3.4 Applications of the Cranking Formula

For concrete calculations the meaning of the collective parameters and the wave functions have to be specified in the cranking formula, which can often be reduced to somewhat simpler forms.

For the *independent particle model* the ground state of the nucleus is the Slater determinant made up of the lowest single-particle orbitals. The operator $\partial/\partial\beta$ in second quantization becomes

$$\frac{\partial}{\partial\beta} = \sum_{\mu\nu} \langle\mu|\frac{\partial}{\partial\beta}|\nu\rangle\, \hat{a}^\dagger_\mu \hat{a}_\nu \tag{9.39}$$

and links the ground state only to the one-particle/one-hole states. We label the single-particle states with Greek letters here to distinguish them from the many-body states $|j\rangle$. The particle–hole excitations thus become the excited states to be summed over. The energy denominator becomes the excitation energy of the particle–hole state,

$$E_j - E_0 \longrightarrow \varepsilon_\mu - \varepsilon_\nu \quad , \tag{9.40}$$

while the matrix element is reduced to

$$\langle j|\frac{\partial}{\partial\beta}|0\rangle \longrightarrow \langle\mu|\frac{\partial}{\partial\beta}|\nu\rangle \quad , \tag{9.41}$$

and the cranking formula is represented by

$$B_{kl} = 2\hbar^2 \sum_{\substack{\mu\ \mathrm{unocc} \\ \nu\ \mathrm{occ}}} \frac{\langle\nu|\partial/\partial\beta_k|\mu\rangle\langle\mu|\partial/\partial\beta_l|\nu\rangle}{\varepsilon_\mu - \varepsilon_\nu} \quad . \tag{9.42}$$

Note that this expression even becomes singular if there is an energy-level crossing at the Fermi surface. This can be avoided only through the use of pairing.

The formula above still does not specify the type of collective coordinate considered. The simplest mass parameter, and the one most easily compared with experiment, is the *moment of inertia*. In this case the collective coordinate is the angle of rotation about an axis not coinciding with the symmetry axis of the nucleus; the x axis is usually chosen. The derivative with respect to the angle can be expressed through the angular-momentum operator $\partial/\partial\beta \rightarrow i J_x/\hbar$, and the moment of inertia becomes

$$\Theta = 2 \sum_j \frac{|\langle j | J_x | 0 \rangle|^2}{E_j - E_0} \quad . \tag{9.43}$$

In this formula the independent-particle case may be inserted as above. It can be shown that under certain conditions the resulting value is just the rigid-body moment of inertia, i.e., the value for a classical solid of uniform density and with the shape corresponding to the nuclear deformation. Comparison with experiment, however, shows that this value is usually too large by a factor of 2. This indicates that not all the nuclear matter may participate in the collective rotation, and again the introduction of pairing provides the solution.

EXERCISE

9.1 The Cranking Formula for the BCS Model

Problem. Derive the cranking formula for the BCS model.

Solution. In this case the nuclear ground state is replaced by the paired ground state, $|0\rangle \rightarrow |\mathrm{BCS}\rangle$. The operator $\partial/\partial\beta$ acts not only on the single-particle states as in (9.39), but also on the pairing occupation probabilities u_μ and v_μ, which will be deformation dependent in general. Denoting this additional contribution by $\overline{\partial/\partial\beta}$, we may write

$$\frac{\partial}{\partial\beta} = \overline{\frac{\partial}{\partial\beta}} + \sum_{\mu\nu} \langle\mu| \frac{\partial}{\partial\beta} |\nu\rangle \hat{a}^\dagger_\mu \hat{a}_\nu \quad . \tag{1}$$

Replacing the operators by quasiparticle operators according to

$$\hat{a}_\mu = u_\mu \hat{\alpha}_\mu + v_\mu \hat{\alpha}^\dagger_{-\mu} \tag{2}$$

and its hermitian adjoint, we get

$$\begin{aligned} \frac{\partial}{\partial\beta} - \overline{\frac{\partial}{\partial\beta}} &= \sum_{\mu\nu} \langle\mu| \frac{\partial}{\partial\beta} |\nu\rangle \; \hat{a}^\dagger_\mu \hat{a}_\nu \\ &= \sum_{\mu\nu} \langle\mu| \frac{\partial}{\partial\beta} |\nu\rangle \; \left(u_\mu \hat{\alpha}^\dagger_\mu + v_\mu \hat{\alpha}_{-\mu}\right) \left(u_\nu \hat{\alpha}_\nu + v_\nu \hat{\alpha}^\dagger_{-\nu}\right) \\ &\rightarrow \sum_{\mu\nu} \langle\mu| \frac{\partial}{\partial\beta} |\nu\rangle \; u_\mu v_\nu \hat{\alpha}^\dagger_\mu \hat{\alpha}_{-\nu} \quad . \end{aligned} \tag{3}$$

Exercise 9.1

The last line shows only that term in the operator contributing to the matrix element. The other combinations yield zero, because the excited state must differ from the ground state in quasiparticle number. Note also that the sum extends over both positive and negative values of μ and ν, whereas in the BCS state only positive values are used. To avoid ambiguity for the latter case $\mu > 0$ etc. will be stated explicitly.

The form of the matrix element shows that only two-quasiparticle states need to be considered. Taking the matrix element between $|\kappa\lambda\rangle = \hat{\alpha}^\dagger_\kappa \hat{\alpha}^\dagger_{-\lambda}|\mathrm{BCS}\rangle$ $(\kappa, \lambda < 0)$ and the BCS ground state results in

$$\begin{aligned}\langle\kappa\lambda|\frac{\partial}{\partial\beta} - \overline{\frac{\partial}{\partial\beta}}|\mathrm{BCS}\rangle &= \sum_{\mu\nu}\langle\mu|\frac{\partial}{\partial\beta}|\nu\rangle u_\mu v_\nu \langle\mathrm{BCS}|\hat{\alpha}_{-\lambda}\hat{\alpha}_\kappa\hat{\alpha}^\dagger_\mu\hat{\alpha}^\dagger_{-\nu}|\mathrm{BCS}\rangle \\ &= \sum_{\mu\nu}\langle\mu|\frac{\partial}{\partial\beta}|\nu\rangle u_\mu v_\nu \left(\delta_{\mu\kappa}\,\delta_{\nu\lambda} - \delta_{\mu-\lambda}\,\delta_{\nu-\kappa}\right) \\ &= u_\kappa v_\lambda \langle\kappa|\frac{\partial}{\partial\beta}|\lambda\rangle + u_\lambda v_\kappa \langle-\lambda|\frac{\partial}{\partial\beta}|-\kappa\rangle \quad . \end{aligned} \tag{4}$$

For the last step $u_\kappa = u_{-\kappa}$ and $v_{-\kappa} = -v_\kappa$ was used.

This result can be simplified further only if the two matrix elements in (4) are related to each other. Obviously this will have something to do with time reversal, since the states involved are related through that operation. Remembering the transformation of matrix elements of any operator $\hat{O}$ under time reversal,

$$\langle A|\hat{O}|B\rangle = \langle\hat{\mathcal{T}}A|\hat{\mathcal{T}}\hat{O}\hat{\mathcal{T}}^{-1}|\hat{\mathcal{T}}B\rangle^* \quad , \tag{5}$$

and assuming that $\partial/\partial\beta$ has definite time reversal properties,

$$\hat{\mathcal{T}}\frac{\partial}{\partial\beta}\hat{\mathcal{T}}^{-1} = \pm\frac{\partial}{\partial\beta} \quad , \tag{6}$$

we get

$$\langle-\lambda|\frac{\partial}{\partial\beta}|-\kappa\rangle = \langle\lambda|\pm\frac{\partial}{\partial\beta}|\kappa\rangle^* = \mp\langle\kappa|\frac{\partial}{\partial\beta}|\lambda\rangle \quad . \tag{7}$$

In the last step the antihermitian nature of $\partial/\partial\beta$ was used (cf. the operator $\partial/\partial x$, which is antihermitian because $\hat{p}_x = -\mathrm{i}\hbar\,\partial/\partial x$ is hermitian). The matrix element has now become

$$\langle\kappa\lambda|\frac{\partial}{\partial\beta} - \overline{\frac{\partial}{\partial\beta}}|\mathrm{BCS}\rangle = \langle\kappa|\frac{\partial}{\partial\beta}|\lambda\rangle(u_\kappa v_\lambda \mp u_\lambda v_\kappa) \quad , \tag{8}$$

the negative sign being valid for time-even $\partial/\partial\beta$.

The final task is evaluating the derivative acting on the occupation probabilities. The definition of the BCS ground state leads directly to

$$\overline{\frac{\partial}{\partial\beta}}|\mathrm{BCS}\rangle = \sum_{\mu>0}\left(\frac{\partial u_\mu}{\partial\beta} + \frac{\partial v_\mu}{\partial\beta}\hat{a}^\dagger_\mu\hat{a}^\dagger_{-\mu}\right)\prod_{\substack{\nu\neq\mu\\ \nu>0}}\left(u_\nu + v_\nu\hat{a}^\dagger_\nu\hat{a}^\dagger_{-\nu}\right)|0\rangle \quad . \tag{9}$$

The overlap with the two-quasiparticle state $|\kappa\lambda\rangle$ will be nonzero only if both κ and λ are equal to μ in each term in the sum, as any other index will meet the unperturbed BCS factor in the product to produce zero. Canceling the products containing the unconcerned index ν both in the two-quasiparticle and the BCS state and using μ for the only remaining index yields

$$\begin{aligned}
\langle\mu\mu|\overline{\frac{\partial}{\partial\beta}}|\mathrm{BCS}\rangle &= \langle 0|\left(u_\mu + v_\mu \hat{a}_{-\mu}\hat{a}_\mu\right)\hat{\alpha}_{-\mu}\hat{\alpha}_\mu\left(\frac{\partial u_\mu}{\partial\beta} + \frac{\partial v_\mu}{\partial\beta}\hat{a}^\dagger_\mu\hat{a}^\dagger_{-\mu}\right)|0\rangle \\
&= \frac{\partial u_\mu}{\partial\beta}\langle 0| - u_\mu^2 v_\mu \hat{a}_{-\mu}\hat{a}^\dagger_{-\mu} - v_\mu^3\hat{a}_{-\mu}\hat{a}_\mu\hat{a}^\dagger_\mu\hat{a}^\dagger_{-\mu}|0\rangle \\
&\quad + \frac{\partial v_\mu}{\partial\beta}\langle 0|u_\mu v_\mu^2\hat{a}_{-\mu}\hat{a}_\mu\hat{a}^\dagger_\mu\hat{a}_\mu\hat{a}^\dagger_\mu\hat{a}^\dagger_{-\mu} + u_\mu^3\hat{a}_{-\mu}\hat{a}_\mu\hat{a}^\dagger_\mu\hat{a}^\dagger_{-\mu}|0\rangle \\
&= -\frac{\partial u_\mu}{\partial\beta}\left(u_\mu^2 v_\mu + v_\mu^3\right) + \frac{\partial v_\mu}{\partial\beta}\left(u_\mu v_\mu^2 + u_\mu^3\right) \\
&= -\frac{\partial u_\mu}{\partial\beta}\left(v_\mu + \frac{u_\mu^2}{v_\mu}\right) = -\frac{1}{v_\mu}\frac{\partial u_\mu}{\partial\beta} \quad . \qquad (10)
\end{aligned}$$

This involved essentially only the usual second-quantization techniques. The expansion of the parentheses in the second step took into account that only terms leaving the particle number unchanged survive, while the last step used normalization:

$$u_\mu^2 + v_\mu^2 = 1 \quad \rightarrow \quad v_\mu\frac{\partial v_\mu}{\partial\beta} = -u_\mu\frac{\partial u_\mu}{\partial\beta} \quad . \qquad (11)$$

The total matrix element now is

$$\langle\kappa\lambda|\frac{\partial}{\partial\beta}|\mathrm{BCS}\rangle = \langle\kappa|\frac{\partial}{\partial\beta}|\lambda\rangle(u_\kappa v_\lambda \mp u_\lambda v_\kappa) - \delta_{\kappa\lambda}\frac{1}{v_\kappa}\frac{\partial u_\kappa}{\partial\beta} \quad , \qquad (12)$$

and it has to be squared and summed over $\kappa, \lambda > 0$ for the cranking formula. The cross terms do not contribute, because the first matrix element vanishes diagonally owing to the normalization of the wave functions:

$$\frac{\partial}{\partial\beta}\langle\kappa|\kappa\rangle = 0 \rightarrow \langle\kappa|\frac{\partial}{\partial\beta}|\kappa\rangle = 0 \quad . \qquad (13)$$

We thus obtain the final formula for the cranking model with pairing:

$$\begin{aligned}
B_{kl} &= 2\hbar^2\sum_{\kappa\lambda}\frac{\langle\kappa|\frac{\partial}{\partial\beta_k}|\lambda\rangle\langle\lambda|\frac{\partial}{\partial\beta_l}|\kappa\rangle}{e_\kappa + e_\lambda}(u_\kappa v_\lambda \mp v_\kappa u_\lambda)^2 \\
&\quad + \hbar^2\sum_\kappa\frac{1}{u_\kappa^2 e_\kappa}\frac{\partial u_\kappa}{\partial\beta_k}\frac{\partial u_\kappa}{\partial\beta_l} \quad . \qquad (14)
\end{aligned}$$

Note the excitation energy of the two-quasiparticle state in the denominator. The negative sign is correct for time-even $\partial/\partial\beta$.

For the moment of inertia the single-particle level structure does not change with rotation, so the second contribution in the cranking formula vanishes, whereas the first one is reduced because the gap increases the energy denominator. On the other hand, for deformation coordinates such as the quadrupole deformation, the term with the derivatives of the occupation probabilities usually dominates.

An additional benefit of pairing is that the gap removes the possibility of a vanishing denominator; the denominator now is at least 2Δ.

Exercise 9.1

This version of the cranking model including pairing was first given by Belyaev [Be59, Be61] and Prange [Pr61].

9.4 Time-Dependent Hartree–Fock

An attractive possibility for the study of large-scale collective motion is provided by the time-dependent Hartree–Fock equations presented in Sect. 8.2.2. If we write the equations in the generic form

$$\mathrm{i}\hbar\frac{\partial}{\partial t}\psi_j(\boldsymbol{r},t) = \hat{H}[\rho]\,\psi_j(\boldsymbol{r},t) \quad , \quad j = 1,\ldots A \quad , \tag{9.44}$$

it is clear that giving initial conditions for the single-particle wave functions determines them for all future times, even though the Hamiltonian depends also, as indicated, on densities summed up over the single particles. The large number of single-particles wave functions and the nonlinearity allow for extremely complicated developments of the wave functions without any need for inventing collective parameters, mass parameters, or such.

As the treatment of the tunneling process in a time-dependent theory is more difficult, TDHF is mostly applied to heavy-ion reactions, which is outside the scope of this volume. The discovery that one could actually determine solutions numerically for interesting cases of heavy-ion collisions stimulated considerable interest [Bo76, Ma76a, Cu76, Cu78, Sa78, Da78] followed by many more papers in succeeding years. Because the method is intimately related to the other theories of collective motion presented here, however, a few more details will be given.

Most practical applications of TDHF have been based on Skyrme forces. The reason is the same as in static Hartree–Fock: the exchange term can be treated simply and everything can be formulated in terms of a few local densities.

In *heavy-ion collisions* the initial wave functions are assumed to be solutions of static Hartree–Fock for the two initial nuclei. These are inserted into the numerical space grid at the position of the corresponding nucleus and then multiplied by plane-wave factors to set them into motion. Thus, if we number the nucleons of the projectile from 1 through A_1 and those of the target from A_1+1 through A_1+A_2, we will have

$$\begin{aligned}\psi_j(\boldsymbol{r},t=0) &= \phi_j(\boldsymbol{r}-\boldsymbol{R}_1)\exp(\mathrm{i}\boldsymbol{k}_1\cdot\boldsymbol{r}) \quad , \quad j = 1,\ldots,A_1 \quad , \\ \psi_j(\boldsymbol{r},t=0) &= \phi_j(\boldsymbol{r}-\boldsymbol{R}_2)\exp(\mathrm{i}\boldsymbol{k}_2\cdot\boldsymbol{r}) \quad , \quad j = A_1+1,\ldots,A_1+A_2 \quad .\end{aligned} \tag{9.45}$$

Here $\boldsymbol{R}_{1,2}$ denote the initial positions of the target and projectile, and $\boldsymbol{k}_{1,2}$ the wave numbers of the nucleons associated with their common velocity.

As time passes the nucleonic wave functions will move with the velocity $\boldsymbol{v} = \hbar\boldsymbol{k}_{1,2}/m$ until the two nuclei overlap, when much more complicated developments begin. Let us briefly examine some of the conceptual features of what happens then.

As can be checked easily, the orthonormalization of the single-particle wave functions is not changed by the time development (the single-particle Hamiltonian is hermitian, leading to a unitary time-development operator). As in static Hartree–Fock, the individual wave functions have no independent physical meaning, only the set of occupied states is significant. The time dependence of the states is quite general. If we expanded them in terms of adiabatic states, i.e., the eigenstates belonging to the density distribution at a given point in time, we would find that they can be highly excited, because the excitation happening near the energy-level crossings is taken into account exactly as it depends on the rapidity of the collective motion.

While all of this makes the method appear very attractive and general, there are also serious drawbacks on second sight. It was found in many calculations that the solutions of TDHF behave almost "classically", in the sense that the spreading of the probability distribution is quite small. This indicates that Slater determinants explore only a tiny part of the wave-function space, and the following simple example shows how much richer the true many-body wave function can be.

Imagine a system of two nuclei interacting and then separating again after the collision. The final fragments may then have different masses in the final state, for example, the reaction ^{16}O + ^{16}O may lead to ^{12}C + ^{20}Ne or even to ^{4}He + ^{28}Si. The TDHF final-state wave function does contain amplitudes for such splitups, but they are coupled by *spurious cross-channel correlation*: although each of the fragments may have a range of different mass numbers with a nonvanishing probability, they are all propagated with *the same average potential*, which practically couples them. Thus a component of the wave function describing an outgoing ^{4}He will find itself inside a potential well appropriate for ^{16}O and all of these fragments will even be forced to move at the same velocity, independently of their mass. Since this is highly inappropriate energetically for fragments with masses differing much from the average, the result is a strong suppression of mass transfer. This shows that the Slater determinant can at best describe only an average behavior of the reaction.

This introduces a further danger: although the TDHF equations can be derived also from a time-dependent version of the variational principle,

$$\delta\langle\Psi|\hat{H} - \mathrm{i}\hbar\frac{\partial}{\partial t}|\Psi\rangle = 0 \tag{9.46}$$

with $|\Psi\rangle$ constrained to a Slater determinant as before, this does not guarantee the quality of the approximation. Even if the initial wave function is a good approximation for the nuclear ground states, the minimization means only that the time derivative will be approximated in the optimum way possible, but that does not imply that the solution will not deviate arbitrarily much from the true solution after a certain amount of time.

For more information about the practical applications of TDHF we refer to the review literature [Ne82, Da85], also the reader will find a code for one-dimensional solution of TDHF in [Ma93].

9.5 The Generator-Coordinate Method

A very general and conceptually attractive foundation of collective motion is provided by the *generator-coordinate method*, which was originally proposed by Hill, Wheeler, and Griffin [Hi53, Gr57a, Gr57b]. It has since been used widely in nuclear theory, and a more recent review can be found in [Re87]. The basic idea is to start with a set of microscopic wave functions $|\Phi(a)\rangle$ which depend parametrically on one or more collective degrees of freedom a. This may for instance be a Slater determinant built out of the occupied single-particle wave functions in the Nilsson model, depending on the deformation of the phenomenological single-particle potential.

A wave function containing collective motion in this collective parameter is then set up as

$$|\Psi\rangle = \int \mathrm{d}a\, f(a)|\Phi(a)\rangle \quad . \tag{9.47}$$

The *weight function* $f(a)$ is yet to be determined. Note that no superfluous coordinate is introduced in this way; the wave function $|\Psi\rangle$ still depends only on the microscopic degrees of freedom.

The weight function is determined from the variational principle

$$\delta\langle\Psi|\hat{H} - E|\Psi\rangle = 0 \quad , \tag{9.48}$$

where the Lagrange multiplier E is necessary to ensure normalization. Carrying out the variation leads to the *Hill–Wheeler–Griffin (HWG) equation*

$$\int \mathrm{d}a'\, \langle\Phi(a)|\hat{H}|\Phi(a')\rangle\, f(a') = E \int \mathrm{d}a'\, \langle\Phi(a)|\Phi(a')\rangle\, f(a') \quad , \tag{9.49}$$

which formally can be written in the shorter form

$$\mathcal{H} f = E\, \mathcal{N} f \quad . \tag{9.50}$$

The new symbols are the *Hamiltonian kernel*

$$\mathcal{H}(a, a') = \langle\Phi(a)|\hat{H}|\Phi(a')\rangle \tag{9.51}$$

and the *overlap kernel*

$$\mathcal{N}(a, a') = \langle\Phi(a)|\Phi(a')\rangle \quad , \tag{9.52}$$

and clearly their application to f is defined as the folding integral over a'.

If the parameter a is discrete, (9.49) becomes an eigenvalue equation, which with suitably changed notation reads

$$\sum_{a'} \mathcal{H}_{aa'} f_{a'} = E \sum_{a'} \mathcal{N}_{aa'} f_{a'} \quad . \tag{9.53}$$

This is the typical eigenvalue equation in a *nonorthogonal basis*, i.e., one in which the basic states are not orthonormal but have overlaps $\mathcal{N}_{aa'}$. Physically this arises from the fact that the Slater determinants for different values of a need not be orthonormal. It appears attractive to reduce the problem by multiplying both sides with the inverse of $\mathcal{N}$,

$$\mathcal{N}^{-1}\,\mathcal{H} f = E\, F \quad , \tag{9.54}$$

leading to a standard eigenvalue problem, but with the complication that the matrix on the left need not be Hermitian.

In practice, however, this is usually not possible, because $\mathcal{N}$ may have vanishing eigenvalues, as the $|\Phi(a)\rangle$ can be linearly dependent. A better way is therefore to perform a Gram–Schmidt orthonormalization, i.e., to construct a new basis $u_k(a)$ which fulfills (switching back to continuous a)

$$\int \mathrm{d}a'\, \mathcal{N}(a,a')\, u_k(a') = n_k\, u_k(a) \tag{9.55}$$

with some norm factors $n_k \geq 0$. Now we still have to worry about those basis functions with $n_k = 0$. This corresponds to a zero norm for the associated wave function $\sum_a u_k|\Phi(a)\rangle$, which again shows that in this case the single-particle wave functions are linearly dependent. The space of such wave functions thus in reality has a lower dimension and such basis functions should simply be omitted. We can thus assume that all the n_k are nonzero.

The remaining wave functions generated with the $u_k(a)$ can now be normalized, forming the *natural states*

$$|k\rangle = \frac{1}{\sqrt{n_k}} \int \mathrm{d}a\; u_k(a)|\Phi(a)\rangle \tag{9.56}$$

which span the Hilbert space of collective states in this method.

Expanding the collective state $|\Psi\rangle$ in the natural states

$$|\Psi\rangle = \sum_k g_k |k\rangle \tag{9.57}$$

leads to an eigenvalue problem of standard form

$$\sum_{k'} H_{kk'}\, g_{k'} = E\, g_k \tag{9.58}$$

with

$$H_{kk'} = \int \mathrm{d}a \int \mathrm{d}a'\, \frac{u_k^*(a)\, \mathcal{H}(a,a')\, u_k'(a')}{\sqrt{n_k\, n_{k'}}} \quad . \tag{9.59}$$

These developments are, of course, purely formal, and one has to see in practice whether the states constructed in this way are good approximations to eigenstates of the Hamiltonian, in the sense that, for example, an integral over Slater determinants with neighboring deformation parameters describes a true collective zero-point vibration or an excited vibrational state.

How could one go about actually solving the HWG equation? A straightforward discretization is problematic. Refining the discretization in a corresponds to a more accurate expansion, but in states with decreasing linear independence because

$|\Phi(a)\rangle$ and $|\Phi(a+\Delta a)\rangle$ become the same wave function with $\Delta a \to 0$. The result is that E and $|\Psi\rangle$ converge, but the weight function does not.

There are some cases in which the overlap kernel can be diagonalized exactly; such a case is presented in Exercise 9.2. The most useful way to proceed, however, is given by the *Gaussian overlap approximation*, which we discuss briefly later. First, however, we find out how to construct a collective Hamiltonian in the space of a.

EXERCISE

9.2 The Harmonic Oscillator in the Generator-Coordinate Method

Problem. Treat the harmonic oscillator in the generator-coordinate method based on an expansion with Gaussian wave packets centered at variable coordinate positions.

Solution. This exercise is not solely of mathematical interest, since we have seen that many types of collective motion can be treated quite well in a harmonic-oscillator approximation. There are a large number of detailed calculations involved, which are laborious but not very instructive, so here only the broad outline is given.

We start with the Hamiltonian

$$\hat{H} = \frac{\hat{p}^2}{2m} + \frac{m\omega^2\hat{q}^2}{2} \tag{1}$$

and use a set of normalized wave functions of the form

$$\langle q|\Phi(a)\rangle = (\pi s^2)^{-1/4} \exp\left[-(q-a)^2/2s^2\right] \quad . \tag{2}$$

Here q stands for the microscopic degree of freedom and a is the "collective parameter"; $|\Phi(a)\rangle$ describes a microscopic wave function centered at the position prescribed by the parameter a. The width of the packets s is a free parameter that will have to be fixed by some additional consideration. Now the kernels can be evaluated by integrations of Gaussians:

$$\begin{aligned} \mathcal{N}(a,a') &= \exp\left[-(a-a')^2/4s^2\right] \quad , \\ \mathcal{H}(a,a') &= \mathcal{N}(a,a')\left\{\frac{\hbar^2}{2m}\left(\frac{1}{2s^2} - \frac{(a-a')^2}{4s^2}\right)\right. \\ &\quad \left. + \frac{m\omega^2}{2}\left(\frac{s^2}{2} + \frac{(a+a')^2}{4}\right)\right\} \quad . \end{aligned} \tag{3}$$

The simplifying factor in this case is that $\mathcal{N}(a,a')$ depends only on $a-a'$, i.e., it is translationally invariant in a space, and the eigenfunctions are plane waves:

$$u_k(a) = \frac{1}{\sqrt{2\pi}} e^{-ika} \quad . \tag{4}$$

Exercise 9.2

The norm of these natural states becomes

$$n_k = \int da\ e^{-ika}\,\mathcal{N}(a,0) = 2\sqrt{\pi s^2}\ e^{-k^2 s^2} \quad . \tag{5}$$

There are no states with $n_k = 0$, but zero is the limiting value for $k \to \infty$.

Since k is a continuous parameter, the Hamiltonian in this basis becomes a distribution:

$$H_{kk'} = \frac{\hbar^2 k^2}{2m}\delta(k-k') - \frac{m\omega^2}{2}\delta''(k-k') \quad . \tag{6}$$

It is interesting to note that this is just the momentum-space version of the harmonic-oscillator Hamiltonian we started with, and that in this special case the free parameter s has disappeared.

The Schrödinger equation for the expansion coefficients $g(k)$ now is

$$\frac{\hbar^2 k^2}{2m} g(k) - \frac{m\omega^2}{2} g''(k) = E\ g(k) \quad , \tag{7}$$

and is almost identical in form to the coordinate-space version, so that the ground-state solution is easy to write down:

$$g_0(k) = \left(\frac{b^2}{\pi}\right)^{1/4} e^{-\frac{1}{2}b^2 k^2} \quad , \quad b = \sqrt{\frac{\hbar}{m\omega}} \quad . \tag{8}$$

This corresponds to the weight function

$$\begin{aligned} f_0(a) &= \int dk\ \frac{g_0(k)}{\sqrt{n_k}}\ u_k(a) \\ &= \frac{1}{\sqrt{2\pi}}\frac{1}{\sqrt{b^2-s^2}}\sqrt{\frac{b}{s}}\ \exp\left[-a^2/2(b^2-s^2)\right] \quad , \end{aligned} \tag{9}$$

which clearly is valid only for $b > s$.

Finally we can calculate the "microscopic" ground-state wave function itself as

$$\langle q|\Psi_0\rangle = \int da\, f_0(a)\langle q|\Phi(a)\rangle = (\pi b^2)^{-1/4}\exp\left(-q^2/2b^2\right) \quad , \tag{10}$$

obtaining the exact ground state of the harmonic oscillator. Is this a surprise or rather trivial? On the one hand, it would have been disconcerting to have a theory that cannot even reproduce the harmonic oscillator; on the other, what happened mathematically was not quite trivial. We have expanded the exact wave function with width b into a superposition of Gaussians of a different width s centered at all possible positions. So the integral over the weight function has performed exactly this expansion, and it is not surprising that this works only for $b > s$: it would be impossible to expand a Gaussian in other Gaussians with a larger width (actually, this case can also be carried through, but with $f_0(a)$ as a distribution and not a regular wave function).

An important advantage of the generator-coordinate method is that a collective Hamiltonian can be derived relatively easily. The goal is an equation of the form

$$\hat{H}_{\rm coll}\, \phi(a) = E\, \phi(a) \quad , \tag{9.60}$$

with $\phi(a)$ the collective wave function, which should be related to the weight function $f(a)$. The main difference is that $\phi(a)$ should be normalized in the standard way

$$\int {\rm d}a\, \phi^*(a)\, \phi(a) = 1 \quad , \tag{9.61}$$

so that the overlap kernel has to be eliminated. This can be done formally by writing

$$\phi(a) = \int {\rm d}a'\, \mathcal{N}^{1/2}(a,a')\, f(a') \tag{9.62}$$

or, more succinctly,

$$\phi = \mathcal{N}^{1/2} f \quad . \tag{9.63}$$

Here the "square root" of $\mathcal{N}$ must fulfil

$$\mathcal{N}(a,a') = \int {\rm d}a''\, \mathcal{N}^{1/2}(a,a'')\, \mathcal{N}^{1/2}(a'',a') \quad . \tag{9.64}$$

Inserting the definition of ϕ into the Hill–Wheeler–Griffin equation yields

$$\mathcal{N}^{1/2} \left(\mathcal{N}^{-1/2}\, \mathcal{H}\, \mathcal{N}^{-1/2} - E \right) \phi = 0 \quad . \tag{9.65}$$

If $\mathcal{N}$ has no singularities or zero eigenvalues, this implies a collective Hamiltonian

$$\hat{H}_{\rm coll} = \mathcal{N}^{-1/2}\, \mathcal{H}\, \mathcal{N}^{-1/2} \quad . \tag{9.66}$$

This formal expression for the collective Hamiltonian is still extremely complicated and in general an integral operator. It can be reduced to a more familiar form in the *Gaussian overlap approximation (GOA)*, which was introduced by Brink and Weiguny [Br68b]. The assumption is that the overlap kernel behaves like a Gaussian, falling off as the collective coordinates differ. The calculations are quite intricate, so here we present a somewhat simplified version; the general case can be found in [Re87].

First we perform a coordinate transformation to a "center-of-mass" system via

$$q = \tfrac{1}{2}(a + a'), \qquad \xi = \tfrac{1}{2}(a - a') \quad , \tag{9.67}$$

with the derivatives transforming as

$$\frac{\partial}{\partial \xi} = \frac{\partial}{\partial a} - \frac{\partial}{\partial a'} \quad , \qquad \frac{\partial}{\partial q} = \frac{\partial}{\partial a} + \frac{\partial}{\partial a'} \quad . \tag{9.68}$$

The Gaussian overlap approximation can now be formulated as

$$\mathcal{N}(q,\xi) = {\rm e}^{-\lambda(q)\,\xi^2} \quad . \tag{9.69}$$

Although the width λ can, in general, depend on the position q, this can be eliminated by a coordinate transformation at least for the one-parameter case, so we shall assume λ to be a constant and $\mathcal{N}(q,\xi) \to \mathcal{N}(\xi)$.

Because of the assumed narrowness of the Gaussian, one may employ second-order approximations; for example, for the Hamiltonian overlap

$$\mathcal{H}(q,\xi) = \left(\mathcal{H}_0(q) - \tfrac{1}{2}\xi^2\mathcal{H}_2(q)\right)\mathcal{N}(\xi) \quad , \tag{9.70}$$

with

$$\begin{aligned} \mathcal{H}_0(q) &= \mathcal{H}(q,\xi=0) \quad , \\ \mathcal{H}_2(q,\xi) &= -\frac{\partial^2}{\partial\xi^2}\mathcal{H}(q,\xi)\bigg|_{\xi=0} \quad , \end{aligned} \tag{9.71}$$

and we will also use

$$\lambda = -\frac{\partial^2}{\partial\xi^2}\mathcal{N}(\xi) \quad . \tag{9.72}$$

Now the task is to rewrite (9.66) using the GOA. First write that equation explicitly with the collective parameters and integrals indicated:

$$\hat{H}_{\rm coll}(a,a') = \int \mathrm{d}x\ \mathrm{d}x'\ \mathcal{N}^{-1/2}(a,x)\,\mathcal{H}(x,x')\,\mathcal{N}^{-1/2}(x',a') \quad . \tag{9.73}$$

Introducing $\bar{q} = \frac{1}{2}(x+x')$ and $\bar{\xi} = \frac{1}{2}(x-x')$, the GOA for the Hamiltonian kernel can be written in the form

$$\mathcal{H}(x,x') = \left(\mathcal{H}_0(\bar{q}) - \frac{\mathcal{H}_2(\bar{q})}{2\lambda^2}\frac{\partial^2}{\partial\bar{\xi}^2} - \lambda\mathcal{H}_2(\bar{q})\right)\mathcal{N}(\bar{\xi}) \quad . \tag{9.74}$$

Note how the second-order expansion was rewritten compared to (9.70) by letting the derivatives act on the overlap kernel.

If we insert the expansion into the collective Hamiltonian of (9.73), the derivatives with respect to x and x' may be replaced by derivatives in a and a' by using partial integration and by noting that

$$\frac{\partial}{\partial a}\mathcal{N} = -\frac{\partial}{\partial a'}\mathcal{N} \quad , \tag{9.75}$$

and analogously for the square root of the kernel. In this way we get

$$\begin{aligned} \hat{H}_{\rm coll}(a,a') = \int & \mathrm{d}x\ \mathrm{d}x'\ \mathcal{N}^{-1/2}(a,x) \\ & \times \left(\mathcal{H}(\bar{q}) - \frac{\mathcal{H}_2(\bar{q})}{2\lambda^2}\frac{\partial^2}{\partial\bar{\xi}^2} - \lambda\mathcal{H}_2(\bar{q})\right)\mathcal{N}(\bar{\xi})\,\mathcal{N}^{-1/2}(x',a') \end{aligned} \quad , \tag{9.76}$$

where the derivative with respect to a should act on the left.

The next step is to deal with the $\bar{q}$ in the integration. We expand all terms consistently to second order about the point q,

$$\begin{aligned} \mathcal{H}_0(\bar{q}) &\approx \mathcal{H}_0(q) + (\bar{q}-q)\frac{\partial\mathcal{H}_0(q)}{\partial q} + \frac{1}{2}(\bar{q}-q)^2\frac{\partial^2\mathcal{H}_0(q)}{\partial q^2} \quad , \\ \mathcal{H}_2(\bar{q}) &\approx \mathcal{H}_2(q) \quad , \end{aligned} \tag{9.77}$$

so that the last remaining task is the evaluation of integrals over $(\bar{q} - q)$. This is straightforward but laborious; for example, the linear term can be rewritten as

$$\begin{aligned}
X &= \int \mathrm{d}x\,\mathrm{d}x'\,\mathcal{N}^{-1/2}(a,x)\,(\bar{q}-q)\,\mathcal{H}_0'(q)\,\mathcal{N}(x,x')\,\mathcal{N}^{-1/2}(x',a') \\
&= \int \mathrm{d}x\,\mathrm{d}x'\,\mathcal{N}^{-1/2}(a,x)\,\tfrac{1}{2}(x-a+x'-a')\,\mathcal{H}_0'(q)\,\mathcal{N}(x,x')\,\mathcal{N}^{-1/2}(x',a') \\
&= \int \mathrm{d}x\,\mathrm{d}x'\,\frac{1}{4\lambda}\left\{\left[\left(\frac{\partial}{\partial x}-\frac{\partial}{\partial a}\right)\mathcal{N}^{-1/2}(a,x)\right]\mathcal{H}_0'(q)\,\mathcal{N}(x,x')\,\mathcal{N}^{-1/2}(x',a')\right. \\
&\quad \left. +\,\mathcal{N}^{-1/2}(a,x)\,\mathcal{H}_0'(q)\,\mathcal{N}(x,x')\left[\left(\frac{\partial}{\partial x'}-\frac{\partial}{\partial a'}\right)\mathcal{N}^{-1/2}(x',a')\right]\right\} \\
&= -\frac{1}{4\lambda}\left(\frac{\partial}{\partial a}+\frac{\partial}{\partial a'}\right)\mathcal{H}_0'(q)\int \mathrm{d}x\,\mathrm{d}x'\,\mathcal{N}^{-1/2}(a,x)\,\mathcal{N}(x,x')\,\mathcal{N}^{-1/2}(x',a') \\
&\quad +\frac{1}{4\lambda}\mathcal{H}_0'(q)\int \mathrm{d}x\,\mathrm{d}x'\,\mathcal{N}^{-1/2}(a,x)\left(\frac{\partial}{\partial x}+\frac{\partial}{\partial x'}\right)\mathcal{N}(x,x')\,\mathcal{N}^{-1/2}(x',a') \\
&= -\frac{1}{4\lambda}\left(\frac{\partial}{\partial a}+\frac{\partial}{\partial a'}\right)\mathcal{H}_0'(q)\,\delta(a-a') \\
&= -\delta(a-a')\,\frac{1}{4\lambda}\,\mathcal{H}_0''(q) \quad . \qquad (9.78)
\end{aligned}$$

The term in $(\bar{q} - q)^2$ can be dealt with analogously, but does not contribute to the order considered. Collecting these results we get

$$\hat{H}_{\mathrm{coll}}(a,a') = \left(\mathcal{H}_0(q) - \frac{1}{4\lambda}\mathcal{H}''(q) - \lambda\mathcal{H}_2(q) - \frac{\mathcal{H}_2(q)}{2\lambda^2}\frac{\partial^2}{\partial\xi^2}\right)\delta(\xi) \quad . \qquad (9.79)$$

In the HWG equation the operator acts as an integral operator on the weight function. In this special form, however, it is proportional to a δ function and thus can be converted to a differential operator by moving the derivatives with respect to a and a' implicit in ξ to the left and right. The result is

$$\hat{H}_{\mathrm{coll}} = \mathcal{H}_0(q) - \frac{1}{4\lambda}\mathcal{H}_0''(q) - \lambda\mathcal{H}_2(q) + \left\{\hat{p},\left\{\frac{\mathcal{H}_2(q)}{8\lambda^2},\hat{p}\right\}\right\} \quad . \qquad (9.80)$$

Here $\hat{p} = -\mathrm{i}\,\partial/\partial q$ is the collective momentum operator, and the braces in the last term denote the anticommutator.

This interesting result shows that the collective Hamiltonian consists of a number of potential terms (the first three, which contain not only the expectation value of the microscopic Hamiltonian at the parameter value q, but also corrections depending on the spread of the microscopic wave function in q space) and a kinetic energy with the usual second-order derivative. The collective mass parameter is clearly prescribed in this formula to be given by

$$\frac{1}{2B} = \frac{\mathcal{H}_2(q)}{8\lambda^2} \quad . \qquad (9.81)$$

This expression is similar but not identical to the one obtained in the much simpler cranking-model approach.

We have thus reached the goal for this chapter: to show that a quantized collective equation of motion can be derived from microscopic theory in a relatively general setting. The crucial approximation was the GOA, which essentially served to reduce the integral equation of motion of HWG type to a differential one similar to the Schrödinger equation.

The generator-coordinate method has been applied to many problems of collective motion, including fission, and has generally been quite successful, although computer limitations have restricted applications mostly to relatively light nuclei. Again the reader is referred to [Re87] for details.

9.6 High-Spin States

9.6.1 Overview

The study of nuclei at large angular momentum has become possible through the use of heavy-ion collisions. Some of the theoretical background will be briefly discussed here, with the emphasis on how to generalize the models presented up to now.

A heavy-ion reaction can populate nuclear states of very high angular momentum, the *high-spin states*, with values of the order of $I \approx 60\hbar$ achievable. The reaction produces such configurations with a considerable internal excitation, but the emission of a few neutrons reduces the excitation energy effectively while not decreasing the spin by much. In this way the nucleus drops down to the *yrast line* ("yrast" is the superlative of Swedish "yr", meaning *dizzy*), which is comprised of the lowest available states for given angular momenta. Along this line we thus expect no internal excitation, i.e., something like the "ground state" for that angular momentum.

One way to describe high-spin states is a proper extension of the Hartree–Fock methods [Qu78, Bo87]. The main theoretical consideration there is how to obtain a wave function with good angular momentum from a theory that violates rotational invariance – remember that the Hartree–Fock ground state is usually deformed – and various projection techniques have been developped for that purpose. Here we will instead regard the more phenomenological methods based on the liquid-drop model and shell corrections. In the next section we will look more closely at the Nilsson model for a rotating nucleus, but before that some idea of the general behavior should be given.

Cohen, Plasil, and Swiatecki [Co74] have calculated the properties of a rotating liquid drop, and this was extended to a more refined version by Sierk [Si86]. Although the model can be treated only numerically, the basic idea is quite simple: for a given surface shape the liquid-drop energy is calculated in the standard way, with surface and Coulomb contributions the dominant ingredients. To these a rotational energy is added, which is determined with the assumption of rigid-body rotation, i.e., the nucleus is assumed to rotate with an internal velocity field of $\boldsymbol{v} = \boldsymbol{\omega} \times \boldsymbol{r}$ at each point $\boldsymbol{r}$ inside and with homogeneous density distribution. The angular momentum is calculated with the same assumption.

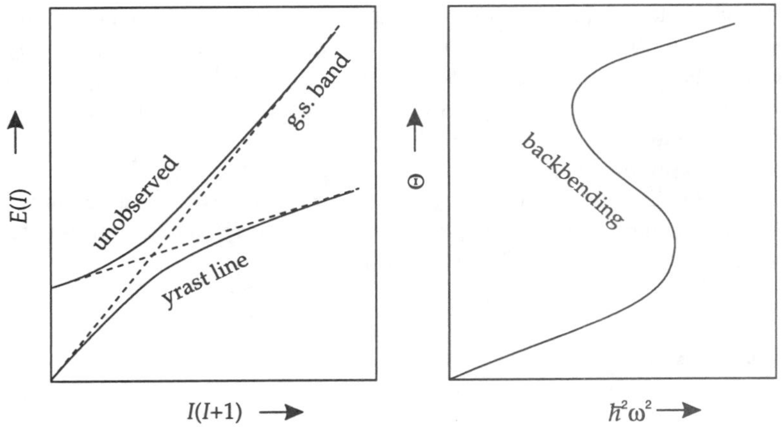

Fig. 9.18 Illustration of the backbending effect. The normal rotational band crosses a band with a broken pair and larger moment of inertia. Near the crossing the usual repulsion mixes the bands and pushes them apart so that the transition becomes gradual. The observable states along the yrast line are the lower ones for each angular momentum. On the right the moment of inertia is plotted as a function $\hbar^2\omega^2$, showing the backbending behavior. Experimentally this transition is usually seen for angular momenta between 10 and 20.

For a given angular momentum, then, one can calculate the energetically most favorable shape of the nucleus. In the ground state, of course, a liquid-drop model produces a spherical shape, which then goes over to an oblate configuration as the nucleons are pushed out by the centrifugal force, so that the nucleus looks like a disk rotating about its short axis. At even higher spin this configuration becomes unstable and the shape starts to elongate along one of the axes, so that a triaxial shape is reached. As this elongation proceeds, the shape becomes closer to a prolate deformation rotating about one of the shorter axes, and finally there is a limiting angular momentum beyond which the nucleus fissions. There is by now evidence for very large prolate deformations – corresponding experimentally to very closely spaced rotational bands – called *superdeformation*.

As mentioned, however, along the yrast line there is no internal excitation, so that single-particle effects and pairing can play an important role just like those near the true ground state. We have already seen that pairing influences the moments of inertia for small spins, and it was found that pairing changes drastically with angular momentum. The Coriolis force tends to break up the nucleon pairs by making it more advantageous for the nucleons to align their angular momenta along the axis of collective rotation instead of coupling them mutally to zero. Without going into the detailed mathematics, it is easy to see what happens: whenever a pair is broken up, the moment of inertia increases, as the angular moment increases for little change in the angular frequency, so that the rotational band gets a smaller spacing. As illustrated in Fig. 9.18, this is a crossing of bands with the yrast band following the lower set of levels. If this is plotted in a suitably dramatic way, the well-known *backbending* effect results. The rotational frequency in these plots is defined classically from the observed energy levels $E(I)$ as

$$\omega = \frac{\mathrm{d}E(I)}{\mathrm{d}J} \quad , \quad J = \sqrt{I(I+1)} \quad , \tag{9.82}$$

so that for discrete states

$$\omega \approx \frac{E(I) - E(I-2)}{\sqrt{I(I+1)} - \sqrt{(I-1)(I-2)}} \quad . \tag{9.83}$$

The moment of inertia is then determined as

$$\Theta = \frac{J}{\omega} \approx \frac{2I - 1}{E(I) - E(I-2)} \quad . \tag{9.84}$$

At sufficiently high angular momenta the pairing is destroyed completely and the rigid-body moment of inertia becomes a good approximation, allowing one to infer the deformations quite straightforwardly. Even in this regime, however, the single-particle structure remains important and shell effects can be studied in terms of a rotating phenomenological shell model as described in the next section.

9.6.2 The Cranked Nilsson Model

A simple and widely used way to describe the change of the single-particle structure with rotation is given by the *cranked Nilsson model* [Be75b, Ne75, Ne76, An76].

We assume that the particles move in a deformed-oscillator potential like in the Nilsson model (Sect. 7.3.2), but that this potential rotates about the x axis. The deformation of the core is described as usual by β and γ, but it turns out that non-axial deformations are very important in this case, so that a restriction to $\gamma = 0$ is not justified. One reason for this is apparent: even if the nucleus is axially symmetric about the z axis in the ground state, the x axis now plays a different role than the y axis and the Coriolis force will definitely violate axial symmetry.

It appears natural to use a system of coordinates rotating with the nucleus. How does this affect the Hamiltonian? Let us transform from space-fixed coordinates (x', y', z') to body-fixed ones (x, y, z) rotating with angular velocity ω about the x'=x axis. The coordinate transformation is

$$\begin{aligned} x &= x' \quad , \\ y &= y' \cos(\omega t) + z' \sin(\omega t) \quad , \\ z &= -y' \sin(\omega t) + z' \cos(\omega t) \quad . \end{aligned} \tag{9.85}$$

The kinetic energy can be transformed easily:

$$\begin{aligned} T &= \tfrac{1}{2} m \left(\dot{x}'^2 + \dot{y}'^2 + \dot{z}'^2 \right) \\ &= \tfrac{1}{2} m \left[\dot{x}^2 + \dot{y}^2 + \dot{z}^2 + 2\omega \left(\dot{y} z - \dot{z} y \right) + \omega^2 \left(y^2 + z^2 \right) \right] \quad . \end{aligned}$$

Since the potential in the corotating frame is just $V(x, y, z)$ – in this frame it is, of course, time independent – the Lagrangian is the standard $L = T - V(x, y, z)$, but the conjugate momenta become a bit more complicated:

$$\begin{aligned} p_x &= \frac{\partial L}{\partial \dot{x}} = m\dot{x} \quad , \\ p_y &= \frac{\partial L}{\partial \dot{y}} = m(\dot{y} - \omega z) \quad , \\ p_z &= \frac{\partial L}{\partial \dot{z}} = m(\dot{z} + \omega y) \quad . \end{aligned} \tag{9.86}$$

Now the Hamiltonian can be set up and becomes

$$\begin{aligned} H_\omega &= xp_x + yp_y + zp_z - L \\ &= \frac{1}{2m}\left(p_x^2 + p_y^2 + p_z^2\right) - \omega(yp_z - zp_y) + V(x,y,z) \\ &= \frac{1}{2m}\left(p_x^2 + p_y^2 + p_z^2\right) + V(x,y,z) - \omega j_x \quad . \end{aligned} \tag{9.87}$$

Here j_x is the angular momentum of the particle about the x axis *defined in the rotating coordinate system.* This corresponds to the primed angular momenta (j_x') in the collective model (the primes are used in the opposite way here to keep in accord with the usage the literature).

Note that the first two terms in (9.87) are just the standard Hamiltonian for the nonrotating situation, so that we can conclude that the Hamiltonian in the rotating system differs from it only by the term

$$H_\omega = H_0 - \omega j_x \quad . \tag{9.88}$$

This derivation works the same way for a system of many particles; j_x then is replaced by the total angular momentum. Also quantization can be carried through trivially and leads to the replacement of j_x by the angular-momentum operator.

Note, though, that for a rotating coordinate system (and generally for time-dependent constraints) the Hamiltonian is not equal to the energy of the system, which instead has to be evaluated as the expectation value of the static Hamiltonian H_0. This also becomes apparent when we interpret (9.88) in a different way: it corresponds to a variational problem

$$0 = \delta\langle\Psi|\hat{H}_\omega|\Psi\rangle = \delta\langle\Psi|\hat{H}_0 - \omega\hat{\jmath}_x|\Psi\rangle \tag{9.89}$$

with the subsidiary condition that j_x is prescribed (ω plays the role of a Lagrange multiplier).

The eigenvalues of $\hat{H}_\omega$ determined from

$$\hat{H}_\omega|\chi_i^\omega\rangle = \varepsilon_i^\omega|\chi_i^\omega\rangle \tag{9.90}$$

are different from the energies of the single-particle states, which have to be calculated as

$$e_i^\omega = \langle\chi_i^\omega|\hat{H}_0|\chi_i^\omega\rangle \quad . \tag{9.91}$$

There are also no angular-momentum quantum numbers left, and even the projection onto the x axis can be calculated as an expectation value only.

The only symmetries still preserved are reflection symmetry, implying conserved parity, and a rotation by 180° about the x axis. The latter leads to the so-called *signature quantum number* α. A rotation by π about the x axis produces a factor of $\exp(-\mathrm{i}\pi j_x)$ for a wave function with good j_x. For half-integer angular momentum the result can be written as

$$\mathrm{e}^{-\mathrm{i}\pi j_x} = \mathrm{e}^{-\mathrm{i}\alpha} \quad \text{where} \quad \alpha = \begin{cases} \frac{1}{2} & \text{for } j_x = \frac{1}{2}, \frac{5}{2}, \frac{9}{2}, \ldots \\ -\frac{1}{2} & \text{for } j_x = \frac{3}{2}, \frac{7}{2}, \ldots \end{cases} \quad . \tag{9.92}$$

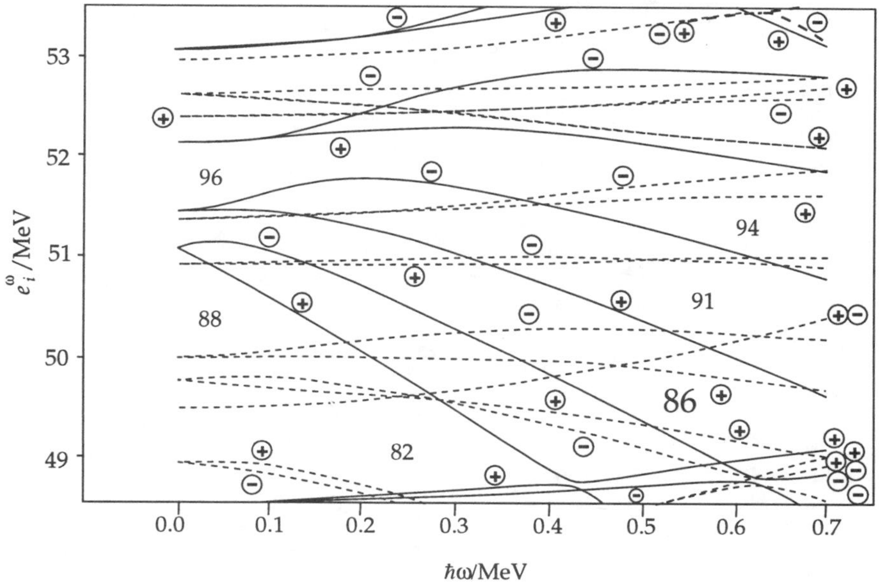

Fig. 9.19 Excerpt from a level scheme in the cranked Nilsson model [Be91]. Shown are neutron orbitals for a quadrupole-deformed shape with $\beta = 0.2$, $\gamma = 0$, and no hexadecupole deformation. The parity is indicated by the type of line (dashed, positive; full, negative parity), while the sign of the signature $\alpha = \pm\frac{1}{2}$ is indicated in circles near the lines. The new magic numbers are indicated in the gaps. Note that because of the lifting of the Kramers degeneracy each orbital can be occupied by one nucleon only.

This fact makes it advantageous to expand the eigenfunctions in a basis with good angular-momentum projection along the x axis, because then only those states with the same value of α will be coupled.

A full calculation within this model requires a larger number of ingredients, most of which have already been discussed. To give an impression of what is involved, we can look at the outline of a computer program published in [Be91]. It uses a Nilsson potential with axial asymmetry but including hexadecupole deformations. For a given deformation and angular frequency ω, the following steps are performed.

1. The single-particle states are determined by diagonalizing $\hat{H}_\omega$.
2. The shell correction is calculated by the procedure outlined in Sect. 9.2.2, using the energy expectation values e_i.
3. Pairing is calculated in the BCS model (Sect. 7.5.4). The correct treatment of pairing at higher angular momenta becomes somewhat more difficult, so that in this code, for example, it is used only for $\omega = 0$.
4. The liquid-drop energy is added. It includes, as usual, the shape-dependent Coulomb and surface energies, but also a liquid-drop rotation energy according to Sect. 9.6.1, which sometimes is shell corrected as well.

The result of this type of calculation is a potential-energy surface depending not only on the various deformation parameters but also on ω. This allows one to

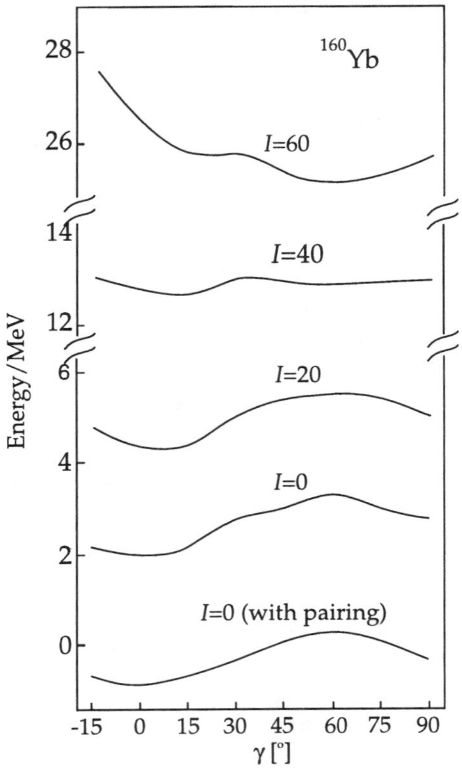

Fig. 9.20 The total energy of the nucleus ^{160}Yb as a function of γ for different spins, from [Be91]. A shift of the minimum from a prolate shape ($\gamma = 0$) to an oblate shape ($\gamma = 60^o$) is apparent. In this case, however, the quadrupole deformation β was kept fixed and not allowed to increase with spin.

study the systematic shift of the potential wells with spin and often shows that shell structure remains quite important even for rotating nuclei at high spin. An example of a level diagram is shown in Fig. 9.19 and clearly demonstrates that shell gaps occur at large values of ω as well as for the nonrotating nucleus. Figure 9.20 shows that the shell corrections are big enough that even together with the liquid-drop energy they can lead to systematic shifts in deformation.

Appendix

Some Formulas from Angular-Momentum Theory

Symmetry properties:

$$
\begin{aligned}
(j_1 j_2 J\,|m_1 m_2 M) &= (-1)^{j_1+j_2-J}(j_2 j_1 J\,|m_2 m_1 M) \\
&= (-1)^{j_1-m_1}\sqrt{\frac{2J+1}{2j_2+1}}\,(j_1 J j_2|m_1 - M - m_2) \\
&= (-1)^{j_1-m_1}\sqrt{\frac{2J+1}{2j_2+1}}\,(J j_1 j_2|M - m_1 m_2) \\
&= (-1)^{j_2+m_2}\sqrt{\frac{2J+1}{2j_1+1}}\,(J j_2 j_1| - M m_2 - m_1) \\
&= (-1)^{j_2+m_2}\sqrt{\frac{2J+1}{2j_1+1}}\,(j_2 J j_1| - m_2 M m_1) \quad , \\
(j_1 j_2 J\,|m_1 m_2 M) &= (-1)^{j_1+j_2-J}(j_1 j_2 J\,| - m_1 - m_2 - M) \quad .
\end{aligned}
\tag{A.1}
$$

All projections zero: coefficients of the type $(j_1 j_2 J\,|000)$. There are two cases:

$$
(j_1 j_2 J\,|000) = 0 \quad \text{if} \quad j_1 + j_2 + J \quad \text{is odd.} \tag{A.2}
$$

Otherwise, define $g = \frac{1}{2}(j_1 + j_2 + J)$. Then

$$
\begin{aligned}
(j_1 j_2 J\,|000) = &\frac{(-1)^{g-J}\sqrt{2J+1}\,g!}{(g-j_1)!(g-j_2)!(g-J)!} \\
&\times \sqrt{\frac{(2h-2j_1)!(2g-2j_2)!(2g-2J)!}{(2g+1)!}} \quad .
\end{aligned}
\tag{A.3}
$$

Coupling with and to zero:

$$
\begin{aligned}
(j_1 j_2 0|m_1 - m_2 0) &= (-1)^{j_1-m_1}\frac{\delta_{j_1 j_2}\,\delta_{m_1 m_2}}{\sqrt{2j_1+1}} \quad , \\
(j_1 0 j_2|m_1 0 m_2) &= \delta_{j_1 j_2}\,\delta_{m_1 m_2} \quad .
\end{aligned}
\tag{A.4}
$$

Coupling with angular momentum of unity:

$$
\begin{aligned}
(l-1\,1l|m-1\,1m) &= \sqrt{\frac{(l+m-1)(l+m)}{2l(2l-1)}} \quad ,\\
(l-1\,1l|m0m) &= \sqrt{\frac{(l+m)(l-m)}{l(2l-1)}} \quad ,\\
(l-1\,1l|m+1\,-1m) &= \sqrt{\frac{(l-m-1)(l-m)}{2l(2l-1)}} \quad ,\\
(l1l|m-1\,1m) &= -\sqrt{\frac{(l-m+1)(l+m)}{2l(l+1)}} \quad ,\\
(l1l|m0m) &= \frac{m}{\sqrt{l(l+1)}} \quad ,\\
(l1l|m+1\,-1m) &= \sqrt{\frac{(l+m+1)(l-m)}{2l(l+1)}} \quad ,\\
(l+1\,1l|m-1\,1m) &= \sqrt{\frac{(l-m+1)(l-m+2)}{(2l+2)(2l+3)}} \quad ,\\
(l+1\,1l|m0m) &= -\sqrt{\frac{(l+m+1)(l-m+1)}{(l+1)(2l+3)}} \quad ,\\
(l+1\,1l|m+1\,-1m) &= \sqrt{\frac{(l+m+2)(l+m+1)}{(2l+2)(2l+3)}} \quad .
\end{aligned}
\tag{A.5}
$$

Some special coefficients for angular momentum 2:

$$
\begin{aligned}
(222|-220) = (222|0\pm 2\pm 2) &= (222|\pm 20\pm 2)\\
&= -(222|000) = \sqrt{\frac{2}{7}} \quad ,\\
(221|m,-m,0) &= (-1)^m \frac{m}{\sqrt{10}} \quad ,\\
(221|m+1,-m,1) &= \frac{(-1)^{m+1}}{2\sqrt{5}}\sqrt{(2-m)(3+m)} \quad ,\\
(221|m-1,-m,-1) &= \frac{(-1)^m}{2\sqrt{5}}\sqrt{(2+m)(3-m)} \quad ,\\
(l2l|m0m) &= \frac{3m^2-l(l+1)}{\sqrt{l(l+1)(2l+3)(2l-1)}} \quad .
\end{aligned}
\tag{A.6}
$$

Rotation matrices:

$$\int \mathcal{D}^{(l)\,*}_{mm'}(\phi,\theta,0)\,\mathcal{D}^{(l)}_{\mu\mu'}(\phi,\theta,0)\,\mathrm{d}\Omega = \frac{4\pi}{2l+1}\,\delta_{m\mu}\,\delta_{m'\mu'} \quad , \tag{A.7}$$

$$d^{(j)}_{mm'}(\theta_2) = (-1)^{j-m} d^{(j)}_{m-m'}(\pi-\theta_2) \quad . \tag{A.8}$$

$$\int \mathcal{D}^{(I_3)}_{M_3K_3}(\boldsymbol{\theta})\mathcal{D}^{(I_2)*}_{M_2K_2}(\boldsymbol{\theta})\,\mathcal{D}^{(I_1)*}_{M_1K_1}(\boldsymbol{\theta})\,\sin\theta_2\,\mathrm{d}\theta_1\,\mathrm{d}\theta_2\,\mathrm{d}\theta_3$$

$$= \frac{8\pi^2}{2I_3+1}\,(I_1I_2I_3|M_1M_2M_3)\,(I_1I_2I_3|K_1K_2K_3) \quad . \tag{A.9}$$

Matrix elements of spherical harmonics:

$$\langle l||Y_L||l'\rangle = \sqrt{\frac{(2l'+1)(2L+1)}{4\pi(2l+1)}}\,(l'Ll|000) \quad . \tag{A.10}$$

References

[Ab64] M. Abramowitz, I. A. Stegun, *Handbook of Mathematical Functions*, (National Bureau of Standards, Washington, DC, 1964).

[An58] P. W. Anderson, Phys. Rev. **112**, 1900 (1957).

[An76] G. Andersson, S. E. Larsson, G. Leander, P. Möller, S. G. Nilsson, I. Ragnarsson, S. Åberg, R. Bengtsson, J. Dudek, B. Nerlo-Pomorska, K. Pomorski, Z. Szymanski, Nucl. Phys. A**268**, 205 (1976).

[An83] M. R. Anastasio, L. S. Celenza, W. S. Pong, C. M. Shakin, Phys. Rep. **100**, 327 (1983).

[Ar75] A. Arima, F. Iachello, Phys. Rev. Lett. **35** (1975) 1069.

[Ar76] A. Arima, F. Iachello, Ann. Phys. (NY) **99** (1976) 253.

[Ar77] A. Arima, T. Ohtsuka, F. Iachello, I. Talmi, Phys. Lett. **B66**, 205 (1977).

[Ba57] J. Bardeen, L. N. Cooper, J. R. Schrieffer, Phys. Rev. **108**, 1175 (1957).

[Ba60] M. Baranger, Phys. Rev. **120**, 957 (1960).

[Ba71] W. H. Bassichis, M. R. Strayer, Ann. Phys. (NY) **66**, 457 (1971).

[Ba83] A. B. Balantekin, I. Bars, F. Iachello, Phys. Rev. **C27**, 1761 (1983).

[Be36] H. A. Bethe, F. Bacher, Rev. Mod. Phys. **8**, 82 (1936).

[Be59] S. T. Belyaev, Kgl. Danske Videnskab Selskab Mat.-Fys. Medd. **31**, No. 11 (1959).

[Be61] S. T. Belyaev, Nucl. Phys. **24**, 322 (1961).

[Be73] H. A. Bethe, Phys. Rev. **167**, 879 (1973).

[Be75a] M. Beiner, H. Flocard, Nguyen Van Giai, P. Quentin, Nucl. Phys. **A238**, 29 (1975).

[Be75b] R. Bengtsson, S. E. Larsson, G. Leander, P. Möller, S. G. Nilsson, S. Åberg, Z. Szymanski, Phys. Lett. **B57**, 301 (1975).

[Be91] T. Bengtsson, I. Ragnarsson, S. Åberg, in *Computational Nuclear Physics 1 – Nuclear Structure*, ed. K. Langanke, J. A. Maruhn, S. E. Koonin, p. 88 (Springer-Verlag, Berlin and Heidelberg 1991).

[Bi82] R. Bijker, A. E. L. Dieperink, Nucl. Phys. **A379**, 221 (1982).

[Bj80] S. Bjornholm, J. E. Lynn, Rev. Mod. Phys. **52**, 725 (1980).

[Bl80] J. P. Blaizot, Phys. Rep. **C64**, 171 (1980).

[Bl87] P. G. Blunden, M. J. Iqbal, Phys. Lett. **B196**, 295 (1977).

[Bl89] V. Blum, J. Fink, P. G. Reinhard, J. A. Maruhn, W. Greiner, Phys. Lett. **B223**, 123 (1989).

[Bo52] A. Bohr, Kgl. Danske Videnskab Selskab Mat.-Fys. Medd. **26** No. 14 (1953).

[Bo53a] A. Bohr, B. Mottelson, Kgl. Danske Videnskab Selskab Mat. Fys. Medd. **27** No. 16 (1953).

[Bo53b] D. Bohm, D. Pines, Phys. Rev. **92**, 609 (1953).

[Bo54] A. Bohr, *Rotational States in Atomic Nuclei* (Thesis, Copenhagen 1954).

[Bo58] N. N. Bogolyubov, Soviet Phys. JETP **7**, 41 (1958); Nuovo Cimento **7**, 843 (1958).

[Bo59] N. N. Bogolyubov, Soviet Phys. Usp. **2**, 236 (1959); Usp. Fiz. Nauk **67**, 549 (1959).

[Bo76] P. Bonche, S. E. Koonin, J. W. Negele, Phys. Rev. **13**, 1226 (1976).

[Bo77] J. Boguta, A. R. Bodmer, Nucl. Phys. **A292**, 414 (1977).

[Bo87] P. Bonche, H. Flocard, P. H. Heenen, Nucl. Phys. **A467**, 115 (1987).

[Br59] G. E. Brown, M. Bolsterli, Phys. Rev. Lett. **3**, 472 (1959).

[Br61] G. E. Brown, J. A. Evans, D. J. Thouless, Nucl. Phys. **24**, 1 (1961).

[Br68a] K. A. Brückner, J. R. Buchler, S. Jorna, R. L. Lombard, Phys. Rev. **171**, 1188 (1968).
[Br68b] D. M. Brink, A. Weiguny, Nucl. Phys. **A120**, 59 (1968).
[Br68c] D. M. Brink, G. R. Satchler, *Angular Momentum* (Clarendon Press, Oxford 1968).
[Br72] M. Brack, J. Damgaard, A. S. Jensen, H. C. Pauli, V. M. Strutinsky, C. Y. Wong, Rev. Mod. Phys. **44**, 320 (1972).
[Br81] P. von Brentano, A. Gelberg, U. Kaup, in *Interacting Boson-Fermion Systems in Nuclei*, ed. F. Iachello, p. 303, (Plenum, New York, 1981).
[Br84] R. Brockmann, R. Machleidt, Phys. Lett. **B149**, 283 (1984).
[Br85] M. Brack, Proc. of a NATO ASI on Density Functional Methods in Physics, p. 331, (Plenum Press, New York 1985).
[Br92] R. Brockmann, J. Frank, Phys. Rev. Lett. **68**, 1830 (1992).
[Ca88] R. F. Casten and D. D. Warner, Rev. Mod. Phys. **60**, 389 (1988).
[Ce86] L. S. Celenza, C. S. Shakin, *Relativistic Nuclear Physics Theories of Structure and Scattering* (World Scientific, Singapore 1986).
[Ch76a] E. Chacón, M. Moshinsky, R. T. Sharp, J. Math. Phys. **17**, 668 (1976).
[Ch76b] S. A. Chin, Phys. Lett. **B62**, 263 (1976).
[Ch77a] E. Chacón and M. Moshinsky, J. Math. Phys. **18**, 870 (1977).
[Ch77b] S. A. Chin, Ann. Phys. **B108**, 301 (1977).
[Co74] S. Cohen, F. Plasil, W. J. Swiatecki, Ann. Phys. (NY) **82**, 557 (1974).
[Cs91] L. P. Csernai, D. D. Strottman, eds., "Relativistic Heavy Ion Physics", Int. Rev. Nucl. Phys. **6** (1991).
[Cu76] R. Y. Cusson, R. K. Smith, J. A. Maruhn, Phys. Rev. Lett. **36**, 1166 (1976).
[Cu78] R. Y. Cusson, J. A. Maruhn, H. W. Meldner, Phys. Rev. **C18**, 2589 (1978).
[Cu82] M. A. Cunningham, Nucl. Phys. **A385**, 204, 221 (1982).
[Da58] A. S. Davydov, B. F. Fillipov, Nucl. Phys. **8**, 237 (1958) .
[Da64a] M. Danos, W. Greiner, Phys. Lett. **8**, 113 (1964).
[Da64b] M. Danos, W. Greiner, Phys. Rev. **134**, B284 (1964).
[Da78] K. T. R. Davies, V. Maruhn-Rezvani, S. E. Koonin, J. W. Negele, Phys. Rev. Lett. **41**, 632 (1978).
[Da85] K. T. R. Davies, K. R. Sandhya Devi, S. E. Koonin, M. R. Strayer, in *Treatise on Heavy-Ion Science*, Vol. 3, ed. D. A. Bromley (Plenum Press, New York 1985).
[Di30] P. A. M. Dirac, Proc. Cambridge Phil. Soc. **26**, 376 (1930).
[Di80] A. E. L. Dieperink, O. Scholten, F. Iachello, Phys. Rev. Lett. **44**, 1747 (1980).
[Du56] H. P. Duerr, Phys. Rev. **103**, 469 (1956).
[Ed60] A. R. Edmonds, *Angular Momentum in Quantum Mechanics*, 2nd Ed. (Princeton University Press, 1960).
[Eh59] H. Ehrenreich, M. H. Cohen, Phys. Rev. **115**, 786 (1959).
[Ei72] J. M. Eisenberg, W. Greiner, *Nuclear Theory. Volume 3: Microscopic Theory of the Nucleus* (North Holland, Amsterdam 1973).
[Ei87] J. M. Eisenberg, W. M. Greiner, *Nuclear Theory. Volume 1: Nuclear Models*, 3rd Ed. (North Holland, Amsterdam 1987).
[El57] J. P. Elliott, B. H. Flowers, Proc. Royal Soc. (London) **24**, (2A), 57 (1957).
[En75] Y. M. Engel, D. M. Brink, K. Goeke, S. J. Krieger, D. Vautherin, Nucl. Phys. **A249**, 215 (1975).
[Fa62a] A. Faessler, W. Greiner, Z. Physik **168**, 425 (1962).
[Fa62b] A. Faessler, W. Greiner, Z. Physik **170**, 105 (1962).
[Fa64a] A. Faessler, W. Greiner, Z. Physik **177**, 190 (1964).
[Fa64b] A. Faessler, W. Greiner, Nucl. Phys. **59**, 177 (1964).
[Fa65a] A. Faessler, W. Greiner, R. K. Sheline, Nucl. Phys. **62**, 241 (1965).
[Fa65b] A. Faessler, W. Greiner, R. K. Sheline, Nucl. Phys. **70**, 33 (1965).
[Fa65c] A. Faessler, W. Greiner, R. K. Sheline, Nucl. Phys. **80**, 417 (1965).
[Fe49] E. Feenberg, Phys. Rev. **75**, 320 (1949).
[Fe53] G. M. Ferentz, M. Gell-Mann, D. Pines, Phys. Rev. **92**, 836 (1953).
[Fe57] R. A. Ferrell, Phys. Rev. **107**, 1631 (1957).
[Fi72] E. O. Fiset, J. R. Nix, Nucl. Phys. **A193**, 647 (1972).
[Fi74] H.-J. Fink, J. Maruhn, W. Scheid, W. Greiner, Z. Physik **268**, 321 (1974).

[Fi89] J. Fink, V. Blum, P.-G. Reinhard, J. A. Maruhn, W. Greiner, Phys. Lett. **B218**, 277 (1989).

[Fl72] K. F. Flynn E. P. Horvitz, C. A. A. Bloomquist, R. F. Barnes, R. K. Sjoblom, P. R. Fields, L. E. Glendenin, Phys. Rev. **C5**, 1725 (1972).

[Fl75] K. F. Flynn, J. E. Gindler, R. K. Sjoblom, L. E. Glendenin, Phys. Rev. **C11**, 1676 (1975).

[Ga88] Y. K. Gambhir, P. Ring, Phys. Lett. **B202**, 5 (1988).

[Gi80a] M. J. Giannoni, P. Quentin, Phys. Rev. **C21**, 2060 (1980).

[Gi80b] M. J. Giannoni, P. Quentin, Phys. Rev. **C21**, 2076 (1980).

[Gn71] G. Gneuss, W. Greiner, Nucl. Phys. A**171**, 449 (1971).

[Go48] M. Goldhaber, E. Teller, Phys. Rev. **74** (1948) 1946.

[Go56] K. Gottfried, Phys. Rev. **103**, 1017 (1956).

[Go59] J. Goldstone, K. Gottfried, Nuovo Cimento **13**, 849 (1959).

[Go73] D. Gogny, *Proc. of the International Conference on Nuclear Physics*, München 1973 (North Holland, Amsterdam 1973).

[Gr57a] J. J. Griffin, J. A. Wheeler, Phys. Rev. **108**, 311 (1957).

[Gr57b] J. J. Griffin, Phys. Rev. **108**, 328 (1957).

[Gr65a] I. S. Gradshteyn, I. W. Ryzhik, *Table of Integrals, Series, and Products*, (Academic Press, New York 1965).

[Gr65b] I. M. Green, S. A. Moszkowski, Phys. Rev. **139**, B790 (1965).

[Gr69] J. Grumann, U. Mosel, B. Fink, W. Greiner, Z. Physik **228**, 371 (1969).

[Gr89] W. Greiner, M. Ivaşcu, D. N. Poenaru, A. Săndulescu, in *Treatise on Heavy Ion Science*, ed. D. A. Bromley, Vol. 8, p. 641 (Plenum, New York 1989).

[Gu66] C. Gustafsson, I. L. Lamm, B. Nilsson, S. G. Nilsson, *Proc. of the International Symposium on Why and How Should We Investigate Nuclides Far off Stability Line* Lysekil, Sweden, 1966, eds. W. Forsling, C. J. Herrlander, H. Ryde (Almquist & Wiksell, Stockholm 1974).

[Gu77] R. K. Gupta, C. Părvulescu, A. Săndulescu, W. Greiner, Z. Physik **A283**, 217 (1977).

[Ha49] O. Haxel, J. H. D. Jensen, H. E. Suess, Phys. Rev. **75**, 1766 (1949).

[Ha62] T. Hamada, I. D. Johnston, Nucl. Phys. **34**, 382 (1962).

[Ha88] R. W. Hasse, W. D. Myers, *Geometrical Relationships of Macroscopic Nuclear Physics* (Springer-Verlag, Berlin and Heidelberg 1988).

[Ha89] J. H. Hamilton, in *Treatise on Heavy Ion Physics*, ed. D. A. Bromley, Vol. 8, p. 3 (Plenum Press, New York 1989).

[He80] P. O. Hess, M. Seiwert, J. A. Maruhn, W. Greiner, Z. Physik **A296**, 147 (1980).

[He81] P. O. Hess, J. A. Maruhn, W. Greiner, J. Phys. **G7**, 737 (1981).

[Hi53] D. L. Hill, J. A. Wheeler, Phys. Rev. **89**, 1106 (1953).

[Ho87] C. J. Horowitz, B. D. Serot, Nucl. Phys. **A464**, 613 (1987).

[Ho95a] S. Hofmann V. Ninov, F. P. Heßberger, P. Armbruster, H. Holger, G. Münzenberg, H. J. Schött, A. G. Popenko, A. V. Yeremin, A. N. Andreyev, S. Saro, R. Jynik, M. Leino, Z. Physik A**350**, 277 (1995).

[Ho95b] S. Hofmann V. Ninov, F. P. Heßberger, P. Armbruster, H. Holger, G. Münzenberg, H. J. Schött, A. G. Popenko, A. V. Yeremin, A. N. Andreyev, S. Saro, R. Jynik, M. Leino, Z. Physik A**350**, 281 (1995).

[Ia79] F. Iachello, O. Scholten, Phys. Rev. Lett. **43**, 679 (1979).

[Ia80] F. Iachello, Nucl. Phys. **A347**, 51 (1980).

[Ia87] F. Iachello, A. Arima, *The Interacting Boson Model* (Cambridge University Press, Cambridge 1987).

[In54] D. R. Inglis, Phys. Rev. **96**, 1059 (1954).

[In56] D. R. Inglis, Phys. Rev. **103**, 1786 (1956).

[Ir75] J. M. Irvine, *Heavy Nuclei, Superheavy Nuclei, and Neutron Stars* (Oxford University Press, Oxford 1975).

[Is81] P. van Isacker, J. Q. Chen, Phys. Rev. **C24**, 684 (1981).

[Is84] P. van Isacker, A. Frank, B. Z. Sun, Ann. Phys. (NY) **157**, 183 (1984).

[It92] M. G. Itkis et al., Sov. J. Nucl. Phys. **52**, 601 (1990).

[Ja74] D. Janssen, R. V. Jolos, F. Dönau, Nucl. Phys. **A224**, 93 (1974).

[Ke56] A. K. Kerman, Kgl. Danske Videnskab Selskab Mat.-Fys. Medd. **30** No. 15 (1956).
[Ke61] A. K. Kerman, Ann. Phys. (NY) **12**, 300 (1961).
[Ke64] I. Kelson, Phys. Rev. **B136**, 1667 (1964).
[Kl91] H. Klein, D. Schnabel, J. Maruhn, W. Greiner, unpublished results.
[Kö71] H. S. Köhler, Y. C. Lin, Nucl. Phys. **A167**, 305 (1971).
[Kö76] H. S. Köhler, Nucl. Phys. **A258**, 301 (1976).
[Kr79] H. J. Krappe, J. R. Nix, A. J. Sierk, Phys. Rev. **C20**, 992 (1979).
[Kr80] H. Krivine, J. Treiner, O. Bohigas, Nucl. Phys. **A336**, 155 (1980).
[Ku61] K. Kumar, R. Bhaduri, Phys. Rev. **122**, 1926 (1961).
[Ku89] K. Kumar, *Superheavy Elements* (Adam Hilger, Bristol and New York 1989).
[La64] A. M. Lane, *Nuclear Theory* (W. A. Benjamin, New York 1964).
[Li60] H. J. Lipkin, Ann. Phys. (NY) **9**, 272 (1960).
[Li73] P. Lichtner, D. Drechsel, J. Maruhn, W. Greiner, Phys. Lett. **B45**, 175 (1973).
[Lo70] K. E. G. Löbner, M. Vetter, V. Hönig, Atomic Data and Nuclear Data Tables **A7**, 495 (1970).
[Lo73] R. J. Lombard, Ann. Phys. (NY) **77**, 380 (1973).
[Lu78] A. Săndulescu, H. J. Lustig, J. Hahn, W. Greiner, J. Phys. **G4**, L279 (1978).
[Lu80] H. J. Lustig, J. A. Maruhn, W. Greiner, J. Phys. **G6**, L25 (1980).
[Ma48] M. G. Mayer, Phys. Rev. **74**, 235 (1948).
[Ma49] M. G. Mayer, Phys. Rev. **75**, 1969 (1949).
[Ma50] M. G. Mayer, Phys. Rev. **78**, 22 (1950).
[Ma55] M. G. Mayer, J. H. D. Jensen, *Elementary Theory of Nuclear Shell Structure* (Wiley, New York 1955).
[Ma72] J. Maruhn, W. Greiner, Z. Physik **251**, 431 (1972).
[Ma74] J. Maruhn, W. Greiner, Phys. Rev. Lett. **32**, 548 (1974).
[Ma75] V. Maruhn-Rezwani, W. Greiner, J. A. Maruhn, Phys. Lett. **57B**, 109 (1975).
[Ma76a] J. A. Maruhn, R. Y. Cusson, Nucl. Phys. **A270**, 471 (1976).
[Ma76b] J. A. Maruhn, W. Greiner, Phys. Rev. **C13**, 2404 (1976).
[Ma85] J. A. Maruhn, W. Greiner, "Relativistic Heavy-Ion Reactions: Theoretical Models", in *Treatise on Heavy-Ion Science, Vol. 4*, ed. D. A. Bromley (Plenum Press, New York 1985).
[Ma93] J. A. Maruhn, S. E. Koonin, in *Computational Nuclear Physics 2 – Nuclear Reactions*, ed. K. Langanke, J. A. Maruhn, S. E. Koonin, p. 88 (Springer-Verlag, Berlin and Heidelberg 1993).
[Mi72] L. D. Miller, L. D. S. Green, Phys. Rev. **C5**, 241 (1972).
[My76] W. D. Myers, Atomic Data and Nuclear Data Tables **17**, 411 (1976).
[Mo55] S. A. Moszkowski, Phys. Rev. **99**, 803 (1955).
[Mo59] B. R. Mottelson, *Cours de l'Ecole d'Eté de Physique Theéorique des Houches 1958*, p. 283 (Dunod, Paris 1959).
[Mo71] U. Mosel, J. A. Maruhn, W. Greiner, Phys. Lett. **43**B, 587 (1971).
[Mo93] S. Mordechai, C. F. Moore, Int. J. Mod. Phys. **E3**, 39 (1994).
[Ne29] J. von Neumann, E. Wigner, Physik. Zeitschrift **XXX**, 467 (1929).
[Ne75] K. Neergård, V. V. Pashkevich, Phys. Lett. **B59**, 218 (1975).
[Ne76] K. Neergård, V. V. Pashkevich, S. Frauendorf, Nucl. Phys. **A262**, 61 (1976).
[Ne82] J. . Negele, Rev. Mod. Phys. **54**, 913 (1982).
[Ni55] S. G. Nilsson, Kgl. Danske Videnskab. Selsk. Mat.-Fys. Medd. **29** (1955).
[Ni61] S. G. Nilsson, O. Prior, Kgl. Danske Videnskab. Selsk. Mat.-Fys. Medd. **32** No. 16 (1961).
[Ni68] J. R. Nix, Los Alamos Report UCRL-17958 (1968).
[Ni69] S. G. Nilsson, C. F. Tsang, A. Sobiczewski, Z. Szymanski, S. Wycech, C. Gustafson, I.–L. Lamm, P. Möller, B. Nilsson, Nucl. Phys. **A131**, 1 (1969).
[No64] Y. Nogami, Phys. Rev. **134**, B313 (1964).
[No68] C. C. Noack, Nucl. Phys. **A108**, 493 (1968).
[Oh78] T. Ohtsuka, A. Arima, F. Iachello, I. Talmi, Phys. Lett. **B76**, 139 (1978).
[Pe86] R. J. Perry, Phys. Lett. **B182**, 269 (1986).
[Po28] B. Podolsky, Phys. Rev. **32**, 812 (1928).
[Po62] S. M. Polikanov et al., Exp. Theor. Phys. **42**, 1464 (1962).

[Po89] D. N Poenaru, M. Ivaşcu, W. Greiner, in *Particle Emissions from Nuclei*, Vol. III, p. 203, ed. D. N. Poenaru and M. Ivaşcu (CRC, Boca Raton, FL, 1989).
[Pr61] R. Prange, Nucl. Phys. **22**, 287 (1962).
[Pr73] H. C. Pradhan, Y. Nogami, J. Law, Nucl. Phys. **A201**, 357 (1973).
[Qu78] P. Quentin, H. Flocard, Ann. Rev. Nucl. Part. Sci. **28**, 523 (1978).
[Ra43] G. Racah, Phys. Rev. **63**, 367 (1943).
[Ra74] J. Randrup, S. E. Larsson, P. Moller, A. Sobiczewski, A. Lukasiak, Phys. Scr. **10A**, 60 (1974).
[Re68] R. V. Reid, Ann. Phys. (NY) **50**, 411 (1968).
[Re86] P. G. Reinhard, M. Rufa, J. Maruhn, W. Greiner, J. Friedrich, Z. Physik **A323**, 13 (1986).
[Re87] P.-G. Reinhard, K. Goeke, Rep. Prog. Phys. **50**, 1 (1987).
[Re89] P. G. Reinhard, Rep. Prog. Phys. **52**, 439 (1989)
[Ro57] M. E. Rose, *Elementary Theory of Angular Momentum* (Wiley, New York 1957).
[Ro68] E. Rost, Phys. Lett. **26B**, 184 (1968).
[Ro84] H. J. Rose, G. A. Jones, Nature **307**, 245 (1984).
[Ru95] K. Rutz, J. A. Maruhn, P.–G. Reinhard, W. Greiner, Nucl. Phys. **A590**, 680 (1995)
[Sa76] A. Săndulescu, R. K. Gupta, W. Scheid, W. Greiner, Phys. Lett. **60B**, 225 (1976).
[Sa78] K. R. Sandhya Devi, M. R. Strayer, Phys. Lett. **77B**, 135 (1978).
[Sa80] A. Săndulescu, D. N. Poenaru, W. Greiner, Sov. J. of Particles and Nuclei **11**, 528 (1980).
[Sa89] A. Săndulescu, J. Phys. **G15**, 529 (1989).
[Sc55] G. Scharff-Goldhaber, J. Weneser, Phys. Rev. **98**, 212 (1955).
[Sc68] W. Scheid, R. Ligensa, W. Greiner, Phys. Rev. Lett. **21**, 1479 (1968).
[Sc78] O. Scholten, F. Iachello, A. Arima, Ann. Phys. (NY) **115** (1978) 325.
[Sc82] O. Scholten, N. Blasi, Nucl. Phys. **A380**, 509 (1982).
[Sc91] O. Scholten, in *Computational Nuclear Physics 1 – Nuclear Structure*, ed. K. Langanke, J. A. Maruhn, S. E. Koonin, p. 88 (Springer-Verlag, Berlin and Heidelberg 1991).
[Se67] P. A. Seeger, R. C. Perisko, Los Alamos Report No. LA-3751 (1967).
[Se68] P. A. Seeger, Los Alamos Report No. LA-DC-8950a (1968).
[Se86] B. D. Serot, J. D. Walecka, Advances in Nuclear Physics **16**, (1986).
[Se89] B. D. Serot, in *Nuclear Matter and Heavy Ion Collisions*, Les Houches, France, Proc. of the NATO Advanced Research Workshop, New York 1989.
[Si86] A. J. Sierk, Phys. Rev. **C86**, 2039 (1986).
[Sk56] T. H. R. Skyrme, Phil. Mag. **1**, 1043 (1956).
[Sk59] T. H. R. Skyrme, Nucl. Phys. **9**, 615 (1959).
[St50] H. Steinwedel, J. H. D. Jensen, Z. Naturforschung **5a**, 413 (1950).
[St67] V. M. Strutinsky, Nucl. Phys. **A95**, 420 (1967).
[St68] V. M. Strutinsky, Nucl. Phys. **A122**, 1 (1968).
[Te86] B. Ter Haar, B. Malfliet, Phys. Lett. **B172**, 10 (1986); Phys. Rev. Lett. **56**, 1237 (1986).
[Th61] D. J. Thouless, Nucl. Phys. **22**, 78 (1961).
[Tr91a] D. Troltenier, J. A. Maruhn, W. Greiner, V. Velazquez-Aguilar, P. O. Hess, J. H. Hamilton, Z. Physik **A338**, 261 (1991).
[Tr91b] D. Troltenier, J. A. Maruhn, P. O. Hess, in *Computational Nuclear Physics 1 – Nuclear Structure*, ed. K. Langanke, J. A. Maruhn, and S. E. Koonin, p. 105 (Springer-Verlag, Berlin and Heidelberg 1991).
[Va58] J. G. Valatin, Nuovo Cimento **7**, 843 (1958).
[Va61] J. G. Valatin, Phys. Rev. **122**, 1012 (1961).
[Va70] D. Vautherin, D. M. Brink, Phys. Lett. **32B**, 149 (1970).
[Va72] D. Vautherin, D. M. Brink, Phys. Rev. **C5**, 626 (1972).
[Va88] D. A. Varshalovich, A. N. Moskalev, and V. K. Khersonskii, *Quantum Theory of Angular Momentum* (World Scientific, Singapore 1988).
[Wa74] J. D. Walecka, Ann. Phys. (NY) **83**, 491 (1974).
[Wa77] A. H. Wapstra, K. Bos, At. Data and Nucl. Data Tables **19**, 177 (1977).
[Wa88] B. Waldhauser, J. Maruhn, H. Stöcker, W. Greiner, Phys. Rev. **C38**, 1003 (1988).

[We35] C. F. von Weizsäcker, Z. Physik **96**, 461 (1935).
[Wi56] L. Wilets, M. Jean, Phys. Rev. **C102** (1956) 788.
[Wi58] L. Wilets, Rev. Mod. Phys. **30**, 542 (1958).
[Wo54] R. D. Woods, D. S. Saxon, Phys. Rev. **95**, 577 (1954).
[Zh86] H. X. Zhang, T. R. Yeh, H. Lancman, Phys. Rev. **C34**, 1397 (1986).

Subject Index

Springer-Verlag and the Environment

We at Springer-Verlag firmly believe that an international science publisher has a special obligation to the environment, and our corporate policies consistently reflect this conviction.

We also expect our business partners – paper mills, printers, packaging manufacturers, etc. – to commit themselves to using environmentally friendly materials and production processes.

The paper in this book is made from low- or no-chlorine pulp and is acid free, in conformance with international standards for paper permanency.